Practical Stress Analysis in Engineering Design

MECHANICAL ENGINEERING

A Series of Textbooks and Reference Books

Editor: L.L. FAULKNER   Columbus Division, Battelle Memorial Institute, and Department of Mechanical Engineering, The Ohio State University, Columbus, Ohio

Associate Editor: S.B. MENKES   Department of Mechanical Engineering, The City College of the City University of New York, New York

1. Spring Designer's Handbook, *by Harold Carlson*
2. Computer-Aided Graphics and Design, *by Daniel L. Ryan*
3. Lubrication Fundamentals, *by J. George Wills*
4. Solar Engineering for Domestic Buildings, *by William A. Himmelman*
5. Applied Engineering Mechanics: Statics and Dynamics, *by G. Boothroyd and C. Poli*
6. Centrifugal Pump Clinic, *by Igor J. Karassik*
7. Computer-Aided Kinetics for Machine Design, *by Daniel L. Ryan*
8. Plastics Products Design Handbook, Part A: Materials and Components; Part B: Processes and Design for Processes, *edited by Edward Miller*
9. Turbomachinery: Basic Theory and Applications, *by Earl Logan, Jr.*
10. Vibrations of Shells and Plates, *by Werner Soedel*
11. Flat and Corrugated Diaphragm Design Handbook, *by Mario Di Giovanni*
12. Practical Stress Analysis in Engineering Design, *by Alexander Blake*
13. An Introduction to the Design and Behavior of Bolted Joints, *by John H. Bickford*
14. Optimal Engineering Design: Principles and Applications, *by James N. Siddall*
15. Spring Manufacturing Handbook, *by Harold Carlson*
16. Industrial Noise Control: Fundamentals and Applications, *edited by Lewis H. Bell*
17. Gears and Their Vibration: A Basic Approach to Understanding Gear Noise, *by J. Derek Smith*

18. Chains for Power Transmission and Material Handling: Design and Applications Handbook, *by the American Chain Association*
19. Corrosion and Corrosion Protection Handbook, *edited by Philip A. Schweitzer*
20. Gear Drive Systems: Design and Application, *by Peter Lynwander*
21. Controlling In-Plant Airborne Contaminants: Systems Design and Calculations, *by John D. Constance*
22. CAD/CAM Systems Planning and Implementation, *by Charles S. Knox*
23. Probabilistic Engineering Design: Principles and Applications, *by James N. Siddall*
24. Traction Drives: Selection and Application, *by Frederick W. Heilich III and Eugene E. Shube*
25. Finite Element Methods: An Introduction, *by Ronald L. Huston and Chris E. Passerello*
26. Mechanical Fastening of Plastics: An Engineering Handbook, *by Brayton Lincoln, Kenneth J. Gomes, and James F. Braden*
27. Lubrication in Practice, Second Edition, *edited by W. S. Robertson*
28. Principles of Automated Drafting, *by Daniel L. Ryan*
29. Practical Seal Design, *edited by Leonard J. Martini*
30. Engineering Documentation for CAD/CAM Applications, *by Charles S. Knox*
31. Design Dimensioning with Computer Graphics Applications, *by Jerome C. Lange*
32. Mechanism Analysis: Simplified Graphical and Analytical Techniques, *by Lyndon O. Barton*
33. CAD/CAM Systems: Justification, Implementation, Productivity Measurement, *by Edward J. Preston, George W. Crawford, and Mark E. Coticchia*
34. Steam Plant Calculations Manual, *by V. Ganapathy*
35. Design Assurance for Engineers and Managers, *by John A. Burgess*
36. Heat Transfer Fluids and Systems for Process and Energy Applications, *by Jasbir Singh*
37. Potential Flows: Computer Graphic Solutions, *by Robert H. Kirchhoff*
38. Computer-Aided Graphics and Design, Second Edition, *by Daniel L. Ryan*
39. Electronically Controlled Proportional Valves: Selection and Application, *by Michael J. Tonyan, edited by Tobi Goldoftas*
40. Pressure Gauge Handbook, *by AMETEK, U.S. Gauge Division, edited by Philip W. Harland*
41. Fabric Filtration for Combustion Sources: Fundamentals and Basic Technology, *by R. P. Donovan*
42. Design of Mechanical Joints, *by Alexander Blake*
43. CAD/CAM Dictionary, *by Edward J. Preston, George W. Crawford, and Mark E. Coticchia*

44. Machinery Adhesives for Locking, Retaining, and Sealing, *by Girard S. Haviland*
45. Couplings and Joints: Design, Selection, and Application, *by Jon R. Mancuso*
46. Shaft Alignment Handbook, *by John Piotrowski*
47. BASIC Programs for Steam Plant Engineers: Boilers, Combustion, Fluid Flow, and Heat Transfer, *by V. Ganapathy*
48. Solving Mechanical Design Problems with Computer Graphics, *by Jerome C. Lange*
49. Plastics Gearing: Selection and Application, *by Clifford E. Adams*
50. Clutches and Brakes: Design and Selection, *by William C. Orthwein*
51. Transducers in Mechanical and Electronic Design, *by Harry L. Trietley*
52. Metallurgical Applications of Shock-Wave and High-Strain-Rate Phenomena, *edited by Lawrence E. Murr, Karl P. Staudhammer, and Marc A. Meyers*
53. Magnesium Products Design, *by Robert S. Busk*
54. How To Integrate CAD/CAM Systems: Management and Technology, *by William D. Engelke*
55. Cam Design and Manufacture, Second Edition; with cam design software for the IBM PC and compatibles, disk included, *by Preben W. Jensen*
56. Solid-State AC Motor Controls: Selection and Application, *by Sylvester Campbell*
57. Fundamentals of Robotics, *by David D. Ardayfio*
58. Belt Selection and Application for Engineers, *edited by Wallace D. Erickson*
59. Developing Three-Dimensional CAD Software with the IBM PC, *by C. Stan Wei*
60. Organizing Data for CIM Applications, *by Charles S. Knox, with contributions by Thomas C. Boos, Ross S. Culverhouse, and Paul F. Muchnicki*
61. Computer-Aided Simulation in Railway Dynamics, *by Rao V. Dukkipati and Joseph R. Amyot*
62. Fiber-Reinforced Composites: Materials, Manufacturing, and Design, *by P. K. Mallick*
63. Photoelectric Sensors and Controls: Selection and Application, *by Scott M. Juds*
64. Finite Element Analysis with Personal Computers, *by Edward R. Champion, Jr. and J. Michael Ensminger*
65. Ultrasonics: Fundamentals, Technology, Applications, Second Edition, Revised and Expanded, *by Dale Ensminger*
66. Applied Finite Element Modeling: Practical Problem Solving for Engineers, *by Jeffrey M. Steele*
67. Measurement and Instrumentation in Engineering: Principles and Basic Laboratory Experiments, *by Francis S. Tse and Ivan E. Morse*

68. Centrifugal Pump Clinic, Second Edition, Revised and Expanded, *by Igor J. Karassik*
69. Practical Stress Analysis in Engineering Design, Second Edition, Revised and Expanded, *by Alexander Blake*

Additional Volumes in Preparation

Mechanical Engineering Software

Spring Design with an IBM PC, *by Al Dietrich*

Mechanical Design Failure Analysis: With Failure Analysis System Software for the IBM PC, *by David G. Ullman*

Practical Stress Analysis in Engineering Design

Second Edition, Revised and Expanded

ALEXANDER BLAKE

Lawrence Livermore National Laboratory
Livermore, California

MARCEL DEKKER, INC.   New York and Basel

Library of Congress Cataloging-in-Publication Data

Blake, Alexander.
　　Practical stress analysis in engineering design.

　　Includes bibliographical references.
　　1. Strains and stresses.　2. Engineering design.
I. Title.
TA648.3.B57　　1990　　　624.1'76　　　89-23370
ISBN 0-8247-8152-X (alk. paper)

This book is printed on acid-free paper

Copyright © 1990 by MARCEL DEKKER, INC. All Rights Reserved

Neither this book nor any part may be reproduced or transmitted in any form or by any means, electronic or mechanical, including photocopying, microfilming, and recording, or by any information storage and retrieval system, without the permission in writing from the publisher.

MARCEL DEKKER, INC.
270 Madison Avenue, New York, New York 10016

Current printing (last digit):
10 9 8 7 6 5 4 3 2 1

PRINTED IN THE UNITED STATES OF AMERICA

To my wife, Iris,
　　daughter, Susan,
　　　and grandchildren, Nathan and Becky

Preface to the Second Edition

Encouraging experience with the first edition clearly indicates that too much emphasis cannot be placed upon the values of practical design methodology. Although the mechanical and structural features of new products have been developing rather quickly, the basic design knowledge has changed very little. In spite of the contemporary trends toward more of the theoretical sciences and less of engineering design, our daily technical decisions often rely on ballpark estimates and that priceless commodity known as "horse sense," which comes only with long-standing practice. For this reason the new edition continues to emphasize closed-form solutions, a hands-on approach to fundamentals, and calculational techniques derived from the actual design activities in the field.

The basic philosophy, style, and intent of the first edition have been retained in order to promote simplicity of presentation, which is still the best strategy and a mark of truth in the face of "rising seas of detail" of modern technology. In line with this premise, the profile readers should include the technical audience of design practitioners, technologists, industrial consultants, students of product engineering, and those members of the teaching profession who recognize the importance of reinstating and maintaining design-orientated curricula.

The entire volume has undergone extensive scrutiny with special regard to the existing and new topics, numerical illustrations, working formulas, and the expanded list of references in concert with the constructive suggestions of the reviewers, informed users of the first edition, and pragmatists of stress analysis in this country and abroad. The author is particularly indebted to his associates at the Lawrence Livermore National Laboratory, Los Alamos National Laboratory, and the supporting industrial organizations who provided constant encouragement and helpful ideas. Special mention in this regard is due to Dr. Richard G. Dong from the Lawrence Livermore National Laboratory for his detailed revision of the selected portions of the manuscript, including dynamic response criteria, fundamentals of seismic design, and mechanics of stress propagation. Last but not least, the author is grateful to Mr. Anthony M. Davito from the Lawrence Livermore National Laboratory for his unwavering support of this book project featuring closed-form calculational tools applicable to bracketing solutions and interpretations of computer models of stress and strain.

It is hoped that, with the major additions and the innumerable minor improvements made throughout this text, the material presented constitutes another forward step in the application of stress analysis principles to engineering design. Since the value of this work cannot be enhanced without a continuous recourse to the latest experience of design practitioners in the field, the author welcomes their further comments and suggestions.

Alexander Blake

Preface to the First Edition

The purpose of this book is to provide a practice-oriented guide for the application of stress analysis principles to mechanical and structural design. This discipline cuts rather broadly across various facets of engineering wherever the response of deformable bodies to external forces is involved. It is based on the laws of Newtonian mechanics, and it relates to a number of traditional topics from machine design, strength of materials, and elasticity. However, the material given in this text also contains several new formulas and approaches derived from recent design experience.

Although the topics selected are intended for a diversified technical audience, the principal profile readers could be described as designers, technologists, industrial consultants, and engineering majors. In addition the discussion of some of the more advanced problems may be of interest to the members of the teaching profession advocating the merits of design-oriented curricula. These considerations, together with recent trends in national productivity goals, demand greater emphasis on practical design methodology to which this volume is dedicated.

Despite a rather extensive use of computer programs in stress calculations at the present time, crude models and "ballpark" estimates are still being made in daily technical decisions, often because of time and budget constraints. By the very nature of a ballpark estimate, the result can depart from what might be termed rigorous and accurate. However, there are many justifiable situations where even a crude estimate is far better than no basis for a conclusion. It is not uncommon for engineering practitioners to subscribe to such a philosophy. To respect this realistic view it is prudent to emphasize the elementary closed form solutions and the art of applying such tools to more complex problems.

The first two parts of this book are intended as a quick refresher course in fundamentals of stress analysis and solid mechanics, including such areas as steady-state stresses, elastic strain energy, dynamic response, elastic stability, fracture control, and thermal effects. The remaining four parts of the text have developed around applications of the theoretical principles and formulas to a great variety of sizing problems. These involve straight members, curved machine elements, circular rings, pipe flanges, support brackets, plates, pressure vessels, and similar structural configurations.

In addition to a number of rigorous equations, the book contains simplified formulas, tables, and charts which can be used directly during the preliminary stress and deflection calculations. Special features seldom found in more traditional texts include out-of-plane response of curved structural members, design of piston rings, practical guide to fracture mechanics, stress propagation criteria, working equations for rib-stiffened flanges, design aids for welded brackets, and sizing rules for circumferential stiffeners. Approximate calculation methods are also given for such problems as the effect of out-of-roundness on external collapse pressure of cylindrical vessels, influence of radial clearance on eyebar strength, actuating pressures in rolling diaphragms, and ultimate loads on Belleville springs, to mention a few.

Each chapter of this book contains its own list of symbols because of a large diversity of the topics and mathematical formulas involved. A brief summary of the principal stress and deflection equations, collated with the text, is given in Appendix A. This appendix has been compiled to give the reader a quick overview of the entire spectrum of the topics and the type of formulas found in the book. The theoretical material is supported by a number of illustrative design problems worked out in full detail. The numerical answers to these problems are given in English and SI units. A condensed outline of some of the principal conversion factors and the application of the SI units to a number of stress formulas are provided in Appendix B.

The growth rate of published technical material appears to be doubling every decade so that only a small fraction of the available stress analysis literature could be quoted in any single volume, including this book. In compiling the references, the goal was to draw upon long-lived and proven methods or formulas suitable for highlighting a particular topic. In this context a number of articles published by the author in *Machine Design, Design News,* and *Product Engineering* have been of special help in making the text more responsive to practical design needs. The author wishes to acknowledge the effect of these publications on the contents and the character of this book.

The predominantly applied nature of this work in general has evolved over many years of the author's industrial experience in this country and England. However, the final concept of the book has crystallized during the last 10 years, during which the author served as a technical advisor and chairman of a number of engineering review panels at the Lawrence Livermore National Laboratory in California.

Various organizations and individuals have influenced the choice of the topics and the methods of their presentation. In particular the author wishes to acknowledge the support and encouragement of the U.S. Department of Energy. Mr. Robert F. Pigeon of the Office of Technical Information, Oak Ridge, Tennessee, was

Preface to the First Edition

involved in the planning stage of the manuscript and continued his interest into the project. The completed manuscript was reviewed by Professor Stephen C. Cowin of Tulane University and Professor Harry Brandt of the University of California, with particular regard to the current requirements of the vocation-oriented mechanical engineering majors.

The preparation of the draft material and the camera-ready version of the text would have not been possible without the support of the Mechanical Engineering Department of the Lawrence Livermore National Laboratory. The original manuscript and its painstaking revisions have been typed and collated by Miss Saundra L. Carey, Mrs. Shirley A. Milani, and Mrs. Michelle I. Jewell. Mr. Henry L. Knoll reviewed a number of practical features of the text for their technical consistency. Finally the arduous and proficient task of composition, typing, and checking of the camera-ready material for the entire volume was accomplished by Mrs. Mildred E. Rundquist. The author is greatly indebted to all the above individuals for their contributions.

Last, but not least, the author would like to thank Industrial Press, the original publisher of his book *Design of Curved Members for Machines*, from which a number of illustrations and tables have been reproduced, and also Robert E. Krieger Publishing Company, from which a new edition of that book is now available, for their cooperation.

Although no effort was spared in citing the appropriate references and in providing due acknowledgment to all who have played a part in the development of this book, it would be too much to hope that nobody has been omitted. It would also be too naive to wish that none of the errors in formulas and the numerical work escaped attention. Therefore, the author welcomes any comments that readers and users of this text may have regarding any matter of the overall concept, technical details, or the physical makeup of the volume.

Alexander Blake

Contents

Preface to the Second Edition   v
Preface to the First Edition   vii

I ELEMENTS OF STATIC STRENGTH

1 Simple Stress and Strain   3

Introduction · Concept of Stress · Hooke's Law in One Dimension · Modulus of Elasticity · Poisson's Ratio · Implications of Hooke's Law · True Stresses and Strains · Symbols

2 Stresses in Shear and Torsion   16

Modulus of Rigidity · Shearing Stress and Strain · Analysis of Shear Stress Distribution · Stresses in Beams under Twist · Angle of Twist · Torsion of Tubes · Torsional Response of Rectangular Cross Sections · Symbols

3 Bending Stress   29

Stress Distribution · Section Modulus · Comments on Bending Theory · Symbols

4 Combined Stresses   34

Introduction · Two-Dimensional Stresses · Strain in Two Dimensions · Mohr's Circle of Stress · Plane Stress and Strain · Symbols

5 Criteria of Mechanical Strength   43

Introductory Comments · Mechanical Properties · Criteria of Brittle Failure · Criteria of Ductile Failure · Symbols

6 Elastic Strain Energy — 50
Basic Relations · Theory of Castigliano · Strain Energy in Bending · Energy Storage Criteria · Symbols

7 Deflection Analysis — 59
Introduction · Comments on Analysis · Strain Energy for a Curved Bar · Exact Deflection for a Thick Curved Bar · Deflection by Double Integration Versus Castigliano Method · Symbols

8 Statically Indeterminate Structures — 66
Problem Definition · Theorem of Least Work · Propped Cantilever · Pin-Jointed Arch · Symbols

9 Stress Concentration — 71
State of Macrostress · Elastic Stress Factors · Common Types of Stress Raisers · Stress Distribution · Plastic Reduction of Stress Factors · Symbols

10 Stability and Buckling Resistance — 80
Introduction · Historical Note · Basic Column Formulas · Concept of Euler Buckling · Inelastic Column Response · Weight Comparison · Buckling of Plates in Elastic Regime · Inelastic Buckling of Plates · Shear Buckling of Panels · Buckling due to Bending · Design of Special Columns · Other Buckling Problems · Symbols

II DYNAMIC AND THERMAL EFFECTS

11 Dynamic Response — 101
Introduction · General Criteria · Dynamic Strength · Response in Natural Mode · Free-Fall Effect · Shock Mitigation · Calculation of Frequency · Elementary Formulas for Dynamics · Symbols

12 Elements of Seismic Design — 115
Introduction · Building Code Method · Preliminary Elastic Design · Structural Damping · Symbols

13 Engineering Aspects of Fatigue — 125
Basic Definitions · Cumulative Damage Criterion · Neuber Effect · Elements of Theoretical Design · Effect of Surface Finish · Effect of Creep · Effect of Corrosion · Effect of Size · Endurance Limit · Low-Cycle Fatigue · Symbols

14 Design Aspects of Fracture Mechanics — 135
Characterization of Materials · Practical Aspects of Fracture Mechanics · Applications of Crack Size Parameter to Design · Implications of Fracture Toughness · Plane Stress Parameter · Plane Stress Criterion for Pressure Vessel Design · Symbols

15 Fundamentals of Fracture Control — 151
Historical Note · Basic Concepts and Definitions · Correlation of Fracture Properties · Practical Use of Crack Arrest Diagrams · Thickness Criteria · Significance of Stress and Strength ·

Contents   xiii

       Developments in Technology of Materials · Closing Remarks · Symbols

16 Mechanics of Stress Propagation   163

Introduction · Basic Concepts · Stress Propagation Theory · Elastic Impact · Axial Impact on Straight Bar · Conditions of Spall · Axial and Radial Modes of Elementary Structures · Response of Buried Structures · Stress Propagation in Granular Medium · Applications to Machinery · Symbols

17 Thermal Stresses and Materials   177

Nature of Thermal Response · Basic Stress Formula · Thermal Effect on Strength · Materials for Special Applications · Thermal Stress Index · Thermal Shock · Thermal Conditions in Piping · Thermal Stress Fatigue · Preliminary Thermal Design · Symbols

III STRAIGHT MEMBERS

18 General Design Criteria   189

Introduction · Practical Rules · Ultimate Strength of Beams · Symbols

19 Axial Response of Straight and Tapered Bars   193

Resilience of Solid Bar · Tapered and Stepped Bars · Tapered Bar Under Own Weight · Discussion of Tapered Bar Formula · Analysis of Composite Bars · Kern Limit · Symbols

20 Uniform Cantilever Beams   211

Applications and Implications · Design Charts and Tables · Effect of Shear · Symbols

21 Variable Section Cantilever Beams   222

Basic Considerations · Constant Parameter Criteria · Analysis of Tapered Cantilevers · Tapered Flat Springs · Design of Multiple-Leaf Springs · Symbols

22 Beams on Simple Supports   233

Basic Assumptions · Mathematical Concepts · Design Tables for Simple Beams · Symbols

23 Beams with Constraint   240

General Considerations · Beam with Sinking Support · Clevis Design · Frame Under Lateral Load · Partial Uniform Load · Special Design Problems · Symbols

24 Special Beam Problems   249

Introduction · Deep Section Beams · Wide Beams · Composite Beams · Beam Columns · Beams on Elastic Foundation · Symbols

IV CURVED MEMBERS

25 Curved Cantilevers   271

Related Considerations · Arched Cantilever Under Vertical Load · Arched Cantilever Under Horizontal Load · Arched Cantilever Under

Uniform Load · Arched Cantilever Under Moment · Complex-Shape Cantilever · Curved-End Cantilever · Symbols

26 Complex-Shape Springs · · · 292

Definitions and Assumptions · Snap Ring · Precurved Cantilever · Three-Quarter Wave Spring · Clip Spring · General U Spring · Instrument-Type U Spring · Symmetrical Wave Spring · Frame Spring Versus Wave Configuration · Symbols

27 Thin Rings and Arches · · · 310

Assumptions · Thin Elastic Ring · Design Charts for Circular Rings · Estimate by Superposition · Ring with Constraint · Analysis of Piston Rings · Analysis of Special Rings · Simply Supported Arch · Pin-Jointed Arch · Built-in Arch · Pinned Arch Under Uniform Load · Symbols

28 Curved Beams and Hooks · · · 337

Early Developments · Correction for Neutral Axis · Experimental Factors in Design · Approximation of Stress Factors · Stresses in Hooks · Design of Curved Beams · Curved Beams with Variable Cross Sections · Symbols

29 Links and Eyebars · · · 356

Introduction · Thick-Ring Theory · Thick-Ring Experiment · Theory of Chain Links · Link Reinforcement · Proof Ring Formulas · Knuckle Joint · Eyebar with Zero Clearance · Thick-Ring Method of Eyebar Design · API Standard · Eyebar with Finite Clearance · Eyebar Experiments · Other Modes of Eyebar Failure · Control of Sling Loads · Symbols

30 Out-of-Plane Response · · · 383

Torsional Strength Factors · Basic Equations · Transversely Loaded Arched Cantilevers · Ring Under Twisting Moments · Transversely Loaded Rings · Gimbal Ring Design · Symbols

V PLATES AND FLANGES

31 Design Fundamentals · · · 399

Simplifying Assumptions · Basic Plate Equations · Bending to Spherical Shape · Theory of Rectangular Panels · Circular Plate · Applications of Plate Theory to Flanges · Approximation of Large Deflection in Plates · Design Charts for Large Deflection of Plates · Special Plate Problems · Symbols

32 Panels and Closures · · · 416

Problem Definition · Design Charts for Panels · Similarities of Rectangular and Elliptical Panels · Circular Openings · Formulas and Charts for Circular Plates · Symbols

33 Flanges and Brackets · · · 430

General Background · Key Stress Criteria · Early Design Methods · Thin-Hub Theory · Thick Hub and Ring Model · Criterion of Flange Rotation · Use of Plate Theory in Flanges · Formula for

Contents   xv

      Hub Stress · German Practice of Flange Design · Waters–Taylor Formula · Circumferential Stress · Apparent Stress Criteria · Plastic Correction · Heavy-Duty Flanges · Equivalent Depth Formula · Load Sharing in Ribbed Flanges · Strength of Flange Ribs · Local Bending of Flange Ring · Correction for Tapered Gussets · Elements of Bracket Design · Theory of Weld Stresses · Selection of Formulas for Brackets · Strength and Stability Considerations · Symbols

34 Special Configurations   475
      Perforated Plates · Reinforced Plates · Pin-Loaded Plates · Belleville Washer · Symbols

VI   PIPING AND VESSELS

35 Internal Pressure   491
      Introduction · Membrane Theory · Thin Cylinders · Radial Growth · Ellipsoidal Shells · Toroidal Vessel · Thick-Cylinder Theory · Thick-Walled Sphere · Design Charts for Thick Cylinders · Ultimate Strength Criteria · Burst Pressure of Cylinders and Spheres · Shrink-Fit Design · Symbols

36 External Pressure   513
      Introduction · Thinness Factor · Stress Response · Stability Response · Mixed Mode Response · Classical Formula for Short Cylinders · Modified Formula for Short Cylinders · Simplified Criterion for Out-of-Roundness · Long Cylinder with Out-of-Roundness · Effective Out-of-Roundness · Empirical Developments · Effect of Axial Stresses on Collapse · Elastic Buckling of Spherical Shells · Corrected Formula for Spherical Shells · Plastic Strength of Spherical Shells · Effect of Initial Imperfections · Experiments with Hemispherical Vessels · Response of Shallow Spherical Caps · Strength of Thick Cylinders · Strength of Thick Spheres · Approximate Stress Criterion for Cylinders · Out-of-Roundness Correction for Stress · Symbols

37 Axial and Bending Response   546
      Introduction · Comment on Sectional Properties · Column Behavior of Pipe · Pipe on Elastic Foundation · Theory of One-Way Buckling · Axial Response of Cylinders · Plastic Buckling in Axial Mode · Analysis of Bellows-Type Buckle · Example of Load Eccentricity · Buckling due to Bending · Theory of Rolling Diaphragm · Symbols

38 Special Problems in Cylinders   569
      Dilation of Closed Cylinders · Nested Cylinders · Design of Ring Stiffeners · Symbols

VII   MISCELLANEOUS STRESS TOPICS IN DESIGN

39 Design of Bolted Joints   579
      Torque Formula · External Load · Effective Area Concept · Elastic Response · Stresses in Bolts · Improvements in Bolt Performance · Analysis of Thread Loading · Joint Rigidity and Leakage · Selection and Testing of Bolt Materials · Symbols

40	**Load Transfer in Mechanical Connections**	595

General Comments · Split-Hub Design · Piping Supports and Branches · Key and Pin Connections · Structural Pin Design · Wire Rope Connection · Symbols

41	**Mechanical Springs**	615

Introduction · Compression Spring · Spring Deformation and Dynamics · Extension Spring · Torsion Spring · Buckling Column Spring · Symbols

42	**Application of Metric Units to Stress Analysis**	629

Introduction · Shear Stress · Beam Deflection · Buckling Stress · Fracture Toughness · Stress Propagation

Appendix: Selection of Practical Stress Formulas	637
References	667
Index	*679*

I
ELEMENTS OF STATIC STRENGTH

1
Simple Stress and Strain

INTRODUCTION

The primary objective of this part of the book is to provide a quick overview of the basic stress analysis concepts which can be readily applied to the sizing and evaluation of the structural integrity of typical load-carrying members. The topics selected are intended to emphasize those aspects of stress analysis which form the ground work for many design decisions tempered only by engineering judgment, knowledge of materials, and loading conditions. Since the performance of various hardware systems and components may be deeply involved in safety and economic considerations, the importance of the rational approach to stress calculations cannot be overemphasized.

The strength criterion for a particular machine or a structural component is judged on the basis of stress, deformation, material's characteristics and on the specific work requirements. These requirements may involve a life-span of the component, thermal response, dynamic environment as well as manufacturing tolerances. For instance, the shaft of a machine is designed to carry twisting and bending loads simultaneously for many millions of revolutions while its transverse deflection has to remain limited by the available working clearances. In another situation the bolts holding a pipe flange, normally sized, say, for the axial loads and temperature effects, may be subject to some additional tensile and bending stresses due to an external bending moment caused by load eccentricity or by a dynamic environment. It is easy to see then that in many practical situations a complex arrangement of interacting structural components and loading conditions can exist. Under such terms of reference the task of detailed and rigorous stress analysis may often be rather difficult, time consuming, and open to question. It is extremely fortunate, however, that in real life the manipulation of even the simplest stress formulas and the application of sometimes crude experiments can save the day. For this reason this and the subsequent sections will be devoted to relatively simple but indispensable concepts of stress analysis.

CONCEPT OF STRESS

The fundamental concept of stress may be defined as the internal force per unit area of a given cross section which resists a change in size or shape within the structural member. Since the effect of an applied external force must depend on the cross-sectional area of the load-carrying component, the most convenient representation of structural behavior is through the theory of stress and strain. This fundamental theory permits a description of the material's characteristics independently of the size or the gage length of the test pieces.

Although this book is not concerned with material evaluation per se, the fact is that testing of materials and structures provides a basis for the verification of theoretical concepts. The mechanical properties of materials are inseparable from the subject of solid mechanics, and their realistic magnitudes are indispensable in testing new design formulas and even very sophisticated computer codes. The mathematics of stress analysis is only as good as our knowledge of engineering materials.

One of the more important ingredients in applying the concept of stress to design is an understanding of the nature, orientation, and magnitude of the external force. In many cases we deal with stationary or dead loads that can cause axial, bending, or torsional response in a structural member. In other situations, a change in velocity can create an inertia force. A change in temperature may initiate internal forces, or a frictional effect can influence the equilibrium of forces acting on a particular structural or machine component.

Experience indicates that, quite often, knowledge of a material's properties and the nature of loading is not satisfactory. This situation may not change the basic concept of stress, but it can certainly affect the confidence level of the results of stress analysis. The overview that follows assumes an adequate knowledge of materials and loads.

HOOKE'S LAW IN ONE DIMENSION

To illustrate the basic concept of stress in a one-dimensional field, consider the simple test piece shown in Fig. 1.1. In materials testing, the original length of the specimen, L, is referred to as a *gage dimension* or *gage distance*. Under the action of an external load W applied axially, the test piece undergoes an increase

Simple Stress and Strain

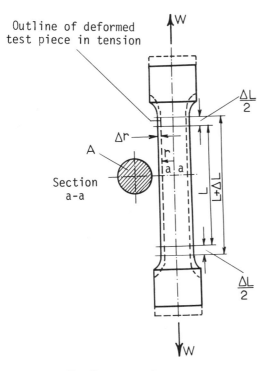

Fig. 1.1 Tensile test specimen.

in length ΔL.* As long as the material behaves elastically, a large portion of the load-deflection diagram obtained from the test will be found to be linear, as shown in Fig. 1.2.

Although Fig. 1.1 refers to axial tension, similar arguments should apply to compression, assuming that the test piece or a machine component is sufficiently rigid not to fail in buckling as a column. Hence, the general response implied by Fig. 1.2 is good for axial tension as well as compression.

The elementary case described here has important design ramifications because it forms the basis for defining the one-dimensional *Hooke's law*, widely used in stress calculations. This law states that the relationship between the stress and strain may be regarded as linear for the great majority of engineering materials. The *axial stress*, then, defined as the intensity of the force on a cross section A, follows directly from Fig. 1.1.

$$S = \frac{W}{A} \tag{1.1}$$

---

*The meanings of symbols and dimensional units involved are listed under "Symbols" at the end of each chapter.

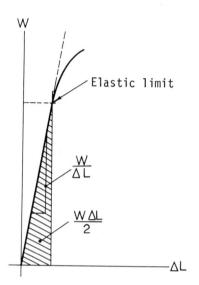

Fig. 1.2 Load-deflection diagram.

The *strain* in the axial direction is

$$\epsilon = \frac{\Delta L}{L} \tag{1.2}$$

The original load-deflection diagram (Fig. 1.2) can be replotted as shown in Fig. 1.3, where the slope of the curve is also constant below the elastic limit of the material. Here the proportionality between the stress and strain determines the fundamental quantity in the mechanics of materials known as the *modulus of elasticity*

$$E = \frac{S}{\epsilon} \tag{1.3}$$

Since S is expressed, for instance, in pounds per square inch and strain ϵ is nondimensional by definition, the dimensions of the modulus of elasticity must be the same as those of the stress.

Scientific and industrial workers are in the process of adopting a version of the metric system, the International Systems of Units (SI), according to which the stress is expressed in terms of megapascals. To reflect the current transition period, this book has adopted a dual system of units for stress, featuring pounds per square inch (psi) and newtons per square millimeter. According to the SI system, 1 megapascal (MPa) is equivalent to 1 newton per square millimeter (N/mm^2). Until recently, various other dimensions were used throughout the world, including tons per square inch, kilograms per square centimeter, and kilograms per square millimeter. See Chap. 42 for details of conversion from English to the SI units pertinent to stress analysis.

Simple Stress and Strain

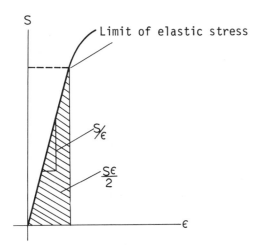

Fig. 1.3 Stress-strain diagram.

MODULUS OF ELASTICITY

In practice, the modulus of elasticity will not have the same value for all the specimens cut out of a plate from a given mill run. Precision tests will always indicate some slight variations. However, these differences are small and the average values of E are quite satisfactory for most engineering design. As far as commonly recognized structural materials are concerned, the modulus can vary as shown in Table 1.1.

Other modulus values, for standard as well as some exotic materials, are readily available from engineering handbooks. It will suffice here to indicate that the range of E values can be rather wide, including, for instance, tungsten at 58,000,000 psi down to about 100,000 psi for a number of selected high polymers commonly referred to as plastics. The lower boundary will, of course, approach zero for a number of viscoelastic materials. The upper boundary for certain exotic materials may be even higher than 58 million psi.

The modulus of elasticity, denoted in most textbooks by E, is often called *Young's modulus*. The latter name honors Thomas Young (1773–1829), who first

Table 1.1 Selected Values of Modulus of Elasticity for General Construction Work (psi)

Structural steel	29,000,000
Wrought iron	27,000,000
Cast iron	15,000,000
Aluminum	10,000,000
Concrete (unreinforced, Portland cement)	2,000,000
Timber (parallel to the grain)	1,200,000

proposed this concept to engineers in England. The proportionality between the uniaxial stress and strain can be extended to biaxial and triaxial states of stress.

For reasons of expediency, the original concept of the elastic modulus has been extended to applications involving plastic behavior of materials and those areas of design where an equivalent modulus of elasticity can be used as a measure of rigidity of a component or material.

Numerous solutions in structural design and mechanics utilize some interpretation of the original concept of Young's modulus. For instance, in studies of the stability and inelastic buckling of structural members it is customary to employ secant and tangent moduli defined in Fig. 10.4. In dynamics, the assessment of the critical velocity of impact involves the notion of a local modulus of elasticity as later shown in Chap. 16. The apparent stress concept in plastic correction and the critical buckling criteria in plates given in Chap. 33 invoke the definitions of the reduced and tangent moduli of elasticity. The design of a perforated plate according to the accepted practice is based on the definition of modified elastic constants described in Chap. 34. The plastic strength of a spherical shell subjected to external pressure is shown to depend on an equivalent modulus in Chap. 36. Finally, the analysis of axial response of a pipe in the plastic regime, discussed in Chap. 37, utilizes the concept of reduced modulus.

The foregoing examples scattered throughout the various chapters of this book emphasize the remarkable utility and flexibility of the original concept of the elastic modulus. These comments, however, would be incomplete without a mention of the use of this concept in more complex situations where the properties may vary with direction in the material [21] or configurational features. For instance, in the process of designing and testing sandwich and honeycomb structures, the elastic and shear moduli become complicated functions of the dimensional relationships and experimental parameters.

POISSON'S RATIO

The longitudinal strain ϵ caused by the uniaxial stress is accompanied by a lateral strain $u = \Delta r/r$, as shown in Fig. 1.1. The constant of proportionality between the lateral and axial strains is known as *Poisson's ratio* and is often denoted by ν. Hence, the ratio between the two strain components is

$$\nu = \frac{u}{\epsilon} \tag{1.4}$$

The concept defined by Eq. (1.4) enters numerous stress formulas, where ν is normally determined experimentally. Knowledge of this constant permits calculation of the change in volume due to strain. For example, in the case of a bar of uniform cross section subjected to tension [1], the *volumetric strain*, equal to the ratio of the change in volume to the original volume, can be derived as follows.

Let the length of the bar be L and its radius r, as shown in Fig. 1.1. Before the bar is stretched, the volume must be equal to

$$V = \pi r^2 L \tag{1.5a}$$

Simple Stress and Strain

The elongation of the bar follows from Eq. (1.2)

$$\Delta L = \epsilon L \tag{1.5b}$$

At the same time, the cross section contracts by an amount defined by Eq. (1.4). Since $\Delta r = ur$, we get

$$\Delta r = \epsilon \nu r \tag{1.5c}$$

Hence, the volume of the bar subjected to the tensile mode becomes

$$V' = \pi(r - \epsilon \nu r)^2 (L + \epsilon L) \tag{1.5d}$$

or

$$V' = \pi r^2 L (1 - \epsilon \nu)^2 (1 + \epsilon) \tag{1.5e}$$

Multiplying throughout and neglecting the terms involving the squares and higher powers of the uniaxial strain gives

$$V' = \pi r^2 L (1 + \epsilon - 2\nu\epsilon) \tag{1.5f}$$

We can now define the volumetric strain as the ratio of the change of volume to the original volume, using Eqs. (1.5a) and (1.5f)

$$\frac{V' - V}{V} = \frac{\pi r^2 L(1 + \epsilon - 2\nu\epsilon) - \pi r^2 L}{\pi r^2 L} \tag{1.5g}$$

Simplifying Eq. (1.5g) and denoting the volumetric strain for this bar by $\Delta V/V$ gives

$$\frac{\Delta V}{V} = \frac{\Delta L}{L}(1 - 2\nu) \tag{1.6}$$

Although Fig. 1.1 illustrates a bar of circular cross section, the same rules apply to bars with other uniform cross sections. For an elemental cube submitted to hydrostatic stress, the volumetric strain is equal to three times the uniaxial strain, provided that the response is elastic.

When engineering materials behave elastically, the values of Poisson's ratio can vary within a relatively narrow range, usually between 0.25 and 0.35. For a material subjected to very high stresses and plastic deformation, the theoretical limit of 0.5 may be reached. The volume of the material under these conditions remains essentially constant, and the change in volume, as shown, for example, by Eq. (1.6), becomes zero. On the other hand, for an ideally brittle material, the volume change would be equal to the linear change in a dimension, as given by Eq. (1.6). In practice, it is unlikely that the minimum value of Poisson's ratio will be less than 0.1. It should also be noted that, by itself, the Poisson's ratio effect creates no additional stresses unless some lateral constraint is introduced

Table 1.2 Typical Values of Poisson's Ratio

Upper theoretical limit (perfectly deformable material)	0.50
Lead	0.43
Gold	0.42
Platinum	0.39
Silver	0.37
Aluminum (pure)	0.36
Phosphor bronze	0.35
Tantalum	0.35
Copper	0.34
Titanium (pure)	0.34
Aluminum (wrought)	0.33
Titanium (alloy)	0.33
Brass	0.33
Molybdenum (wrought)	0.32
Stainless steel	0.31
Structural steel	0.30
Magnesium alloy	0.28
Tungsten	0.28
Granite	0.28
Sandstone	0.28
Thorium (induction-melted)	0.27
Cast iron (gray)	0.26
Marble	0.26
Glass	0.24
Limestone	0.21
Uranium (D-38)	0.21
Plutonium (alpha phase)	0.18
Concrete (average water content)	0.12
Beryllium (vacuum-pressed powder)	0.027
Lower theoretical limit (perfectly brittle material)	0.000

which obviates the natural tendency of the material to obey Poisson's law. The absolute theoretical limits for Poisson's ratio are zero and 0.5 for perfectly brittle and ductile materials, respectively. Some typical design values of this ratio for a variety of materials are given in Table 1.2.

Design Problem 1.1

A stainless steel bar having a rectangular cross section of 2 in. by 0.5 in. is subjected to an axial pull of 25,000 lb. Estimate the decrease in the cross-sectional dimensions on the assumption that the modulus of elasticity is 29×10^6 psi.

Solution

From Table 1.2, the design value of Poisson's ratio is $\nu = 0.31$. The tensile stress is

$$S = \frac{W}{A} = \frac{25,000}{2 \times 0.5} = 25,000 \text{ psi}$$

Simple Stress and Strain

The corresponding longitudinal strain follows from Eq. (1.3)

$$\epsilon = \frac{S}{E} = \frac{25,000}{29 \times 10^6} = 0.000862$$

From Eq. (1.4)

$$u = \nu\epsilon = 0.31 \times 0.000862 = 0.000267$$

Hence, multiplying the lateral strain by the two sides of the cross section gives the magnitudes of contraction

$$2 \times 0.000267 = 0.000534 \text{ in.} \quad (0.013564 \text{ mm})$$

and

$$0.5 \times 0.000267 = 0.0001335 \text{ in.} \quad (0.003391 \text{ mm}) \quad \blacklozenge$$

Design Problem 1.2

A short, solid cylinder made out of wrought aluminum is to sustain a compressive stress of 50,000 psi. Calculate the necessary diameter of the cylinder in order not to exceed its radial expansion equal to $\Delta r = 0.0025$ in., assuming the modulus of elasticity to be 10×10^6 psi. Indicate the magnitude of the compressive load that this cylinder can carry within the assigned stress limit. It is assumed that this cylinder will not buckle as a column.

Solution

From Table 1.2, the design value of Poisson's ratio is $\nu = 0.33$. From Eq. (1.4), the necessary radial strain is

$$\frac{\Delta r}{r} = \nu\epsilon \tag{1.7a}$$

whereas, from Eq. (1.3), the longitudinal strain is

$$\epsilon = \frac{S}{E} \tag{1.7b}$$

Combining Eqs. (1.7a) and (1.7b) and solving for r gives

$$r = \frac{E(\Delta r)}{\nu S} \tag{1.7c}$$

Hence, substituting the numerical values into Eq. (1.7c) yields

$$r = \frac{10 \times 10^6 \times 0.0025}{0.33 \times 50,000} = 1.515 \text{ in.}$$

The maximum allowable compressive load is

$$W = \pi r^2 S$$
$$= \pi \times 1.515^2 \times 50{,}000 = 360{,}533 \text{ lb} \quad (1{,}603{,}723 \text{ N}) \quad \blacklozenge$$

Design Problem 1.3

A gray cast-iron bar with a square cross section of 1.5 in. side carries a tensile load of 100,000 lb. Assume the modulus of elasticity to be equal to 20×10^6 psi. Calculate the absolute change in volume corresponding to the maximum stretch assuming the initial length of the bar to be 60 in.

Solution

From Table 1.2, Poisson's ratio is $\nu = 0.26$. Combining Eqs. (1.1) to (1.3) with Eq. (1.6) gives

$$\frac{\Delta V}{V} = \frac{W(1-2\nu)}{AE}$$

Hence, substituting the numerical data yields

$$\frac{\Delta V}{V} = \frac{100{,}000(1 - 2 \times 0.26)}{1.5 \times 1.5 \times 20 \times 10^6} = 0.001$$

The original volume of the bar is

$$V = 1.5 \times 1.5 \times 60 = 1.35 \text{ in.}^3$$

The absolute change in volume is then

$$0.001 \times 135 = 0.135 \text{ in.}^3 \quad (2212 \text{ mm}^3) \quad \blacklozenge$$

IMPLICATIONS OF HOOKE'S LAW

Design Problems 1.1 to 1.3 illustrate application of the simplest form of Hooke's law in terms of the axial tension or compression within the limits of the elastic response. The theory of elasticity is built on the premise of Hooke's law and deals with structural materials that can be characterized by the proportionality of stress and strain, elasticity, isotropy, and homogeneity. The modulus of elasticity and Poisson's ratio involved in this law are obtainable from experiments. For the purpose of the analysis, the material parameters are considered to be independent of time and spatial coordinates.

When loads are applied along the x and y axes, the relevant deformations depend on the stresses S_x and S_y which act along these axes. These deformations can be used to calculate the stresses if E and ν are known.

$$S_x = E\frac{\epsilon_x + \nu\epsilon_y}{1-\nu^2} \qquad (1.8)$$

Simple Stress and Strain

and

$$S_y = E\frac{\epsilon_y + \nu\epsilon_x}{1 - \nu^2} \tag{1.9}$$

For the specific case of equal strains along the two axes, we have $\epsilon_x = \epsilon_y$, so that Eqs. (1.8) and (1.9) reduce to the form

$$S_x = S_y = E\frac{\epsilon_x}{1 - \nu} \tag{1.10}$$

Suppose now that a cubic inch of material is subjected to a single tensile stress S_x. Figures 1.2. and 1.3 show that the areas under the load-deflection curve and the corresponding stress-strain curve represent the amount of the work done on this material. Assuming that, within the elastic limit, the amount of this work is equal to the elastic energy stored, we can state that

$$U' = \frac{(S_x + 0)\epsilon_x}{2} \tag{1.11a}$$

In Eq. (1.11a), the stress along the y axis is made equal to zero. Since ϵ_x now follows from Eq. (1.3)

$$\epsilon_x = \frac{S_x}{E} \tag{1.11b}$$

combining Eqs. (1.11a) and (1.11b) gives the principal formula for the unit strain energy

$$U' = \frac{S_x^2}{2E} \tag{1.12}$$

Following identical reasoning, the unit strain energy can be obtained for the case when a unit volume of material is acted upon by two tensile stresses, S_x and S_y, along the orthogonal axes x and y. This should yield

$$U' = \frac{S_x^2 + S_y^2 - 2\nu S_x S_y}{2E} \tag{1.13}$$

A typical three-dimensional state of stress is found, for instance, in the walls of a pressure vessel. This involves hoop, meridional, and radial components of stress which resist pressure and temperature loading. The extension of Hooke's law to the triaxial state of stress indicates the following basic relations:

$$\epsilon_x = \frac{1}{E}(S_x - \nu S_y - \nu S_z) \tag{1.14}$$

$$\epsilon_y = \frac{1}{E}(S_y - \nu S_x - \nu S_z) \tag{1.15}$$

$$\epsilon_z = \frac{1}{E}(S_z - \nu S_x - \nu S_y) \tag{1.16}$$

Hence, the stresses can be explicitly defined in terms of the unit deformations measured along the x, y, and z axes. This procedure yields

$$S_x = E\frac{\nu(\epsilon_y + \epsilon_z) + (1 - \nu)\epsilon_x}{(1 + \nu)(1 - 2\nu)} \tag{1.17}$$

$$S_y = E\frac{\nu(\epsilon_x + \epsilon_z) + (1 - \nu)\epsilon_y}{(1 + \nu)(1 - 2\nu)} \tag{1.18}$$

$$S_z = E\frac{\nu(\epsilon_x + \epsilon_y) + (1 - \nu)\epsilon_z}{(1 + \nu)(1 - 2\nu)} \tag{1.19}$$

When a cubic inch of material is stressed in tension along the three orthogonal axes, the unit elastic strain energy becomes

$$U' = \frac{1}{2E}\left[S_x^2 + S_y^2 + S_z^2 - 2\nu(S_xS_y + S_yS_z + S_zS_x)\right] \tag{1.20}$$

It is clear from Eqs. (1.8) and (1.9), and (1.14) to (1.16) that when a cube of material is elongated in a given direction, it is at the same time shortened by the Poisson's ratio effect of one or two of components acting at right angles to the first. For the case of uniform compression acting along the three axes x, y, and z, it is possible to calculate the unit decrease in volume caused by the hydrostatic compression. These considerations lead to the following expression for the bulk modulus [7]:

$$K = \frac{E}{3(1 - 2\nu)} \tag{1.21}$$

Formula (1.21) represents the ratio of the external pressure, acting on a submerged solid, to the amount of the relative change in volume of this solid. For instance, if the volumetric change is 0.001 in.^{3}/in. of the material when the external pressure is 20,000 psi, the bulk modulus calculates to be $20,000/0.001 = 20,000,000$ psi. It may be of interest to note that for a perfectly brittle material exhibiting $\nu = 0$, $K = E/3$. However, for a plastic material with ν approaching the theoretical limit of 0.5, the bulk modulus based on Eq. (1.21) becomes very large indeed. This corresponds to what is known as the incompressible feature of the material subjected to the triaxial compression.

TRUE STRESSES AND STRAINS

The ideal case of uniaxial tension is defined by Fig. 1.1. However, tests of relatively ductile materials show that as the load is increased the elongation tends to localize, forming a neck in the specimen where the failure eventually occurs. The engineering stress given by Eq. (1.1) is normally used for design purposes because it is convenient to base the external load on the original cross-sectional area. When it is required to better understand the actual behavior of the material stressed beyond the elastic limit, the concept of the true stress is introduced by dividing the force W by

Simple Stress and Strain

the actual cross-sectional area. As we load the test piece further, the true stress continues to increase and in the fully plastic regime it becomes closer to reality.

The concept of true strain can be related to engineering strain given by Eq. (1.2) in a way similar to the relation of true stress to engineering stress [29]. The true strain at any stage of loading may be obtained by dividing a small increment of deflection by the actual gage length. This strain can be calculated as

$$\epsilon_t = \ln(1 + \epsilon) \tag{1.22}$$

For instance, when engineering strain is 25%, the true strain obtained from Eq. (1.22) is 22%. Note that the natural logarithm is used in the foregoing equation.

It should be pointed out in general that when the strains are small, it is of little consequence whether the stress is calculated from the original or the actual dimensions of the structural member under load. When the strains are expected to be large, such as in special metal forming processes, the differences between the engineering and the so-called true strain values can be appreciable.

SYMBOLS

A	Cross-sectional area, in.^{2} (mm^2)
E	Modulus of elasticity, psi (N/mm^2)
K	Bulk modulus, psi (N/mm^2)
L	Length, in. (mm)
ΔL	Change in length, in. (mm)
r	Radius of solid bar, in. (mm)
Δr	Change in radius, in. (mm)
S	General symbol for stress, psi (N/mm^2)
S_x	Stress along x axis, psi (N/mm^2)
S_y	Stress along y axis, psi (N/mm^2)
S_z	Stress along z axis, psi (N/mm^2)
U'	Unit strain energy, lb-in. (N-mm)
u	Lateral strain, in./in. (mm/mm)
V	Volume of stressed material, in.^{3} (mm^3)
V'	Original volume, in.^{3} (mm^3)
ΔV	Change in volume, in.^{3} (mm^3)
W	External load, lb (N)
ϵ	Engineering strain, in./in. (mm/mm)
ϵ_t	True strain, in./in. (mm/mm)
ϵ_x	Strain in x direction, in./in. (mm/mm)
ϵ_y	Strain in y direction, in./in. (mm/mm)
ϵ_z	Strain in z direction, in./in. (mm/mm)
ν	Poisson's ratio

2
Stresses in Shear and Torsion

MODULUS OF RIGIDITY

So far, only the direct stresses have been considered, as these are fundamental to understanding of the basic concepts of the two most commonly used design constants, E and ν. There are, however, many other cases where certain shearing action takes place, as illustrated, for instance, in Fig. 2.1. It is clear that the rivet shown resists load Q in such a manner that its total cross section is subject to transverse shearing action. An element of the rivet, cut out in the form of a rectangular block, is shown with greatly exaggerated displacement of the upper edge of the element with respect to the lower edge. For the case of a relatively small elastic deformation, the experimental evidence indicates that there is a linear relationship between the shearing stress τ and the distortion angle γ. As shown in Fig. 2.1, $\tan \gamma$ is nearly equal to γ in radians for all small values of δ_0, and by analogy to the case of direct tension, we can state the following general principle:

$$G = \frac{\tau}{\gamma} \qquad (2.1)$$

The elastic constant G defined in Eq. (2.1) is known in engineering as the *modulus of rigidity* or the *shearing modulus of elasticity*, and bears a definite numerical relation to the modulus of elasticity and the Poisson's ratio of the material. The

Stresses in Shear and Torsion

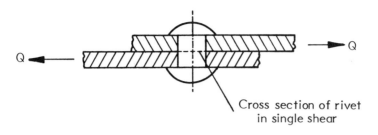

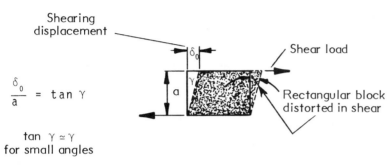

Fig. 2.1 Direct shear concept and notation.

three basic elastic constants are connected by the following relation:

$$G = \frac{E}{2(1+\nu)} \tag{2.2}$$

Equation (2.2) is often used in experimental work for the determination of Poisson's ratio since the values of E and G can be obtained from the stress-strain characteristics of the material.

In practical applications the ratio G/E is sometimes taken as 0.4. Solving Eq. (2.2) for ν gives

$$\nu = \frac{1}{2(G/E)} - 1 \tag{2.3}$$

Substituting $G/E = 0.4$ into Eq. (2.3) gives $\nu = 0.25$. Using several values of Poisson's ratio between the two theoretical extremes of 0 and 0.5, Eq. (2.2) yields Table 2.1.

Table 2.1 Ratio of Elastic Constants

ν	0	0.1	0.2	0.3	0.4	0.5
G/E	0.500	0.455	0.417	0.385	0.357	0.333

SHEARING STRESS AND STRAIN

By analogy to Eq. (1.2) illustrating axial strain, the shearing strain may be defined as

$$\gamma = \frac{\delta_0}{a} \tag{2.4}$$

Strictly speaking, then, shearing displacement δ_0 may be thought of as an angle. This concept is helpful in the analysis of torsional problems arising in machine shafts, aircraft fuselages, eccentrically loaded beams, and similar structures. The direct shearing stress in general can be computed as the shearing force Q divided by the cross-sectional area A, and for many practical applications, where generous factors of safety are involved, this approach should be entirely acceptable. Figure 2.1, pertaining to the general concept of shearing strains and stresses, gives no clue as to the actual distribution of the stresses across the cross section.

The first step in the treatment of any shear problem is to define the magnitude and direction of the external load involved. It is assumed that the reader is familiar with the elements of statics. For instance, the magnitude of the vertical shear load may be determined from the resolution of all the external forces acting, say, on either the left- or right-hand portion of the beamlike component. In Fig. 2.1, the rivet behaves as a short, structural beam, and Q represents the external force.

Before turning our attention to some of the basic aspects of the distribution of shear stresses in beams, it is well to note that the shear stress, acting on a given cross section, produces sliding of the transverse elements. This action provokes a response from the longitudinal fibers, which develop a complementary stress of equal magnitude to the transverse stress. This effect becomes quite apparent, for example, in loading wooden beams to destruction. Since wood is weaker along the grain parallel to the neutral axis, shearing takes place along that grain. The concept of a complementary stress therefore promotes the conclusion that the shearing stresses must occur in pairs.

ANALYSIS OF SHEAR STRESS DISTRIBUTION

Suppose that we consider an element of the beam defined by the two cross sections, distance dx apart and loaded by Q, M, and $M + dM$, shown in Fig. 2.2. These forces are needed for the static equilibrium of the beam element indicated. To obtain an expression for the shear stress at the layer located a distance a from the neutral axis, first consider a bending stress at distance y

$$S_b = \frac{My}{I} \tag{2.5}$$

Equation (2.5) represents the well-known beam formula used by practicing design engineers.

Stresses in Shear and Torsion

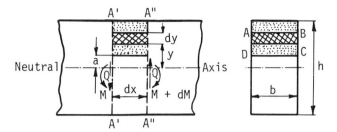

Fig. 2.2 Detail of beam element for shear stress analysis.

Horizontal force on the shaded element of beam cross section $ABCD$ at the left plane $A'A'$ is

$$F' = \frac{Myb\,dy}{I} \tag{2.6a}$$

Similarly, for the plane $A''A''$ as shown, we get

$$F'' = \frac{(M+dM)yb\,dy}{I} \tag{2.6b}$$

The net horizontal force parallel to the neutral axis must be equal to $F'' - F'$, so Eqs. (2.6a) and (2.6b) give

$$F'' - F' = \frac{dM}{I} by\,dy \tag{2.6c}$$

The total horizontal force between $y = a$ and $y = h/2$ is

$$F = \int_a^{h/2} \frac{dM\,by\,dy}{I} \tag{2.6d}$$

Finally, the shear stress at $y = a$ is

$$\tau = \frac{\frac{dM}{I}\int_a^{h/2} by\,dy}{b\,dx} = \frac{dM}{I\,dx}\int_a^{h/2} y\,dy \tag{2.6e}$$

or

$$\tau = \frac{Q}{I}\int_a^{h/2} y\,dy \tag{2.6f}$$

In going from Eq. (2.6e) to (2.6f) we have defined the shear force Q as the first derivative of the bending moment. The term I stands for the moment of inertia for the entire beam cross section. The proof of the statement $dM/dx = Q$ will be found in the section on mathematical concepts in Part III of this book.

For the total rectangular cross section, the shear stress can be found from Eq. (2.6f) by integrating between the limits of 0 and $h/2$. This gives

$$\tau = \frac{Q}{I} \int_0^{h/2} y\, dy = \frac{Q}{I} \left. \frac{y^2}{2} \right|_0^{h/2} = \frac{3Q}{2bh} \tag{2.6g}$$

Two important observations can now be deduced from the foregoing analysis. The shear stress distribution is parabolic for the rectangular beam cross section with the maximum value attained at the neutral axis. The maximum stress is 3/2 times the mean stress, which follows directly from Eq. (2.6g)

$$\tau = \frac{3Q}{2A} \tag{2.7}$$

Similar analysis for a beam with circular cross section shows that the shear stress distribution is again parabolic. The maximum value of the shear stress for the circular geometry works out to be 4/3 times the average stress. The values of 3/2 and 4/3 are known as the *shear distribution factors*. Examples of the maximum shear stresses for these and other important cross-sectional geometries are given in Table 2.2. Note that τ_{av} in this table is obtained by dividing the shear load by the total beam cross section.

Structural beams with open cross sections such as those of the I or T type transmit the shear loads largely through the webs, and the maximum stress approximates closely that obtained by dividing the shear load by the area of the web. The effect of the flange on the shear stress distribution is found to be extremely small. In rapidly changing cross-sectional geometry, however, some engineering judgment would be required in deciding which portions of the section are likely to behave as flanges and which should be treated as typical webs.

The fact that the shear stresses vanish at the edges of all the cross-sectional configurations is certainly a very convenient feature of structural behavior. It should be stipulated, however, that the foregoing shear considerations apply only to those shear loads which act through the center of twist coinciding with the centroid of the cross section. This type of loading assures that no torsional moments will be produced. The foregoing condition is fulfilled by cross-sectional areas having two areas of symmetry.

For a general shape of beam cross section, the basic expression for the derivation of the shear distribution factor may be given as follows:

$$\tau = \frac{Q}{IB} \int_a^c by\, dy \tag{2.8}$$

The reason for retaining both width dimensions B and b in this expression is that the width of the section may not be uniform. The limits of integration a and c refer to some arbitrary point, distance a removed from the neutral axis, and to the maximum distance from the neutral axis, respectively. In the case of Fig. 2.2, $c = h/2$, and for the specific case calculated for the rectangular cross section, the distance a is made equal to zero,.

Stresses in Shear and Torsion

Table 2.2 Examples of Maximum Shear Stresses

Type of Section	Type of Shear Stress Distribution	Formula for Maximum Shear Stress
Rectangle (A, h, b)	τ_{av}, τ_{max}	$\tau_{max} = 1.5\tau_{av}$
Solid circle (A, r)	τ_{av}, τ_{max}	$\tau_{max} = 4\tau_{av}/3$
Thin tube (A, r, t)	τ_{av}, τ_{max}	$\tau_{max} = 2\tau_{av}$
I-section (h, b)	τ_{av}, τ_{max}	$\tau_{max} = \dfrac{Qbh}{2I}\left(1 + \dfrac{h}{4b}\right)$ Q = transverse load, lb I = Moment of inertia about x-x axis, in.^{4}

The general expression, Eq. (2.8), can also be integrated for the case of a hollow circular cross section having outer and inner radii R and r, respectively [117]. Using notation from Table 2.2, the relevant formula for the shear distribution factor can be stated as

$$\frac{\tau_{max}}{\tau_{av}} = \frac{4(R^2 + Rr + r^2)}{3(R^2 + r^2)} \tag{2.9}$$

When the inner radius r approaches the magnitude of the outer radius R, we have the condition applicable to a thin tubular member, for which Table 2.2 indicates the value of 2. On the other hand, when the inner radius becomes very small, the geometry of a solid cylinder is attained, for which both Table 2.2 and

Table 2.3 Shear Distribution Factors for Tubular Members

r/R	0	0.2	0.4	0.6	0.8	1.0
τ_{max}/τ_{av}	1.333	1.590	1.793	1.922	1.984	2.000

Eq. (2.9) give a value of 4/3. Some intermediate values for the tubular members with the various wall thicknesses are given in Table 2.3.

STRESSES IN BEAMS UNDER TWIST

When a shaft or structural beam is subjected to a pure twisting moment, every cross section finds itself in a state of pure shear. The induced shearing action causes the development of a moment of resistance equal and opposite to the applied torque. In this type of analysis the following basic assumptions are made:

1. Material is homogeneous and elastic.
2. Hooke's law is applicable.
3. Shear stress at a point is proportional to the shear strain at that point.
4. Stresses are below the elastic limit.
5. Radial lines of the cross section remain radial after twist.
6. Plane sections remain plane for circular shafts.

Assumption 5 has a special meaning because it implies that the stresses and strains are directly proportional to the radius. Hence, the maximum shear stress is found at the outer surface. This behavior is quite different for the case of the transverse shear, where the maximum value is found at the neutral axis. For an element of the cross-sectional area dA, located a distance x from the central axis, we get

$$\tau_x = \frac{x}{r}\tau_{max} \tag{2.10a}$$

The corresponding shear force on the elementary area dA is

$$F_x = \frac{x}{r}\tau_{max}\,dA \tag{2.10b}$$

The elementary reactive moment of this force about the central axis is

$$dM_r = xF_x \tag{2.10c}$$

The total moment of resistance M_r can then be obtained by the process of integration extended over the complete cross-sectional area as follows:

$$M_r = \frac{\tau_{max}}{r}\int x^2\,dA \tag{2.10d}$$

Stresses in Shear and Torsion

Remembering that M_r must be equal to the applied torque M_t and recognizing that the above integral actually defines the polar moment of inertia I_p, we get

$$M_t = \frac{\tau_{max} I_p}{r} \qquad (2.11)$$

Hence, the maximum shear stress at the surface of a circular shaft is

$$\tau_{max} = \frac{r M_t}{I_p} \qquad (2.12)$$

The stress at any other radius within the cross section can be stated as follows:

$$\tau_x = \frac{x M_t}{I_p} \qquad (2.13)$$

The quantity I_p/r in the foregoing equations may be designated as the section modulus for torsion. For a solid shaft subjected to twist, $Z_t = \pi r^3/2$.

ANGLE OF TWIST

For the case of a solid circular shaft subjected to torque as shown in Fig. 2.3, the relationship between the shear stress, angle of distortion, and modulus of rigidity is given by Eq. (2.1). It is clear from Fig. 2.3 that the shearing displacement δ_0 can be related to the cross-sectional parameters as follows:

$$\delta_0 = \gamma L = \phi r \qquad (2.14a)$$

Hence, from Eq. (2.14a) we get

$$\gamma = \frac{\phi r}{L} \qquad (2.14b)$$

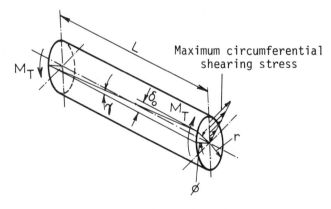

Fig. 2.3 Solid circular shaft under twist.

From Eq. (2.1), the maximum stress is

$$\tau_{max} = G\gamma \qquad (2.14c)$$

Combining Eqs. (2.12) and (2.14c) gives

$$G\gamma = \frac{rM_t}{I_p} \qquad (2.14d)$$

Substituting next Eq. (2.14b) into Eq. (2.14d) shows that

$$\frac{G\phi r}{L} = \frac{rM_t}{I_p} \qquad (2.14e)$$

Hence, simplifying Eq. (2.14e) and solving for ϕ gives the following important design formula for the angle of twist

$$\phi = \frac{M_t L}{G I_p} \qquad (2.15)$$

Also, from Eq. (2.14c) we have

$$\gamma = \frac{\tau_{max}}{G} \qquad (2.14f)$$

Hence, combining Eqs. (2.14b) and (2.14f) yields another useful stress formula

$$\tau_{max} = \frac{Gr\phi}{L} \qquad (2.16)$$

Although elementary in nature, Eqs. (2.12), (2.15), and (2.16) have numerous practical applications in engineering design and materials testing. These formulas are good for the solid and hollow circular shafts when the appropriate polar moments of inertia are taken into account.

TORSION OF TUBES

In considering various elementary concepts of shear stresses, torsion of thin-walled tubes of arbitrary cross-sectional geometry should be mentioned for which *Bredt's approximate formula* gives quite satisfactory results and is extremely simple in use [3]

$$\tau = \frac{M_t}{2A_0 t} \qquad (2.17)$$

For a tube of circular cross section, Eq. (2.17) becomes

$$\tau = \frac{M_t}{2\pi r^2 t} \qquad (2.18)$$

Equation (2.17) indicates that A_0 in this particular case denotes the whole area described by mean radius r, with wall thickness t. It is emphasized that A_0 is not the actual cross-sectional area of the tube, which in this instance would have been equal to $2\pi rt$.

It is well to point out that Bredt's formula also applies to other than circular cross sections, provided only that closed sections are involved. However, the error of applying this formula to split tubes of any geometry can be very substantial. It may also be of interest to note that since the area enclosed by the circular contour is a maximum for a given length of a perimeter, a circular tube is more efficient in carrying the twisting moment than any other form of a hollow cross section. This feature has been exploited in many designs of frames and support structures as well as in machine elements where torsional strength and rigidity are of importance.

For instance, if a thin-walled tube of square cross section is compared with a circular tube of the same wall thickness and weight, we will find from Bredt's formula, Eq. (2.17), that the shearing stress for the circular tube geometry is 21% lower. If, instead, we select the enclosed areas as our criterion, the corresponding weight of the circular tube would be 11% lower, although the theoretical values of the maximum shear stresses are still the same for both designs. These differences in stresses and weights do not seem like very much until we begin to consider the economical factors in large frameworks, for instance, transmission towers, where weight reduction, manufacture, and assembly costs can become significant.

A simple comparison of Eqs. (2.12) and (2.18) also brings out another interesting point—that the sectional property in torsion for a circular tube can be defined as I_p/r or $2\pi r^2 t$. It will suffice to state here that other simplified formulas for circular cross sections can be obtained in terms of the average tube radius and wall thickness for a surprisingly wide range of r/t ratios before any significant design errors, caused by this simplifying process, are encountered.

TORSIONAL RESPONSE OF RECTANGULAR CROSS SECTIONS

In analyzing torsional stresses and rigidity of structural members having rectangular cross sections, it becomes necessary to introduce two design factors. In the case of the torsional stresses, we find the concept of a *section modulus for torsion*, denoted in this book by K_s. It is recalled that in the section on stresses in beams under twist, this parameter for a circular geometry depicted the ratio I_p/r. In dealing with the torsional rigidity, it is convenient to introduce the *torsional shape factor*, denoted by K. By analogy to the section modulus and the moment of inertia properties listed in engineering handbooks, the K_s and K factors are expressed in in.[3] and in.[4], respectively. Using this notation, the two principal formulas for the rectangular cross sections can be stated as follows:

$$\tau = \frac{M_t}{K_s} \tag{2.19}$$

Table 2.4 Torsional Parameters for Rectangular Sections

b/h	1.0	1.2	1.5	2.0	2.5	3.0	4.0	5.0	10
K/bh^3	0.141	0.166	0.196	0.229	0.249	0.263	0.281	0.291	0.312
K_s/bh^2	0.208	0.219	0.231	0.246	0.258	0.267	0.282	0.291	0.312

and

$$\phi = \frac{M_t L}{GK} \qquad (2.20)$$

Also, combining Eqs. (2.19) and (2.20) gives

$$\tau = \frac{\phi G K}{L K_s} \qquad (2.21)$$

If we now denote by b and h the longer and shorter sides of the rectangular cross section, respectively, the appropriate numerical values of the torsional parameters can be obtained from Table 2.4.

The majority of torsional parameters, and particularly those that relate to more complex cross sections, have been deduced experimentally using membrane analogy. The technique is first to cut a hole, the shape of the section, in a flat plate positioned horizontally. The hole is filled with a membrane of constant tension using, for example, soap film. This film is then pressurized slightly on one side, and the shape of the film is used as a guide in developing the approximate forms of solution for the shear stress distribution in thin-walled open sections and other complex geometries [38].

Some of the structural shapes can be approximated by combining a number of rectangular sections. The torsional parameter K or K_s for such a composite section is roughly equal to the sum of the K values determined for individual rectangular areas. The maximum torsional stress in a rectangular cross section occurs at the midpoint of the longer side.

Design Problem 2.1

Determine the maximum stress in a structural steel member of rectangular cross section when the twisting moment of 4000 lb-in. produces an angle of twist equal to 0.0018 rad. The long and the shorter dimensions of the cross section are 4.2 and 1.4 in., respectively. The length of the steel component is 16 in. Assume that $G = 0.4E$.

Solution

Since $b = 4.2$ in. and $h = 1.4$ in., we get $b/h = 4.2/1.4 = 3$. Hence, Table 2.4 gives

$$\frac{K}{bh^3} = 0.263$$

Stresses in Shear and Torsion

and

$$\frac{K_s}{bh^2} = 0.267$$

The relevant torsional parameters are

$$K = 0.263 \times 4.2 \times 1.4^3 = 3.03$$
$$K_s = 0.267 \times 4.2 \times 1.4^2 = 2.20$$

From Table 1.1, the modulus of elasticity for the structural steel is $E = 29,000,000$ psi. Therefore,

$$G = 0.4 \times 29,000,000 = 11,600,000 \text{ psi}$$

Since $\phi = 0.0018$, using Eq. (2.21) yields

$$\tau = \frac{0.0018 \times 11,600,000 \times 3.03}{16 \times 2.20} = 1797 \text{ psi} \quad (12.4 \text{ N/mm}^2) \quad \blacklozenge$$

SYMBOLS

A	Cross-sectional area, in.² (mm²)
A_0	Total enclosed area, in.² (mm²)
a	Arbitrary distance, in. (mm)
B	Additional width dimension, in (mm)
b	Width of cross section, in. (mm)
c	Arbitrary distance, in. (mm)
E	Modulus of elasticity, psi (N/mm²)
F, F', F''	General symbols for forces, lb (N)
F_x	Shear force on elementary area, lb (N)
G	Modulus of rigidity, psi (N/mm²)
h	Depth of cross section, in. (mm)
I	Moment of inertia, in.⁴ (mm⁴)
I_p	Polar moment of inertia, in.⁴ (mm⁴)
K	Torsional shape factor, in.⁴ (mm⁴)
K_s	Section modulus for torsion, in.³ (mm³)
L	Length, in. (mm)
M	General symbol for bending moment, lb-in. (N-mm)
M_r	Moment of resistance, lb-in. (N-mm)
M_t	Twisting moment, lb-in. (N-mm)
Q	Shear load, lb (N)
R	Outer radius, in. (mm)
r	Radius of solid or hollow shaft, in. (mm)
S_b	Bending stress, psi (N/mm²)
t	Wall thickness of tube, in. (mm)
x	Arbitrary distance, in. (mm)
y	Arbitrary distance, in. (mm)

Z_t	Section modulus for torsion, in.³ (mm³)
γ	Angle of distortion due to shear, rad
δ_0	Shearing displacement, in. (mm)
ν	Poisson's ratio
τ	General symbol for shear stress, psi (N/mm²)
τ_x	Shear stress at radius x, psi (N/mm²)
τ_{av}	Average shear stress, psi (N/mm²)
τ_{max}	Maximum shear stress, psi (N/mm²)
ϕ	Angle of twist, rad

3
Bending Stress

STRESS DISTRIBUTION

When a structural beam is subjected to a pure bending moment M_b such as that shown in Fig. 3.1, the longitudinal fibers in the upper portion of the beam become extended, producing a convex surface. At the same time, the fibers in the lower part are compressed so that the lower surface adopts the concave shape. This action results in the development of the tensile and compressive stresses indicated in Fig. 3.1 as varying linearly across the beam cross section. The maximum value of the stress is denoted here by S_b. The stress at any other point of the cross section must be lower than S_b. When all the elementary forces parallel to the length of the beam are multiplied by their respective distances from the neutral axis, the summation forms a quantity called the *moment of internal resistance*. This quantity must be equal and opposite to the applied bending moment M_b if no other effects are present.

The analysis of bending results in one of the most frequent calculations made by engineers at work. Stress distribution for a straight member with a longitudinal plane of symmetry can be defined with the aid of a simple flexure formula given previously by Eq. (2.5). The most important stress is in the extreme outer fiber, which can be estimated as follows:

$$S_b = \frac{M_b C}{I} \tag{3.1}$$

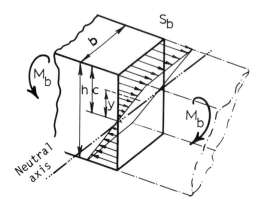

Fig. 3.1 Stress distribution in pure bending.

This classical formula has many applications and is predicated on knowledge of the moment of inertia, referred to sometimes as the *second moment of area* about a given axis.

For a beam of a rectangular cross section, shown in Fig. 3.1, the moment of inertia about the neutral axis is given by the well-known formula of solid mechanics

$$I = \frac{bh^3}{12} \tag{3.2}$$

For the outermost fiber, then, the maximum bending stress is found for $C = h/2$, which becomes

$$S_b = \frac{6M_b}{bh^2} \tag{3.3}$$

In Eq. (3.3) the term $bh^2/6$ is known as the *section modulus for bending* of a beam of a rectangular cross section. The general formula for beams of other cross-sectional geometry has the form of

$$S_b = \frac{M_b}{Z} \tag{3.4}$$

As indicated by the foregoing formulas and Fig. 3.1, the calculation of the elastic stresses due to bending for a simple cross-sectional geometry should not produce any unusual difficulties provided that either I or Z is easy to define. However, when the applied bending moment M_b is sufficiently high to cause some yielding of the material at the extreme fibers, the relation between the stress and strain attains nonlinear characteristics. The onset of yield in a simple tension or compression test is well defined because all the portions of the cross section respond simultaneously. This is not so in the case of bending, where the outer fibers yield first and the process of yielding gradually extends toward the neutral axis of the beam. However, for purposes of engineering analysis, we assume that a particular fiber obeys a stress-strain relationship identical with that obtained in a simple tensile or comprehensive test.

Bending Stress

This assumption implies that the stress at a specific point remains essentially constant for a considerable increase in strain, and it is compatible with the concept of a bilinear stress-strain characteristic. Here the linear portion of the elastic response defined by Hooke's law is followed by a line of a constant stress. The theoretical, bilinear diagram is well established in practice by the behavior of a common mild steel. The total strain for this type of a material is on the order of 10 times the strain observed at the onset of yield.

SECTION MODULUS

The concept of section modulus Z is very convenient in many engineering calculations involving straight beams. It is also applicable to curved members for which cross-sectional dimensions are small compared with the total length of the member. According to some conventions adopted in books on strength of materials, the bending moment is considered positive when it produces tension. Hence, the upper portion of the beam shown in Fig. 3.1 is in a state of tension which decreases linearly to zero and transforms into compression below the neutral axis. Since the cross section in this particular case is symmetrical, there can only be one value of section modulus causing the magnitudes of the tensile and compressive stresses to be equal. It is obvious that this is not so in the case illustrated in Fig. 3.2, where the centroid of the section is located closer to the base of the beam, giving rise to two distinctly different numerical values of the section modulus

$$Z_{\min} = \frac{I}{C_2} \tag{3.5}$$

and

$$Z_{\max} = \frac{I}{C_1} \tag{3.6}$$

Using Eqs. (3.4) to (3.6), we can calculate the two independent values of the bending stresses, one of which must be of the tensile and the other of the compressive nature. Since the term I in Eqs. (3.5) and (3.6) denotes the moment of inertia of this cross section about the neutral axis, the difference in the numerical values of these stresses depends only on the relative magnitudes of C_1 and C_2 for a given external bending moment. Such a difference, however, may be quite important in practical design situations. For instance, when the material is such that the ultimate strengths

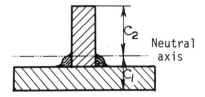

Fig. 3.2 T-type cross section.

in tension and compression are markedly different, the calculated stresses for the two extreme values of the section modulus will give two distinct factors of safety, one of which may not be acceptable. In another case where complicated sectional geometry is involved, the fibers in compression may give rise to the onset of local flange buckling, although the factor of safety based on the stress alone may appear to be quite satisfactory. It is therefore sound practice in many design situations to calculate the two extreme values of the section modulus.

COMMENTS ON BENDING THEORY

The analysis of bending stresses in beamlike members is generally well treated in standard textbooks dealing with strength of materials and elasticity, and handbooks of engineering design. This brief chapter on bending stresses is restricted to some of the fundamental concepts and those practical aspects of design which can be subject to some shaky interpretations. One of such obvious areas can be found in connection with the concept of neutral axis in bending of straight and curved members.

The term *neutral axis*, related to the essentially straight beams, has the same meaning as the "central" or "centroidal axis." The location of such an axis is determined from the elementary principle of statics, which states that the distance of the centroid of a section from a given axis is equal to the first moment of area of the section about the same axis divided by the total cross-sectional area. The absolute position of the centroid is obtained with reference to two reference axes perpendicular to each other. Of the above three terms, the term "neutral axis," directly implies that the normal stress at that axis vanishes. The normal stress is assumed to be that which acts parallel to the longitudinal dimension of the beam. The position of the neutral axis for all straight and curved structural members can be assumed to be the same as long as the depth of the beam cross section is small compared with the length or radius of curvature. However, in the case of curved members having sharp curvature, the neutral axis does not coincide with the central axis and is always displaced toward the center of curvature. The amount of this displacement depends on the ratio of the mean radius of curvature to the depth of the cross section. This problem will be examined in greater detail in Part IV.

The elementary beam theory characterized by Eq. (3.1) has a number of limitations, which can be summarized as follows:

1. Transverse sections of the beam remain plane before and after bending.
2. Young's modulus of the beam material is the same in tension and in compression.
3. The radius of curvature of the deflected beam is large compared with the beam's depth.
4. The effect of lateral stresses on the distribution of normal stresses is small.
5. The sum of internal forces normal to the beam cross section is zero.

It follows from the distribution of the elementary forces and the corresponding stresses that the material near the neutral surface of the beam lends very little help in resisting the applied external moments. Hence, for theoretical reasons, a beam cross section should have the greatest portion of its area placed as far away

Bending Stress

from the neutral axis as possible. This principle has been adopted in engineering practice by employing the I, T and channel sections in beams subjected to bending loads.

Although the bending stress formula given by Eq. (3.1) does not take into account the effect of shearing forces, which nearly always accompany the customary beam loading, it is sufficient to calculate the maximum fiber stress, because the transverse shearing stress is zero at the extreme beam fibers. Conversely also, the maximum shearing stress is found where the bending stress is essentially zero. The practice also indicates that the relatively long beams, designed on the basis of bending stresses alone, are sufficiently strong at all points along the span where other loading conditions might exist.

When the plane of loading does not coincide with one of the axes of symmetry of the beam cross section, we can resort to the method of superposition utilizing the relevant components of loading. Such components are expressed as a function of the angle measured between the plane of loading and the plane of symmetry. In dealing with unsymmetrical cross sections for which the location of the neutral axis can be established, the point of the maximum stress is the point in the cross section characterized by the greatest distance from the neutral axis.

It was noted in connection with the calculation of the section moduli in the preceding section that the allowable stresses for a particular material could be different for the tensile and compressive loading. A typical example here is concrete, which is quite strong in compression but rather weak in tension. The "no tension" theory has been used rather generally in structural design which resulted in the development of reinforced concrete beams. The assumptions included perfect adhesion between the steel and the concrete, with all the tensile stress carried by the reinforcement.

SYMBOLS

b	Width of cross section, in. (mm)
C, C_1, C_2	Distances to extreme fibers, in. (mm)
h	Depth of cross section, in. (mm)
I	Moment of inertia of cross section, in.^{4} (mm^4)
M_b	Bending moment, lb-in. (N-mm)
S_b	Bending stress, psi (N/mm^2)
y	Arbitrary distance from neutral axis, in. (mm)
Z	Section modulus, in.^{3} (mm^3)
Z_{min}	Minimum section modulus, in.^{3} (mm^3)
Z_{max}	Maximum section modulus, in.^{3} (mm^3)

4
Combined Stresses

INTRODUCTION

Thus far we have discussed the various stresses arising in structural members one at a time, and it should be stated with a considerable degree of certainty that in many practical design situations elementary calculations suffice. Numerous technical decisions are made solely on the basis of such ballpark figures. However, the machine and structural components may be subjected to several types of loads, simultaneously, such as, for instance, in the case of bending and torsion of a transmission shaft, bending and tension in an eccentrically loaded member, or internal pressure and axial tension in a cylindrical pressure vessel or piping. In some cases the direct algebraic superposition of individually calculated stresses is all that is required for obtaining the acceptable final result. However, where shear and normal loads act on a section simultaneously, the stresses must be treated in a different manner. Here considerable care should be exercised in combining the individual components of stresses utilizing either numerical or graphical methodology.

The analysis of combined stresses and strains has been presented in many textbooks on elasticity, illustrating the required calculational skills. Although such material may appear to be an exercise in pure mathematics, it should be appreciated that it has a practical bearing on the analysis of strength problems because it is impossible to make materials tests under all feasible load combinations.

Combined Stresses

TWO-DIMENSIONAL STRESSES

The general two-dimensional stress system is illustrated in Fig. 4.1 adopting the usual sign convention [4]. In this case both external stresses S_x and S_y have been shown to be tensile and therefore positive. The state of stress on a plane inclined at an angle θ can be defined by the normal and shear components, S and τ, respectively. From the consideration of equilibrium of forces acting on this element, the stress components become

$$S = \frac{1}{2}(S_x + S_y) - \frac{1}{2}(S_x - S_y)\cos 2\theta + \tau_{xy}\sin 2\theta \tag{4.1}$$

$$\tau = \frac{1}{2}(S_x - S_y)\sin 2\theta + \tau_{xy}\cos 2\theta \tag{4.2}$$

To find the maximum value of the normal stress, one can differentiate Eq. (4.1) with respect to θ and set the derivative equal to zero. This procedure should give

$$\tan 2\theta = -\frac{2\tau_{xy}}{S_x - S_y} \tag{4.3}$$

The required formulas for the maximum and minimum principal stresses can now be developed either with the aid of the trigonometric transformations utilizing Eq. (4.3), or directly from the equilibrium of the element shown in Fig. 4.1.

$$S_1 = \frac{1}{2}(S_x + S_y) + \frac{1}{2}\sqrt{(S_x - S_y)^2 + 4\tau_{xy}^2} \tag{4.4}$$

$$S_2 = \frac{1}{2}(S_x + S_y) - \frac{1}{2}\sqrt{(S_x - S_y)^2 + 4\tau_{xy}^2} \tag{4.5}$$

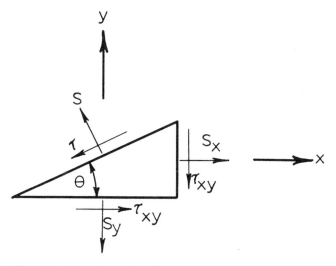

Fig. 4.1 Two-dimensional stress systems.

To find the maximum value of the shearing stress, Eq. (4.2) is differentiated with respect to θ, and the derivative is set equal to zero. Subsequent elimination of θ from Eq. (4.2) with the help of Eqs. (4.4) and (4.5) gives

$$\tau_{max} = \frac{S_1 - S_2}{2} \tag{4.6}$$

In the above equations the stresses are positive and correspond to the sign convention shown in Fig. 4.1. If one or both stresses happen to be compressive in the actual design case, the appropriate change in signs should be introduced. The sign of the shear stress presents no problem because this term is eventually squared. It is also obvious from the foregoing equations that S_1 and S_2 represent the maximum and the minimum principal stresses, respectively. The planes on which these stresses act are called the *principal planes*. These planes are defined as those on which no shearing stresses exist.

STRAIN IN TWO DIMENSIONS

When Hooke's law and the concept of Poisson's ratio, established for the uniaxial loading, are applied to the case of a two-dimensional stress at a given point in a solid body, the method of superposition can be used as long as the material is elastic. An element of elastic solid is shown in Fig. 4.2, where the two mutually perpendicular stresses S_x and S_y are applied. The state of stress on the element is shown diagrammatically as the sum of the two individual stresses acting at right angle to each other. Assuming the tensile strain to be positive and the compressive strain negative, the resultant strains in the x and y directions, respectively, become

$$\epsilon_x = \frac{S_x - \nu S_y}{E} \tag{4.7}$$

$$\epsilon_y = \frac{S_y - \nu S_x}{E} \tag{4.8}$$

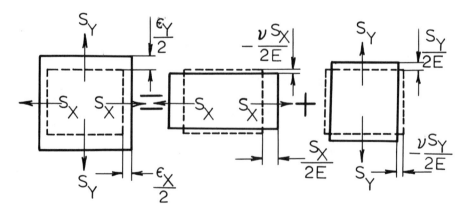

Fig. 4.2 Symbolic representation of strain superposition in a two-dimensional field.

Combined Stresses

If the shearing stresses on the element are present, the additional strain equation for a two-dimensional field is

$$\gamma_{xy} = \frac{\tau_{xy}}{G} \qquad (4.9)$$

For the principal directions, where shearing strains vanish, the principal strain formulas are

$$\epsilon_1 = \frac{S_1 - \nu S_2}{E} \qquad (4.10)$$

$$\epsilon_2 = \frac{S_2 - \nu S_1}{E} \qquad (4.11)$$

The system of stresses and strains in a two-dimensional field lends itself to a geometrical interpretation by means of a Mohr's circle [5]. Although this approach leads to a relatively simple method of stress analysis, it appears that few practicing engineers and designers resort to the graphical solutions. This is caused perhaps by the availability of the various numerical charts and calculators.

MOHR'S CIRCLE OF STRESS

The example of a Mohr's circle of stress given in Fig. 4.3 represents Eqs. (4.4) to (4.6). It may be helpful to recall that in this particular geometrical interpretation of principal stresses the maximum shearing stress is defined by the radius of the circle, while the principal stresses are described by the points of intersection of the horizontal stress axis with the circle. As shown in Fig. 4.3, both applied stresses S_x and S_y are positive and therefore are both measured from the origin 0 to the right. If one or both of these stresses are compressive, the construction of the circle is the same but the relevant distances are set out to the left of the origin. For example, if the applied stress along the y axis becomes compressive, negative S_y is situated to the left of the origin in Fig. 4.3, while the remaining stress components are the same. The new Mohr's circle for this case is shown by the dashed line. When a straight bar of uniform cross section is subjected to axial tension implied by Fig. 4.4, further Mohr's circle techniques can be analyzed. Suppose that we wish to know the normal and shear stresses on a plane inclined at an angle θ to the direction of axial force. Since the applied stress is also the principal stress S_1, making $S_1 = S$ and denoting the normal stress on the inclined plane by S_n, the solution can be illustrated in the upper portion of Fig. 4.4. The method shows that the maximum shearing stress for an axially loaded bar is found for $2\theta = 90°$. This result is often encountered in engineering work. It has also been observed in practice that yielding of a ductile material in tension is due to the slip along the planes of the maximum shearing stress. These slip lines appear roughly at 45° to the longitudinal axis of the test specimen subjected to tension. The pattern of the fine lines appearing on a well-polished specimen in tension was first observed by Lüder in 1854.

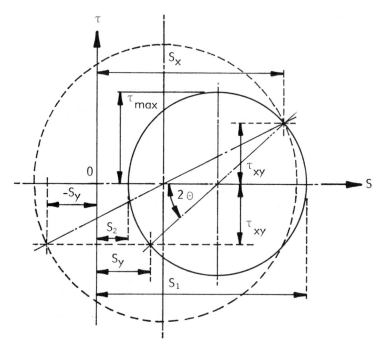

Fig. 4.3 Mohr's circle for finding the principal stresses.

The normal stress component follows from a simple trigonometric relation depicted in Fig. 4.4.

$$S_n = \frac{S(1 - \cos 2\theta)}{2} \tag{4.12}$$

It is evident from Eq. (4.12) that the normal stress varies from 0 to S, depending on the magnitude of θ. Furthermore, the relation between the horizontally distributed stress along the inclined plane (Fig. 4.4) and the stresses S_n and τ can be defined by vectorial resolution, so that

$$S_n = S' \sin \theta \tag{4.13}$$

and

$$\tau = S' \cos \theta \tag{4.14}$$

Since the area on which S' acts is $A/\sin \theta$, and since the forces must satisfy static equilibrium, we get

$$W = \frac{S'A}{\sin \theta} \tag{4.15}$$

Combined Stresses

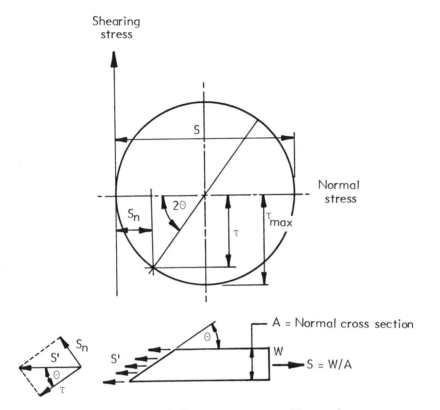

Fig. 4.4 Mohr's circle for finding stresses on an arbitrary plane.

Making $W = SA$ and equating this result to that of Eq. (4.15), we get $S' = S \sin \theta$, from which Eq. (4.13) yields

$$S_n = S \sin^2 \theta \tag{4.16}$$

By a trigonometric transformation we can show that this result is equivalent to Eq. (4.12), which also follows from the Mohr's circle given in Fig. 4.4. Similarly, it can be deduced that the shearing stress is

$$\tau = \frac{S \sin 2\theta}{2} \tag{4.17}$$

It can be seen from Fig. 4.4 that the normal stress $S_n = S$ and $\tau = 0$ when $\theta = 90°$.

The specific results given by Eqs. (4.12) and (4.17) were only used to illustrate the general versatility of the Mohr's circle method. They could also have been obtained from the previously quoted formulas, Eqs. (4.1) and (4.2), by making $S_x = S$ and $S_y = \tau_{xy} = 0$.

There are also other methods of geometric representation of the two-dimensional state of stress available in the technical literature. These include such concepts as the ellipse of Lamé, the dyadic circle, or the polar method of stress and strain

Table 4.1 Examples of Mohr's Circle for Special Cases

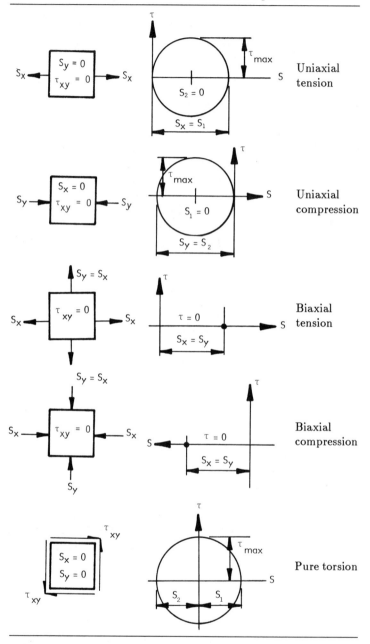

Combined Stresses

analysis [6]. However, the Mohr's circle approach to the graphical description of the state of stress and strain at a given point is perhaps the most practical. The applications of this method easily extend to some of the special stress cases such as those indicated in Table 4.1.

It is not the primary purpose of this section to go into the many details of the Mohr's stress analysis. In general, the approach involves the sketch of an element of the particular structure, showing the disposition of the normal and shearing stresses. Since the center of the Mohr's circle is established by the relation $(S_x + S_y)/2$ with the corresponding radius of $(1/2)\sqrt{(S_x - S_y)^2 + 4\tau_{xy}^2}$, any two given stress values are sufficient to determine one point of the circle. The combined stresses can then be defined to complete the resultant stress distribution [3,7].

PLANE STRESS AND STRAIN

In dealing with the two-dimensional problems discussed in this chapter it may be well to note the nature of certain general elastic properties and conditions of loading characterized as plane stress or plane strain.

The concept of a plane stress criterion is illustrated in Fig. 4.5. Here the assumption is made that for given set of forces applied externally the interior effects of this loading are uniformly distributed across the thickness. The state of plane stress is presumed to exist when the stresses in the z direction vanish on both faces of the slice and when the three components S_x, S_y, and τ_{xy} do not depend on z. This is obviously a convenient approximation for the analysis of thin plates on the premise that the stresses can vary gradually as functions of x and y.

In describing the concept of plane strain consider, for example, a long cylinder with the axis aligned along the z coordinate. Assume next that all the surface and mass forces act perpendicular to this axis and are independent of z. Let the ends

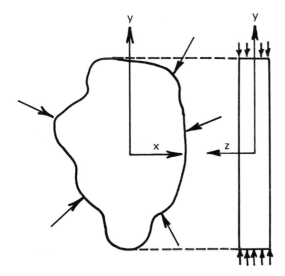

Fig. 4.5

of the cylinder be constrained in the z direction but otherwise free. If we now cut a slice from the cylinder defined by two planes $z =$ constant and if we invoke the appropriate boundary conditions, the basic stress relationship describing the state of plane strain becomes

$$S_z = \nu(S_x + S_y) \tag{4.18}$$

The foregoing expression simply states that the stress along the z direction depends on x and y only and therefore it must also act at the cylinder ends. When in practice such end effects are often neglected the theory must be regarded as approximate at best. The plane stress theory is used in the studies of elasticity, inertial fields, gravity, and other areas where a two-dimensional stress element and the equations of equilibrium are required. In the case of brittle fracture of ductile materials the conditions of plane strain constraint are often involved. For instance, the local deformation prior to fracture under plane strain response is smaller than that for the plane stress.

SYMBOLS

A	Cross-sectional area, in.² (mm²)
E	Modulus of elasticity, psi (N/mm²)
G	Modulus of rigidity, psi (N/mm²)
S	General symbol for stress, psi (N/mm²)
S'	Stress component on an inclined plane, psi (N/mm²)
S_n	Normal stress, psi (N/mm²)
S_1	Maximum principal stress, psi (N/mm²)
S_2	Minimum principal stress, psi (N/mm²)
S_x	Stress in direction of x axis, psi (N/mm²)
S_y	Stress in direction of y axis, psi (N/mm²)
S_z	Stress along z axis, psi (N/mm²)
W	Concentrated load, lb (N)
x	Coordinate axis, in. (mm)
y	Coordinate axis, in. (mm)
γ_{xy}	Shearing strain in two dimensions
ϵ_1, ϵ_2	Principal strains, in./in. (mm/mm)
ϵ_x	Strain in direction of x axis, in./in. (mm/mm)
ϵ_y	Strain in direction of y axis, in./in. (mm/mm)
θ	Angle at which stress plane is inclined, rad
ν	Poisson's ratio
τ	General symbol for shear stress, psi (N/mm²)
τ_{xy}	Shear stress in two dimensions, psi (N/mm²)
$\tau_{\max}$	Maximum shear stress, psi (N/mm²)

5
Criteria of Mechanical Strength

INTRODUCTORY COMMENTS

The preceding chapters were concerned with the basic mathematical formulations related to the structural response of mechanical components in the most likely modes, such as tension, compression, shear, torsion, or bending. However, the task of the designer cannot be completed without reliable information about the mechanical properties of the proposed material and the likely modes of failure.

When a metal part is subjected to a perfect axial load and Hooke's law applies, it is a simple matter to predict the point of failure or to design a part in such a way as to avoid reaching the yield stress. The problem here is simple because we are dealing with only one principal stress. However, when the design task involves such configurations as pressure vessels, rotating disks, lifting hooks, and other components subjected to two- or three-dimensional stress systems, the problem of determining the point of failure can be rather complex. The method of building each component and testing it up to the point of yield would be very uneconomical. Hence, many investigators during the past 100 years have been preoccupied with the process of establishing the theoretical criteria of the mechanical behavior which would allow to relate the complex stress systems to the yield point in a simple tension or compression.

The range of acceptable calculational accuracy in practical stress analysis should be a matter of engineering judgment in a particular design situation. This judgment comes from experience and it relates to the accuracy with which we can develop and measure the conventional mechanical properties. There is ample evidence in the open literature suggesting that we seldom know the values of properties with the accuracy better than 5 to 10%.

MECHANICAL PROPERTIES

One of the major tasks of a manufacturer in the metal industry is to produce the required alloy from the raw material in accordance with the specifications. The method of uniaxial testing, referred to in Chap. 1 in connection with the definition of Hooke's law, is the most widely used in industry to determine the design properties of the materials. The basic uniaxial test gives the stress-strain curve for the material, elongation, modulus of elasticity, yield point, and ultimate strength. Most materials display linear stress-strain relationships, although there are some exceptions to this rule.

The shape of the stress-strain curve is important is characterizing the material's behavior beyond the elastic range. As indicated previously, the stress-strain curve can sometimes be approximated with the aid of two or more portions of straight line, with the bilinear stress-strain curve being the most popular in stress analysis.

A number of materials, such as the light alloys, do not exhibit a clearly defined yield point. In such situations the yield point is established by the *offset method*. The amount of offset depends on the amount of preload above the limit of proportionality, and it can be measured in terms of the inelastic deformation obtained after the stage of unloading the test piece. The actual magnitude of the offset is derived from experience and it is often defined as 0.2%. In other words, the yield point is assumed to be that stress at which the material exhibits a permanent set of 0.2%.

Several generic shapes of the stress-strain curves are shown in Fig. 5.1. This diagram illustrates the approximate differences in mechanical characteristics for some of the typical materials found in construction. The offset stress S_y corresponds to the 0.2% strain where the yield point is difficult to define. The two points Y and Y' given in Fig. 5.1 are determined on the basis of different criteria. For instance, point Y represents a methodology where the stress-strain curve does not contain a unique yield point or the limit of proportionality. Point Y' on the other hand indicates an abrupt transition at the onset of yield such as one would expect to find in mild steel or a similar construction material.

The initial portion of the stress-strain curve, as stated before, determines the elastic modulus shown by the line OA. The slope of this line is expressed in psi since strain on horizontal axis is dimensionless. The location of the yield point S_y, such as that indicated for the alloy steel, is found by drawing the line BC parallel to OA. The BC line then must contain the yield point Y and the strain point C corresponding to 0.2% offset. This elementary but important convention is often used in structural design where higher-strength materials are involved or where the yield point transition is unknown.

After the initial yield point is passed, an increasing stress produces further straining and the material undergoes the process of strain hardening. This mechanical property is quite important. The material becomes harder and a higher yield point can be obtained by a number of loading cycles into the plastic range. However, the material has a natural limit to which it can be strain-hardened. This point is reached when the slope of the nominal stress-strain curve becomes zero.

Criteria of Mechanical Strength

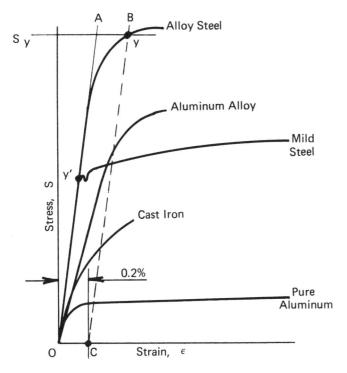

Fig. 5.1

The "nominal" stress is obtained when the load on a bar in tension is divided by the original cross section of the bar. It should be pointed out here that the increase in yield strength caused by strain hardening is attained at the expense of elongation.

In the case of a standard tensile test the ductility can be demonstrated either by elongation or reduction in the cross-sectional area. These two mechanical properties are dimensionless and are always quoted by the manufacturers. Although elongation depends to some extent on the geometrical form of the test piece, standards have been developed over the years so that elongation can be regarded as a useful index of ductility.

The various mechanical properties are greatly influenced by the temperature and chemical composition. The literature dealing with this topic is very extensive and the serious student of this aspect of engineering is encouraged to consult specialists in the field. It will suffice here to mention some of the basic elements of this knowledge as a general reminder for the design audience. The strength of steel, for instance, increases with the decrease in temperature, but the material can also become very brittle. This aspect of design is explored further in Part II. The modulus of elasticity generally decreases with increased temperature. The effect of temperature on the strength and ductility is more complicated, and the exact variations should be analyzed on an individual basis. The effects of chemical composition and heat treatment on the mechanical properties of metals are profound. The strength of steel increases with an increase in the carbon content while the ductility drops off. The steel can be hardened and strengthened by heating it to a high temperature followed by rapid cooling in a cold liquid. The same process can, however, drastically reduce the material's ductility, to the point of inherent brittleness. This problem can be circumvented in many cases by skillful design of a heat-treatment process tailored to a specific material, but the nature of the various effects on the mechanical strength should always be kept in mind. Furthermore, in addition to the chemical composition and heat-treatment parameters, some of the mechanical properties can be affected by the various processes of cold or hot working of the metal parts. This is certainly a valid area of concern in many design situations.

CRITERIA OF BRITTLE FAILURE

Brittleness, as a material's property, can be defined as the lack of ability to deform plastically. The conventional materials falling into this category include ceramics, glass, concrete, and certain forms of cast iron, to mention a few. Some structural steels, however, particularly in the welded state, can also exhibit highly brittle behavior below critical temperatures. The fractures in such materials can develop at and below the elastic limit, which also serves as a common point for the yield and ultimate strength. The corresponding stress-strain curve for a brittle material ends abruptly, as shown in Fig. 5.1. The plane of the fractured part adopts normal orientation to the direction of the tensile stress that precipitates this fracture. On a microscopic scale the brittle fracture can be characterized as a separation of individual crystals.

The mechanical strength criterion required for the prediction of brittle failure can be based on the maximum principal stress theory which determines a general state of stress. This theory, often attributed to Rankine, states that under a complex system of stress involving three principal components, the material should yield when the largest of the principal stresses attains a value equal to the yield strength of the material. This critical value can be in either tension or compression, depending on the type of the material. In a two-dimensional stress system, Eqs. (4.4) and (4.5) may be used in the calculations. It is recalled that these equations can be derived on the basis of Mohr's theory.

Design Problem 5.1

Calculate the induced shear stress in a cast-iron component subjected to a normal stress $S_x = 20,000$ psi for a maximum principal stress not to exceed $S_1 = 30,000$ psi. Assume no transverse load acting on this component. The material is expected to fail in a brittle manner.

Solution

From Eq. (4.4)

$$S_1 = \frac{S_x}{2} + \frac{1}{2}(S_x^2 + 4\tau_{xy}^2)^{1/2}$$

Criteria of Mechanical Strength

Hence

$$2S_1 - S_x = (S_x^2 + 4\tau_{xy}^2)^{1/2}$$

Squaring both sides and solving for τ_{xy} gives

$$\tau_{xy} = (S_1^2 - S_1 S_x)^{1/2}$$

Substituting the prescribed numerical values yields

$$\tau_{xy} = (30{,}000 \times 30{,}000 - 30{,}000 \times 20{,}000)^{1/2} = 17{,}300 \text{ psi} \quad (119.3 \text{ N/mm}^2) \quad \blacklozenge$$

CRITERIA OF DUCTILE FAILURE

Ductility in general can be defined as the ability of the material to undergo a significant amount of plastic deformation before the onset of failure. This process is accompanied by shear-sliding of the crystal structure on planes oriented at about 45° to the direction of the tensile stress. In the case of a cylindrical tensile specimen, the geometry of failure can be described as the cup-and-cone pattern of the fracture. The flat portion of the observed break in the tensile test specimen is not considered to be an indication of brittle separation. It is the result of the triaxial stress system set up in the neck of the test bar. The flat portion, when examined under the microscope, is characterized by a mass of tiny shear planes inclined 45° to the axis of the bar. The lip found along the edge of the break is also of the typical shear type.

The theory of ductile failure was established independently by Maxwell, von Mises, and Hencky. In modern technical literature this finding is referred to as the *von Mises–Hencky criterion of ductile failure*, or the *distortion energy theory*. The theory proposes that yielding starts when the total strain energy in shear reaches the value of yielding in simple tension. In other words, this yield criterion is established on an energy basis. If an element of the material is subjected to the principal stresses S_1, S_2, and S_3, then the von Mises–Hencky criterion becomes

$$(S_1 - S_2)^2 + (S_2 - S_3)^2 + (S_3 - S_1)^2 = 2S_0^2 \tag{5.1}$$

For a two-dimensional system, which is most likely to be needed in design calculations, we get

$$S_1^2 + S_2^2 - S_1 S_2 = S_0^2 \tag{5.2}$$

Experiments conducted under complex stress conditions have shown that the von Mises–Hencky criterion is a valid index of ductile behavior. Other theories of strength of materials are not recommended for the analysis of ductile metals.

In many practical cases we deal with the normal and shear stresses in a component, such as S_x and τ_{xy}. For instance, a machine shaft can be subjected to bending and torsion simultaneously. The maximum and minimum principal stresses in this situation can be defined as follows:

$$S_1 = \frac{S_x}{2} + \frac{1}{2}(S_x^2 + 4\tau_{xy}^2)^{1/2} \tag{5.3a}$$

and

$$S_2 = \frac{S_x}{2} - \frac{1}{2}(S_x^2 + 4\tau_{xy}^2)^{1/2} \tag{5.3b}$$

Substituting Eqs. (5.3a) and (5.3b) into Eq. (5.2) gives

$$S_0 = (S_x^2 + 3\tau_{xy}^2)^{1/2} \tag{5.4}$$

Equation (5.4) is very useful in preliminary stress analysis involving ductile materials. In this expression S_0 defines the resultant stress for a given uniaxial stress S_x and a transverse shear stress τ_{xy}.

Design Problem 5.2

A cylindrical component made of structural steel has a mean radius R of 12 in. and a wall thickness T of 0.15 in. It is subjected to an internal pressure of 600 psi and a torsional stress of 10,000 psi. Assuming that the material is inherently ductile, calculate the resultant stress in accordance with the von Mises–Hencky criterion.

Solution

The two principal stresses can be calculated on the assumption that the membrane stresses are as follows:

$$S_h = \frac{PR}{T} = \frac{600 \times 12}{0.15} = 48{,}000 \text{ psi}$$
$$S_l = \frac{PR}{2T} = 24{,}000 \text{ psi}$$

Then using Mohr's formulas, Eqs. (4.4) and (4.5), gives

$$S_1 = \frac{48{,}000 + 24{,}000}{2} + \frac{1}{2}\left[(48{,}000 - 24{,}000)^2 + 4 \times 10{,}000^2\right]^{1/2}$$
$$= 36{,}000 + 15{,}620 = 51{,}620 \text{ psi}$$

and

$$S_2 = 36{,}000 - 15{,}620 = 20{,}380 \text{ psi}$$

Substituting the above results into Eq. (5.2) and solving for S_0 gives

$$S_0 = (S_1^2 + S_2^2 - S_1 S_2)^{1/2}$$
$$= (51{,}620^2 + 20{,}380^2 - 51{,}620 \times 20{,}380)^{1/2} = 45{,}000 \text{ psi} \quad (310 \text{ N/mm}^2) \quad \blacklozenge$$

Criteria of Mechanical Strength

In a simple case of uniaxial tension, the elastic failure may be expected to take place when the stress reaches the elastic limit of the material. However, in more complex stress systems, such as those illustrated in Design Problems 5.1 and 5.2, the elastic failure does not necessarily happen when the maximum principal stress reaches the yield point of the material. The resultant stress S_0 can either be greater or smaller than the uniaxial yield S_y, depending on the signs of the relevant applied stresses and the type of material. Although various theories of elastic failure have been developed over the years, it is sufficient to keep in mind the Rankine (maximum principal stress) and the von Mises–Hencky (shear strain energy) theories for the analysis of brittle and ductile materials, respectively.

SYMBOLS

P	Internal pressure, psi (N/mm^2)
R	Mean radius of cylinder, in. (mm)
S_1, S_2, S_3	Principal stresses, psi (N/mm^2)
S_h	Hoop stress, psi (N/mm^2)
S_l	Longitudinal stress, psi (N/mm^2)
S_x	Stress in x direction, psi (N/mm^2)
S_0	Resultant stress, psi (N/mm^2)
S_y	Stress at uniaxial yield, psi (N/mm^2)
T	Wall thickness, in. (mm)
τ_{xy}	Shear stress, psi (N/mm^2)

6
Elastic Strain Energy

BASIC RELATIONS

One of the very important principles of mechanics concerns the amount of internal work stored in a body due to the externally applied forces. This work manifests itself as the elastic energy of deformation, which together with the principle of the conservation of energy forms the basis for several stress analysis methods.

The work done by an external force is simply the average magnitude of this force multiplied by the displacement of its point of application in the direction of this force. If the material is elastic (Fig. 1.2) the area under the load-deflection curve is $W \Delta L/2$ for the case of uniaxial stresses. The corresponding stress-strain diagram gives $S\epsilon/2$, which represents the energy stored in an elastic body (Fig. 1.3). Utilizing Hooke's law [Eq. (1.3)], we obtain a rather elementary but indispensable formula for the quantity of strain energy, stored in a unit volume of the material

$$U' = \frac{S^2}{2E} \tag{6.1}$$

Although Eq. (6.1) does not directly indicate the dependence of the stored energy on the load-deflection relation, it defines the material's ability to absorb the energy without any permanent set. This type of energy is recoverable upon the removable of external loading. If it is desired to express the total elastic strain energy for a prismatic bar of uniform cross section in terms of the uniaxial load, then substituting $S = W/A$ in Eq. (6.1) and multiplying the result by the volume AL

Elastic Strain Energy

gives

$$U = \frac{W^2 L}{2AE} \tag{6.2}$$

Since the work of the external force $W \Delta L/2$ is by the definition equal to the elastic strain energy given by Eq. (6.2), it follows by equating the two quantities, that

$$\Delta L = \frac{WL}{AE} \tag{6.3}$$

This formula is recognized by many analysts and practitioners throughout the engineering world.

THEORY OF CASTIGLIANO

It should be noted that Eq. (6.3) can be used in all cases where the external force is applied gradually. Also the algebraic transformation leading to the formula points to the possibility of utilizing the concept of the elastic strain energy in calculating the deflections. Indeed Eq. (6.3) defines the elongation of a bar of uniform cross section, and it can be obtained by differentiating the energy expression, Eq. (6.2), with respect to W, on the premise that the other terms in the equation are constant. This elementary observation is of special importance to practical stress analysis and dates back over 100 years, to the ingenious work of Castigliano [8, 9].

Suppose that a force W is applied to an elastic body giving a displacement Y, as shown in Fig. 6.1. The work done by the external force W is equal to the area

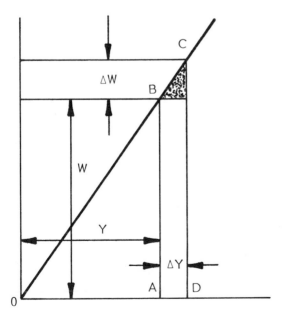

Fig. 6.1 Load-deflection relation for Castigliano's theory.

under the curve. Hence, we have

$$\text{Area } OBA = \frac{WY}{2} \tag{6.4a}$$

If an additional small force ΔW is now applied, producing an additional displacement ΔY, the extra work added is

$$\text{Area } ABCD = W\,\Delta Y + \frac{\Delta W\,\Delta Y}{2} \tag{6.4b}$$

The total work done in the elastic regime is equal to the total elastic strain energy stored. Using Eqs. (6.4a) and (6.4b) gives

$$U = \frac{WY}{2} + W\,\Delta Y + \frac{\Delta W\,\Delta Y}{2} \tag{6.4c}$$

Since the work done does not depend on the order in which the forces are applied, it follows from Fig. 6.1 that

$$U = \frac{(W + \Delta W)(Y + \Delta Y)}{2} \tag{6.4d}$$

Hence, equating Eqs. (6.4c) and (6.4d), and neglecting the products of small quantities, yields

$$W\,\Delta Y = Y\,\Delta W \tag{6.4e}$$

Since area $ABCD$, Eq. (6.4b), represents an increment of elastic strain energy ΔU, we get

$$\Delta U = W\,\Delta Y \tag{6.4f}$$

Now, combining Eqs. (6.4e) and (6.4f) gives

$$\frac{\Delta U}{\Delta W} = Y \tag{6.4g}$$

For very small increments of W and Y, we can write

$$\frac{\partial U}{\partial W} = Y \tag{6.4h}$$

Equation (6.4h) is the proof of a very important hypothesis of Castigliano, which states that the partial derivative of the strain energy with respect to a force gives the displacement corresponding to that force.

In the usual mathematical notation, then, the elongation of a bar in tension may be stated as follows:

$$\frac{\partial U}{\partial W} = \frac{WL}{AE} \tag{6.5}$$

Elastic Strain Energy

It may be recalled that the rules of partial differentiation with respect to a given variable require the same procedure as that for ordinary derivatives. Furthermore, the partial derivative has essentially the same geometric interpretation as that for the derivative of a function of one variable [10]. Although the concept of the elastic strain energy and the associated mathematical operations are at times considered to be beyond the scope of interest to practical engineers, the record of development of some of the more popular formulas and methods should be an integral part of the engineer's armory. The classical formula for uniaxial loading, Eq. (6.2), as well as other general expressions for strain energy indicate that the relevant functions involve the second degree of the external forces or displacements. The only necessary conditions for the energy concepts to apply include compliance with Hooke's law, restriction of displacements to small values, and the assumption that these displacements do not alter the action of the external forces [5].

In exceptional cases where the material of the machine component obeys Hooke's law but the displacements are not proportional to the loads, Castigliano's formula, Eq. (6.5), does not apply. This situation occurs when the deformations affect the action of the external loads and the expression for strain energy becomes no longer a second-degree function.

STRAIN ENERGY IN BENDING

In many stress analysis problems, the deflections of beams are calculated with the aid of strain energy considerations. For the case of pure bending of a bar that is built in at one end and carries a bending couple M_b at the other end (Fig. 6.2), the work done during the deflection is simply the area under the bending moment curve. Since the angular displacement is $\phi_0 = M_b L/EI$ and the energy stored in the bar is $M_b \phi_0 / 2$, we get

$$U = \frac{M_b^2 L}{2EI} \tag{6.6}$$

The more general form of Eq. (6.6), applicable to all straight bars and all different bending moment distributions along the length of the bar, is obtained directly from the considerations of the elastic energy stored in an element of the bar dx and summing-up the energy for the complete length L

$$U = \int_0^L \frac{M_b^2 \, dx}{2EI} \tag{6.7}$$

The term EI, defining the product of the modulus of elasticity and the moment of inertia, is the *flexural rigidity*, used universally in many deflection formulas. It is also clear that since the bending moment is expressed in lb-in. units and dx is a linear dimension, the resultant dimensional value must be

$$\frac{\text{lb}^2 \times \text{in.}^2 \times \text{in.}}{\frac{\text{lb}}{\text{in.}^2} \times \text{in.}^4} = \text{lb-in.}$$

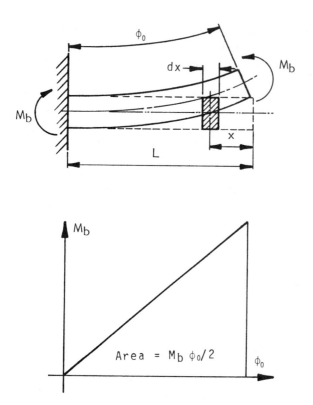

Fig. 6.2 Case of pure bending.

This is in compliance with all the expressions and geometric interpretations of the elastic strain energy, such as that given by Eq. (6.2) or the area under the bending moment curve shown in Fig. 6.2. As stated before, it represents the internal energy stored, which is equal to the amount of the external work done on the body. It must, in accordance with the principle of mechanics, have units of lb-in.

When an initially curved member is subjected to a bending moment, Eq. (6.7) can be modified with the aid of the notation indicated in Fig. 6.3. Replacing dx by $R\,d\theta$ and changing the limits of integration to 0 and some finite angle θ subtended by the curved member gives

$$U = \int_0^\theta \frac{M_b^2 R\,d\theta}{2EI} \tag{6.8}$$

The application of Eqs. (6.7) and (6.8) to stress calculations requires finding the derivative of the total strain energy with respect to a given load. Again according to the well-known principle of Castigliano, the partial derivative of the total elastic strain energy with respect to the selected force gives the displacement at the point and in the direction of the force. It is noted here that the bending moment is a function of the force, and the strain energy is a function of the bending moment. Hence, to find $\partial U/\partial W$ the differentiation is carried out for the function of a func-

Elastic Strain Energy

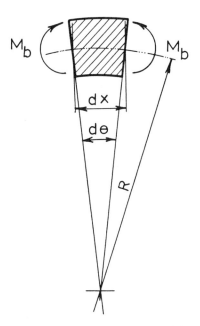

Fig. 6.3 Segment of a curved beam.

tion. This method of attack is known sometimes as the *chain rule*, which in our particular case can be stated as follows:

$$\frac{\partial U}{\partial W} = \frac{\partial U}{\partial M_b}\frac{\partial M_b}{\partial W} \tag{6.9}$$

Since by the usual rule of differentiation $\partial U/\partial M_b = 2M_b$, Eqs. (6.7) and (6.8) give

$$\frac{\partial U}{\partial W} = \int_0^L \frac{M_b}{EI}\left(\frac{\partial M_b}{\partial W}\right)dx \tag{6.10}$$

and

$$\frac{\partial U}{\partial W} = \int_0^\theta \frac{M_b}{EI}\left(\frac{\partial M_b}{\partial W}\right)R\,d\theta \tag{6.11}$$

ENERGY STORAGE CRITERIA

It is well known that a variety of mechanical springs are used to absorb energy, which in turn depends heavily on the geometrical and material parameters. It has also been useful [264] to compare the mechanical springs and other components on the basis of volume efficiency and form coefficients. The volumetric efficiency can be obtained by dividing the energy capacity by the volume, say, of a spring material or that of a structural component. Such a term then carries units of in.-lb/in.^{3}. The form coefficient on the other hand is a nondimensional quantity which attains a maximum value of 1.0 for a solid bar under uniform axial stress. On the premise that the material obeys Hooke's law, the quantity of strain energy stored in a unit volume is given by Eq. (6.1). It follows then that for components of identical geometry but of dissimilar material the best energy storage is obtained from the lowest elastic modulus. In other words, the part is superior as an energy absorber when it can be both stressed and elongated to a high level. The same principle, of course, applies to an element in compression provided no buckling occurs.

When a prismatic element of material is acted upon by a shear load, as shown for instance in Fig. 2.1, the energy storage per unit volume becomes

$$U' = \frac{\tau^2}{2G} \tag{6.12}$$

In many practical situations, however, the stress distribution across a given section may not be uniform, making the volumetric efficiency smaller than that for a solid bar. In other words, the energy storage is directly proportional to the form factor K_f, which in a number of cases is smaller than 1.0. In general we tend to subdivide all energy-absorbing devices into two major classes such as E- or G-governed systems. It can also be shown that the form factor $K_f = 1.0$ can be attained by a thin-walled cylindrical member under internal pressure or by an elastic ring made of the so-called shock cord. The shock cord design consists of a multiple-strand elastic rubber with braided cotton cover. This type of an elastic ring had, over the years, several industrial applications. The ring is characterized by a low elastic modulus (less than 2000 psi) resulting in elongation on the order of 100 to 200%. In the G class of energy storage the conventional shear mount has a form factor $K_f = 1.0$. The second best storage in this class is for a thin tube in torsion with $K_f = 0.9$. The mechanical springs of helical and torsional type appear to have K_f on the order of 0.25 to 0.35. Some of the lowest form factors are found in the stacks of Belleville washers which are characterized by small deflections, nonlinear behavior, and sharply varying compressive stresses in circumferential direction. The K_f values for these components can be as low as 0.05.

One of the more unique energy storage systems is known as a "ring-spring" consisting of inner and outer rings with matching tapered surfaces. The outer rings expand in uniform tension while the inner rings compress uniformly, resulting in a telescoping action. The dynamic response of this unique spring system is greatly influenced by the angle of the taper as well as the friction angle. Maier [264] gives a simple formula for estimating the compression and extension form factors K_f for this spring as follows:

$$K_f = \frac{\tan(\beta + \rho)}{\tan \beta} \tag{6.13}$$

In the compression mode the K_f values using Eq. (6.13) were found to be in the range of 1.4 to 1.7. For the case of extension the plus in Eq. (6.13) becomes a minus sign, leading to K_f numbers between 0.63 and 0.35. Standard taper angle for this design is usually $\beta = 14°$. The friction angle ρ falls within the range of $5°$ to $9°$.

Elastic Strain Energy

It is also of interest to mention the energy storage efficiency in bending for a cantilever beam with a rectangular cross-section. Because of the change in stress from tension to compression over the cross section, the total energy storage is found to be relatively small resulting in a form factor K_f of only 0.11.

In summary, then, the efficiency of an energy-storing material is greatly enhanced through a combination of low elastic modulus and high stress capacity. Such combinations of mechanical properties are possible, for instance, in music wire, chrome-silicon steel, and clock spring steel. However, shock cord rubber filaments or glass-fiber reinforced plastics can also claim exceptional energy-storing efficiency. Hence modern nonmetallic compounds with the right mechanical properties should continue to offer advantages in this area of materials technology.

Design Problem 6.1

Calculate the energy storage efficiency of a cantilever beam of a rectangular cross-section where b and h denote width and depth, respectively. The cantilever of length L carries concentrated end load W and the response is elastic. Utilize the quantity of energy stored in a solid bar in tension as the criterion for determining the form factor K_f.

Solution

The elastic work done follows directly from the linear load-deflection curve of the cantilever beam. This work is equal to $Wy/2$, which is the area under the load-deflection curve where y is cantilever deflection given by the usual formula

$$\frac{Wy}{2} = \frac{W}{2} \times \frac{WL^3}{3EI} = \frac{W^2L^3}{6EI}$$

The volumetric efficiency is the work divided by the volume of the material in cantilever

$$\frac{12 \times W^2 L^3}{6 \times bh^3 E \times bhL} = \frac{2W^2L^2}{Eb^2h^4}$$

The maximum cantilever stress is

$$S = \frac{6WL}{bh^2}$$

From this

$$WL = \frac{Sbh^2}{6}$$

Substituting this product term into the formula for the volumetric efficiency derived above, we get

$$\frac{2}{Eb^2h^4} \times \frac{S^2b^2h^4}{36} = \frac{S^2}{2E} \times \frac{1}{9}$$

Hence by a direct reference to Eq. (6.1) and the foregoing result we have to conclude that the form factor K_f for the cantilever beam under consideration is 1/9 or 0.11. ◆

SYMBOLS

A	Cross-sectional area, in.^{2} (mm^2)
b	Width of cross section, in. (mm)
E	Modulus of elasticity, psi (N/mm^2)
G	Modulus of rigidity, psi (N/mm^2)
h	Depth of cross section, in. (mm)
I	Moment of inertia of cross section, in.^{4} (mm^4)
K_f	Form factor
L	Length, in. (mm)
ΔL	Increment length, in. (mm)
M_b	Bending moment, lb-in. (N-mm)
R	Radius of curvature, in. (mm)
S	General symbol for stress, psi (N/mm^2)
U	Total elastic strain energy, lb-in. (N-mm)
U'	Unit elastic strain energy (lb-in./in.^{3}) (N-mm/mm^3)
W	Concentrated load, lb (N)
x	Arbitrary distance, in. (mm)
β	Taper angle, deg
θ	Angle at which bending moment is considered, rad
ρ	Friction angle, deg
τ	Shear stress, psi (N/mm^2)
ϕ_0	Angular displacement, rad

7
Deflection Analysis

INTRODUCTION

The analytical expressions given as Eqs. (6.10) and (6.11) are extremely useful in developing formulas for the deflection of straight and curved bars on the assumption that the strain energy due to the bending alone is taken into account. It should be emphasized that in practice the great majority of stress problems involve slender bars, where the effect of the bending moment is predominant. To illustrate briefly the method of application of this important method, the following examples may be of interest.

Design Problem 7.1

It is required to find the formula for the deflection of the free end of a straight cantilever beam illustrated in Fig. 7.1 utilizing Eq. (6.10). Consider the effect of bending alone.

Solution

The bending moment due to the end load W is

$$M_b = Wx \tag{7.1a}$$

Since W is considered here as the variable as far as the bending moment is concerned, keeping x constant for the purpose of finding the variation of the moment with respect to

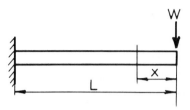

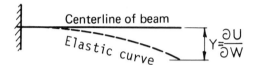

Fig. 7.1 Straight bar with end load.

the load W gives

$$\frac{\partial M_b}{\partial W} = x \tag{7.1b}$$

Substituting the results, Eqs. (7.1a) and (7.1b), into Eq. (6.10) yields

$$\begin{aligned} Y = \frac{\partial U}{\partial W} &= \int_0^L \frac{Wx}{EI} x \, dx \\ &= \frac{W}{EI} \int_0^L x^2 \, dx \\ &= \frac{WL^3}{3EI} \quad \blacklozenge \end{aligned} \tag{7.1c}$$

COMMENTS ON ANALYSIS

Equation (7.1c) is easily recognized as the standard formula for a maximum deflection of a cantilever beam of constant cross section. It is noted that the term *flexural rigidity*, EI, can be taken outside the integral sign before the integration is completed. As a rule the modulus of elasticity is considered to be constant, and therefore it is practically never involved in integration terms. The moment of inertia can, however, vary along the length of the beam and in such a case it must be included in integration.

It is well at this point of the discussion to review the character of variables involved in a problem using Eq. (6.10) in stress analysis and design. In elementary studies of beam deflections, the student is accustomed to think of the term x as the only variable involved. The external load W is considered to be constant, and hence the bending moment is directly proportional to x. Although this relationship does not appear explicitly in Eq. (6.10), it must and does enter the detailed expression, as shown in Design Problem 7.1, and the integration is carried out with respect to x. Hence, the term $\partial M/\partial W$ is only an intermediate step which helps to avoid

Deflection Analysis

more cumbersome algebra and integration. This is especially important when the expression for the bending moment consists of several terms. In the case of the cantilever beam with a single load the analysis can be performed just as easily with the aid of Eq. (6.7). The procedure in such a case is to make the direct substitution of $M_b = Wx$, squaring the term, and then integrating with respect to x.

Design Problem 7.2

Calculate the vertical displacement of the curved bar shown in Fig. 7.2, assuming constant rigidity and an elastic strain energy due to bending. Use Eq. (6.11) in the analysis.

Solution

The bending moment equation follows directly from Fig. 7.2:

$$M_b = WR\sin\theta \tag{7.2a}$$

The partial derivative of the moment with respect to the load W is

$$\frac{\partial M_b}{\partial W} = R\sin\theta \tag{7.2b}$$

Substituting the results, Eqs. (7.2a) and (7.2b), in Eq. (6.11) gives for the respective limits of integration of 0 and $\pi/2$

$$\begin{aligned} Y &= \int_0^{\pi/2} \frac{WR\sin\theta}{EI} \times R\sin\theta \times R\,d\theta \\ &= \frac{WR^3}{EI} \int_0^{\pi/2} \sin^2\theta\,d\theta \\ &= \frac{\pi WR^3}{4EI} \quad \blacklozenge \end{aligned} \tag{7.2c}$$

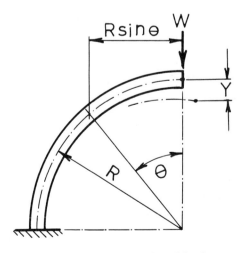

Fig. 7.2 Curved bar with end load.

The deflection formula (Eq. (7.2c)) is often used in practical design and can also be easily obtained from Eq. (6.8) because of the simplicity of the bending moment expression. In setting up the bending moment equation, it is advisable to check the bracketing assumptions. For instance, when $\theta = 0$, $M = 0$ and for $\theta = \pi/2$, $M_b = WR$, which is in agreement with the standard definition.

STRAIN ENERGY FOR A CURVED BAR

When a curved bar of a relatively deep cross section is analyzed, the total strain energy is expressed by the relation [5]

$$U = \int_0^\theta \left(\frac{M_b^2}{2AER\delta} + \frac{N^2}{2AE} + \frac{\xi Q^2}{2AG} - \frac{M_b N}{AER} \right) R\, d\theta \qquad (7.3)$$

In Eq. (7.3), M_b, N, and Q denote bending moment, normal force, and shearing force on a bar cross section, respectively. The distance between the neutral and central axis δ is featured in the denominator of the first term. The shear distribution factor defined by ξ depends entirely on the cross-sectional geometry as mentioned previously in reviewing the case of simple shear. The terms E and G refer to the moduli of elasticity and rigidity. The cross-sectional area of the bar and radius of curvature are denoted by A and R, respectively.

In suggesting the use of Eq. (7.3), a basic practical problem arises, how to distinguish between the thin and thick members. This question may be of importance in designing proving rings, machine frames, rolling-element bearing rings, hoist shackles, hooks, chain links, and the like. The usual rule often quoted states that when the mean radius of curvature is about 10 times the depth of the cross section (or more), the member is considered thin and such formulas as Eq. (6.8) and (6.11) are sufficiently accurate. Bearing in mind the accuracy of engineering calculations, improved knowledge of materials, and certain experimental evidence [11], the rule "10 times the depth" may well be lowered to "six times the depth" without undue error. Such a relaxation of this rule of thumb appears to be justified in many practical situations, especially where a reasonably good factor of safety can be applied.

EXACT DEFLECTION FOR A THICK CURVED BAR

In more rigorous design and research investigations, the deflection of a thick member is sometimes calculated from the following expression derived from Eq. (7.3):

$$Y = \int_0^\theta \left[\frac{M_b}{AER\delta} \left(\frac{\partial M_b}{\partial W} \right) + \frac{N}{AE} \left(\frac{\partial N}{\partial W} \right) + \frac{\xi Q}{AG} \left(\frac{\partial Q}{\partial W} \right) \right.$$
$$\left. - \frac{M_b}{AER} \left(\frac{\partial N}{\partial W} \right) - \frac{N}{AER} \left(\frac{\partial M_b}{\partial W} \right) \right] R\, d\theta \qquad (7.4)$$

It is emphasized that Eq. (7.4) applies only to in-plane bending of curved bars. The evaluation of the deflection using such a complex expression is seldom justified in

Deflection Analysis

practical design situations, and Eq. (7.4) is quoted here only as an illustration of how rapidly the degree of complexity of the calculation increases when the terms in addition to the bending energy are taken into account. The essence of practical stress analysis is to learn how simplified formulas and procedures can be utilized to provide rational answers to specific questions without undue mathematical complexity. In this area, applications of simplified expressions for the elastic strain energy seem to be most appropriate.

DEFLECTION BY DOUBLE INTEGRATION VERSUS CASTIGLIANO METHOD

In this regard let us return to Design Problem 7.1 and the method of strain energy used in the derivation of the deflection formula for a simple cantilever beam. Let us obtain the same formula using the well-known double-integration method using the differential equation of the deflection curve, treated in most college texts and engineering handbooks

$$EI \frac{d^2 y}{dx^2} = M_b \tag{7.5}$$

Since the bending moment equation is $M_b = Wx$, as before, the first integration gives

$$EI \frac{dy}{dx} = \frac{Wx^2}{2} + B_1 \tag{7.6}$$

Equation (7.6) involves a constant of integration B_1 and represents the equation of the slope. Second integration yields

$$EIY = \frac{Wx^3}{6} + B_1 x + B_2 \tag{7.7}$$

Equation (7.7) represents the equation of the deflection curve and involves two constants of integration. In accordance with the usual procedure, the integration constants are evaluated from a knowledge of boundary conditions. This is an important statement which often leads to difficulties when the exact boundary conditions are not well defined. In our elementary case, of course, we know that the slope is zero for $x = L$ (Fig. 7.1), provided that the beam is rigidly clamped there. From Eq. (7.6), this condition gives $B_1 = -WL^2/2$. The second boundary condition, which probably can be fulfilled more precisely, gives $Y = 0$ at $x = L$, so that from Eq. (7.7), we obtain $B_2 = WL^3/3$. Hence, substituting both constants in Eq. (7.7) yields

$$Y = \frac{W}{EI} \left(\frac{x^3}{6} - \frac{L^2 x}{2} + \frac{L^3}{3} \right) \tag{7.8}$$

The formula given by Eq. (7.8) is very useful because it gives the deflection at any point of the beam, although the derivation required specific information on the state of the boundary conditions. It is seen from Design Problem 7.1 that the advantage of using the energy approach is demonstrated by the absence of integration constants. The formula obtained in Design Problem 7.1 follows directly from Eq. (7.8) by putting $x = 0$. It is also well to point out that the dimensional check should always be made after the derivation. In the case of Eq. (7.8) it is obvious that the product of various individual linear dimensions such as x and L must be inches to the third power, which applies to all the deflection problems involving concentrated loads.

The advantage of the double-integration method is evident from the foregoing discussion and concerns the generality of the formula given by Eq. (7.8). It will be recalled, however, that the general formula can also be obtained using the strain energy approach, which will be illustrated by the following example. Consider the case of a simple cantilever beam shown in Fig. 7.3. Let W and F represent the real and imaginary loading, respectively. Since the displacement here is required at a point other than at which the external force is applied, the strain energy is expressed in terms of all the real and fictitious quantities. Next the partial differentiation is carried out with respect to the fictitious force F, and finally this force is made equal to zero. The remaining terms yield the desired deflection formula. For instance, for the case illustrated in Fig. 7.3, the bending moment equation must be written for the two separate regions a and $L - a$. It will become obvious during the derivation that only the second region need be considered, since in the first interval $\partial M_b / \partial F$ must be zero.

$$M_b = Wx + F(x - a) \qquad (7.9)$$

$$\frac{\partial M_b}{\partial F} = x - a \qquad (7.10)$$

Hence, substituting Eqs. (7.9) and (7.10) into the general deflection formula, Eq. (6.10), and extending the integration over the length $L - a$ gives

$$Y = \int_a^L \frac{Wx + F(x - a)}{EI}(x - a)\,dx \qquad (7.11)$$

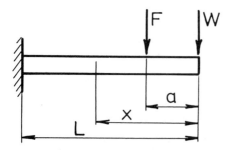

Fig. 7.3 Cantilever beam model for deflection at any point.

Deflection Analysis

From Eq. (7.11), after the integration and substitution of limits, we obtain

$$Y = \frac{W}{6EI}(2L^3 - 3aL^2 + a^3) \tag{7.12}$$

Equations (7.8) and (7.12) are, of course, identical for $x = a$. It should be pointed out here that the use of the fictitious force method, illustrated above, constitutes a very powerful tool of stress analysis.

When the magnitudes of statically indeterminate reactions are required, the total strain energy of a load-carrying member is expressed in terms of the unknown redundant reactions. The partial derivatives of the strain energy with respect to each of the redundant reactions are then equated to zero to obtain the necessary number of simultaneous equations.

Other methods of deflection analysis such as those applicable to beams of various sections and more complex examples of out-of-plane loading in curved beam analysis will be presented in further discussion as the need arises. For additional methods involving semigraphical solutions and superposition, the reader is referred to standard texts of strength of materials. One of the aims of this book is to emphasize the continuing need for reviewing closed-form solutions and the physical meaning of the various mathematical models employed.

SYMBOLS

A	Cross-sectional area, in.^{2} (mm^2)
a	Arbitrary distance, in. (mm)
B_1, B_2	Integration constants
E	Modulus of elasticity, psi (N/mm^2)
F	Fictitious concentrated force, lb (N)
G	Modulus of rigidity, psi (N/mm^2)
I	Moment of inertia of cross section, in.^{4} (mm^4)
L	Length of beam, in. (mm)
M_b	Bending moment, lb-in (N-mm)
N	Normal force on beam cross section, lb (N)
Q	Shear force, lb (N)
R	Radius of curvature, in. (mm)
U	Total elastic strain energy, lb-in. (N-mm)
W	Concentrated load, lb (N)
x	Distance along x axis, in. (mm)
Y	Beam deflection, in. (mm)
δ	Displacement of neutral axis, in. (mm)
θ	Angle at which bending moment is considered, rad
ξ	Shear distribution factor

8
Statically Indeterminate Structures

PROBLEM DEFINITION

A structure becomes statically indeterminate when there is at least one redundant quantity (i.e., force or couple) which is not essential for maintaining static equilibrium. The majority of the conventional stress formulas are derived by considering a portion of the loaded structure as a body in static equilibrium under the action of external forces. The mathematical analysis is based on Hooke's law and the elastic constants are determined experimentally. The *principle of superposition* states that the effect produced on an elastic system by any final loading is the same whether the forces are applied simultaneously or in any given sequence. It is obvious then that the accuracy of stress analysis is often limited due to certain assumptions as to the material properties, geometry, boundary conditions, or design procedure. Nevertheless, most of the appropriate procedures are satisfactory for practical purposes, and the aim of this chapter is to illustrate the basic approach to the problem. In particular, then, we deal with certain solutions in which the equations of equilibrium of forces (normally sufficient for solving statically determinate problems) must be supplemented with the expressions for elastic strain energy.

THEOREM OF LEAST WORK

One of the more convenient methods of dealing with a statically indeterminate structure involves the *theorem of least work*, which states that the stress distribution

Statically Indeterminate Structures

in a given structure is such as to make the strain energy a minimum. The strain energy of the entire system given as a function of the unknown redundant quantities can be partially differentiated with respect to a given redundant reaction and then set equal to zero. These supplementary relations, together with the basic equations of equilibrium, are finally used in the evaluation of all the external forces and couples.

Before considering any stresses or deflections in a given structure, it is essential to establish whether we are dealing with a statically determinate or indeterminate system. To illustrate the method of tackling such problems, two elementary cases will now be examined in some detail.

PROPPED CANTILEVER

A beam of length L shown in Fig. 8.1 is built in at one end and simply supported at the other. It carries a concentrated load of a given magnitude W at a distance a from the right support, producing unknown reactions R_1 and R_2 and a fixing couple M_0. Applying the equations of static equilibrium to this beam gives

$$W = R_1 + R_2 \tag{8.1}$$
$$M_0 = R_2 L - Wa \tag{8.2}$$

Obviously, these conditions are not sufficient for the solution because there are three unknown quantities in the two equations. It is therefore necessary to supplement the equations of equilibrium with the expression for elastic strain energy due to bending. The effect of shear may generally be ignored when beam thickness is small compared with the length between the supports.

The elastic equation can be set up on the basis that at a rigid support the displacement in direction of R_1 must be equal to zero. Consequently, the expression for strain energy of bending for a portion of the beam is

$$U = \int_0^x \frac{M_b^2}{2EI} dx \tag{8.3}$$

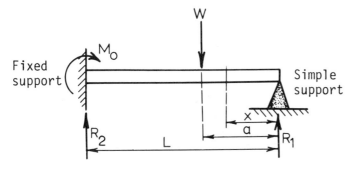

Fig. 8.1 Model of propped cantilever.

The displacement under a given load W follows directly from Eq. (6.10) in accordance with the first principle of Castigliano.

Taking the origin at the right support, as shown in Fig. 8.1, the relevant bending moments are

$$M_{1b} = R_1 x \quad \text{for } 0 < x < a$$
$$M_{2b} = M_{1b} - W(x - a) \quad \text{for } a < x < L$$

Since one of the quantities, for example R_1, must be considered to be statically indeterminate, the following expression is obtained:

$$\int_0^a M_{1b} \frac{\partial M_{1b}}{\partial R_1} dx + \int_a^L M_{2b} \frac{\partial M_{2b}}{\partial R_1} dx = 0 \tag{8.4}$$

where

$$\frac{\partial M_{1b}}{\partial R_1} = \frac{\partial M_{2b}}{\partial R_1} = x$$

Integrating Eq. (8.4) yields

$$R_1 = 0.5W(k^3 - 3k + 2)$$

Finally, inserting the foregoing value of the reaction into the equations of static equilibrium (8.1) and (8.2) gives

$$R_2 = 0.5W(3k - k^3)$$

and

$$M_0 = 0.5W(k^3 L - a)$$

where

$$k = \frac{a}{L}$$

PIN-JOINTED ARCH

Another statically indeterminate system is shown in Fig. 8.2. In this case the value of the vertical reaction follows immediately from considerations of statics on account of symmetry. However, the horizontal thrust H is still unknown and its value must be determined from the condition of elastic strain energy. Since the arch is pin-jointed, there are no fixing couples at the supports, and the redundant reaction H can be calculated on the basis that the displacement of the arch at

Statically Indeterminate Structures

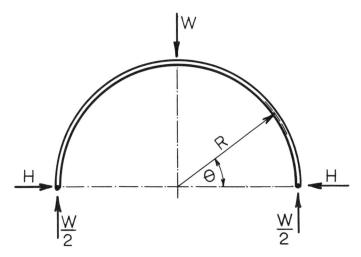

Fig. 8.2 Model of pin-jointed arch.

the point of application of H is equal to zero. The *principle of least work* (often referred to as *Castigliano's second theorem*) requires that

$$\int_0^{\pi/2} M_b \frac{\partial M_b}{\partial H} R\, d\theta = 0 \tag{8.5}$$

Since the bending moment at an angle θ is

$$M_b = HR\sin\theta - \frac{W}{2}R(1 - \cos\theta)$$

and

$$\frac{\partial M_b}{\partial H} = R\sin\theta$$

on substitution of the foregoing expression into Eq. (8.5) and on integration, we obtain

$$H = 0.318W$$

It is assumed in this case that the arch radial thickness is small in comparison with R, so that it will be sufficient to take into account the bending energy alone. The numerical value of the bending moment at any section defined by θ can now be calculated and the required bending stresses obtained.

The subject of statically indeterminate structures has been treated by many analysts and experimenters, and is almost as old as engineering science itself. The examples described in this brief section are limited to the most elementary applications. Since only one redundant quantity was involved, only one extra design equation was needed to effect the solution. This is not so in real life, where more complicated structural designs are encountered every day. Fortunately for the designer, many complex problems can be simplified so that a number of bracketing values of stresses and deflections can be obtained with a good deal of confidence as a basis for selecting the appropriate margins of safety. Serious students of structural indeterminacy are referred to the specialized publications on the subject [12,13]. Other examples of dealing with indeterminate quantities may be found in Part IV of this book.

SYMBOLS

a	Arbitrary distance, in. (mm)
E	Modulus of elasticity, psi (N/mm^2)
H	Horizontal constraint, lb (N)
I	Moment of inertia of beam cross section, in.^{4} (mm^4)
$k = a/L$	Length ratio in beam analysis
L	Length of beam, in. (mm)
M_b	Bending moment, lb-in. (N-mm)
M_0	Fixing moment, lb.-in (N-mm)
R	Radius of curvature, in. (mm)
R_1, R_2	Beam reactions, lb (N)
U	Total elastic strain energy, lb-in. (N-mm)
W	Transverse beam load, lb (N)
x	Arbitrary distance, in. (mm)
θ	Angle at which bending moment is considered, rad

9
Stress Concentration

STATE OF MACROSTRESS

The concept of *stress concentration*, as it is known in design, refers to the macroscopic state of stress, and it has a unique meaning only for plane problems involving the definition of the average stress. For instance, if a small hole were drilled in the test specimen shown in Fig. 1.1, the stress would remain essentially the same at appreciable distances from the hole, but the tangential stress at the edge of the hole would increase substantially. By the concept of a *macroscopic stress* we understand the average calculated stress related to the material's volume, characterized by a very fine structure. In terms of the practical requirements then this assumption is sufficiently accurate for the great majority of design situations. The factor of stress concentration K is generally expressed as a function of the average stress on the net section

$$S_{\max} = KS'' \tag{9.1}$$

For a bar of a rectangular section with a central hole subjected to uniform tensile stress, the concept of the factor K is illustrated in Fig. 9.1. Note that $S'' > S$ because the net section in tension has been decreased by drilling the hole. However, this decrease alone does not account for the difference between S'' and $S_{\max}$.

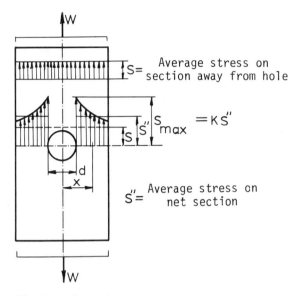

Fig. 9.1 Example of stress concentration in tension.

ELASTIC STRESS FACTORS

As a rule, stress concentrations arise due to the various local changes in shape, such as sharp corners, screw threads, abrupt changes in thickness, and even curved members of sharp curvature. This phenomenon is characteristic of elastic behavior. On the other hand, plastic yielding accompanies high stresses and tends to mitigate stress concentrations even in relatively brittle materials. This is a very important practical rule to keep in mind in developing rational designs. Particularly in the case of ductile response under static conditions, such as rivet holes in structural steel members, high local stresses based on the elastic theory can, indeed, be tolerated.

Under the conditions of static loading applied to the parts made of brittle materials, stress raisers cannot be ignored. This is also true in the case of some inherently ductile materials, which, at lower temperatures, fail due to the acquired brittle characteristics.

The stress concentration in any kind of cyclic loading should be avoided or at least mitigated. Furthermore, tests show that a single isolated hole or a notch appears to have a worse effect than that due to a number of similar stress raisers placed relatively close together.

The elastic stress concentration factors can be obtained either analytically or experimentally [14, 15]. Extensive design tables for stress concentration factors have been published [16] and design against fatigue failure in the presence of stress raisers has been discussed extensively in a number of books [17–20]. Useful design data for the selected stress concentration problems have also been presented in chart form [21].

Stress Concentration

COMMON TYPES OF STRESS RAISERS

Design experience indicates that there are at least two groups of questions which frequently come up during structural reviews. One concerns the effect of holes in plate and shell members. The other involves stress concentration due to the fillets and grooves under various conditions of loading. The first group can be best illustrated in Table 9.1, based on the rules established quite some time ago [22–24]. The results should also apply to curved surfaces, provided that the local curvature is not too sharp.

In the second group of stress concentration problems, one question frequently encountered concerns the difference in the effect of the type of loading on a circular shaft and rectangular bar with the transversely drilled central holes, as shown in Fig. 9.2. The upper curve is based on the case of uniaxial tension from Table 9.1, using, however, a more exact empirical formula [25, 26]. It is of interest to note that both curves start at $K = 3$ and that as a conservative guide a factor of 3 can be used in many circumstances. In the case of a hole drilled near the free edge, however, as shown in Table 9.1, good practice is to make the dimension e equal to at least two hole diameters. Additional comparison between the rectangular and round bars with grooves, indicated in Fig. 9.3, shows that round bars are less susceptible to the effect of stress raisers. Again the ultimate values of K do not appear to exceed the value of 3.0 [27]. A brief comparison of stress concentration for bending and torsion is shown in Fig. 9.4.

Where sharp grooves and notches are involved, the theoretical values of stress concentration can be very high, and for this reason the theory should be corrected for small radii of curvature. Under repeated loading, sharp notches can be especially detrimental. The highest stress concentration will develop when the notch depth is large while the notch radius and the angle are small.

Table 9.1 Effect of Circular Hole on Direct Stress for Flat Plates and Rectangular Bars

Uniaxial tension central hole	$K = \dfrac{3b}{b + d}$ (approximate formula)	
Uniaxial tension of center hole	e/d | 0.67 0.77 0.91 1.07 1.39 1.56 K | 4.37 3.93 3.61 3.40 3.25 3.16	
Biaxial tension (d/b small)	$K = 2$	
Biaxial tension and compression (d/b small)	$K = 4$	

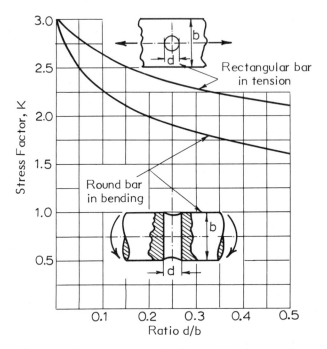

Fig. 9.2 Stress concentration in round and square bars with holes.

Stress concentration in the presence of a groove produces the effect of a combined stress pattern, decreasing the shear stress, for instance, in the middle of a grooved cylindrical specimen. This effect results in a cup-and-cone type of failure of a tensile specimen, so that the ductile material appears to have the characteristics of brittle failure on the inside.

From the point of view of design of steel structures [28], several practical considerations merit special attention. In addition to the problems of grooves or sharp reentrant corners, nonuniform stress distribution can be found in welded joints such as those shown in Fig. 9.5. To produce a more uniform stress gradient Fig. 9.5a indicates an elongated weld in the direction of the applied load. By comparison a conventional type of transverse weld (Fig. 9.5b) shows a more drastic transition between the two component parts.

STRESS DISTRIBUTION

In discussing sharp transitions and the effect of holes on the stress field, the question is often raised as to the character of stress distribution. Consider, for example, Fig. 9.1, illustrating the stress concentration in tension. The theoretical and photoelastic studies suggest that in this case the distribution of stresses can be represented by the equation

$$S_{max} = S'' \left[1 + \frac{1}{8}\left(\frac{d}{x}\right)^2 + \frac{3}{32}\left(\frac{d}{x}\right)^4 \right] \tag{9.2}$$

Stress Concentration

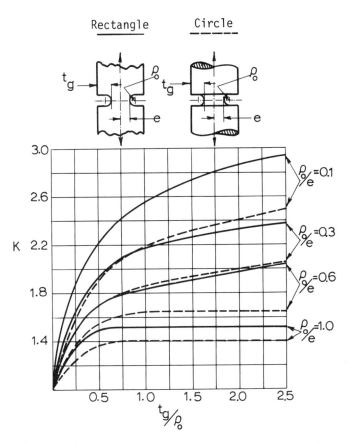

Fig. 9.3 Comparison of stress concentration for round and rectangular bars with grooves.

When $x = d/2$, $S_{max} = 3S'''$; that is, we obtain the maximum theoretical stress at the edge of the hole. On the other hand, when x becomes rather large in comparison with the hole diameter, we get the average stress on the cross section of the beam. Equation (9.2) shows that stress disturbance is highly localized. Practical rules often state that the maximum theoretical stress concentration for a plate in tension is encountered when the width of the plate is more than about four times the diameter of the hole. Putting $b = 4d$ into an approximate formula, given in Table 9.1, we get $K = 2.4$. Only when d becomes very small is the theoretical value of 3.0 attained.

In general the effect of open holes in beams is not easy to evaluate despite the various theoretical and experimental tools available. For example, when holes are present in the flange, the problem of location of the neutral axis can lead to many interesting speculations. Furthermore, the effect of a hole in the tension flange of a beam is difficult to assess if the beam does not fracture and the compression flange carries the significant share of the load. On this basis it would seem that the effect of holes in flanges can often be ignored, particularly when rivets are used. Under these circumstances the American Institute of Steel Construction allows us

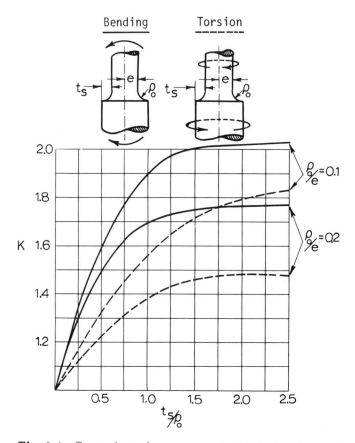

Fig. 9.4 Comparison of stress concentration in bending and torsion.

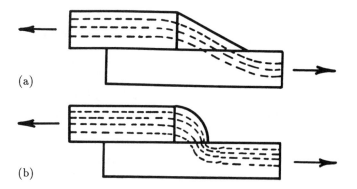

Fig. 9.5 Approximate stress transfer lines in a welded joint.

Stress Concentration

to neglect the reduction of beam area and girder flanges of up to 15% of the gross area.

PLASTIC REDUCTION OF STRESS FACTORS

So far the material treated in this chapter dealt primarily with the more conventional elastic criteria of stress concentration starting with the classical case illustrated by Fig. 9.1 and Eq. (9.2). Much of the earlier work in this area was abstract and mathematical in nature, following the scientific progress during the first half of the nineteenth century. Although practical men of that period did not feel at home with the theoretical solutions, they still managed to build magnificent structures, many of which we can admire today. The design factors of safety were naturally rather high and almost independent of weight and cost.

As the industry developed and the environmental conditions became more restrictive the accident rate increased and the elastic stress concentration problems came under more intense scrutiny. It was then quite natural that the first step toward better understanding of the unexpected structural failures was to develop a practical model of the more common geometric irregularities and defects. At about that time Inglis [72] proposed a simple formula for estimating the increase in stress due to a finite discontinuity such as an elliptical-shape opening in a plate, porthole, or hatchway. The design formula, related to Fig. 9.6, was

$$S_{\max} = S\left[1 + 2\left(\frac{L}{r}\right)^{1/2}\right] \qquad (9.3)$$

The application of this expression can be extended to the geometry of cracks, notches, scratches, and similar stress raisers as the L/r ratio is increased. The corresponding $S_{\max}/S$ ratio becomes the conventional stress concentration factor when symbol S denotes the nominal stress existing at a point away from the discontinuity. Note that when $L = r$ in Fig. 9.6, the case reduces to that for a plate

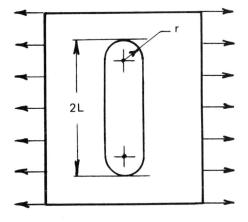

Fig. 9.6 Notation for Inglis model.

with a symmetrically placed circular hole for which $K = 3$ under uniaxial tension. This is a remarkable result considering that Eq. (9.3) was derived 75 years ago. What is even more remarkable that about the same time Kirsch in Germany and Kolosoff in Russia made similar predictions and yet little notice was taken of these results in shipbuilding and other industries where the stress failures appeared to be more frequent.

While the Inglis formula planted some interesting ideas in the minds of engineers it became necessary to invoke the principles of ductile response in the face of disturbing theoretical results derived from the purely elastic considerations. This process has led to the development of a plastic reduction procedure [30, 254] for the elastic stress factors. It became obvious that in the case of a truly ductile material under static loading the conventional elastic factor should be modified with the aid of the appropriate stress-strain diagram of the material [29]. One of the simplest approaches to the correction of any type of elastic stress concentration factor [265] is as follows:

$$K_p = 1 + (K - 1)\left(\frac{E_s}{E}\right) \tag{9.4}$$

Here E_s is the secant modulus of the material in the vicinity of the discontinuity (crack, notch, hole, etc.) while E denotes the well-known modulus of elasticity. For the case of a circular hole in a wide plate Eq. (9.4) yields

$$K_p = 1 + 2\left(\frac{E_s}{E}\right) \tag{9.5}$$

This method provides the opportunity for rounding off the calculated higher peaks of the elastic stresses and in this manner assuring a more reasonable value of the design stress factor. The magnitude of the plastic stress concentration factor depends then on the shape of the stress-strain curve while the conventional K factor is a function of the geometry of the part alone.

It should be stated in closing that stress concentrations in general are virtually inevitable in real structures and machines due to the presence of grooves, fillets, holes, threads, and similar discontinuities. The worst situations, of course, include machining errors, gravel nicks, nonmetallic inclusions, and microvoids, which may be difficult or even impossible to detect. The stress intensities due to the cracks in inherently brittle materials and some of the ductile materials displaying brittle behavior under specific environmental conditions are further subject to fracture mechanics control, discussed in Chaps. 14 and 15.

SYMBOLS

b		Width of rectangular bar and diameter of round bar, in. (mm)
d		Diameter of bolt or rivet hole, in. (mm)
e		Hole-to-edge distance, groove-to-bar center distance, in. (mm)
E		Modulus of elasticity, psi (N/mm^2)
E_s		Secant modulus, psi (N/mm^2)

Stress Concentration

K	Stress concentration factor
K_p	Plastic stress concentration factor
L	Length, in. (mm)
r	Radius, in. (mm)
S	Average tensile stress, psi (N/mm^2)
S''	Average stress on net cross section, psi (N/mm^2)
S_{max}	Maximum stress, psi (N/mm^2)
t_g	Depth of groove, in. (mm)
t_s	Depth of shoulder, in. (mm)
W	Concentrated load, lb (N)
x	Distance from bar center, in. (mm)
ρ_0	Fillet or groove radius, in. (mm)

10
Stability and Buckling Resistance

INTRODUCTION

Although in the conventional stress problems involving axial loading no differentiation is made between the yield point in tension and compression for the metal components, the mechanical properties in compression do not seem to be as clearly defined as those in tension. This is particularly true as the length of the compression members increases to the point where failure can take place by lateral bending or buckling at axial stresses far below the yield point. The notion of equal yield in tension and compression, however, should be qualified for special materials such as concrete, where yield strength in tension may be as low as 8% of that in compression.

HISTORICAL NOTE

For the past 65 years, and particularly since the time when the American Society of Civil Engineers authorized a special committee of Steel Column Research in 1923, the technical literature on stability and buckling resistance has developed to a sophisticated degree, involving many authorities in the field. Nevertheless, despite the classical works available to engineers and educators in this country, and abroad [31–34], various topics and particularly those involving shells with complex boundary conditions are still open to further scrutiny. Although confronted with the vast theoretical literature on the subject, one rarely deals with the ideal solutions

Stability and Buckling Resistance

in practice. Furthermore, everyday questions arising from design seldom warrant lengthy theoretical and experimental investigations. For this reason many simplified and ingenious concepts have been proposed from time to time [29] and various compilations of the formulas and references have been presented [35]. Out of this collection of charts and formulas the designer has often the task of selecting the optimum solution within the shortest possible time.

BASIC COLUMN FORMULAS

Fortunately, apart from highly specialized research and design, certain general rules exist for the three basic structural elements—columns, plates, and thin shells—which help us to pinpoint at least some of the important variables and modes of response. For example, the critical buckling load for a slender column is expressed as

$$W_{CR} = K_c \frac{EI}{L^2} \tag{10.1}$$

Here the moment of inertia of the column cross section denoted by I is calculated for that axis about which the buckling of the bar is likely to take place. It is well to note that this load is proportional to the modulus of elasticity but independent of the mechanical strength. The column factor K_c is a function of the manner of loading and support, and once its value is adopted for the particular case at hand, other critical loads follow, since they are inversely proportional to the second power of column length.

In real life all column action is nearly always a combination of direct axial compression and bending. One of the established rules for column design is known as the *secant formula* [36] developed on the basis of extensive full-scale column tests

$$W_{CR} = \frac{A\sigma_y}{1 + 0.25 \ \sec\left[\sqrt{\frac{W_{CR}}{AE}}\left(\frac{L'}{2\rho}\right)\right]} \tag{10.2}$$

Although Eq. (10.2) has both rational and experimental justification, it is seldom used, because the solution involves successive approximations of W_{CR}/A for a given value of L'/ρ, where L' defines the effective column length in inches and ρ is the radius of gyration, to be calculated as

$$\rho = \left(\frac{I}{A}\right)^{1/2} \tag{10.3}$$

To avoid tedious computations but still to make rational assumptions for the column ends, working design formulas can be developed for specific applications. Examples of such formulas for steel construction are given in Tables 10.1 and 10.2 [37]. These column unit stresses are based on a factor of safety of 1.80. The columns, defined as having pinned ends, are not considered to be completely unrestrained because there is always some frictional resistance present in hinges and pinned supports.

Table 10.1 Columns with Pinned Ends

Yield Strength (psi)	Range of Values for L/ρ	Critical Buckling Stress
33,000	0–140	$S_{CR} = 15{,}000 - 0.325(L/\rho)^2$
33,000	140–200	$S_{CR} = \dfrac{15{,}000}{0.5 + \frac{(L/\rho)^2}{15{,}860}}$
45,000	0–120	$S_{CR} = 20{,}500 - 0.605(L/\rho)^2$
45,000	120–200	$S_{CR} = \dfrac{20{,}500}{0.5 + \frac{(L/\rho)^2}{11{,}630}}$
50,000	0–110	$S_{CR} = 22{,}500 - 0.738(L/\rho)^2$
50,000	110–200	$S_{CR} = \dfrac{22{,}500}{0.5 + \frac{(L/\rho)^2}{10{,}460}}$
55,000	0–105	$S_{CR} = 25{,}000 - 0.902(L/\rho)^2$
55,000	105–200	$S_{CR} = \dfrac{25{,}000}{0.5 + \frac{(L/\rho)^2}{9{,}510}}$

Table 10.2 Columns with Riveted Ends (Partially Fixed Supports)

Yield Strength (psi)	Ranges of Values for L/ρ	Critical Buckling Stress
33,000	0–140	$S_{CR} = 15{,}000 - 0.253(L/\rho)^2$
33,000	140–200	$S_{CR} = \dfrac{15{,}000}{0.5 + \frac{(L/\rho)^2}{20{,}370}}$
45,000	0–135	$S_{CR} = 20{,}500 - 0.471(L/\rho)^2$
45,000	135–200	$S_{CR} = \dfrac{20{,}500}{0.5 + \frac{(L/\rho)^2}{14{,}930}}$
50,000	0–125	$S_{CR} = 22{,}500 - 0.574(L/\rho)^2$
50,000	125–200	$S_{CR} = \dfrac{22{,}500}{0.5 + \frac{(L/\rho)^2}{13{,}400}}$
55,000	0–120	$S_{CR} = 25{,}000 - 0.702(L/\rho)^2$
55,000	120–200	$S_{CR} = \dfrac{25{,}000}{0.5 + \frac{(L/\rho)^2}{12{,}220}}$

Stability and Buckling Resistance

Similarly, there is no such thing as a fully fixed-ended column. These assumptions are accounted for in the formulas through the following substitution:

$$L' = 0.85L \quad \text{(pinned ends)}$$
$$L' = 0.75L \quad \text{(riveted ends)}$$

It should also be emphasized that while the classical type of a formula, Eq. (10.1) is independent of material strength, the working-type equations, allowing for unavoidable eccentricity of loading and support, involve the yield strength in compression. The extent of eccentricity assumed in Eq. (10.2) is

$$\frac{ac}{\rho^2} = 0.25$$

where a defines the eccentricity of load application in inches and c is the distance from a given centroidal axis to the most stressed fiber in compression, also expressed in inches. The design of a column with the specific value of eccentricity involves, unfortunately, the method of successive approximations using the following equation for the maximum stress:

$$S_{max} = \frac{W_{CR}}{A}\left\{1 + \frac{ac}{\rho^2}\sec\left[\sqrt{\frac{W_{CR}}{AE}}\left(\frac{L'}{2\rho}\right)\right]\right\} \qquad (10.4)$$

The application of column Tables 10.1 and 10.2 is illustrated by the following example.

Design Problem 10.1

Calculate the critical buckling load for a structural column riveted at both ends and having cross-sectional dimensions as shown in Fig. 10.1. The total length of the column is 10 ft and the yield strength of the material is given as 50,000 psi. Assume the factor of safety to be 2.00. Neglect the effect of the welds.

Solution

The moments of inertia with respect to the two major axes are

$$I_{xx} = 49.60 \text{ in.}^4$$
$$I_{yy} = 18.02 \text{ in.}^4$$

The cross-sectional area is

$$A = 7.88 \text{ in.}^2$$

From Eq. (10.3),

$$\rho = \left(\frac{18.01}{7.88}\right)^{1/2} = 1.512 \text{ in.}$$
$$\frac{L}{\rho} = \frac{120}{1.512} = 79.4$$

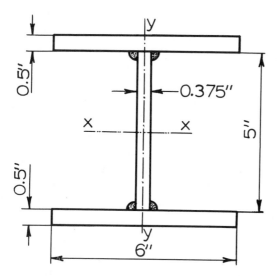

Fig. 10.1 Section dimensions for Design Problem 10.1.

Hence, the recommended formula from Table 10.2 is

$$S_{CR} = 22{,}500 - 0.574 \left(\frac{L}{\rho}\right)^2$$

$$= 22{,}500 - 0.574 \times 79.4 \times 79.4 = 18{,}880 \text{ psi} \quad (130.2 \text{ N/mm}^2)$$

Since the tabulated formulas are given for a factor of safety of 1.8, the critical buckling load for the column becomes

$$W_{CR} = 7.88 \times 18{,}880 \times \frac{1.8}{2.0} = 133{,}900 \text{ lb} \quad (595{,}614 \text{ N}) \quad \blacklozenge$$

CONCEPT OF EULER BUCKLING

The solution to the primary problem of column buckling was obtained some 200 years ago [29, 38] by Euler, who first developed the concept of flexural rigidity EI, although he assumed erroneously that the stiffness varied as the square of the thickness. Examples of the theoretical formulas involved are shown in Table 10.3. The table indicates how sensitive the theoretical buckling load is to changes in fixity at the supports. Many different cases of columns have been worked out [31] which have intermediate fixity at the supports. Those given in Table 10.3 are probably more frequently employed in the preliminary analysis. It should be realized, however, that Euler buckling loads represent the theoretical upper limits of compressive loading in the elastic range. For example, referring to the Design Problem 10.1 and using the buckling load factor from Table 10.3 corresponding to the pin-jointed condition, the critical buckling load is 370,000 lb. A comprehensive discussion of the

Stability and Buckling Resistance

Table 10.3 Euler Buckling Loads for Typical Columns

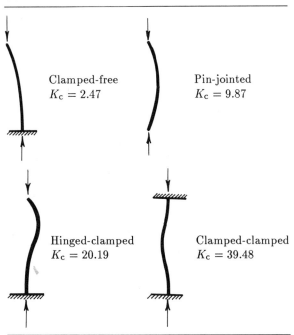

Note: $W_{CR} = K_c EI/L^2$; L, length between supports.

elastic and inelastic effects, falling between the concepts described by Eqs. (10.1) and (10.4), has been presented by Shanley [29].

INELASTIC COLUMN RESPONSE

The analysis of inelastic column behavior is based on Engesser's theory [32], which involves the *tangent modulus* E_t, defined as the local slope of the stress-strain diagram

$$E_t = \frac{dS}{d\epsilon} \tag{10.5}$$

It is advisable to develop a complete curve for E_t instead of depending on one value obtained by graphical differentiation. Once the value of E_t is decided upon, the critical inelastic load can be found directly from the Euler equation by substituting E_t for E. Despite some discussion over the past 50 years, the concept of the tangent modulus remains in force and it is believed today that the Engesser load, as it became known, corresponds to the upper practical limit of column strength. The shape of the inelastic portion of the column curve, based on this concept, depends on the shape of the stress-strain curve for a given material. The increase in the critical load is a function of the improvement of the base metal by alloying and

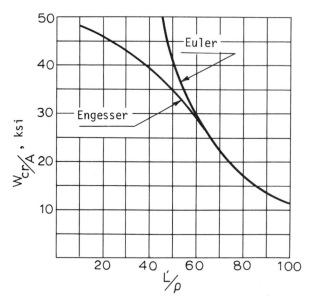

Fig. 10.2 Comparison of Euler and Engesser critical column stresses for lower values of L'/ρ.

heat treatment. However, where rolled shapes are involved, any residual stresses present would tend to lower the inelastic load. Since the critical buckling stresses are often expressed in terms of the L'/ρ values, one way of increasing the buckling stress without changing the material's characteristics in design is to reduce the ratio L'/ρ. In this connection, when the allowable working stress is specified in design, it may be more convenient to use the secant formula Eq. (10.4) for the elastic as well as the inelastic range.

It is instructive to keep in mind Shanley's comparison [39] between the Euler and Engesser tangent modulus column stresses, as shown in Fig. 10.2. The comparison indicates how conservative Euler prediction can be for smaller values of L'/ρ. A survey of experimental data on various columns also shows [40] that practically all test points fall below the Euler curve. This fact proves Euler's remarkable foresight in predicting the physical behavior of a structure by purely mathematical deductions and provides an example of the great value of a closed-form approach to the solution.

WEIGHT COMPARISON

In those cases where the dimensional similarity, optimization, and testing are involved, it may be in order to quote a few useful rules. When the ratio of W_{CR}/L^2 is the same for the two column designs whose cross sections are geometrically similar, the columns are subjected to the same axial stress. The only scatter in the results in real column tests will be due to the material properties and manufacturing tolerances. It also follows that for any two column structures having the same

Stability and Buckling Resistance

proportions, the ratio of the weights can be defined as

$$\frac{\text{Weight 2}}{\text{Weight 1}} = \left(\frac{W_{\text{CR2}}}{W_{\text{CR1}}}\right)^{3/2} \tag{10.6}$$

The analysis of structural columns discussed so far has been characterized by the assumption that the entire load-carrying member is involved. For instance, the pin-jointed column listed in Table 10.3 develops a well-defined curvature along its entire length. The solution obtained by Euler indicates that the shape of the deflection curve is sinusoidal. More advanced studies of stability problems in plates and shells utilize energy methods in conjunction with an assumed shape of the deflected structural member. These solutions give generally good results provided that the assumed shape is reasonably correct [31, 32].

BUCKLING OF PLATES IN ELASTIC REGIME

A special class of problems in the area of elastic design involves relatively thin and flat sections in which local instability can take place under the action of a compressive load. The buckling strength of a plate component is not necessarily lost entirely when such a local distortion occurs. Because of the significant residual strength left in the member, the design analysis can therefore be based on a twofold approach. In one instance we may select not to have any buckling deformation because of appearance requirements. Under yet another set of circumstances, local buckling may be permitted if the structural integrity of the total system is not compromised.

The general form of the stress equation that applies to all types of plate buckling is

$$S_{\text{CR}} = K_{\text{p}} E \left(\frac{t}{b}\right)^2 \tag{10.7}$$

The buckling stress coefficient K_{p} is similar to the column factor K_{c} entering Eq. (10.1), and depends to a large degree on edge constraint. It is a nondimensional quantity and it is sometimes called the *plate coefficient*. The notation applicable to Eq. (10.7) is shown in Fig. 10.3.

When the critical stress, calculated from Eq. (10.7) is less than the yield strength of the material, the buckling is considered to be elastic and the modulus of elasticity remains constant. The design value of S_{CR}, given by the various formulas of the foregoing type, should be regarded as an upper limit because the stresses actually measured are found to be smaller by an appreciable margin. The discrepancies between the theory and experiment are mainly due to the geometrical irregularities, and this effect is known to increase with the decrease in plate thickness.

When the critical stress exceeds the yield strength of the material, the plate buckling is termed *inelastic* and the modulus of elasticity decreases with the stress. However, the yield strength of the material is considered to be a natural limit for the critical stress.

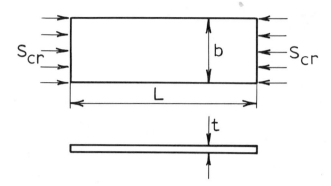

Fig. 10.3 Rectangular plate in compression.

In the case of most practical problems, the ratio of the plate length to width is greater than 5. For this ratio the buckling strength of the plate is found to be independent of length. For lower length-to-width ratios, the buckling coefficient K_p increases somewhat, but it is a common, conservative practice to disregard this change and to consider the nature of the edge supports as the only controlling factor. Here again the choice of a value for K_p depends to a large extent upon engineering judgment in a particular design situation. The most common values of K_p are given in Table 10.4. These values are intended for material characterized by a Poisson's ratio of 0.3. Again, as in the case of structural columns, the complete fixity of both edges, represented by condition 5 in Table 10.4, is practically never realized. Unless the weight requirements are such that the fixed-end condition must

Table 10.4 Buckling Stress Coefficients for Edge-Loaded Flat Plates (Poisson's Ratio = 0.3)

Edge 1	Condition	Edge 2	K_p
Simple support	1.	Free	$K_p = 0.38$
Fixed support	2.	Free	$K_p = 1.15$
Simple support	3.	Simple support	$K_p = 3.62$
Simple support	4.	Fixed support	$K_p = 4.90$
Fixed support	5.	Fixed support	$K_p = 6.30$

Note: All loaded edges are simply supported and plates are considered to be relatively long. Loading is perpendicular to the plane of the paper.

Stability and Buckling Resistance

be satisfied, the most simple and practical solution in design is to assume either both simple supports (case 3) or the simple support-free condition (case 1).

In using the design formula for the critical compressive stress, the common practice is to place the dimension b between the rivet or weld lines. In rolled sections made by cold forming or pressing, the distance b is normally measured from the edges of the fillets.

When a very long plate of width b is supported at the two long sides and it is loaded in compression, case 3 in Table 10.4 can be used as a criterion for estimating the elastic buckling stress. The theory also shows that when K_p is plotted as a function of L/b, the panel with simply supported edges tends to buckle into a series of equivalent square panels when L/b is an integer [29].

When both edges of the plate remain free, they are not compelled to remain straight under load, and the plate behaves as a column. All cases illustrated in Table 10.4, however, provide some degree of edge restraint, so that the plate is forced to bend about two different axes. Here, therefore, lies the principal distinguishing feature between the behavior of a wide column and a panel. Under the conditions of developing a double curvature, the elements of the panel located off center are subject to twist as well as bending. This provides a rational explanation of why the buckling load of a panel is markedly higher than the corresponding buckling load of a wide column.

INELASTIC BUCKLING OF PLATES

To account for inelastic effects, Eq. (10.7) modifies to

$$S_{\text{CR}} = K_{\text{p}} \eta E \left(\frac{t}{b}\right)^2 \tag{10.8}$$

Here η denotes the inelastic reduction factor [29], which is obtained by dividing the effective modulus by the elastic modulus. Many studies of this subject have been made and various formulas recommended. The approximate values for inelastic plate buckling in compression are given in Table 10.5.

The relation of secant to tangent modulus is illustrated in Fig. 10.4. The secant modulus E_s is the relationship between stress and total strain at a particular point, consisting of elastic and plastic components ϵ_E and ϵ_p, respectively. The tangent modulus E_t may be regarded as a measure of instantaneous resistance of

Table 10.5 Approximate Factors for Inelastic Buckling of Plates

Edge Conditions	Inelastic Reduction Factor, η
Both edges simply supported	$(E_t/E)^{1/2}$
Both edges fixed	$(E_t/E)^{1/2}$
One edge free, the other fixed	$(E_t/E)^{1/2}$
One edge free, the other supported	E_s/E

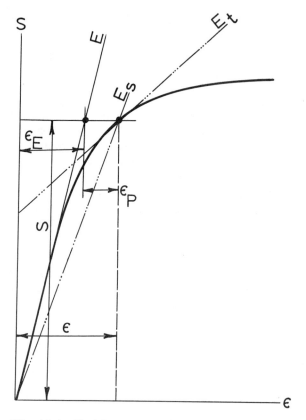

Fig. 10.4 Modulus concepts for inelastic analysis.

the material against increase in strain. This value diminishes quite rapidly as the total strain increases.

SHEAR BUCKLING OF PANELS

When relatively thin panels are employed to carry transverse loading, such as in airplane fuselage structure, the following two equations can be used to calculate the critical shear stress for simply supported and fixed edges of slender panels, respectively:

$$\tau_{\text{CR}} = \frac{4.8 G_s E}{G} \left(\frac{t}{b}\right)^2 \tag{10.9}$$

$$\tau_{\text{CR}} = \frac{8.1 G_s E}{E} \left(\frac{t}{b}\right)^2 \tag{10.10}$$

In Eqs. (10.9) and (10.10), G_s is the secant modulus of the shear stress strain diagram and G is the elastic modulus of rigidity defined in Chap. 2. When modulus of rigidity values are not available, the approximate correction factor (E_t/E) can

Stability and Buckling Resistance

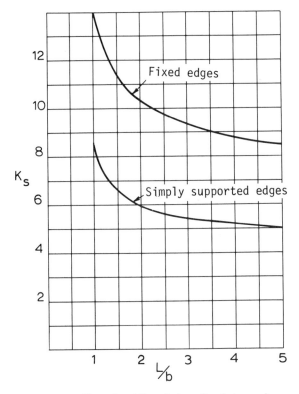

Fig. 10.5 Shear buckling factors for flat panels.

be used in the analysis at a stress level of 2τ. The shear panels tend to buckle in a series of parallel wrinkles. This response, however, does not mean collapse of a flat shear web or a panel. When the panels are not slender, the critical shear stress can be computed from the following equations and Fig. 10.5.

$$\tau_{\mathrm{CR}} = K_s \eta E \left(\frac{t}{b}\right)^2 \tag{10.11}$$

BUCKLING DUE TO BENDING

Occasionally, the question is asked as to the mechanism of buckling of beams due to the in-plane bending. In the case of a narrow rectangular bar or a plate, lateral buckling will occur when the bending moment will reach a critical value. Such a failure then involves a combined effect of bending and twist, which is reflected in the torsional shape factor, having the same dimensions as the moment of inertia. The shape factor is denoted in this chapter by K_t and it should be employed in the product GK_t, known as the *torsional rigidity*. A summary of the more commonly encountered formulas in this area of structural stability is given in Table 10.6 [38]. The table covers cantilever and simply supported beams with bending about the

Table 10.6 Buckling Due to Bending for Beams of Rectangular Cross Section

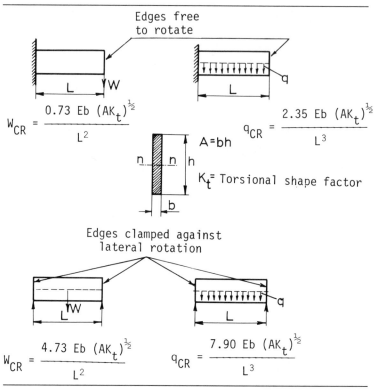

n-n axis, as shown. In the case of simple supports, edges are clamped against the lateral rotation. For a study of more complex cross sections, the reader is referred to other works [32, 38, 40]. For the wide flange and I-beam geometry, a semiempirical design formula for lateral buckling is recommended [41]

$$S_{\text{CR}} = \frac{0.7 E b t}{h L} \qquad (10.12)$$

In Eq. (10.12), b denotes the total width of the flange, h is the total depth of beam, and t is the thickness of the compression flange. The factor 0.7 represents an average test value. In applying the factor of safety to Eq. (10.12), the "working stress" is obtained by dividing the particular yield strength of the material by the required factor. Equation (10.12) is, however, applicable to simply supported beams subjected to uniform bending under end moments only.

DESIGN OF SPECIAL COLUMNS

Certain structural members sized to carry axial loads may have variable cross sections and other features where the standard column formulas, such as so far given

Stability and Buckling Resistance

in this chapter, may not apply. Research organizations and aerospace industry in particular have been concerned with this problem for a long time now and have developed excellent practical design tools for various column geometries. The expression for the critical buckling load for a column of variable cross section can be postulated with direct reference to the general formula given by Eq. (10.1).

$$W_{\text{CR}} = \phi \frac{EI}{L^2} \tag{10.13}$$

In this model ϕ denotes the buckling coefficient, which is a complicated function of the column geometry, bending rigidity, and end constraints. The ϕ parameter is often represented by elaborate design curves involving the ratios of flexural rigidities and lengths. The special class of columns also includes stepped and tapered transitions, which can be defined in terms of the appropriate dimensionless parameters. Consider for instance the case of a stepped column shown in Fig. 10.6. The critical load for this case [21] is

$$W_{\text{CR}} = \phi \frac{EI_2}{L^2} \tag{10.13a}$$

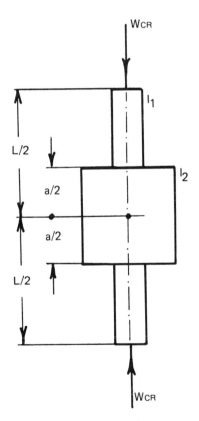

Fig. 10.6 Stepped column.

where

$$\phi = \frac{9.87}{m + k(1-m) - 0.32(k-1)\sin(\pi m)} \tag{10.13b}$$

The buckling coefficient here is expressed in terms of $k = I_2/I_1$ and $m = a/L$, consistent with the notation given in Fig. 10.6. When $m = 0$ and $k = 1$, we obtain the case of the classical column with pin-jointed ends as shown in Table 10.3.

The design of a column with one end fixed and consisting of two components of different size is shown in Fig. 10.7. The notation is such that the critical buckling load can be obtained from Eqs. (10.13a) and (10.13b) directly. This is generally consistent with the conventional cases given in Table 10.3. We note that a column with a fixed end and having a length of $L/2$ behaves as a pin-jointed column of length L. This is not surprising because of the symmetry involved. The slope of a pinned column shown in Fig. 10.6 is zero at the half-length. In other words, the buckled shape of a classical Euler column with one end fixed is identical with that of a complete pin-jointed column.

When a structural member of uniform cross section illustrated in Fig. 10.8 is loaded by its own weight q, the problem of finding the exact solution is not simple because it involves the use of Bessel functions. However, for most practical purposes the approximate solution can be obtained using the strain energy method [21] and the assumed deflection mode under the uniformly distributed axial load. The critical unit weight for the case shown in Fig. 10.8 can be calculated as

$$q_{CR} = 7.89 \frac{EI}{L^3} \tag{10.14}$$

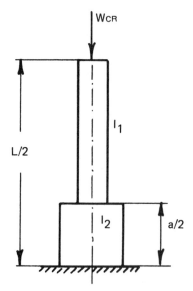

Fig. 10.7 Pinned column.

Stability and Buckling Resistance

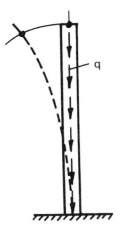

Fig. 10.8 Column with uniform cross-section.

Design Problem 10.2

Estimate the critical length of a thin-walled aluminum pipe of uniform cross section, at which the pipe is expected to fail as Euler column. Assume mean cross-sectional radius of $r = 2$ in. and Young's modulus of $E = 10 \times 10^6$ psi. Check the axial stress at the instant of elastic instability for the specific weight of aluminum of $\gamma = 0.098$ lb/in.^{3}, and wall thickness $t = 0.2$ in.

Solution

For a relatively thin pipe of radius r and wall thickness t, the following approximations apply

$$q_{CR} = 2\pi r t \gamma$$

and

$$I = \pi r^3 t$$

Utilizing Eq. (10.14) yields

$$2\pi r t \gamma = 7.89 \frac{E \pi r^3 t}{L^3}$$

Hence simplifying and solving for L gives

$$L = 1.58 \left(\frac{E r^2}{\gamma} \right)^{1/3}$$

and

$$L = 1.58(10 \times 10^6 \times 4/0.098)^{1/3} = 1172 \text{ in.} = 97.7 \text{ ft}$$
$$= 29.77 \text{ m}$$

The total weight of the pipe is

$$W = 2\pi rt\gamma L$$

Hence

$$S = \frac{W}{A} = \frac{2\pi rt\gamma L}{2\pi rt} = \gamma L$$

and

$$S = 0.098 \times 1172 = 114.9 \text{ psi} \quad (0.79 \text{ N/mm}^2) \quad \blacklozenge$$

It is of interest to note that this case is strictly governed by stability rather than stress. However, the critical length calculated here represents an ideal case which does not take into account eccentricity or any other geometric irregularity likely to exist in a real structure. Any such effect should lower the allowable calculated value of the critical length.

One of the highly involved theoretical and experimental problems is concerned with a pin-connected column supported by an elastic foundation. This problem is found in the railroad industry, where high compressive forces tend to buckle the rails laterally against a rather rigid underlying support. The reaction offered by the foundation is proportional to the deflection. Several aspects of this and other related cases in the area are discussed in Chapters 24 and 37.

Table 10.7 Critical Buckling Loads for Arches and Rings

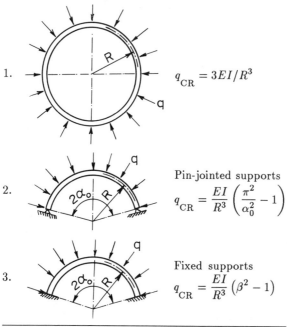

1. $q_{CR} = 3EI/R^3$

2. Pin-jointed supports
$$q_{CR} = \frac{EI}{R^3}\left(\frac{\pi^2}{\alpha_0^2} - 1\right)$$

3. Fixed supports
$$q_{CR} = \frac{EI}{R^3}(\beta^2 - 1)$$

Note: For values of β, see Fig. 10.9.

Stability and Buckling Resistance

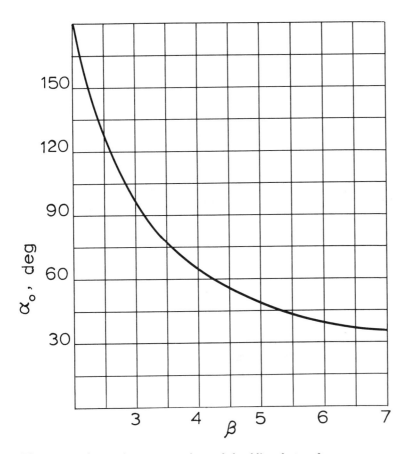

Fig. 10.9 Approximate curve for arch buckling factor β.

OTHER BUCKLING PROBLEMS

The problem of buckling of circular rings and arches can be treated analytically [31], and a brief summary of the more important conventional cases is given in Table 10.7. The approximate arch buckling factor β, illustrated in Fig. 10.9, follows from the relation $\beta \tan \alpha_0 \cot \alpha_0 \beta = 1$. It is noted that for $\alpha_0 = \pi$, case 3 from Table 10.7 reduces to case 1. The formula for case 1 is well known in the study of the behavior of cylindrical shells under uniform external pressure. There are wide areas of application of stability and strength formulas to shells in engineering design.

In closing this brief review of some of the more frequently encountered aspects of stability and strength, it is noted that in the design of certain light structures, stability alone can determine the structural integrity. Elastic buckling of some of the components often occurs without precipitating damage of the total structural system. The classical works of Timoshenko [31] and Bleich [32] have gone a long way toward removing the reluctance to using elasticity and buckling strength criteria in design.

SYMBOLS

A	Area of cross section, in.² (mm²)
b	Width of flange or plate, in. (mm)
c	Distance from CG to outer fiber, in. (mm)
E	Modulus of elasticity, psi (N/mm²)
E_t	Tangent modulus of elasticity, psi (N/mm²)
E_s	Secant modulus of elasticity, psi (N/mm²)
G	Modulus of rigidity, psi (N/mm²)
G_s	Secant modulus of rigidity, psi (N/mm²)
h	Total depth of beam, in. (mm)
I	General symbol for moment of inertia, in.⁴ (mm⁴)
I_1, I_2	Moments of inertia in stepped columns, in.⁴ (mm⁴)
I_{xx}	Moment of inertia about x axis, in.⁴ (mm⁴)
I_{yy}	Moment of inertia about y axis, in.⁴ (mm⁴)
K_c	Elastic column factor
K_p	Plate coefficient
K_s	Shear buckling factor
K_t	Torsional shape factor, in.⁴ (mm⁴)
k	Ratio of moments of inertia
L	Length of beam, column or plate, in. (mm)
L'	Effective column length, in. (mm)
m	Length ratio
q	Uniform load, lb/in. (N/mm)
q_{CR}	Critical uniform load, lb/in. (N/mm)
R	Radius of curvature, in. (mm)
S	General symbol for stress, psi (N/mm²)
S_{CR}	Critical elastic buckling stress, psi, (N/mm²)
S_{max}	Maximum buckling stress, psi (N/mm²)
t	Thickness of flange or wall, in. (mm)
W	Weight, lb (N)
W_{CR}	Critical buckling load, lb (N)
α	Eccentricity of load application, in. (mm)
α_0	Half-angle subtended by circular arch, rad
β	Arch buckling factor
γ	Specific weight, lb/in.³ (N/mm³)
ϵ	General symbol for strain, in./in. (mm/mm)
ϵ_E	Elastic strain, in./in. (mm/mm)
ϵ_p	Plastic strain, in./in. (mm/mm)
η	Inelastic reduction factor
ρ	Radius of gyration, in. (mm)
σ_y	Yield strength of material, psi (N/mm²)
τ	General symbol for shear stress, psi (N/mm²)
τ_{CR}	Critical shear stress, psi (N/mm²)
ϕ	Buckling coefficient

II
DYNAMIC AND THERMAL EFFECTS

11
Dynamic Response

INTRODUCTION

Everyday engineering design, concerned with sizing and shaping structural and machine elements, is for the most part concerned with the elastic, static response of the systems. There are many good reasons for following such a plan. In the first place, static, elastic analysis gives conservative and satisfactory results without undue refinement. The load application is usually found to be static, and when some dynamic effects are present, it is convenient to compute the static stresses and to multiply the results by, say, a factor of 2. Nevertheless, questions arise concerning the dynamic behavior, justifying at least a brief review of some of the practical concepts as part of the general stress analysis. The dynamic effects have been, for a long time now, special topics of the various theoretical books written for mathematicians and advanced students of engineering. The mathematical complexity of problems in this field is certainly beyond the scope of the design office charged with the primary duty of turning out working drawings and specifications.

In the area of the dynamic behavior, the rate of load application and the corresponding method of analysis vary depending on the history and severity of load application. For example, in fatigue analysis the stress level and the number of applied load cycles must be known before the remaining useful life of a component can be estimated. The stress level also implies that thermal effects may have to be included in the analysis.

GENERAL CRITERIA

Before embarking on a tedious course of dynamic analysis, however, it is well to keep in mind certain principles and rules of thumb. Assuming that we can estimate the time of load application for a particular design case, and assuming that we can either measure or calculate the natural period of vibration of a component, the loading can be considered as static if

$$\frac{\text{Time of load application}}{\text{Natural period of vibration}} > 3$$

Although in the intermediate range of dynamic considerations the static stress analysis is still applicable, the mechanical properties of the material may be sufficiently altered to be included in the calculations. As a rough guide this condition corresponds to

$$1.5 \leq \frac{\text{Time of load application}}{\text{Natural period of vibration}} \leq 3$$

Finally, when the structure is subjected to impact or shock, static analysis becomes pretty much meaningless and rigorous solutions of the problems, if possible at all, become exceedingly complex. An example of this is a bullet fired at a plate, where the dynamic properties of both materials involved must be considered. This occurs when

$$\frac{\text{Time of load application}}{\text{Natural period of vibration}} \leq 0.5$$

The concepts of fatigue and inertia loading are discussed in theoretical books somewhat differently, because the loads are developed from the equations of motion and then stresses are calculated on the assumption of static behavior [21].

DYNAMIC STRENGTH

It is well to note that yield strength of a low-carbon steel increases with increasing the rate of strain comparatively more than that for a high-strength steel. This point is illustrated in Fig. 11.1. The problem is, however, that the best data obtained so far have been derived from uniaxial rather than combined stress conditions. The characteristics shown in Fig. 11.1 also apply to most materials when tested at higher temperatures. The effect of strain rate on the compressive strength of concrete and similar brittle materials has also been measured [42]. It is also known that each material has its own value of the critical normal fracture strength under purely dynamic conditions [43]. Examples of such data are given in Table 11.1.

Materials data of this type are not readily available in reference handbooks. Also, some of the extreme values shown are used only in explosive forming design where significant shock pressures must be postulated. The development of the type of data indicated in Table 11.1 involves rather advanced testing techniques [44].

Dynamic Response

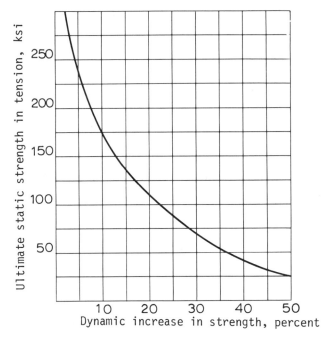

Fig. 11.1 Effect of rapid loading on strength.

RESPONSE IN NATURAL MODE

As stated in the introductory paragraphs of this chapter, proper estimates of the expected loads can be made if the natural period of vibration of an elastic member can be estimated. The maximum elastic response of a structure is normally associated with the maximum stress. It is customary in the dynamic analysis of various components to represent the particular part by a spring-mass model [45]. Such considerations of a single degree of freedom lead to a very useful, elementary

Table 11.1 Dynamic Strength of Metals

Material	Static Strength (psi)	Dynamic Strength (psi)	Impact Velocity (ft/sec)	Reference
2024 Al (annealed)	65,200	68,600	> 200	94
Magnesium alloy	43,800	51,400	> 200	94
Annealed copper	29,900	36,700	> 200	94
302 Stainless steel	93,300	110,800	> 200	94
SAE 4140 steel	134,800	151,000	175	94
SAE 4130 steel	80,000	440,000	235	44
Brass	39,000	310,000	216	44

criterion for the natural period of vibration expressed in seconds

$$T = 2\pi \left(\frac{\delta_{st}}{g}\right)^{1/2} \tag{11.1}$$

where δ_{st} denotes the static elastic displacement and g is the acceleration of gravity. When δ_{st} is expressed in inches, Eq. (11.1) reduces to

$$T = 0.32(\delta_{st})^{1/2} \tag{11.2}$$

The corresponding frequency in the natural mode of vibration is equal to the reciprocal of the period given by Eq. (11.2)

$$f = \frac{3.1}{(\delta_{st})^{1/2}} \tag{11.3}$$

Design Problem 1.1

A fixed-ended steel beam of length $L = 20$ ft is designed to carry a central load $W = 1000$ lb. Calculate the moment of inertia of the beam in the plane of loading to give the fundamental frequency of 30 cycles per second. Neglect the beam's own weight.

Solution

The conventional deflection formula for this case is

$$\delta_{st} = \frac{WL^3}{192EI}$$

Utilizing Eq. (11.3) the formula for the moment of inertia becomes

$$I = 0.00054 \frac{WL^3 f^2}{E}$$

Taking the modulus of elasticity to be 30×10^6 psi and substituting other numerical values specified for this problem we obtain

$$I = \frac{0.00054 \times 1000 \times 240^3 \times 30^2}{30 \times 10^6} = 224 \text{ in}^4. \quad \blacklozenge$$

FREE-FALL EFFECT

The concepts of the elastic strain energy stored and the change in the potential energy of the system can be used to calculate the maximum stresses and dynamic response of simple structural members subjected to impact. This approach is illustrated by developing an approximate formula for the relation between the dynamic and static stresses for a transversely loaded beam when a given weight W is dropped

Dynamic Response

through a distance a, as shown in Fig. 11.2. The effect of the kinetic energy lost during the impact will be ignored.

Let us assume that the unknown, effective force producing the total displacement δ_t is W_d. Hence, the elastic work done in deforming the beam is

$$U_e = \frac{W_d \delta_t}{2} \tag{11.4a}$$

The deflection for a simply supported and centrally loaded beam is

$$\delta_t = \frac{W_d L^3}{48 EI} \tag{11.4b}$$

Combining Eqs. (11.4a) and (11.4b) by eliminating W_d gives

$$U_e = \frac{24 EI \delta_t^2}{L^3} \tag{11.4c}$$

The change of the potential energy of the system neglecting the weight of the beam is

$$U_p = W(a + \delta_t) \tag{11.4d}$$

Equating U_e and U_p, Eqs. (11.4c) and (11.4d), and rearranging terms gives the quadratic equation in δ_t

$$\delta_t^2 - \frac{WL^3 \delta_t}{24 EI} - \frac{WL^3 a}{24 EI} = 0 \tag{11.4e}$$

The static displacement under weight W is

$$\delta_{st} = \frac{WL^3}{48 EI} \tag{11.4f}$$

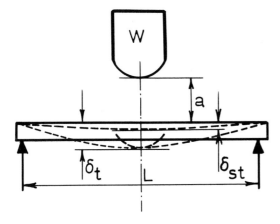

Fig. 11.2 Drop weight problem.

Hence, combining Eq. (11.4f) with Eq. (11.4e) and solving for δ_t yields

$$\delta_t = \delta_{st} + \left(\delta_{st}^2 + 2\delta_{st}a\right)^{1/2} \tag{11.4g}$$

Since the maximum beam stresses are proportional to deflections, dividing both sides of Eq. (11.4g) by δ_{st} gives the required formula

$$S_{dyn} = S_{st}\left[1 + \left(1 + 2\frac{a}{\delta_{st}}\right)^{1/2}\right] \tag{11.5}$$

It is of interest to note that for $a = 0$, Eq. (11.5) gives the familiar design rule which states that for the case of sudden loading, the ratio of dynamic to static stress is 2.

When the impact is horizontal instead of a free fall under gravity, such as that shown in Fig. 11.2, the stress formula (for $g = 386.4$ in./sec^2) becomes

$$S_{dyn} = S_{st}\frac{0.051V}{(\delta_{st})^{1/2}} \tag{11.6}$$

where V denotes the velocity of the moving body in inches per second. The objective of Eqs. (11.5) and (11.6) is to estimate the dynamic stress when the static stress can be calculated from the conventional equations.

It should be stated in all fairness that the stresses due to impact can not be determined accurately. The materials involved are never perfectly elastic. Furthermore, when a body strikes another object simultaneous contact is not realized at all points, and the distribution of stresses and strains under impact loading is not the same as that under static loading, particularly at higher velocities of impact. Last but not least, some kinetic energy of the moving body is dissipated during impact. This loss can be approximated with the aid of Table 11.2 for a number of elementary design cases [35]. The procedure here is to multiply the theoretically calculated energy by the dissipation factor C_e from Table 11.2.

Design Problem 11.2

A steel cantilever of length $L = 10$ ft and the moment of inertia $I = 600$ in.^{4} is designed to the static factor of safety of 8 while carrying a weight of $W = 500$ lb. If the same weight were now to be dropped on the free end of the beam, estimate the height of the maximum allowable drop a to maintain a minimum factor of safety of 1.8.

Solution

The standard deflection formula for this case is

$$\delta_{st} = \frac{WL^3}{3EI}$$

which gives

$$\delta_{st} = \frac{500 \times 120^3}{3 \times 30 \times 10^6 \times 600} = 0.016 \text{ in.}$$

Dynamic Response

Table 11.2 Energy-Loss Factors for Impact Loading

#	Diagram	C_e
1.	Stationary part with M_1, M_2; Moving M part	$C_e = \dfrac{4(M_1 + 3M_2 + 3M)M}{3(M_1 + 2M_2 + 2M)^2}$
2.	Simply supported beam with M_1, M_2, and falling M	$C_e = \dfrac{64M(17M_1 + 35M_2 + 35M)}{35(5M_1 + 8M_2 + 8M)^2}$
3.	Cantilever beam with M_1, M_2, and falling M	$C_e = \dfrac{16M(33M_1 + 140M_2 + 140M)}{35(3M_1 + 8M_2 + 8M)^2}$
4.	Fixed-fixed beam with M_1, M_2, and falling M	$C_e = \dfrac{4M(13M_1 + 35M_2 + 35M)}{35(M_1 + 2M_2 + 2M)^2}$

At a safety factor of 1.8, the total stress under dynamic response must be 4.44 times the static value. Hence, using this result in Eq. (11.5) yields

$$4.44 - 1 = \left[1 + 2\left(\frac{a}{\delta_{\text{st}}}\right)\right]^{1/2}$$

from which

$$a = 5.42\delta_{\text{st}} = 0.087 \text{ in. } (2.2 \text{ mm}). \quad \blacklozenge$$

Design Problem 11.3

Develop an approximate design formula for estimating the dynamic load W_d when a vehicle traveling with velocity V strikes a steel barrier oriented at 90° to the direction of travel. The weight of the vehicle is W. The barrier is made of a heavy-wall pipe having the moment of inertia I and length L acting as a beam on simple supports. Use the elastic strain energy of the barrier as a design criterion.

Solution

The elastic strain energy U_e is assumed to be equal to the area under the curve when W_d is plotted against δ. For a linear-elastic behavior

$$U_e = \frac{W_d \delta}{2}$$

where

$$\delta = \frac{W_d L^3}{48EI}$$

The kinetic energy of the vehicle is

$$U_k = \frac{WV^2}{2g}$$

Assume that

$$U_e = U_k$$

Substituting the above relations and solving for W_d yields

$$W_d = \frac{0.35V}{L}\left(\frac{WEI}{L}\right)^{1/2}$$

The dimensions applicable to this formula include pounds, inches, and seconds. If the vehicle velocity is specified in mph then the conversion gives 1 mph = 17.6 in./sec. The modulus of elasticity E is in lb/in.^{2}, moment of inertia represents in.^{4}, and the length of the barrier is given in inches. ♦

Design Problem 11.4

Assuming that the barrier discussed in Design Problem 11.3 is supported by a crushable material at each anchor, develop a formula for the amount of crush if the absorbed energy can be based on a constant crush strength S_c. Neglect the elastic strain energy of the barrier and let the cross-sectional area of each crushable support be A. Estimate the magnitude of the dynamic load W_d when A is 4 ft by 4 ft and $S_c = 40$ psi.

Solution

Denoting the energy and the amount of compression of the material by U_c and δ_c respectively gives

$$U_c = 2A\delta_c S_c$$

Making $U_c = U_k$ and solving for δ_c, we obtain

$$\delta_c = 0.00065 \frac{WV^2}{AS_c}$$

and

$$W_d = 2AS_c$$
$$= 2 \times 48 \times 48 \times 40 = 184{,}320 \text{ lb} \quad (0.82 \times 10^6 \text{ N}) \quad ♦$$

Dynamic Response

SHOCK MITIGATION

When shock mitigation criteria are part of engineering design [46], one of the basic questions which often arises concerns the level of a negative acceleration that a given body experiences when a physical stop is provided in the path of the motion. If the kinetic energy of the moving body can be completely transformed into the elastic strain energy of the resisting structure, then the dynamic stress can be deduced from Eq. (11.5). This formula represents a very general case of an elastic system where a reasonable approximation to a spring constant can be made. In a particular instance of a packaged item experiencing a half-sinusoidal pulse during a drop the maximum acceleration in terms of g units is sometimes expressed as

$$\Omega_d = 0.113 \frac{(a)^{1/2}}{T_r} \tag{11.7}$$

In the above formula a denotes the height of the drop in inches and T_r is the rise time in seconds. In this case the rise time is assumed to be equal to $1/4f$, where f denotes the fundamental frequency of the structure mitigating the motion. The derivation of the design formula given by Eq. (11.7) can be accomplished when the momentum of the package dropped from a given height a is made equal to the impulse developed during the time interval T_r.

In a number of practical cases a rigid body subjected to a free-fall motion can encounter a resisting medium which causes deceleration over a distance δ_c shown in Fig. 11.3. The corresponding static deflection δ_{st} can be only a small portion of δ_c. The resisting medium, for instance, can be a granular matter such as soil or a similar material which has unique load-deflection characteristics. If we postulate that the total kinetic energy of the moving object at the moment of impact is transformed into the work done in compressing the medium, then a number of different formulas

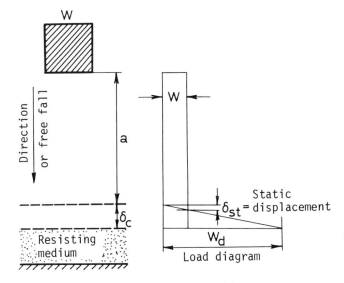

Fig. 11.3 Simplified deceleration model.

can be derived depending on the character of the load-deflection curve assumed. For a linear relationship corresponding to a dynamic load W_d, Fig. 11.3, the number of g units can be expressed as

$$n = \frac{2(a + \delta_c)}{\delta_c} \tag{11.8}$$

Based on the above formula the maximum dynamic force acting on the object would be nW where W denotes the weight of the object. When δ_c is relatively small the dynamic ratio n becomes inversely proportional to the amount of penetration. On the other hand, when $a = 0$, Eq. (11.8) reduces to $n = 2$, which is consistent with the standard case of sudden loading when the support is suddenly removed. The same result, of course, was shown to follow from Eq. (11.5). Although Eq. (11.8) is at best approximate it does indicate that by increasing δ_c we are effectively providing more cushioning. The approximate nature of Eq. (11.8) stems from the fact that the $(\delta_c - \delta_{st})$ portion of deflection shown in Fig. 11.3 is assumed to be linear.

Design Problem 11.5

A section of steel casing of mean radius R, wall thickness h and length $L = 40$ ft falls under gravity from a height $a = 500$ ft on a thick layer of sand. The fall is assumed to be perfectly axial, and the depth of penetration into the sand is found to be equal to $\delta_c = 30$ in. Calculate the approximate axial stress in compression caused by a sudden deceleration on impact. Assume the specific weight of steel to be $\gamma = 0.283$ lb/in.^{3}.

Solution

The weight of the casing is

$$W = 2\pi R h L \gamma$$

The dynamic force is

$$F = nW$$

The corresponding axial stress is

$$S_{dyn} = \frac{nW}{2\pi R h} = \frac{2\pi R h L \gamma n}{2\pi R h} = L\gamma n$$

Hence, employing Eq. (11.8) gives

$$S_{dyn} = \frac{2L\gamma(a + \delta_c)}{\delta_c}$$

On substitution, the above equation yields

$$S_{dyn} = \frac{2 \times 40 \times 12 \times 0.283(500 \times 12 + 30)}{30} = 54,608 \text{ psi} \quad (377 \text{ N/mm}^2) \quad \blacklozenge$$

Dynamic Response

Cursory view of input to Design Problem 11.5 may cause one to expect that cross-sectional dimensions should enter the onset of the analysis. The actual solution, however, quickly dispels such a notion when the radius and wall thickness cancel out during the process of deriving the formula for stress.

CALCULATION OF FREQUENCY

It was noted in the various sections of this chapter that a number of technical decisions in elementary dynamics hinge upon the knowledge of the fundamental mode of vibration. The first frequency and the natural period of vibration are shown to depend on the deflection of a structural member as indicated by Eqs. (11.1) through (11.3). This is an important consideration in dynamic analysis which means that the deflection curve under dynamic loading has essentially the same shape as that under static conditions. The critical vibrational stress is proportional to the maximum amplitude, and the stress during impact can be obtained by multiplying the stress under the static application of the load by a deflection ratio such as δ_t/δ_{st}. Although a simplification, this theory has served engineering purposes well as shown by experiments carried out as far back as 1922 [5]. A more rigorous theory of transverse impact, involving local deformations, was developed by Clebsch more than 100 years ago. For practical purposes, however, the analysis of impact based on the overall deflection mode is still difficult to fault.

As far as the vibration theory is concerned, the transverse and axial frequencies vary as the square of the beam length while the torsional frequency is inversely proportional to the length.

The method of frequency calculation widely used in engineering practice is known as the energy method. It works on the premise that in a simple periodic motion the maximum values of the kinetic and potential energies of the vibrating system are equal. Consequently, when the amplitude of vibration of the component or a system is known, the corresponding frequency in a simple periodic motion can be found.

In order to illustrate the practical aspects of this method consider the fundamental vibration mode of a cantilever beam. The square of the frequency for the uniform beam is

$$f^2 = \frac{g}{(2\pi)^2} \frac{\int_0^L y\,dx}{\int_0^L y^2\,dx} \tag{11.9}$$

The bending moment due to the gravity loading at any arbitrary point x is

$$M_x = \frac{A\gamma x^2}{2} \tag{11.9a}$$

From Eq. (7.5) we have

$$EI\frac{d^2y}{dx^2} = \frac{A\gamma x^2}{2} \tag{11.9b}$$

The two consecutive integrations lead to

$$EI\frac{dy}{dx} = \frac{A\gamma x^3}{6} + C_1 \tag{11.9c}$$

and

$$EIy = \frac{A\gamma x^4}{24} + C_1 x + C_2 \tag{11.9d}$$

The relevant boundary conditions are

$$\frac{dy}{dx} = 0 \quad \text{when } x = L$$

and

$$y = 0 \quad \text{when } x = L$$

Hence using Eqs. (11.9c) and (11.9d) gives

$$C_1 = -\frac{A\gamma L^3}{6} \tag{11.9e}$$

and

$$C_2 = \frac{A\gamma L^4}{8} \tag{11.9f}$$

Substituting Eqs. (11.9e) and (11.9f) into (11.9d) yields

$$y = \frac{A\gamma}{24EI}(x^4 - 4L^3 x + 3L^4) \tag{11.10}$$

Employing this result leads to the two relations required for the solution of Eq. (11.9)

$$\int_0^L y\, dx = \frac{A\gamma L^5}{20EI} \tag{11.11}$$

and

$$\int_0^L y^2\, dx = \frac{13 L^9 A^2 \gamma^2}{3240 E^2 I^2} \tag{11.12}$$

Finally, substituting Eqs. (11.11) and (11.12) into Eq. (11.9) results in the fundamental frequency of a cantilever beam

$$f = \frac{11.04}{L^2}\left(\frac{EI}{\gamma A}\right)^{1/2} \tag{11.13}$$

Table 11.3 Free Fall

$$H = \frac{gt^2}{2} = \frac{Vt}{2} = \frac{V^2}{2g}$$

$$t = \frac{V}{g} = \left(\frac{2H}{g}\right)^{1/2}$$

$$V = gt = (2gH)^{1/2}$$

For the case of a propped cantilever (one end fixed and one supported) the energy method gives

$$f = \frac{48.31}{L^2}\left(\frac{EI}{\gamma A}\right)^{1/2} \tag{11.14}$$

When the cantilever beam is replaced by a beam with both ends built in, the result is

$$f = \frac{70.19}{L^2}\left(\frac{EI}{\gamma A}\right)^{1/2} \tag{11.15}$$

In many situations the tables of formulas for frequency calculation are available. However, if a particular case at hand is markedly different from the standard models, then the energy method of derivation is certainly one of the elegant approaches to the problem.

ELEMENTARY FORMULAS FOR DYNAMICS

The basic motion of a body, without a specific reference to the forces causing that motion, is concerned with the distance, velocity, and acceleration as a function of time. In order to assist the designer with a quick estimate of certain kinematic parameters, Tables 11.3 and 11.4 provide a reference to the free-fall and constant-acceleration conditions, respectively. These two cases are probably the most common in preliminary calculations. In Table 11.3, H denotes the height of the free fall. Table 11.4 includes the initial velocity of motion V_0, and it also defines the acceleration in terms of the g values.

Table 11.4 Constant Acceleration

$$x = V_0 t + \frac{ngt^2}{2} = \frac{(V+V_0)t}{2} = \frac{V^2 - V_0^2}{2ng}$$

$$t = \frac{V - V_0}{ng} = \frac{(V_0^2 + 2ngx)^{1/2} - V_0}{ng}$$

$$V = V_0 + ngt = (V_0^2 + 2ngx)^{1/2}$$

SYMBOLS

A	Cross-sectional area, in.^{2} (mm^2)
a	Arbitrary distance, in. (mm)
C_e	Energy loss factor for impact
E	Modulus of elasticity, psi (N/mm^2)
F	Dynamic force, lb (N)
f	Frequency, Hz
g	Acceleration of gravity, in./sec^2 (mm/sec^2)
H	Free-fall height, in. (mm)
h	Thickness of wall, in. (mm)
I	Moment of inertia of cross section, in.^{4} (mm^4)
L	Length of beam, in. (mm)
M, M_1, M_2	Stationary and moving masses, lb-sec^2/in. (N-sec^2/mm)
M_x	Bending moment, lb-in. (N-mm)
n	Number of g units
R	Mean radius of pipe, in. (mm)
S_c	Crush strength, psi (N/mm^2)
S_{dyn}	Dynamic stress, psi (N/mm^2)
S_{st}	Static stress, psi (N/mm^2)
T	Period of vibration, sec
T_r	Rise time, sec
t	Time, sec
U_c	Energy absorbed in crushing, lb-in. (N-mm)
U_e	Elastic strain energy, lb-in. (N-mm)
U_k	Kinetic energy, lb-in. (N-mm)
U_p	Potential energy, lb-in. (N-mm)
V	Velocity of moving body, in./sec (mm/sec)
V_0	Initial velocity, in./sec (mm/sec)
W	Static load, lb (N)
W_d	Dynamic load, lb (N)
x	Coordinate, in. (mm)
γ	Specific weight, lb/in.^{3} (N/mm^3)
δ	General symbol for deflection, in. (mm)
δ_c	Clearance; also amount of crush, in. (mm)
δ_{st}	Static deflection, in. (mm)
δ_t	Total deflection, in. (mm)
Ω_d	Deceleration during impact, in/sec^2 (mm/sec^2)

12
Elements of Seismic Design

INTRODUCTION

Recent earthquake events, as well as a good deal of research on shock-resistant structures developed since the end of World War II, have stimulated considerable interest in the field of dynamics among practicing engineers and students of science. With special regard to seismic design, a good deal has been published [47] for the purpose of general education. Although the field of seismic knowledge is highly specialized, a brief review of the relevant engineering aspects is included here because of increased emphasis on earthquake design. A comprehensive outline of seismic problems, directed primarily to civil engineers, is available [48]. The work contains description of dynamic effects, characteristics of earthquake motion, behavior of materials, and design concepts. Considerable attention is paid to the probabilistic nature of seismic design.

BUILDING CODE METHOD

The problem of structural design in an earthquake environment has not yet been solved to the full satisfaction of seismologists, geologists, and engineers [49–51]. It is difficult to predict the character and intensity of the earthquake for design purposes, and hence the actual calculations have to be based on rather crude approximations. One particular area of difficulty involves the plastic response of structures to irregular ground motion. The application of rigorous theoretical methods of design to

115

this problem appears to be impractical [45]. Current design procedure is based on the regulatory codes derived from experience by observation of structural behavior in earthquakes. This experience is coupled with some of the established theoretical concepts of dynamic response and vibratory motion, on the assumption that only the first mode of vibration is important. The characteristic shape of the deflection curve is assigned to the first fundamental mode of a particular structure, and then the equivalent static force is calculated. For example, based on a commonly accepted code in California, the following practical formulas are used:

$$Q_b = WCK_d \tag{12.1}$$
$$C = \frac{0.05}{T^{1/3}} \tag{12.2}$$
$$\delta_{max} = \frac{0.49}{f^{5/3}} \tag{12.3}$$

The system to which these equations apply is represented in Fig. 12.1. This can be a building or some other structure resting on a firm base which is submitted to seismic motion. Here Q_b denotes the shear load at the base under dynamic conditions and W is the total weight of the structure. The coefficient C can be multiplied by the weight to obtain the maximum horizontal inertia force. The numerical factor K_d varies between 0.7 and 1.5 and accounts for the ability of the structure to deform plastically.* The maximum displacement, given by Eq. (12.3), is expressed in inches and frequency f in hertz [45].

The beam member supporting weight W in Fig. 12.1 can flex and it represents the effective spring constant of the system. It is important to note that the natural period of vibration T enters the calculations.

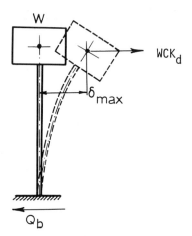

Fig. 12.1 Simplified seismic model.

---

*Low values of K_d are normally assigned to structural members that are able to deform plastically. Higher values of this factor are reserved for relatively brittle materials such as concrete.

PRELIMINARY ELASTIC DESIGN

In general, there appear to be two schools of thought among engineers concerned with the theory of earthquake-resistant design. One thesis is that a building should be perfectly rigid, so that in the event of seismic motion the top and bottom positions of this building would move an identical amount during the same time interval. Such a response would, of course, tend to induce rather large stresses in the building structure. The other thesis also goes to an extreme, by maintaining that a building should be as flexible as possible in order to, literally, sway during an earthquake. In an extreme case, then, one would have to construct a building from vulcanized rubber, which would certainly be resistant to seismic shock. However, such a building would not protect the contents very well because of the possibility of large displacements.

In practice, builders are limited by the availability and cost of construction materials, by design codes, and by soil conditions, so that none of the extreme design criteria outlined above are actually utilized. Good earthquake-resistant design approach is a compromise in which both stresses and deflections should be evaluated. Furthermore, the criterion of flexibility should not imply flimsy construction. Some designers also suggest that a reasonable compromise could be reached if we were to construct rigid buildings on soft ground and flexible buildings on rock.

Over the years a rule of thumb developed which states that a well-designed, earthquake-resistant building should be able, at any level, to withstand a horizontal force equal to one-tenth of its weight above that level. However, to the surprise of all concerned, more recent seismological data indicated a maximum acceleration at times exceeding three-tenths of gravity. It also became clear that the damage was not always proportional to the maximum acceleration, and that a more detailed knowledge of the vibrational modes of a structure was important in determining a realistic seismic response.

All this knowledge prompted the development of the various rational design methods. One such approach is based on a conservative elastic analysis utilizing spectral velocity charts developed for the discrete seismic regions [52].

Before illustrating the spectral-velocity method in some detail, it may well be useful to draw a general plan. It will be assumed here that a structure has a finite flexibility and can respond in one of the fundamental modes of vibration. The first task, therefore, is to calculate the fundamental periods of motion. Knowing this and having a spectral velocity chart for a particular seismic region, the probable levels of g loading on the structure can be estimated. This information can then be used to calculate the stresses using the conventional principles of strength of materials. By analogy to code practice, such as that defined by Eq. (12.1), we can use the concept of the equivalent static force and its effect on the stresses and deformations of a structure.

In applying seismological data, such as that shown in Fig. 12.5, we refer to the *Richter scale* Q. Although all of us have probably heard of it, some readers may not be familiar with its definition and relation to the measurement of seismic damage. Essentially, the concept behind the scale is based on the correlation between the seismic ground motion and the possible energy of the source causing this motion. In this manner, therefore, the Richter scale attempts to describe the strength of an earthquake by allotting to it a number Q on a numerical scale from 1 to 10. In mathematical terms, Q is defined as the logarithm (to the base 10) of the maximum amplitude measured in microns (mm × 10^{-3}) and traced on a standard seismograph at a distance of 100 km from the epicenter. Here the epicenter is defined as the point on the earth's surface directly above the focus of the earthquake. Recent determinations also indicate that a 1-unit increase in Richter's scale Q is equivalent to a 32-fold increase in energy of the earthquake source. However, the exact calculations of the absolute amount of energy involved remain rather uncertain and can vary by as much as a factor of 10. Practical experience also shows that based on this scale earthquakes of magnitude 5 or greater are potentially destructive to buildings.

Richter's scale is useful in classifying the extent of seismic disturbances. However, it should be realized that Q is a magnitude derived from the response of a seismic instrument and, as such, it must be influenced by the sensitivity of the available instrument. Furthermore, because of the nonuniformity of the earth's crust and random orientations of the geological faults Q cannot be a precise measure of the energy released by an earthquake.

The design procedure that follows is based on the spectral velocity data applicable to San Fernando, Parkville, El Centro, and similar types of ground motion. These cases have been selected mainly because of the superior quality of the records involved. In particular, the 1940 El Centro experience indicates the level of acceleration of $0.33g$, which appears to be one of the larger values known at the time [52].

The spectral velocity method can best be described with reference to a specific design problem involving a package W supported on beam-like members (Fig. 12.2 and 12.3) as follows:

1. Two extreme structural modes of behavior are selected, such as the single and double cantilevers shown in Figs. 12.2 and 12.3, respectively.
2. Spring constants are calculated separately for the two selected modes. Let the corresponding values of the spring constants be 60,000 lb/in. and 6600 lb/in.
3. For the case of a single degree of freedom in a mass-spring model, the relevant static displacement is

$$\delta = \frac{W}{k_s}$$

where W = total weight of the package supported by the frame, lb

k_s = spring constant, lb/in.

For the two extreme modes of deformation indicated in Figs. 12.2 and 12.3, the displacements under an assumed load of 7,200 lb are

$$\delta = \frac{7200}{60,000} = 0.12 \text{ in.} \quad \text{(single cantilever)}$$

Elements of Seismic Design

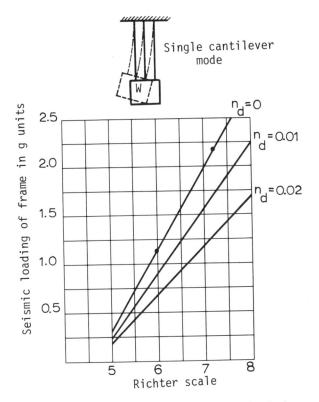

Fig. 12.2 Seismic loading on space frame in single-cantilever mode.

$$\delta = \frac{7200}{6600} = 1.09 \text{ in.} \quad \text{(double cantilever)}$$

4. The periods of the fundamental modes are obtained, for example, from Eq. (11.1):

$$T = 2\pi \left(\frac{\delta}{g}\right)^{1/2}$$

where

$$g = 386.4 \text{ in./sec}^2$$

Hence,

$$T = 6.28 \left(\frac{0.12}{386.4}\right)^{1/2} = 0.11 \text{ sec} \quad \text{(single cantilever)}$$

$$T = 6.28 \left(\frac{1.09}{386.4}\right)^{1/2} = 0.33 \text{ sec} \quad \text{(double cantilever)}$$

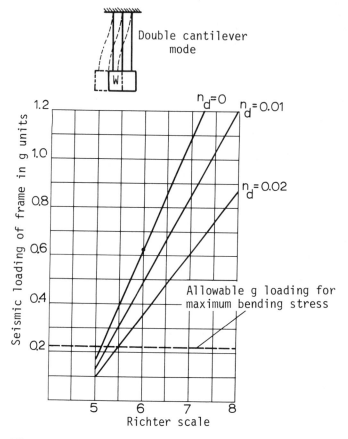

Fig. 12.3 Seismic loading on space frame in double-cantilever mode.

5. For the selected spectral velocity chart approximating, for example, the El Centro type of record (Fig. 12.4) and the relevant numerical value of periods calculated in step 4, the required velocities can be obtained from the appropriate curves. In this instance we take the following:

$$S_v = 15 \text{ in./sec} \quad \text{(single cantilever)}$$
$$S_v = 25 \text{ in./sec} \quad \text{(double cantilever)}$$

6. The spectral velocity chart given in Fig. 12.4 has been sketched in very roughly for the purpose of illustration only. In actual design situations, engineers should consult more precise seismological data [52].

The required formulas for the calculation of g loading corresponding to a particular bending mode of the system given for this example can be

Elements of Seismic Design

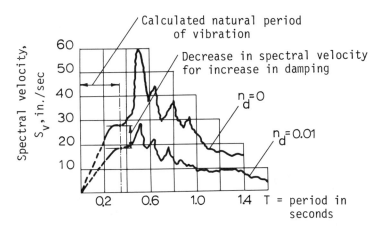

Fig. 12.4 Spectral velocity chart.

obtained as follows. Let

S_v = spectral velocity, in./sec

T = period, sec

Seismic loading on a structure can be now calculated in terms of g units with the aid of the following relations:

$$\Omega = \omega S_v \frac{\text{rad}}{\text{sec}} \times \frac{\text{in.}}{\text{sec}} = \text{in.}/\text{sec}^2$$

$$\omega = \frac{2\pi}{T} \text{ sec}^{-1}$$

$$T = 2\pi \left(\frac{\delta}{g}\right)^{1/2} \text{ sec}$$

Hence, substituting yields the necessary nondimensional parameter

$$\frac{\Omega}{g} = \frac{\omega S_v}{g} = \frac{2\pi S_v}{Tg} = \frac{2\pi S_v}{g 2\pi (\delta/g)^{1/2}} = \frac{S_v}{(\delta g)^{1/2}}$$

The maximum seismic loading on the frame can be now estimated directly using the final design formula from above, values of static deflection from step 3, and spectral velocity numbers from step 5. This gives the following seismic loading on the frame in g units:

$$\frac{\Omega}{g} = \frac{15}{(0.12 \times 386.4)^{1/2}} = 2.20 \quad \text{(single cantilever)}$$

$$\frac{\Omega}{g} = \frac{25}{(1.09 \times 386.4)^{1/2}} = 1.22 \quad \text{(double cantilever)}$$

Since the results are based on the El Centro spectrum, the corresponding maximum ground acceleration is known to be $0.33g$. The dynamic response of the frame in the two extreme modes analyzed here is represented by 2.20 and 1.22 seismic loading in g units, respectively. Assuming that the El Centro earthquake measured about 7.2 on the Richter scale, the points corresponding to this calculation can be found in Figs. 12.2 and 12.3 on lines $n_d = 0$. Linear extrapolation leads to the entire characteristics of seismic loading for zero damping provided that a functional relationship between the ground acceleration and the Richter scale can be established.

7. Such a functional relationship [52] is illustrated in Fig. 12.5, where the design curves are marked by solid lines. Based on the San Fernando and Parkville experiences one might extrapolate the design curves as shown by the dashed lines. Assuming the distance to the major fault line to be about 25 miles, as shown in Fig. 12.5, the remaining seismic loading characteristics depicted in Figs. 12.2 and 12.3 are developed as follows.

Take, for instance, the case of zero damping in a single-cantilever mode, depicted in Fig. 12.2. The level of $2.2g$ corresponds to seismic loading on the frame at a maximum of 0.33 ground acceleration. For an intermediate point such as that of $0.17g$, corresponding to the 25-mile distance to the fault and a Richter scale reading of 6, as indicated in Fig. 12.5, the calculated seismic

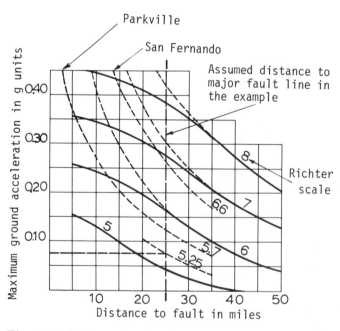

Fig. 12.5 Maximum acceleration versus distance to fault. Dashed lines indicate extrapolation to maximum observed values of ground acceleration: $1.06g$, San Fernando; $0.50g$, Parkville.

Elements of Seismic Design

loading on the frame becomes

$$\frac{0.17}{0.33} \times 2.20 = 1.13g$$

This value, denoted by a small circle in Fig. 12.2, agrees with the line indicated for zero damping. Similarly, for the double-cantilever mode we get

$$\frac{0.17}{0.33} \times 1.22 = 0.63g$$

The point on the chart compatible with this numerical result is also indicated in Fig. 12.3. In principle, a similar procedure applies to other seismic loading characteristics for which damping is other than zero. For instance, for the same period and a damping coefficient n_d of 0.01, a lower spectral velocity must be assumed in the calculations, as shown in Fig. 12.4. In this manner, utilizing the El Centro velocity spectrum, the seismic characteristics for $n_d = 0.01$ and $n_d = 0.02$ have been obtained for Figs. 12.2 and 12.3.

8. The allowable g loading on the space frame under consideration may be determined on the basis of either strength or stability. In the actual case described, the maximum bending stress in a double-cantilever mode was used as a criterion for the design chart in Fig. 12.3. The dashed line corresponds to the allowable g loading on the frame.
9. Interpretation of the design charts may be conducted in the following way. Consider, for example, the critical double-cantilever mode shown in Fig. 12.3. The point of intersection between the dashed line and the line denoted by $n_d = 0.01$ yields the maximum allowable earthquake intensity of 5.25 on the Richter scale, assuming the El Centro type of spectral response and about 25 miles distance to the major fault line. Interpolating in Fig. 12.5, we get a corresponding maximum ground acceleration of about $0.075g$.

The procedure described above contains only the elementary concepts of stress analysis. Yet from a practical point of view, this approach will be found useful, because it reduces the nonsteady response of a structure to a statical problem which can then be handled with ease. Because of the assumptions involving linear elasticity, the results obtained should be sufficiently conservative for most practical needs.

When the seismic input is such that the engineering structure is plastically deformed, it is not particularly appropriate to determine a design stress, because it will always be at the yield level. In such a case the maximum displacement rather than the maximum stress should be taken as a criterion. The practice also indicates that the potential for producing a failure in a ductile structure is relatively low for the majority of seismic spectra available.

STRUCTURAL DAMPING

Even a small amount of damping can significantly decrease the dynamic response of a structure. For instance, as indicated in Fig. 12.4, the peak response for a damping

Table 12.1 Representative Damping Values for Various Types of Structures and Induced Stresses

	No More Than One-Half Yield Stress (%)	At or Just Below Yield (%)
Piping	1 to 2	2 to 3
Welded steel	2 to 3	5 to 7
Prestressed concrete	2 to 3	5 to 7[a]
Well-reinforced concrete	2 to 3[b]	7 to 10
Bolted or riveted steel	5 to 7	10 to 15
Bolted wood	5 to 7	10 to 15
Nailed wood	5 to 7	15 to 20

[a]Without complete loss of prestress.
[b]Only slight cracking.

factor $n_d = 0.01$ is roughly one-half that for zero damping. This attenuation exists in all structures and its magnitude varies with the type of the material and construction. It also depends on how close to yield are the induced stresses. Some of the damping values encountered in practice are given in Table 12.1 [275, 276].

SYMBOLS

C Seismic coefficient
f Frequency, Hz
g Acceleration of gravity, in./sec² (mm/sec²)
K_d Numerical factor for plastic response
k_s Spring constant, lb/in. (N/mm)
M Mass of moving body, lb-sec²/in. (N-sec²/mm)
Q Richter scale number
n_d Damping factor
Q_b Shear load at base, lb (N)
S_v Spectral velocity, in./sec (mm/sec)
T Natural period of vibration, sec
W Weight of structure, lb (N)
δ Displacement, in. (mm)
δ_{max} Maximum displacement, in. (mm)
ω Angular velocity, rad/sec
Ω Acceleration or deceleration, in./sec² (mm/sec²)

13
Engineering Aspects of Fatigue

BASIC DEFINITIONS

Various theories and design methods have been proposed in relation to the mechanism of fatigue failure [53–60]. With reference to engineering design, the relevant fatigue cycle notation and definitions are illustrated graphically in Figs. 13.1 and 13.2. It is of interest to note that when the relevant typical stress σ_{max} is plotted against the number of stress cycles, there is a fairly rapid change in the shape of the curve at about 10^6 to 10^7 cycles, beyond which a constant stress is approached. This value is defined as the *endurance limit*.

The majority of fatigue results are obtained for completely reversed stress in pure bending marked by the significant amount of scatter. For tension-compression or push-pull tests, an endurance limit of approximately 75% of that obtained in bending may be assumed for the correlation purposes. However, because of the significant scatter encountered in fatigue testing, statistical analysis is recommended for data reduction.

CUMULATIVE DAMAGE CRITERION

When several different stress amplitudes and periods of operation are involved throughout the lifetime of a particular component, cumulative damage concept can be used in the calculations [59]. Although this method should be employed

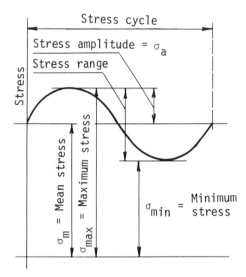

Fig. 13.1 Fatigue cycle notation

with caution (no allowance is made for other variables), a fairly satisfactory approximation can be obtained utilizing the following criterion:

$$\sum \frac{n}{N} = 1 \tag{13.1}$$

where n denotes the number of cycles at a particular working stress and N is the number of cycles to failure at the same stress level. The term n/N represents the cycle ratio. The value of N must be determined from a given fatigue diagram such as that shown in Fig. 13.2. To illustrate the method of approach, consider the following example.

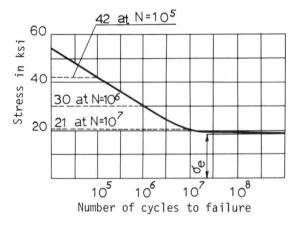

Fig. 13.2 Example of $\sigma - N$ fatigue diagram.

Engineering Aspects of Fatigue

Design Problem 13.1

Perform a safety check for a structure undergoing stress fatigue, assuming the following data

$\quad$ 80,000 cycles at $\quad \sigma_{\max} = 42,000$ psi
$\quad$ 50,000 cycles at $\quad \sigma_{\max} = 30,000$ psi
$\quad$ 100,000 cycles at $\quad \sigma_{\max} = 21,000$ psi

Refer to the fatigue diagram in Fig. 13.2.

Solution

The required values of N are obtained directly from the diagram in Fig. 13.2 for the assigned stress levels.

$\quad \sigma_{\max} = 42,000$ psi $\quad N = 100,000$
$\quad \sigma_{\max} = 30,000$ psi $\quad N = 1,000,000$
$\quad \sigma_{\max} = 21,000$ psi $\quad N = 10,000,000$

Hence, using the design formula for cumulative damage, Eq. (13.1), we get

$$\sum \frac{n}{N} = \frac{80,000}{100,000} + \frac{50,000}{1,000,000} + \frac{100,000}{10,000,000} = 0.8 + 0.05 + 0.01 = 0.86$$

Since the sum of the cycle ratios is less than unity, the structure is presumed safe. In practice, a limiting value smaller than unity is selected in order to be on the safe side. For instance, $\sum n/N < 0.8$ can be taken as a suitable design criterion. $\blacklozenge$

NEUBER EFFECT

It was pointed out previously that in ductile materials, stress concentration may be relieved by inelastic behavior. In fatigue, however, the full value of a theoretical elastic stress concentration K must be taken for design. Numerous fatigue experiments conducted with notched bars and sharp radii led to the establishment of the *Neuber effect* [14]. According to this finding, there is a small limiting value of the notch radius below which no additional stress increase in fatigue is expected. The ratio between the apparent increase and that predicted by the elastic theory has been termed the *notch sensitivity factor* [16, 26].

$$q = \frac{K_f - 1}{K - 1} \tag{13.2}$$

In Eq. (13.2), K_f denotes the fatigue stress concentration factor derived from tests. The particular structural member is considered to have no notch effect when $q = 0$. On the other hand, $q = 1$ defines the maximum theoretical notch sensitivity. Design tables are available [26] which show how the parameter q varies with the notch or hole radius. However, the theory of fatigue suggests that the actual size of hole or depth of notch may prove to be more influential than the geometrical parameters.

ELEMENTS OF THEORETICAL DESIGN

The design for fatigue is relatively straightforward when the experimental curve, such as that shown in Fig. 13.2, is available. However, when such data are not at hand, various theoretical estimates can still be made [58].

A conservative prediction of the maximum and minimum stresses for uniaxial loading follows from *Soderberg's law* [21]

$$\sigma_{max} = \sigma_e + \sigma_m \left(1 - \frac{\sigma_e}{\sigma_y}\right) \qquad (13.3)$$

$$\sigma_{min} = -\sigma_e + \sigma_m \left(1 + \frac{\sigma_e}{\sigma_y}\right) \qquad (13.4)$$

In these formulas, σ_e denotes the endurance limit, such as that shown in Fig. 13.2. The yield strength of the material is σ_y. The mean stress σ_m, together with σ_{max} and σ_{min} is defined in Fig. 13.1. When a structural component is subjected to a combined stress system, the endurance limit can be calculated on the basis of the distortion energy theory [21, 58] utilizing Eqs. (13.3) and (13.4) or similar models. Such problems are rather complicated and it is often necessary to run experiments in support of the theory. For example, interesting studies of this nature have been conducted on cylinders (with side holes) subjected to pulsating internal pressures [21].

EFFECT OF SURFACE FINISH

Although the applied stress is the most important single factor in fatigue performance, geometrical discontinuities, surface effects, and environmental variables should also be taken into account. As far as surface effects are concerned, stress raisers due to roughness, residual stresses, and nonuniformity of the material properties between the surface and the core must influence the fatigue life.

Under the normal conditions of mechanical finishing of the surface, highly polished specimens have the fatigue life higher by at least 10% compared with that of relatively rough machined samples [62]. The effect of finish on the endurance limit for shorter fatigue life appears to be less pronounced [63]. There is also sufficient evidence to indicate that surface sensitivity increases with the tensile strength [64]. Furthermore, the magnitude of the residual stresses in high-strength steels, produced by milling and grinding, can be very high. In one case of a poor practice, for instance, a residual tensile stress in the surface layer of a high-strength steel was found to be close to 200,000 psi [65]. Although this is rather unusual, values of 100,000 psi are not uncommon. Fortunately such stresses can often be mitigated by stress relieving, abrasive tumbling, or shot peening.

In a particular application of machine shafts where bending and torsion can combine to create high stress gradients, hardening of the surface by carburizing, flame hardening, or nitriding produces beneficial effects. In this manner the surface stronger than the core can be provided. If, in addition to this treatment, we can

Engineering Aspects of Fatigue

assure such a residual system on the surface that the mean stress in fatigue is reduced, longer component life can be expected. One of the rather simple ways to achieve such a stress system is to apply shot peening [66]. In this technique the tensile strength of the surface is increased by means of the compressive residual stresses developed in the skin of the metal.

The quality of a machined surface can be affected by a number of commercial processes and it can be defined in terms of the surface roughness measured in microinches. However, a definition of roughness alone does not convey the entire picture; the process of material removal is extremely complex and may result in detrimental surface layer alterations. This can lead to such problems as grinding burns, stress corrosion, cracks, distortion, and increased residual stresses. These features, in turn, influence the fatigue resistance and performance of hardware. Especially harmful can be abusive grinding. In such cases the decrease in endurance limit can be much more than 10%. Where high-temperature nickel-base and titanium alloys are involved the loss in fatigue resistance can become as high as 30%. This is especially noticeable in high-cycle fatigue.

In addition to the purely mechanical means of surface finishing electrical discharge machining (EDM) can seriously affect the fatigue strength, particularly in the case of highly stressed components. In the event of EDM roughing, the outer layer of metal can undergo microstructural changes, local overheating, and cracking, which may extend further, into the adjacent layers of the material. Other methods of chemical and electrochemical machining appear to have a relatively smaller effect on the endurance limit. The surface finish can, of course, be improved by mitigating heat-affected zones, cracks, and other detrimental layers in critically stressed components [61].

EFFECT OF CREEP

There is virtually no thermal effect on the endurance limit of a conventional structural material up to about 650°. The development of creep, even for carbon steels, is also insignificant at these moderate temperatures. However, where this effect becomes more pronounced at a higher temperature, a simple method for combining endurance limit and creep is shown in Fig. 13.3 [67]. The fatigue strength at a completely reversed cycle and the static creep strength are plotted on the two axes for a given design temperature, defining the two points of the straight-line approximation, which is considered to be relatively conservative. The area under the curve denotes safe region, whereas any combination of alternating and creep stresses falling above the line is not acceptable. It should be added that the experimental results fall closer to an elliptical curve, drawn between points A and B in Fig. 13.3.

EFFECT OF CORROSION

A mechanism that can lower the endurance limit is corrosion. The combination of fatigue and corrosion, called corrosion fatigue, is a serious problem because the corrosion products can act as a wedge, opening the crack. This results in about a

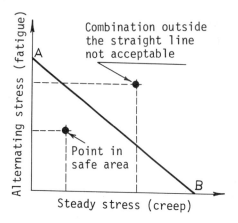

Fig. 13.3 Simplified method for combining creep and fatigue.

20% reduction of the endurance limit for a conventional carbon steel. In salt water and other more corrosive media, the reduction can be even more pronounced. To combat this effect, alloying elements, as well as protective coatings, are used. The rate of cycling can also lower the fatigue life, but it is seldom considered in normal design procedures because such data are poorly defined ahead of time.

EFFECT OF SIZE

A rather important factor influencing the fatigue life is illustrated in Fig. 13.4 [58]. The principle involved here is that a bar with a smaller diameter has better resistance to fatigue failure because it has a relatively lower volume of material affected by a high stress. This principle is acceptable in the design of members of various thicknesses if the endurance limit is known for, at least, some of the thicknesses. For instance, in the case of fatigue data for steel given in Fig. 13.4 the endurance limit can be plotted as a function of thickness, as shown in Fig. 13.5.

ENDURANCE LIMIT

Use of design equations (13.3) and (13.4) is predicated on a knowledge of the endurance limit σ_e after all the influencing factors, such as those indicated above, had been considered. Although this problem is rather complex, the first estimate, based on the available extensive test results [58], can be made during the preliminary materials selection. Some of the recommended ratios for the endurance limit to the ultimate strength of materials are compiled in Table 13.1. In this table, σ_u denotes the ultimate tensile strength of the material and the term σ_e is the endurance limit, defined in Fig. 13.2 for the particular example considered.

Engineering Aspects of Fatigue

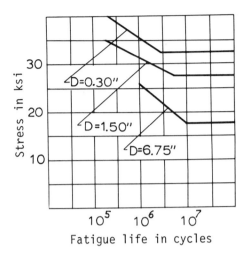

Fig. 13.4 Effect of bar diameter on fatigue life.

LOW-CYCLE FATIGUE

In most practical applications, the designer deals with a fatigue life involving millions of cycles. On the other hand, in an extreme case a simple tensile test corresponds to one-half cycle. It is customary to refer to "low-cycle fatigue" when the total number of cycles loading a given structure beyond the elastic limit is less than 10,000. An empirical relation predicting low-cycle fatigue strength is given below as Eq. (13.5) [21, 68]. The formula agrees with the experimental data on steel, copper, aluminum, nickel, stainless steel, and titanium.

$$\sigma_a = \sigma_e + \frac{CE}{2N^{1/2}} \tag{13.5}$$

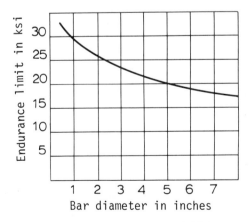

Fig. 13.5 Example of effect of thickness on fatigue.

Table 13.1 Recommended Fatigue Strength (σ_e/σ_u) Ratios for Preliminary Design

Cast aluminum, 220-T4	0.17
Cast aluminum, 108	0.52
Cast aluminum, F132,-T5	0.38
Cast aluminum, 360-T6	0.40
Wrought aluminum, 2014-T6	0.29
Wrought aluminum, 6061-T6	0.45
Beryllium copper, HT	0.21
Beryllium copper, H	0.34
Beryllium copper, A	0.47
Naval brass	0.35
Phosphor bronze	0.32
Gray cast iron (No. 40)	0.48
Malleable cast iron	0.56
Magnesium, AZ80A-T5	0.29
Titanium alloy, 5Al, 2.5Sn	0.60
Steel, A7-61T	0.50
Steel, A242-63T	0.50
Spring steel, SAE 1095	0.36
Steel, SAE 52100	0.44
Steel, SAE 4140	0.42
Steel, SAE 4340	0.43
Stainless steel, Type 301	0.30
Tool steel, H.11	0.43
Maraging steel, 18 Ni	0.45

Note: T, heat-treated; H, hard; HT, hardened; A, annealed.

where

$$C = \frac{1}{2} \ln \left(\frac{100 - \%\text{R.A.}}{100} \right) \qquad (13.6)$$

In Eq. (13.5) E is the modulus of elasticity of the material and N denotes the number of cycles to failure at the stress amplitude σ_a. The factor C depends on the well-known mechanical property termed the reduction of area (R.A.) in a conventional tensile test.

Design Problem 13.2

Calculate the mean and maximum fatigue stresses in uniaxial loading for a machine part made of steel with an endurance limit of 35,000 psi, expected to work for 20,000 cycles. Let the yield strength be 80,000 psi and the ultimate strength 100,000 psi with a reduction of area R.A. = 30%.

Solution

By the definition, from Fig. 13.1

$$\sigma_{max} = \sigma_m + \sigma_a \qquad (13.7a)$$

From Eq. (13.3)

$$\sigma_m + \sigma_a = \sigma_e + \sigma_m\left(1 - \frac{\sigma_e}{\sigma_y}\right) \qquad (13.7b)$$

or

$$\sigma_m = \sigma_y\left(1 - \frac{\sigma_a}{\sigma_e}\right) \qquad (13.7c)$$

From Eq. (1.36)

$$C = \frac{1}{2}\ln\left(\frac{100-30}{100}\right) = -0.1783$$

From Eq. (13.5)

$$\sigma_a = 35,000 - \frac{0.1783 \times 30 \times 10^6}{2 \times (20,000)^{1/2}} = 16,000 \text{ psi} \quad (110.3 \text{ N/mm}^2)$$

Hence, using formula (13.7c), the mean stress is found to be

$$\sigma_m = 80,000\left(1 - \frac{16,000}{35,000}\right) = 43,400 \text{ psi} \quad (299.2 \text{ N/mm}^2)$$

The corresponding maximum stress follows from Eq. (13.7a):

$$\sigma_{max} = 43,400 + 16,000 = 59,400 \text{ psi} \quad (409.6 \text{ N/mm}^2) \quad \blacklozenge$$

The stress calculated in Design Problem 13.2 is often referred to as the *pseudoelastic limit*, because low-cycle fatigue is expected to extend beyond the elastic limit. When $\sigma_a = 0$, $\sigma_m = \sigma_y$, which gives the following condition for the number of cycles to failure:

$$N = \frac{C^2 E^2}{4\sigma_y^2} \qquad (13.8)$$

For the numerical data given in Design Problem 13.2, the lowest number of cycles to failure at a mean stress equal to 80,000 psi would be 1120. Hence, the lowest number of cycles to failure at yield depends only on the strain at yield and the reduction of area based on a standard tensile test. Equations (13.6) and (13.8) also indicate that, as the brittleness of the material increases, the number of cycles to failure falls off rather rapidly. The ideally brittle material, based on Eq. (13.6), is that at which C tends to zero.

SYMBOLS

C	Area reduction factor
D	Bar diameter, in. (mm)
E	Modulus of elasticity, psi (N/mm^2)
K	Stress concentration factor in static analysis
K_f	Stress concentration factor in fatigue
N	Number of cycles to failure
n	Number of working cycles
q	Notch sensitivity factor
$R.A.$	Reduction in area, %
σ_a	Stress amplitude, psi (N/mm^2)
σ_e	Endurance limit, psi (N/mm^2)
σ_m	Mean stress, psi (N/mm^2)
σ_{min}	Minimum working stress, psi (N/mm^2)
σ_{max}	Maximum working stress, psi (N/mm^2)
σ_y	Yield strength, psi (N/mm^2)

14
Design Aspects of Fracture Mechanics

CHARACTERIZATION OF MATERIALS

Since the primary mission of fracture mechanics can be defined as the stress analysis of components sensitive to crack propagation and brittle failure, it may be instructive to reflect briefly upon the ductile and brittle behavior. The design engineer is always faced with the problem of recognizing the most likely mode of material's response under stress.

The distinction between the ductile and brittle behavior of engineering materials has never been too clear. At normal working temperatures mild steel has always been considered to be a ductile material, whereas cast iron, for instance, has often served as an example of a typical brittle material. The dividing boundary between the two types of material could be designated by the amount of elongation equal to about 5%. However, this limit is probably on the low side, and it may be open to challenge in terms of the existing strength theories. It has been generally agreed that brittle failure could be related to the maximum principal stresses, whereas the failure phenomenon in a ductile material is best interpreted with the aid of the von Mises–Hencky criterion [5]. These rules cannot be applied rigidly, however, since the basic problem of the mechanical strength of a component is generally rather complex, involving various geometrical and material's parameters. Hence, the designer is often constrained to make the final decision even when the information available is somewhat incomplete. Under these circumstances the necessary margin between failure and success can be assured only by a sufficiently high factor of safety.

The complexity of the design process has recently been increased due to the requirements of fracture control. It is now generally recognized that the metallurgical phenomenon of a fracture toughness transition with temperature is exhibited by a number of low- and medium-yield-strength steels. This transition results from the interactions among temperature, strain rate, microstructure, and the state of stress. One of the more perplexing aspects of this behavior is that the customary elongation property of the material appears to have virtually no relation to the degree of fracture toughness available. For example, a well-known mild steel such as the ASTM Grade A36, having an elongation greater than 20%, exhibits brittle behavior not only at lower, but also at room temperatures.

It appears, therefore, that it may be advisable to characterize the materials with respect to their brittle tendencies before selecting the method of stress analysis. Traditional mechanical properties—in the form of yield point, ultimate strength, elongation, and elastic constants—must be supplemented with the thermomechanical data. The response of a stressed component, particularly at lower working temperatures, may be impossible to predict without a knowledge of fracture mechanics and the material's toughness.

The application aspects of fracture mechanics given in this chapter are treated in a most elementary fashion. The aim of the presentation is simply to alert the design audience to some potential problem areas and to indicate the nature of modern trends in stress analysis and fracture control. It points to the necessity of characterizing the material's behavior under stress in terms of new parameters.

PRACTICAL ASPECTS OF FRACTURE MECHANICS

The classical theory of fracture mechanics, which goes beyond the traditional concepts of stress analysis, has been further developed during the past 30 years for glasslike materials and higher-strength metals [69–83], although the critical phenomena of fracture in solids were studied as far back as 1920 [73]. It deals with brittle behavior, which can be triggered off by even a low nominal stress and virtually no visible signs of deformation. Once initiated, such a brittle process can propagate at a high velocity to the point of complete failure.

As noted above, classical fracture mechanics is applicable to high-strength metals. As a general guide, steels with the yield strength, say, above 180 ksi, titanium alloys above 120 ksi, and aluminum alloys above 60 ksi fall in this category and should be evaluated on the basis of fracture toughness rather than pure yield strength and elongation. It should again be emphasized, however, that brittle failure can also occur in lower-strength materials. This important thermomechanical aspect is discussed in greater detail in Chap. 15.

Extensive experimentation and some recent developments in continuum mechanics have been aimed at defining a quantitative relationship between stress, the size of a crack, and the mechanical properties. It is well to point out that the advent of fracture mechanics does not negate the traditional concepts of stress analysis, which allow us to design for stresses exceeding the yield strength in the vicinity of such structural discontinuities as holes, threads, or bosses, provided that the material deforms plastically and redistributes the stresses. This concept is still valid

Design Aspects of Fracture Mechanics

unless the material contains critical flaws that produce unstable crack propagation below the design value of the yield strength. The difficulty with introducing the correction for flaws in design is that in many cases flaws cannot be easily detected. It becomes necessary, therefore, to develop a procedure that would define the maximum crack length permissible at a particular level of stress. According to the theory of fracture mechanics, this stress level is inversely proportional to the flaw size. The stress at which crack propagation is expected to occur is given by the relation

$$\sigma = \frac{K_{1C}}{(\pi a)^{1/2}} \tag{14.1}$$

Fracture toughness K_{1C}, which is a mechanical property, approaches a limiting minimum value as the specimen thickness increases. Its magnitude in Eq. (14.1) constitutes an absolute minimum corresponding to a plane strain condition at which the fractured surface has a brittle appearance. This type of failure is associated with the very limited plastic deformation and is typical of fractures in heavy sections. The dimension of K_{1C} is psi(in.)$^{1/2}$, and a denotes the half-length of the crack in the opening mode. Equation (14.1) can be utilized, for instance, in calculations involving through-cracks in relatively large containers and pressure vessels. If the calculated stress of fracture is found to be higher than the operating stress based on a minimum acceptable factor of safety, the component should be satisfactory for service.

It should be emphasized that the nominal stress featured by Eq. (14.1) must be of a tensile nature. This stipulation is of special importance in applications of fracture mechanics principles to design. The key formula, Eq. (14.1), may be used under static or dynamic loading conditions. For conventional structural steels having a static yield strength of up to 100,000 psi, the corresponding dynamic yield can be estimated as follows:

$$\sigma_{yd} = \sigma_y + 30{,}000 \tag{14.1a}$$

According to Lange [256], the upper limit for this prediction can be as high as 140,000 psi while at the same time the numerical term in Eq. (14.1a) is gradually reduced from 30,000 to zero.

The concept of K_{1C} refers to brittle fracture and plane strain conditions. In mathematical terms, *plane strain* is defined as the state of zero plastic flow parallel to a crack front. Thick materials normally develop plane strain fracture characteristics and the broken surface is essentially flat. As the material's thickness decreases, the degree of constraint decreases, creating a *plane stress* condition and a maximum amount of plastic flow associated with the fracture.

The effect of temperature on conventional strength and crack resistance is illustrated in Figs. 14.1 and 14.2 for two high-strength steels. These examples indicate poor resistance to brittle fracture within the specific temperature ranges associated with either a tempering process or testing. It is seen that the conventional tensile property gives essentially no clue as to the brittle behavior of the materials discussed.

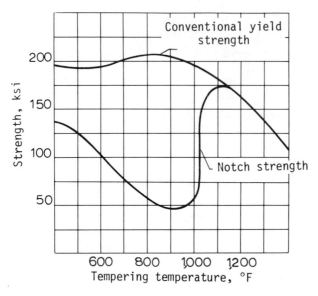

Fig. 14.1 Effect of tempering temperature on high-strength steel (12Mo-V stainless).

APPLICATIONS OF CRACK SIZE PARAMETER TO DESIGN

In considering the application of fracture mechanics principles to typical production defects, involving surface cracks of depth b and half-length a, the following expression can be used to calculate the maximum gross stress σ associated with the

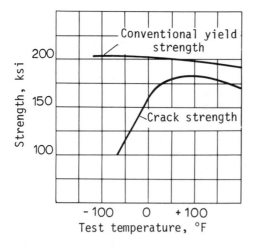

Fig. 14.2 Effect of test temperature on high-strength steel.

Design Aspects of Fracture Mechanics

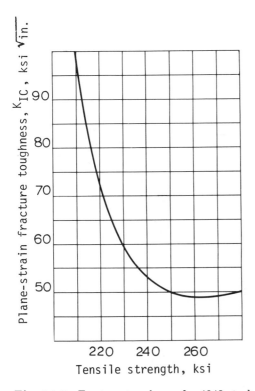

Fig. 14.3 Fracture toughness for 4340 steel.

critical area of a component:

$$\sigma = \frac{K_{1C}(\psi)^{1/2}}{\left(3.77b + 0.21\frac{K_{1C}^2}{\sigma_y^2}\right)^{1/2}} \quad (14.2)$$

The plane strain fracture toughness number K_{1C} and the parameter ψ must be known before the required estimate can be made. An example of typical fracture toughness data is shown in Fig. 14.3 for a steel sheet and plate fabricated according to AISI 4340 material specification [83]. The crack shape parameter ψ is known to depend on the b/a ratio [84]. For convenience of design calculations, this parameter has been plotted in Fig. 14.4.

Design Problem 14.1

Calculate the maximum stress at failure for a large structural component made of 4340 steel plate having an ultimate tensile strength of 220,000 psi. The surface flaw discovered is 0.2 in. deep and 2 in. long.

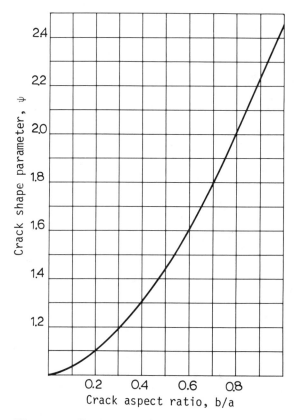

Fig. 14.4 Crack shape parameter chart.

Solution

For the specified tensile strength of 220,000 psi, the fracture toughness parameter K_{1C} follows from Fig. 14.3

$$K_{1C} = 73,000 \text{ psi(in.)}^{1/2}$$

Since $a = 1$ in.

$$\frac{b}{a} = 0.2$$

Hence, Fig. 14.4 gives

$$\psi = 1.09$$

Assuming the critical fracture stress to be equal to the yield strength, we have $\sigma/\sigma_y = 1$. Therefore, using Eq. (14.2), the required stress at failure becomes

$$\sigma = \frac{K_{1C}(\psi - 0.21)^{1/2}}{(3.77b)^{1/2}} = \frac{73,000(1.09 - 0.21)^{1/2}}{(3.77 \times 0.2)^{1/2}}$$
$$= 78,860 \text{ psi} \quad (544 \text{ N/mm}^2) \quad \blacklozenge$$

Design Aspects of Fracture Mechanics

Design Problem 14.2

Select structural material to tolerate the following maximum flaw dimensions, given as 0.1 in. in depth with a total crack length 2a equal to 1.0 in. The maximum applied stress is expected to be equal to the yield strength of the material, approaching 220,000 psi for the 4340 grade of steel. Assume the ratio of ultimate to yield strength to be 1.13.

Solution

From Eq. (14.2),

$$K_{1C} = 1.92 \sigma_y \left(\frac{b}{\psi - 0.21} \right)^{1/2}$$

Utilizing the specified crack dimensions, we get

$$\frac{b}{a} = 0.2$$

From Fig. 14.4, $\psi = 1.09$. Hence, the required fracture toughness is

$$K_{1C} = 1.92 \times 220,000 \left(\frac{0.1}{1.09 - 0.21} \right)^{1/2} = 142,390 \text{ psi(in.)}^{1/2} \quad [4948 \text{ N(mm)}^{-3/2}]$$

From Fig. 14.3, the plane strain fracture toughness is only 50,000 psi $(\text{in.})^{1/2}$. The conclusion, therefore, is that we have to either select material with a higher fracture toughness or decrease the corresponding working stress. ◆

Design Problem 14.3

Estimate the maximum allowable radial interference Δ in a press-fit assembly consisting of a solid shaft fitted into a cylinder of inner radius $R_i = 0.8$ in. and outer radius $R_o = 2$ in. on the premise that the parts are made of high-strength stainless steel with a minimum yield of 210 ksi and the K_{1C} value of 70 ksi$(\text{in.})^{1/2}$. Assume the initial flaw depth of 0.04 in. to extend in radial direction from the inner surface of the cylinder, and take $E = 30,000$ ksi. Compare this design with the results based on the conventional hoop stress criterion when the initial flaw does not exist.

Solution

When the cylinder is subjected to internal pressure only Eq. (35.27) gives

$$S_t = P_i \frac{(R_o^2 + R_i^2)}{(R_o^2 - R_i^2)}$$

The conventional formula for a shrink-fit assembly* using the same materials for outer and inner parts [21] is

$$\Delta = \frac{2P_i R_o^2 R_i}{(R_o^2 - R_i^2)E}$$

Eliminating the contact pressure term P_i between the foregoing equations yields

$$S_t = \frac{E\Delta}{2R_i}\left[1 + \left(\frac{R_i}{R_o}\right)^2\right]$$

For the case of a notch of depth a the classical expression of fracture mechanics given by Eq. (14.1) becomes [255]

$$K_{1C} = 1.12\sigma(\pi a)^{1/2}$$

from which

$$\sigma = \frac{K_{1C}}{2(a)^{1/2}}$$

Making $\sigma = S_t$, and solving for Δ results in a design equation representing the fracture mechanics criterion for our case

$$\Delta_c = \frac{K_{1C} R_i}{E(a)^{1/2}[1 + (R_i/R_o)^2]}$$

Substituting the appropriate numerical values gives

$$\Delta_c = \frac{70 \times 0.8}{30{,}000(0.04)^{1/2}[1 + (0.8/2)^2]} = 0.008 \text{ in.} \quad (0.2 \text{ mm})$$

From the hoop stress criterion

$$\Delta = \frac{2R_i S_t}{E[1 + (R_i/R_o)^2]}$$

which on substitution yields

$$\Delta = \frac{2 \times 0.8 \times 210}{30{,}000[1 + (0.8/2)^2]} = 0.0097 \text{ in.} \quad (0.25 \text{ mm})$$

In this instance, then, the fracture mechanics prediction is conservative by about 21%. It should be pointed out, however, that the degree of conservatism may depend on the relative magnitudes of the three major parameters for this shrink-fit assembly as follows.

$$\frac{\Delta_c}{\Delta} = \frac{K_{1C}}{2(a)^{1/2} S_t}$$

It may be of interest to note that the shrinkage allowance Δ can also be approximated using Design Problem 36.3 together with Figs. 35.8 and 36.12. ♦

---

*For additional formulas applicable to shrink-fit design, see Chapter 35.

Design Aspects of Fracture Mechanics

Design Problem 14.4

Provide a quick estimate of the leak-before-break internal pressure for a high-strength alloy steel cylinder of mean radius $R = 36$ in. and wall thickness $t = 2.5$ in. on the premise that the minimum plane strain fracture toughness is 93 ksi $(\text{in.})^{1/2}$. Neglect end effects and assume that the conventional membrane theory of stress analysis applies.

Solution

For a through-thickness crack Eq. (14.1) becomes

$$\sigma = \frac{K_{1C}}{(\pi t)^{1/2}}$$

The conventional hoop stress is S_t

$$S_t = \frac{PR}{t}$$

Hence making $\sigma = S_t$ and solving for pressure yields

$$P = \frac{tK_{1C}}{R(\pi t)^{1/2}}$$

Finally, substituting the numerical values, the required leak-before-break pressure becomes

$$P = \frac{2.5 \times 93}{36(\pi \times 2.5)^{1/2}} = 2300 \text{ psi} \quad (15.9 \text{ N/mm}^2) \quad \blacklozenge$$

The foregoing examples illustrate the quantitative approach to the prediction of stresses for a given maximum size of flaw. Since, however, the fracture toughness parameter K_{1C} is a compound quantity involving stress and the crack size, it becomes necessary to rely on empirically obtained curves of K_{1C} before calculating the strength of a component for a given size of defect.

When a structural element of brittle nature is subjected to a combined loading and lack of symmetry it is difficult to rely on the size and orientation of the flaw. For example, in the case of glass it may be necessary to describe the fracture mechanics property in terms of the strain energy release rate. This can be defined as the quantity of energy released per unit area of crack surface as the crack extends [257]. The relevant parameter in this case is

$$G_c = \frac{\eta \sigma^2 a}{E} \tag{14.2a}$$

On the premise that a denotes one-half crack length for a through-thickness flaw in a semi-infinite plate in tension and σ is the nominal prefracture stress, Eq. (14.1) gives

$$K_{1C} = \sigma(\pi a)^{1/2} \tag{14.2b}$$

Eliminating the (πa) term between Eqs. (14.2a) and (14.2b) gives

$$K_{1C}^2 = EG_c \tag{14.2c}$$

The parameter G_c may be defined as a measure of the force driving the crack. Once this quantity is established it should be possible to obtain a preliminary estimate of the critical crack size.

Design Problem 14.5

Calculate the approximate critical length of the crack in a glass panel subjected to tension on the premise that the modulus of elasticity is $E = 10 \times 10^6$ psi, ultimate strength is 40,000 psi, and the strain energy release rate amounts to $G_c = 0.08$ (in.-lb/in.^{2}).

Solution

The equivalent fracture toughness from Eq. (14.2c) is

$$K_{1C} = (10 \times 10^6 \times 0.08)^{1/2} = 894 \text{ psi(in.)}^{1/2}$$

Next, using Eq. (14.2b) yields

$$a = \frac{(K_{1C})^2}{\pi (\sigma)^2} = \frac{894^2}{\pi \times 40,000^2} = 0.00016 \text{ in.} \quad (0.0041 \text{ mm})$$

This calculation shows that the critical crack dimension in glass can be extremely small. The result is consistent with the expected behavior of a highly brittle material where even best nondestructive techniques cannot pinpoint a minute discontinuity. ◆

IMPLICATIONS OF FRACTURE TOUGHNESS

The plane strain fracture toughness parameter (K_{1C}) has received rather wide attention because of its role in fracture mechanics and fracture-safe design. As a distinct mechanical property, K_{1C} depends on the temperature and the rate of strain. Experience also shows that in the case of higher-strength materials, which tend to be relatively brittle, the determination of K_{1C} does not present any special problems. However, when using a medium or a lower material strength combined with high toughness, the testing procedure becomes more complex and requires a higher loading capacity for the test equipment. In some situations, such as that involving the Navy-quality material known as HY80, the required high specimen thickness for the test becomes almost impractical. The reason for this is that to assure a valid K_{1C} number in a tough material, the test specimen must satisfy the plane strain conditions as noted previously in the section dealing with practical aspects of mechanics.

For a long time now the status of characterization of fracture properties, particularly in relation to intermediate- and lower-strength steels, has been in a relatively poor condition. This is not surprising, however, when one considers the short development time of fracture mechanics criteria and the complexity of the problem. The basic difficulty is that we are always inclined to look for a sweeping generalization using a limited number of standard tests. A good example of this is the history

Design Aspects of Fracture Mechanics

of the popular and relatively simple Charpy V-notch test. It may be recalled that this test is conducted on a specimen containing a centrally placed sharp notch. The specimen is broken by impact in a beamlike configuration with a three-point loading. The amount of impact energy at the temperature of interest, sometimes denoted by CVN, is usually expressed in ft-lb. Since a K_{1C} value is involved in the fracture mechanics concept described by Eq. (14.1), numerous attempts have been made to develop simple correlations between CVN and K_{1C}. These correlations have generally been of the type

$$K_{1C} = A(\text{CVN})^n \tag{14.3}$$

where A and n denote the specific numerical constants for a selected material and design conditions. Although this correlation may be suitable for design analysis in the elastic range, a continuous debate of the merits and limitations of CVN created an impasse in the various committees that deal with specifications and test standards. In the meantime, other test methods have evolved with the specific charter of plane strain fracture characterization, resulting in a standard document known as ASTM-E399 [85]. Whatever means the designer plans to adopt for the decision regarding the value of K_{1C}, he or she would be well advised to remember that no sweeping generalizations are available within the bounds of the current state of the art in this field and that each design case should be treated individually.

Several typical minimum values of K_{1C} shown in Table 14.1 have been selected from a recent source [254]. The average ratios of the maximum to minimum K_{1C} numbers for the materials quoted in the table were generally on the order of 1.1 to 1.3. Hence the use of these minimum parameters should lead to conservative predictions. The information given in Table 14.1 is intended only as an illustration of the order of magnitude of K_{1C} values, because of the unavoidable variations in chemical composition, heat treatment, mechanical working of structural materials, as well as test temperature and methods of K_{1C} determination. Nevertheless, this fracture mechanics parameter is sufficiently versatile because it is a material's constant which, when properly measured and tailored to the specific design conditions, can represent a conservative criterion. Knowledge of K_{1C} data should help in material selection and in judging the performance of a component when the extent of a defect can be defined. Minute cracks are inherently present in any structural component, and their actual size is often below the limit of the sensitivity of nondestructive test equipment. Linear elastic fracture mechanics recognizes this fact of life and provides the tools for making a conservative prediction. One basic lesson should be learned from the concept of K_{1C} and Eq. (14.1): the higher the value of K_{1C}, the greater the resistance of the material to brittle failure and the greater the stress required to produce such a failure.

PLANE STRESS PARAMETER

The concept of plane strain fracture toughness, denoted by K_{1C} and briefly outlined in this chapter, can be looked upon as nonarrestable instability of the material in the presence of a crack. Under these conditions the brittle fracture can propagate

Table 14.1 Yield and Toughness of Engineering Materials

Material	Condition	Form	Min. Yield (ksi)	Min. K_{1C} [ksi(in.)$^{1/2}$]
Alloy steel				
18 Ni maraging (200)	Aged 900°F, 6 hr.	Plate	210	100
18 Ni maraging (250)	Aged 900°F, 6 hr.	Plate	259	78
18 Ni maraging (300)	Aged 900°F	Plate	276	44
4330 V	Tempered at 525°F	Forging	203	77
4330 V	Tempered at 800°F	Forging	191	93
4340	Tempered at 400°F	Forging	229	40
4340	Tempered at 800°F	Forging	197	71
Stainless steel				
PH13-8 Mo	H1000	Plate	210	78
PH13-8 Mo	H950	Forging	210	70
Titanium alloys				
Ti-6A1-4V	Annealed	—	120	81
Ti-6A1-6V-2Sn	Annealed	—	144	45
Ti-6A1-6V-2Sn	Solution treated and aged	—	179	29
Aluminum alloys				
2014	T651	Plate	57	22
2021	T81	Plate	61	26
2024	T851	Plate	59	19
2124	T851	Plate	64	22
7049	T73	Forging	61	29
7049	T73	Extrusion	73	28
7075	T651	Plate	70	25
7075	T7351	Plate	53	31

at velocities on the order of several thousand feet per second. The amount of plastic energy needed to propagate the crack is assumed to be rather small, so that the process of crack extension would largely be governed by the release of the elastic strain energy.

As the material's thickness decreases, lateral constraint relaxes and the size of the plastic zone around the crack tip experiences a sudden growth. In the science of fracture mechanics we describe this phenomenon as *crack-tip blunting*, indicating that a relatively large volume of metal has deformed. In a sense this process can be compared to the behavior of the neck region in a tensile test specimen. The velocities of crack propagation in this instance drop off very drastically and the crack becomes arrested, until we force further crack extension by increasing the stresses beyond the yield point of the material. We therefore approach the state of arrestable instability, at which the formerly defined fracture toughness parameter K_{1C} no longer applies. Arrestable instability can be described by the plane stress parameter denoted by K_C. This stress intensity factor can be correlated with the corresponding value of K_{1C}, the yield strength of the material σ_y, and the nominal material thickness B using the following expression:

$$K_C = K_{1C}\left[1 + \frac{1.4}{B^2}\left(\frac{K_{1C}}{\sigma_y}\right)^4\right]^{1/2} \tag{14.4}$$

Since the energy required to propagate the fracture under plane stress conditions must be high because of the arrestable crack characteristics, it follows that the critical flaw size under plane stress should be higher than that under plane strain conditions. According to the science of fracture mechanics, the relation between the two crack lengths may be stated as follows:

$$a_C = a_{C1}\left[1 + \frac{1.4}{B^2}\left(\frac{K_{1C}}{\sigma_y}\right)^4\right] \tag{14.5}$$

This expression can be obtained from Eq. (14.4) by making the crack ratio directly proportional to the square of the ratio of the respective stress intensity factors K_C and K_{1C}. Note that a_C and a_{C1} can represent the total as well as the half-length of the cracks.

Although the concept of K_C is relatively easy to adapt to the analysis of thin-walled components, the task of experimental determination of the K_C values is not without some serious limitations. The fracture corresponding to the K_C parameter is of a mixed-mode type involving large amounts of crack-tip plastic flow. Also, the critical length of the crack is difficult to establish because of the limitations of the instrumentation. Nevertheless, approximate values of K_C can be derived experimentally in such cases as, for example, wide panels and sheets. Such a process may be based on the initial crack lengths and the stresses to failure. In the case of heavier sections in low-strength materials, however, more complex correlation techniques are needed before the results can be considered as applicable to design.

PLANE STRESS CRITERION FOR PRESSURE VESSEL DESIGN

The concept of K_C lends itself to the application studies involving typical cylindrical vessels intended for internal pressure. In particular, we may wish to predict the burst pressure of the vessel for a given surface defect, such as a part-through, longitudinal flaw. The choice of the longitudinal orientation of the flaw relates well to the nature of loading in a pressurized cylinder where the critical membrane stress is likely to be in the hoop direction. The task, then, is to estimate the internal pressure to failure when the dimensions of the flaw and a specific value of the plane stress parameter K_C are known. The basic question in this type of analysis is concerned with the essential design criterion of leak-before-break. This situation has, over the years, provoked a number of scientific investigations and it has always had an important issue of industrial safety attached to it. Ideally, given a through-the-wall crack, we would like to have an assurance that the vessel would leak rather than fail suddenly due to unstable crack propagation through the plate proper or in the weld region of the vessel. The designer should also know what kind of a specific relation exists between the critical crack length and the stress for a given material characterization and the material's thickness intended in pressure vessel applications.

Assuming that plane stress conditions exist in the wall of the vessel, the relevant fracture toughness parameter for a longitudinal through-the-wall crack in a cylindrical geometry can be expressed by the equation

$$K_C^2 = \frac{\pi(1+5\nu)a\sigma_m^2}{2(1+\nu)\cos(\pi\sigma_m/2\sigma_u)}\left[1+\frac{1.7a^2}{Rt}(1-\nu^2)^{1/2}\right] \quad (14.6)$$

This formula is based on the classical flat plate theory and it involves a number of corrections to conform to vessel geometry and the nature of critical stress [191]. The corrections include the effect of curvature in going from a flat plate to a cylinder, the influence of the plastic zone at the tip of the defect, and the allowance for the effect of the biaxial state of stress. The design parameter σ_m denotes the nominal circumferential stress in the pressure vessel defined by PR/t, where P is the internal pressure in psi. In the customary notation R stands for the mean radius of the vessel and t is the wall thickness, both dimensions being expressed in inches. The half length of the longitudinal crack is denoted here by a and the ultimate strength of the material is σ_u. Poisson's ratio ν does not vary significantly for different metallic materials, particularly in the case of the various steel and aluminum alloys used in the pressure vessel field. The plane stress parameter K_C is given in psi(in.)$^{1/2}$ when the circumferential stress σ_m is expressed in psi.

Suppose now that we are dealing with a cylindrical vessel of thickness t which contains a part-through longitudinal flaw having a maximum depth d. The flaw is assumed to be here symmetrical and in the form of a hacksaw slot. The presence of this flaw reduces the wall thickness locally and without full penetration. Hence, by this definition t/d must be greater than unity. If we shall denote the actual area of the part-through flaw by A_f, its corresponding equivalent length can be defined as A_f/d. The original half-length of the crack entering Eq. (14.6) under conditions of full penetration is, of course, a. Hence, the approximate nominal stress σ_m for the part-through flaw can be estimated from Eq. (14.6), provided that we introduce $a = A_f/2d$. On substituting this equivalent quantity into Eq. (14.6), we obtain

$$K_C^2 = \frac{\pi(1+5\nu)A_f\sigma_m^2}{4d(1+\nu)\cos(\pi\sigma_m/2\sigma_u)}\left[1+\frac{0.425A_f^2}{Rtd^2}(1-\nu^2)^{1/2}\right] \quad (14.7)$$

The membrane stress to failure in a cylindrical vessel containing a part-through flaw can now be expressed on the basis of experimental evidence [191] in the following way:

$$\sigma_f = \frac{(t-d)\sigma_u^2}{t\sigma_u - d\sigma_m} \quad (14.8)$$

The method of predicting the failure stress for a thin vessel with a part-through, longitudinal flaw can be established with the aid of Eqs. (14.7) and (14.8). For instance, knowing the actual area of a part-through flaw A_f and the maximum

Design Aspects of Fracture Mechanics

depth d, we can calculate the parameter σ_m from Eq. (14.7). For this calculation, involving trial solutions, we have to know K_C, ν, σ_u, R, and t. If required, the plane stress parameter K_C may be approximated from Eq. (14.4), provided that the relevant K_{1C} value, material thickness B, and yield strength of the material σ_y are known.

In summary, then, the leak-before-break criterion can be verified analytically by comparing the two critical stress levels, σ_f and σ_m. If the stress to failure σ_f, obtained from Eq. (14.8) and intended for a given surface defect, exceeds σ_m calculated from Eq. (14.7) for an equivalent length of through-the-wall crack, we can expect the flaw to propagate, leading to a structural failure of the vessel. However, if the failure stress σ_f, calculated for the surface defect such as a part-through flaw, proves to be lower than σ_m, the corresponding vessel may be expected to leak before a catastrophic break.

It should be emphasized that although quite useful, the foregoing method of fracture mechanics does not provide complete answers for all design situations. For instance, the effect of non-applied loads, such as residual stresses induced by welding, is very difficult to interpret. The design parameters, such as K_C or K_{1C}, have to be determined with reference to the direction of the working stresses during the manufacture as well as in service. Other special considerations, such as neutron radiation, may also enter the picture. Fracture toughness of ferritic steels is known to be reduced by such a process. Furthermore, the design criteria may be based on the choice of the flaw size, its geometry, and its orientation with respect to a working stress field different from that required for a particular design case. Highly stressed regions such as nozzle junctions and similar transitions pose separate problems of interpretations of test results, inspection techniques, stress analysis methods, and fracture mechanics criteria which are certainly beyond the scope of this introductory treatment of fracture analysis. Last, but not least, the design factors of safety will be affected by all the technical issues noted above, together with considerations of production economics and the potential consequences of failure.

SYMBOLS

a	Half-length of crack, in. (mm)
a_C	Critical length of crack in plane stress, in. (mm)
a_{C1}	Critical length of crack in plane strain, in. (mm)
A	Numerical constant
A_f	Actual area of part-through flaw, in.^{2} (mm^2)
B	Material thickness, in. (mm)
b	Depth of crack, in. (mm)
CVN	Charpy V-notch energy, ft-lb (N-mm)
d	Maximum depth of part-through crack, in. (mm)
E	Modulus of elasticity, psi (N/mm^2)
G_c	Strain energy release rate, (in.-lb)/in^2 (N-mm/mm^2)
K_C	Plane stress fracture toughness, psi(in.)$^{1/2}$ [N(mm)$^{-3/2}$]
K_{1C}	Plane strain fracture toughness, psi(in.)$^{1/2}$ [N(mm)$^{-3/2}$]
n	Numerical constant
P	Internal pressure, psi (N/mm^2)

P_i	Internal shrink-fit pressure, psi (N/mm^2)
R	Mean radius of vessel, in. (mm)
R_o	Outer radius, in. (mm)
R_i	Inner radius, in. (mm)
S_t	Hoop stress, psi (N/mm^2)
t	Wall thickness, in. (mm)
Δ	Radial interference, in. (mm)
Δ_c	Critical radial interference, in. (mm)
ν	Poisson's ratio
σ	General symbol for stress, psi (N/mm^2)
σ_f	Failure stress for part-through crack, psi (N/mm^2)
σ_y	Yield strength, psi (N/mm^2)
σ_m	Circumferential stress, psi (N/mm^2)
σ_u	Ultimate strength, psi (N/mm^2)
σ_{yd}	Dynamic yield, psi (N/mm^2)
ψ	Crack shape parameter

15
Fundamentals of Fracture Control

HISTORICAL NOTE

The recognition of the basic problem of "ductile-to-brittle transition" in metallic materials dates back more than 40 years to the time when the welded fabrication of World War II ships was plagued by catastrophic failures. The incidents were characterized by almost instantaneous fractures of the entire ships. Although Navy records indicate that World War II ship steel exhibited some 40% elongation, there was obviously no beneficial effect of this property on the structural integrity of the ship plate in the particular environment. This experience has not been limited to the Navy materials and structures. At times, other costly failures were observed, such as a sudden burst of a multimillion-gallon storage tank or unexpected break of a main aircraft spar in flight. The problems were very serious, of course, and, over the years, many large-scale investigations have been sponsored by the U.S. government and private industry to develop remedial measures. In particular, the Naval Research Laboratory has been very active in the studies of fracture phenomena. This work has provided an excellent theoretical and experimental background for fracture-safe design [86–89], with special regard to low- and medium-strength materials. Normally, this implies "fracture control" in design utilizing steels with the yield strength lower than about 120,000 psi. In this category we encounter the majority of quenched and tempered steels used in modern applications involving rolling stock, shipping, bridges, lifting gear, storage tanks, and automotive components, to mention a few.

High-alloy steels having high resistance to fracture at room temperature include HY-80, HY-100, HY-130, and A-543 ASTM grades, and may be acceptable up to the thicknesses of about 3.0 in. In the low-alloy class, A-514 and A-517 ASTM grades serve as examples. For thicker sections, in the range of 6 to 12 in., the alloying elements have to be increased. However, experience with thick pressure vessels indicates that even good-quality quenched and tempered steels may become highly brittle. One of the best fracture-tough materials has been HY-80 steel, specified by the Navy, which even in welded regions performed successfully in hull structures of both surface ships and submarines since the mid-1950s.

Current manufacturing specifications are designed to give full assurance that the HY-80 weldment system is a well-proven structural material. However, even in this and similar cases, special attention must be paid to any changes in the fabrication processes that may significantly affect the quality of the material, as well as the meaning of a standard test procedure such as Charpy V notch. This is especially important during the process of quality control and the correlation of fracture-tough parameters for design purposes.

Fracture mechanics and its natural derivative, fracture control, are the elements of a relatively new science affecting the work of modern designers. It will be some time before the theoretical and experimental developments of fracture mechanics become fully understood and applied by the various manufacturing industries. General industrial acceptance will, most likely, come from the economic necessity of complying with the more stringent rules of product safety at all levels of the government and in the private sector of our society.

BASIC CONCEPTS AND DEFINITIONS

The art and science of fracture control is related to the basic concepts of fracture mechanics, as it utilizes the stress field parameter K_{1C}, defined in Chap. 14. This parameter has been designed primarily to measure very fine differences in fracture toughness related to the brittle state of the material, and represents the elastic stress field. For this reason, however, it cannot represent the full transition range of fracture toughness between the brittle and plastic behavior, unless a suitable correlation between the K_{1C} and other test parameters can be developed. A number of difficulties encountered with the various correlation methods, referred to in Chap. 14, can, however, be overcome by conducting a *dynamic tear* (DT) test, in which a specimen, featuring a deep sharp crack, is broken in a pendulum-type machine. The upswing of the pendulum following the break indicates the level of energy absorbed. This requires that the size and geometry of the test piece be standardized. The methods of deriving the K_{1C} parameter from the dynamic tear data are described in considerable detail in the literature [87–89].

The dynamic tear concept provides a relatively inexpensive method of characterizing the particular material with regard to its fracture toughness as a function of test temperature, as shown in Fig. 15.1. The solid line in the figure is known as the *crack arrest temperature* (CAT) curve. The point on the curve denoted by NDT denotes the *nil ductility transition* temperature. As the temperature is decreased, a critical transition point NDT is reached which characterizes the ini-

Fundamentals of Fracture Control

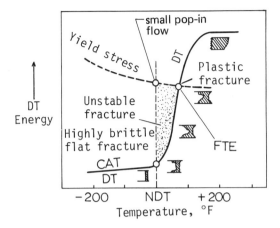

Fig. 15.1 Transition characteristics for low-strength steel with high shelf fracture toughness. CAT, crack arrest temperature; NDT, nil ductility transition temperature; DT, dynamic tear; FTE, fracture transition elastic point. (Adapted from Ref. 87.)

tiation of the elastic fracture in the presence of a dynamically loaded small crack. As the temperature is increased above the NDT, we reach the point called FTE (*fracture transition elastic*). This point defines the highest possible temperature for unstable crack propagation driven by the elastic stress. The shaded portion of the diagram, then, represents unstable fracture conditions. To the right of the crack arrest temperature or dynamic tear curve, the material is stable and will not propagate the crack. The diagram also shows a superimposed yield stress curve, indicating a small downward trend with the increase in temperature. To the left of the dashed vertical line is the region of highly brittle flat fracture. The CAT and DT curves have distinct S shapes. The horizontal branches of the curve are called the lower and upper shelves of fracture toughness for the left and right portions of the curve, respectively. The material characterized in Fig. 15.1 indicates a relatively high-shelf fracture toughness. The types of fracture expected all along the S curve are shown to the right of the solid curve.

The brittle failure is reflected in a flat fracture, indicating a condition of plane strain. On the other hand, a plastic shear-type failure is characteristic of plane stress behavior. The condition of plane strain exists when the elastic portion of the structure envelops and maintains in equilibrium a smaller volume of localized plastic flow. In other words, the elastic part provides a triaxial constraint to plastic flow. This can also be explained in terms of the degree of through-toughness lateral contraction which is developed during the course of testing a dynamic tear sample. In the event of plane strain, this contraction is very small. Hence, in the case of plane stress the flow of metal during the test must be less restricted and considerable notch blunting should be expected to occur. The plane stress fracture, therefore, must be associated with a rather high energy absorption in ductile response, because a relatively large plastic zone is being formed continuously ahead of the propagating crack.

CORRELATION OF FRACTURE PROPERTIES

The practice of fracture control involves a number of specification and procurement decisions affecting the material's toughness. Since specifications, codes, and fabrication techniques frequently change it is necessary to assure the most reliable and current information.

The transition characteristics for a low-strength steel shown in Fig. 15.1 were derived from the appropriate measured values of DT energy. In a similar fashion the transition temperature curve can be obtained from the CVN tests. These tests are still popular in industry during the process of evaluation and certification of the candidate steels despite the findings [258] that the two approaches can give different results along the energy and temperature axes. The primary intent in developing the DT tests is to assure a more sensitive fracture resistance criterion for the elastic-plastic regimes over a broad range of materials.

The use of the various tests and parameters has led to the development of certain correlation techniques. For instance, in the case of a high-strength steel casing, found in oil field explorations, the correlation along the energy axis between the CVN and DT data was of the type

$$(\text{CVN}) = 0.12(\text{DT}) + 15 \tag{15.1}$$

The discrepancy along the temperature axis for the casing material, using the CVN and DT techniques, was in this case on the order of 60°F in such a manner as to make the CVN results less conservative. Similarly, when dealing with the conventional bridge and other steels in the range of 35 to 65 ksi, the shift of the DT energy curve was toward the higher temperature. The corresponding formula suggested by Barsom [259] is

$$T_s = 215 - 1.5 S_y \tag{15.2}$$

In this expression T_s is given in °F and the yield strength at room temperature is inserted in ksi. The formula should be good for the yield strength values between 36 and 140 ksi.

There is ample evidence [260] that CVN energy increases with the tensile strength of steels for a number of fabricated products such as plates, forgings, and welded parts. The general form of correlation between the appropriate K_{1C} and CVN parameters is of the type given by Eq. (14.3). This consideration, together with the available empirical relations between fracture mechanics properties and the transition temperature test energy, has led to the following useful correlation:

$$\left(\frac{K_{1C}}{S_y}\right)^2 = 5\left[\left(\frac{\text{CVN}}{S_y}\right) - 0.05\right] \tag{15.3}$$

Here K_{1C} is given in ksi (in.)$^{1/2}$, S_y is the 0.2% yield strength in ksi, and CVN defines the Charpy V-notch energy in ft-lb. Several other mathematical relationships are available for this purpose in the open literature [260].

Fundamentals of Fracture Control

PRACTICAL USE OF CRACK ARREST DIAGRAMS

Analysis of fracture safety can be reduced to the study of one basic curve defining the relationship between the stress at the instant of crack arrest and the temperature of the material. This relationship, known as the *CAT curve* (Fig. 15.1), represents a practical, conservative criterion on which the design can be based. In the literature, the CAT curve is always shown as part of a more complex picture of material behavior known as a *generalized fracture analysis diagram* [86]. The complete diagram contains three basic variables: nominal stress, crack size, and temperature. By adopting a conservative view, only stress and temperature need be considered. We assume, therefore, that our particular structure is flaw-free and we further stipulate that it is not necessary to know the actual fracture property of the material, such as K_{1C}. This is a very convenient assumption for the entire process of fracture safe design and material certification.

The principal features of this aspect of fracture-safe design are illustrated in Fig. 15.2. The line $FABC$ represents a portion of the crack arrest temperature curve designated as CAT, and it is the most important element of the fracture analysis diagram compatible with our conservative philosophy. This line divides the two main regions of the material's behavior. The shaded area under the curve represents, so to speak, a "safe" region with respect to fracture initiation and propagation. Suppose now that ΔT_W will denote the temperature increment above the NDT at which we expect our particular structure to operate. We tacitly assume here that the curve $FABC$ has been developed experimentally for the selected material and the temperature range containing NDT. The limiting stress corresponding to ΔT_W is S_{CA} This point is denoted by B in Fig. 15.2. Let us also assume that the anticipated working stress is S_W. Hence, we have a positive margin of safety when $S_{CA} > S_W$. When the actual working temperature is lower than (NDT + ΔT_W), such as that corresponding to, for example, B', a fracture may, or may not, propagate at a working stress level of S_{CA}. The region above the curve $FABC$ is always difficult to analyze without knowledge of the actual crack size. It can only be stated

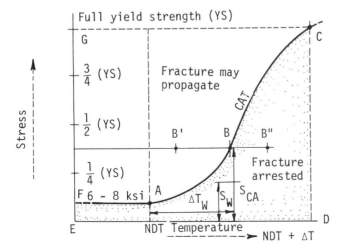

Fig. 15.2 Crack arrest diagram (CAT).

in general that for a given temperature of NDT $+ \Delta T$, the crack propagation stress should increase with the decrease in crack size. Figure 15.2 also suggests that any increase in ΔT_W at a constant stress level shifts our working point to B'', progressing further into the "safe" operational region $FABCDE$. From the point of view of fracture safety, however, the upper region $FGCBA$ should be considered as more "uncertain" for most applications.

The fracture arrest relationship, expressed in the form of the $FABC$ curve, relates to the brittle behavior of the material for various levels of the applied stress. The FA portion of this curve represents the lower stress limit in fracture propagation criteria. Numerically, it corresponds to a level of 5 to 8 ksi fracture extension stress in the plane strain region.

A combination of crack-tip blunting and low nominal stresses should prevent rapid crack propagation above NDT, regardless of crack size. Experience with ship structures, involving weld residual stresses, indicates that when the crack moves out of the welded region into a stress field of about 5 to 8 ksi, there is insufficient elastic energy to propagate brittle fracture.

The concept of the lower-bound stress indicated in Fig. 15.2 dates back to the work of Robertson [261]. It offers a practical and conservative approach to fracture-safe design particularly in those areas of material control where the existing structures cannot be certified as fracture resistant. This area then includes off-the-shelf items without prior history of satisfactory fracture toughness as well as new designs in structural steel which exhibits K_{1C} values of not more than about 25 ksi(in.)$^{1/2}$. This level of plane strain fracture toughness represents a practical lower limit of K_{1C} for what may be termed a "garden variety steel." The corresponding limit for the aluminum is taken here as 15 ksi(in.)$^{1/2}$.

When the appropriate edge-notch criteria of linear elastic fracture mechanics are invoked, the lower-bound nominal stress becomes a function of the material's thickness and fracture toughness. This leads to a useful design chart for steel and aluminum shown in Fig. 15.3. As long as the designer limits the particular working stress to the area below the appropriate curve any existing crack should not propagate catastrophically under the usual conditions of loading and geometry.

The lower-bound region of toughness for steel and aluminum, which formed the basis of design limits in Fig. 15.3, is consistent with fracture control studies of load-carrying members including welds and heat-affected zones [262]. For other materials, the lower-bound fracture toughness should be obtained from empirical sources. The exact role of the lower shelf stress field, such as 5 to 8 ksi, requires further clarification with respect to the behavior of structures made of lower-strength steels. However, it does appear to offer an approach to establishing a useful design criterion for the non-fracture-tough steels employed at ambient and lower temperatures encountered in field construction.

As we progress along the CAT curve through points A, B, and C in Fig. 15.2, the highest temperature of fracture propagation for purely elastic stresses is reached at point C. In the literature of fracture-safe design, this point is referred to as "fracture transition elastic." The corresponding stress level reaches the yield strength of the material. The generalized fracture analysis diagram developed over the years indicates that point C coincides with NDT $+ 60°F$, which is the more frequently quoted temperature consistent with the last, purely elastic point on the CAT curve. This correlation has been developed primarily for use with steels. By restricting

Fundamentals of Fracture Control

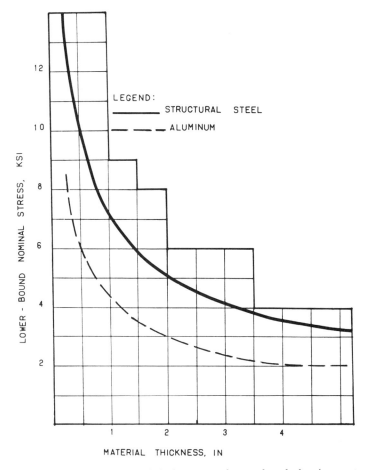

Fig. 15.3 Lower-bound design curve for steel and aluminum structures.

our service use to above the temperature NDT + 60°F, the problems of crack size becomes unimportant, provided that the level of stress does not exceed the yield strength of the selected steel. Under these conditions the fractures can neither initiate nor propagate.

THICKNESS CRITERIA

Plane strain fracture toughness K_{1C} is a key parameter in fracture mechanics. However, it only determines the degree of brittleness. The complete range of fracture toughness between the plane strain and plane stress conditions should be defined by the dynamic tear, and therefore the general relationship between DT and K_{1C} is usually required. It is also important to note that the K_{1C} value alone is not a sufficient index of fracture toughness unless it is related to the yield strength of the material. In this particular case, the relationship K_{1C}/S_y is an important parameter. This relation, in turn, can be calculated if the test specimen thickness

B is known. The thickness B is normally required for the measurement of plane strain fracture toughness as recommended by the American Society for Testing and Materials [85].

Design factors affecting overall fracture safety must include the material's properties, temperature, flaw size, and geometry, as well as thickness. The criterion of material thickness lends itself to a relatively simple correlation with the plane strain fracture toughness K_{1C} and yield strength of the material S_y. The following are the key relationships recommended for fracture safe design:

$$B \leq \left(\frac{K_{1C}}{S_y}\right)^2 \tag{15.4}$$

and

$$B \geq 2.5 \left(\frac{K_{1C}}{S_y}\right)^2 \tag{15.5}$$

Design thickness B, meeting or exceeding the foregoing criteria, provides an assurance that through-thickness flaws will not propagate unless the nominal stresses exceed the yield strength of the material. Here Eq. (15.4) defines the yield criterion and Eq. (15.5) corresponds to the theoretical plane strain limit. In general, high K_{1C}/S_y ratios require large section size and large flaws for plane strain fracture initiation. On the other hand, low K_{1C}/S_y ratios correspond to small section size and small cracks for plane strain fracture. Theoretically, the K_{1C}/S_y ratios can vary between 0.5 and 2.0. The value in excess of the upper limit of 2.0 is unattainable irrespective of the section or flaw size, because metal ductility becomes too high to allow the state of plane strain.

In thicker sections, the level of the mechanical constraint is indicated by the extent of through-thickness contraction adjoining the fracture surface. In thinner sections the imposed mechanical constraint must be small. These considerations have been explored and reported extensively [89].

For practical design needs, Eqs. (15.4) and (15.5) provide a satisfactory guide for the development of safe fracture control plans. The only difficulty encountered throughout the various phases of applications of fracture mechanics and fracture-safe design principles lies in the basic stress analysis used to determine the nominal stresses. It is well to emphasize that a fracture critical component is defined as being a tension-loaded member which in the presence of geometrical discontinuities may have significant stress gradients. The accuracy of the analysis of such gradients and local stress concentrations should determine the final factors of safety based on the thickness criteria given by Eqs. (15.4) and (15.5).

SIGNIFICANCE OF STRESS AND STRENGTH

The nominal engineering stress in fracture control is normally of the P/A type. However, if the observed or anticipated flaw is located in a structural transition, such as that near a sharp corner, the relevant elastic stress concentration factor should be included to obtain, say, a KP/A stress level. Moreover, as far as the mechanism of crack propagation is concerned, this stress would have to be of a type and direction such as to cause the crack to open. Obviously, the resultant tensile mode is most significant in this regard and it is compatible with the stress value S_{CA} indicated in Fig. 15.2.

According to the basic criteria involving the transition temperature, the designer can aim at either fracture-tough or fracture-safe design. From the stress point of view, both approaches constitute the major elements of the fracture control process. The object of fracture-tough design would be to select a material that could prove to be insensitive to crack propagation and brittle failure throughout the working range of temperatures, material thicknesses, and design stresses. In fracture-safe design with known CAT curve (Fig. 15.2), the working stress S_W should be less than S_{CA} at a selected temperature. For instance, at temperature NDT+ΔT_W corresponding to point B, the maximum stress that can be arrested is on the order of 37.5% of the yield strength of the material depicted by S_y on the vertical axis of Fig. 15.2. When the operating temperature is less than NDT, the rule of 5000 to 8000 psi applies. The basic problem with such a low shelf value is that a weight penalty may be involved in developing a low-stressed system. Unless the stress is low in this design region (left of the dashed vertical line in Fig. 15.2), the material is essentially brittle and a plane strain failure can be initiated in the presence of a very small flaw.

In many cases structural members are designed to perform in an as-welded but not stress-relieved condition. The residual stresses in this instance can be as high as the yield strength, and they are oriented in the direction parallel to the weld. Such stresses result from the longitudinal shrinkage during cooling of the weld area, which is restrained by the adjacent colder metal. In practice, these peak stresses extend to about one to two weld widths and therefore they can contain only relatively small cracks. We therefore have a highly localized residual stress region in the weld or heat-affected zone, which may initiate fracture due to a small crack for a wide range of stresses and when the service temperature falls below the NDT.

The brittle condition in the heat-affected zone is further aggravated by the formation of a coarse grain structure, which is inherently more brittle than the finer grain size of the parent metal. Additional aggravation is encountered if the rapid cooling of the weld results in the formation of martensite, a hard, brittle microconstituent. However, at temperatures above the NDT, such a fracture will not occur because of the requirement for extensive plastic deformation. This theory has been verified experimentally [90].

When the transition curve corresponds to a low-strength steel, it normally exhibits a high shelf fracture toughness. On the other hand, a flat S curve, indicating no significant transition temperature, is characteristic of high-strength steels. This is an important practical consideration, because increasing service temperature for a high-strength material may not necessarily assure that the new system will become much more fracture-safe.

The technological progress puts increased emphasis on the development of improved specifications for fracture testing and fracture-safe design, particularly in the areas of structures carrying dynamic loads. The prediction of elastic stresses in the context of fracture mechanics still remains difficult despite the progress in analytical methods. The basic reason for this is that only the simplest, statically determined loads and stresses can be recognized and calculated with any reasonable degree of confidence. Residual and locally induced stresses are especially difficult to handle analytically; therefore, the majority of publications in fracture mechanics and stress analysis are seldom design oriented.

DEVELOPMENTS IN TECHNOLOGY OF MATERIALS

The evolution of fracture control principles and materials research in the United States has been largely driven by the needs of the Navy concerned with fracture resistance of the high-yield family of steels. Opening comments of this chapter refer to HY-80, HY-100, HY-130, and similar alloy steels selected for their excellent resistance to fracture.

It may be of interest to recall that the forerunner of HY-80 steel was the so-called modified Krupp steel, in a quenched and tempered condition, which was developed at the start of this century. This particular German product was characterized as a low-carbon material with addition of modest amounts of nickel, chromium, and molybdenum to ensure superior toughness properties in a quenched and tempered state.

The availability of the high-yield family of steels had to be matched by the developments in electrode materials for reliable welds, an accomplishment of a tall order considering the number of variables and imponderables involved in the process. Among the complexities which are difficult to control are structural constraints, joint preparation, weld accessibility, cooling rate, transformation, and shrinkage. The development process had to be painfully slow for technical, economic, and legal reasons and was punctuated by special investigations and disputes such as those concerned with the formulation and use of the high-strength material known as T-1 steel.

One of the more recent developments in the area of fracture-tough materials with superior weldability is the low-carbon and low-alloy steel known as HSLA-80. This type of material offers considerable potential for providing fracture-safe welded structures when stressed to the nominal yield strength in the presence of flaws, down to about 0°F service temperature. The welding process is rather straightforward and essentially does not require preheating or postweld heating. Since HSLA-80 is a low-carbon steel there is no risk of developing brittle martensitic constituents near defects or along the flame-cut surfaces. A similar material is known in industry as ASTM A-710, grade A, class 3. Both materials represent an excellent choice for the designer of critical structural members. These basic, ferritic alloy steels are able to achieve a high strength level through grain refinement and precipitation hardening while deriving toughness and weldability benefits from the limited carbon content. With additional amounts of manganese and molybdenum, and proper fabrication techniques to assure predominantly bainitic microstructure, these materials are likely to be used in plate thicknesses of up to 8 in. [263]. Structures of the future demand improved characteristics in design, fabrication, and performance. The applications of these materials will be found in construction equipment, railroad

Fundamentals of Fracture Control

tank cars, mining and dredging, offshore platforms, oceangoing vessels, and truck trailers, to mention a few.

CLOSING REMARKS

The material presented in this chapter only scratches the surface of the volume of published literature on the subject of fracture mechanics and fracture control. It is hoped, however, that this discussion will alert the reader to the modern era of stress analysis, which was born from the necessity of dealing with catastrophic failures of engineering materials. The primary message here is that we are making progress in characterizing the brittle behavior of materials and that the knowledge of a few new parameters can help us to design safer mechanical systems.

In summary, the following specific conclusions can be added as an aid to designers:

1. Standard yield strength and ductility of a material provide no clue as to the fracture resistance.
2. Ductile materials can fail in a brittle manner when the operating temperature is lower than NDT.
3. Fracture mechanics and fracture-safe design relate only to the tensile mode of failure.
4. When small flaws are present, the flaw and not the section size may control the initiation and progress of fracture.
5. The effect of very large flaws is influenced by the section size; the procedure shifts the CAT-DT curve toward higher temperatures.
6. Cracks will not propagate under normal conditions when the nominal stress in the member is less than the appropriate lower-bound design values, such as those given in Fig. 15.3.
7. Crack mitigation can be expected when the crack enters either a lower-stressed region, a region of higher fracture toughness, or both.
8. A safe metal thickness can be selected when K_{1C} and S_y are known.
9. Increasing the section thickness may not provide any additional safety margin unless the K_{1C}/S_y ratio indicates an improvement.
10. Design modification directed toward lowering the NDT parameter is likely to increase the cost of the project.

Whatever method is adopted for the development of a fracture control plan, it is safe to assume that real engineering materials must contain some manufacturing imperfections. We have, therefore, the task of designing around a potential problem area using both linear elastic fracture mechanics and experimental data on crack arrest characteristics.

SYMBOLS

A Cross-sectional area, in.^{2} (mm^2)
B Material thickness, in. (mm)
CAT Crack arrest temperature, °F

CVN	Charpy V-notch energy, ft-lb (N-mm)
DT	Dynamic tear, ft-lb (N-mm)
FTE	Fracture transition elastic
K	Stress concentration factor
K_{1C}	Plane strain fracture toughness, psi(in.)$^{1/2}$ [N(mm)$^{-3/2}$]
NDT	Nil ductility transition temperature, °F
P	Nominal axial force, lb (N)
S_y	Yield strength, psi (N/mm^2)
S_{CA}	Limiting stress, psi (N/mm^2)
S_W	Working stress, psi (N/mm^2)
T_s	Temperature shift, °F
ΔT	Temperature margin above NDT, °F
ΔT_W	Temperature margin above NDT corresponding to working stress, °F

16
Mechanics of Stress Propagation

INTRODUCTION

The problem of unsteady load application, reviewed briefly in Chaps. 11–13, is a frequent source of concern in modern engineering design. Consequently, recent studies of shock, impact, and seismic stress waves have attracted new attention of both theoreticians and practical engineers. Although the elastic strain propagation phenomena have been known for some time now, relatively few practical books on structural analysis, strength of materials, or dynamics include the subject of wave propagation. The reasons for this omission is the mathematical sophistication required in developing the various theoretical relationships and perhaps the lack of practical formulas of sufficient generality which could be applied to various design situations. In addition, the majority of books and technical articles on this subject are not easy to follow, with perhaps one exception [91]. The reader should be warned, however, that even in the case of easier presentations, considerable mathematical background is needed in dealing with the problem.

The material in this chapter contains only the simplest relations and formulas. It falls into the class of the dynamic response, where the ratio of the time of load application to the natural period of vibration is less than about 0.5. It also implies that in this particular region of dynamic behavior the properties of materials should be taken into account.

BASIC CONCEPTS

The elementary illustration of a wave propagation phenomenon can be based on the analogy of a locomotive starting to pull a long string of stationary freight cars, or a similar train running into a barrier of fixed buffers. In the first case, each car starts up the one behind it, while the last car is, so to speak, "unaware" of the load applied to the front sections of the train. This analogy applies also to a rod to which an axial force is suddenly applied, sending a tensile stress wave along its axis. Any section of the rod, other than that experiencing a wave propagation phenomenon, remains unstressed. By analogy to a freight train, the particles of the rod at the impacted end are displaced and create a wave which begins to travel from one end of the bar to the other.

By reference again to the train analogy, we note that each car resists the motion by the inertia of the car behind it, except the last car in the string, which will run after the train faster than the cars in front of it, initiating a compressive type of wave. This wave is expected to proceed until the front of the train is reached. On the other hand, when a freight train runs into a barrier of fixed buffers, each car is brought to rest in turn. However, the last car rebounds and initiates a tensile wave which begins to travel up the locomotive.

The foregoing simplified illustration of mechanics of wave propagation leads to the following basic conclusions:

1. A compression wave reaching a free end transforms into a tensile wave and vice versa.
2. A wave is reflected from a fixed end without a stress reversal.

STRESS PROPAGATION THEORY

The general theoretical relationship between a stress wave and the velocity of propagation in one-dimensional impact is usually expressed as follows:

$$\sigma(x,t) = \rho C V(x,t) \tag{16.1}$$

where σ denotes the stress normal to the wave front and V is the particle velocity at a particular point in space and time [250]. For a straight bar, for instance, the positions in space and time are denoted by x and t, respectively. The parameter C, known as the *sonic velocity*, is defined as

$$C = \left(\frac{E}{\rho}\right)^{1/2} \tag{16.2}$$

where E stands for the modulus of elasticity and ρ is the mass density of the material in which the stress wave propagation takes place. This formula defines the velocity of stress wave propagation in the elastic range. Similarly, in the case of plastic deformation, the velocity of propagation of a given plastic strain is a

Mechanics of Stress Propagation

function of the slope of the stress-strain characteristics at that strain.

$$C_\mathrm{p} = \left[\left(\frac{d\sigma}{d\epsilon} \right) \Big/ \rho \right]^{1/2} \tag{16.3}$$

When $d\sigma/d\epsilon$ approaches zero, as is the case with the point on the stress-strain curve corresponding to the ultimate strength of the material, Eq. (16.3) suggests that a sudden change takes place. The stress propagation theory, therefore, can predict the velocity of impact corresponding to the instant of material's rupture. The relevant velocity is sometimes called the *critical impact velocity*. In a practical manner, this velocity for a particular metal can be estimated in the following way:

1. Acquire the stress-strain curve for the given material by the usual methods of mechanical testing under static conditions.
2. Plot the strain ϵ versus the parameter $[(d\sigma/d\epsilon)/\rho]^{1/2}$.
3. Measure the area under this curve between the points corresponding to zero and maximum strain, respectively. This area expressed in correct dimensions is equal to the required critical impact velocity. The procedure is illustrated graphically in Fig. 16.1.

This concept of velocity has been developed on the basis of classical mechanics [92]. It may be defined as that velocity at which the energy of absorption and strain begin a sharp decrease. The critical velocity can be stated in mathematical terms as

$$V_\mathrm{CR} = \int_0^\epsilon \left[\left(\frac{d\sigma}{d\epsilon} \right) \Big/ \rho \right]^{1/2} d\epsilon \tag{16.4}$$

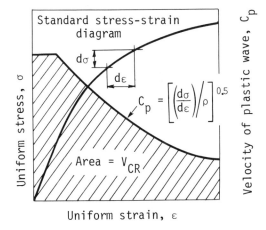

Fig. 16.1 Graphical method for finding critical impact velocity.

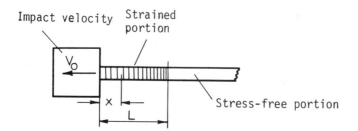

Fig. 16.2 Model of uniform bar under tensile impact.

ELASTIC IMPACT

Consider next the case of elastic impact on a bar of uniform cross section, as shown in Fig. 16.2. Provided that no break is experienced, the stress wave will move a distance equal to the elongation of the strained portion of length L, shown by the shaded area in Fig. 16.2. The stress wave will move with acoustic velocity C a distance L during a time interval t. When the distance traveled by the end of the bar is $V_0 t = \epsilon L$, and $L = Ct$, a useful formula is obtained

$$V_0 = \epsilon C \qquad (16.5)$$

Utilizing Hooke's law $\epsilon = \sigma/E$ and Eq. (16.1) gives the dynamic stress immediately after impact

$$\sigma = \frac{EV_0}{C} \qquad (16.6)$$

When $x = 0$ and $t = 0$, Fig. 16.2 shows that $V_0 = V$. Also note that Eqs. (16.1) and (16.6) have essentially the same form. For a steel bar with $E = 30 \times 10^6$ psi, the appropriate formula from Table 16.1 gives $C = 16,870$ ft/sec so that the velocity corresponding to the yield strength of the material of, say, 80,000 psi is $V_0 = 45$ ft/sec. It should be emphasized here that a significant numerical difference exists between the particle velocity and the sonic velocity of stress propagation. In

Table 16.1 Formulas for Sonic Velocity

Uniform bar of infinite length	$C = \left(\dfrac{E}{\rho}\right)^{1/2}$
Infinite slab or plate	$C = \left[\dfrac{E}{(1-\nu^2)\rho}\right]^{1/2}$
Cylinder	$C = \left[\dfrac{E}{(1-\nu^2)\rho}\right]^{1/2}$
Elastic continuum	$C = \left[\dfrac{E(1-\nu)}{(1+\nu)(1-2\nu)\rho}\right]^{1/2}$

Mechanics of Stress Propagation

the case of Fig. 16.2, the applied impact velocity V_0 and the particle velocity at $x = 0$ are assumed to be the same.

When dealing with the dynamic response of the bars of uniform cross section, the theoretical sonic velocity can be directly obtained from Eq. (16.2). However, it should be appreciated that certain limitations on the use of this equation are imposed when considering other geometries. The corresponding formulas are compared in Table 16.1. The general case of elastic continuum refers to an unbounded isotropic medium through which the wave propagates in a three-dimensional system.

Since ν in Table 16.1 denotes the Poisson's ratio, it is easy to see how the sonic velocity increases from the uniform bar to the continuum geometry by about 20% for the case of a typical metal. The parameter denoted by C in Table 16.1 is called the *dilatational velocity*, to distinguish it from the shear or the rotational type of propagation velocity.

Based on the data in Table 16.1, a quick estimate of the upper-bound elastic stress can be made utilizing the relation

$$\sigma = \rho C V \qquad (16.7)$$

This calculation however, depends on the choice of the value of a particle velocity in the stressed region.

In considering the reflection of the stress wave at the end of a rod, the maximum theoretical stress in the uniform rod with a free end can be no more than the value given by Eq. (16.7). On the other hand, when the remote end of the rod is built in as in a rigid support, there can be no displacement or velocity at such a point, and a tensile wave, for instance, is reflected back as a tensile wave. Vector addition of the two tensile waves, then, gives 2σ, where σ follows from Eq. (16.7). This is then the maximum theoretical value of the longitudinal impact stress on a bar or other structure in the elastic regime of the material.

AXIAL IMPACT ON STRAIGHT BAR

An interesting case of a compressive impact stress [1] can be analyzed with reference to the sketch in Fig. 16.3, where a rigid body of mass M, traveling with velocity V, strikes axially the free end of the rod at $x = 0$; the particle velocity at this point must also be equal to V. Since the motion of the mass M is resisted and slowed down by the reaction created between the rigid body and the cantilever beam, the impact velocity v of mass M, after a short time interval t, must be less than V. The corresponding equation of motion becomes

$$\sigma_0 A = -M \frac{dv}{dt} \qquad (16.8)$$

Since v also represents the particle velocity at the point of contact, Eq. (16.7) is applicable. Eliminating sonic velocity from Eq. (16.7) with the help of Eq. (16.2)

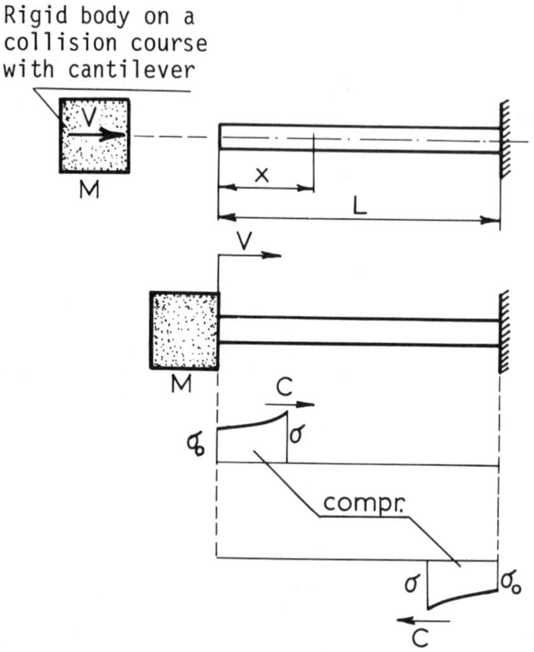

Fig. 16.3 Model of uniform bar under compressive impact.

yields

$$v = \frac{\sigma_0}{(\rho E)^{1/2}} \tag{16.9}$$

If the ρE term is assumed to be constant for the problem considered, differentiating both sides of Eq. (16.9) with respect to time gives

$$\frac{dv}{dt} = \left(\frac{d\sigma_0}{dt}\right) \bigg/ (\rho E)^{1/2} \tag{16.10a}$$

Combining Eqs. (16.8) and (16.10a) gives

$$\frac{M}{(\rho E)^{1/2}} \frac{d\sigma_0}{dt} + A\sigma_0 = 0 \tag{16.10b}$$

Solution of Eq. (16.10b) is

$$\sigma_0 = Be^{-(A\sqrt{\rho E}/M)t} \tag{16.10c}$$

When $t = 0$, $\sigma_0 = \sigma$ and $B = \sigma$, so that Eq. (16.10c) becomes

$$\sigma_0 = \sigma e^{-(A\sqrt{\rho E}/M)t} \tag{16.11}$$

Mechanics of Stress Propagation

Hence, the intensity of the impact compressive stress at $x = 0$ decreases exponentially while the wave front continues to travel on toward the fixed end of the rod at $x = L$. At this point the compressive stress is suddenly reflected with no change in sign, as shown in Fig. 16.3. At this instant also the compressive stress attains the value of $2\sigma_0$; this value can be used in the preliminary calculations. It should be emphasized, however, that in general the complete process of impact may involve the analysis of several stress wave reflections.

CONDITIONS OF SPALL

The phenomenon of spall is a direct result of a high-amplitude compressive wave encountering a free surface. The basic features of spall, for instance, may be observed at the ground surface disturbed by an underground explosion or at the ruptured surface opposite the point of impact of a projectile striking a thick plate. A compressive wave reaching a free surface generates a reflected tensile wave of the same amplitude. If this stress amplitude happens to exceed the tensile strength of the material, a layer near the free boundary will spall off. The corresponding plane of failure must be below the surface since the peak tensile stress can occur only after the reflected wave starts to travel back toward the original point of impact.

The thickness of the spalled layer is known to depend on the wave amplitude, wave shape, and tensile strength of the material.

The mechanical model of spall is complicated and it requires a rather sophisticated analysis of the various phases of stress propagation where the impinging compressive wave interacts with the reflected tensile wave throughout the thickness of the particular medium. The reader interested in theoretical ramifications of the spall problem is advised to consult the specialized literature on the subject [250]. From the practical point of view it will suffice here to state that spall can occur in materials whose tensile strength is numerically smaller than the compressive strength, or when the amplitude of the impinging stress wave is significantly higher than the tensile strength of the material.

Design Problem 16.1

A rigid object free-falls over a distance of 10 ft and strikes the top of a steel column built in at the base. Assuming that the impinging object produces a purely axial response of the column, calculate the maximum compressive stress at the base if the sonic velocity for the column material is 16,820 ft/sec, with the specific weight of steel of 0.283 lb/in.^{3}.

Solution

The required stress in this case will be estimated on the premise that the free-fall velocity of the object at the instant of impact is equal to the particle velocity at the point of contact.

The elementary formula for a free-fall velocity given in Table 11.3 is

$$V = (2gH)^{1/2}$$

Hence the propagated compressive stress can be obtained from Eq. (16.7)

$$\sigma = \rho C (2gH)^{1/2}$$

where

ρ = mass density of column material, (lb-sec^2)/in.^{4}
C = sonic velocity, in./sec
g = acceleration of gravity, in./sec^2
H = height of drop, in.

Substituting the appropriate numerical values gives

$$\sigma = \frac{0.283 \times 16{,}820 \times 12}{386.4}(2 \times 386.4 \times 10 \times 12)^{1/2}$$
$$= 45{,}017 \text{ psi} \quad (310.5 \text{ N/mm}^2)$$

At the fixed end the stress wave is reflected with no change in sign so that the maximum value becomes 90,034 psi or 621 N/mm^2. ♦

Design Problem 16.2

A rigid object weighing 5000 lb collides with the free end of a steel cylinder which is supported in the manner shown in Fig. 16.3. The impact induces the stress wave equal to $\sigma_0 = 6000$ psi. Assuming that the cylinder is 28 ft long and has a cross-sectional area of 8 in.^{2}, estimate the compressive stress in the cylinder wall at 0.001 sec after the impact and indicate the location of the wave front. The specific weight and the elastic modulus for the cylinder material can be taken as 0.283 lb/in.^{3} and 28×10^6 psi, respectively.

Solution

The mass density is

$$\rho = \frac{0.283}{386.4} = 0.00073 \text{ lb-in.}^2/\text{in.}^4$$

The mass of the striking object is

$$M = \frac{5000}{386.4} = 12.94 \text{ lb-sec}^2/\text{in.}$$

Hence using Eq. (16.11) gives

$$\sigma = \frac{\sigma_0}{e^{-[A(\rho E)^{1/2}/M]t}}$$

Since

$$\frac{A(\rho E)^{1/2}}{M} = \frac{8(0.00073 \times 28 \times 10^6)^{1/2}}{12.94} = 88.39$$

Mechanics of Stress Propagation

then

$$\sigma = \frac{6000}{e^{-88.39 \times 0.001}} = 6557 \text{ psi} \quad (45.2 \text{ N/mm}^2)$$

The sonic velocity in steel is

$$C = \left(\frac{28 \times 10^6}{0.00073}\right)^{1/2} = 195,847 \text{ in./sec}$$

Therefore the distance traveled by the wave front is

$0.001 \times 195,847 = 195.8$ in.

or

$$\frac{195.8}{28 \times 12} = 0.58L$$

where L denotes the length of the cylinder. ♦

AXIAL AND RADIAL MODES OF ELEMENTARY STRUCTURES

A continuous elastic system can be simplified for the purpose of dynamic analysis if its mass can be lumped together at a particular central point and when its rigidity is represented by a mechanical spring. In the literature, this approach is referred to as a *lumped mass-spring model*, and it permits computation of the response of a structure to a specific input of acceleration-time history. The general consequence of the lumped parameter simplification is the development of a working formula for the calculation of the fundamental frequency. Since the mass density and the modulus of elasticity of the structure material always enter the formulation of the vibrational problem, reference to Eq. (16.2) indicates that the final formulas can be expressed in terms of the sonic velocity of a given material. This can be accomplished quite independently of the type of the vibrational response, be that longitudinal or purely radial response. In the case of simple shells, for instance, we often talk about the *breathing mode*. This mode implies radial expansion and contraction cycles of, say, a cylindrical component at a specific fundamental frequency dependent on the dimensional proportions and the physical properties of the material.

To assist with the choice of the first modes of dynamic response, the fundamental frequencies of several common structures are given in Table 16.2 in terms of the characteristic dimensions and the sonic velocity of the material. The formulas for cylinders and spheres define the breathing modes [94, 95].

The equations compiled in Table 16.2 may be utilized in the following design situations:

1. In the field of machines and structures a forced frequency of external loading may be coinciding with the natural frequency of a particular mechanical component. Such external excitation can push the component at the right

Table 16.2 Fundamental Frequency Data as a Function of Sonic Velocity (Axial and Radial Modes)

Direction of impulse

Shape	Condition	Formula
	Free/free	$f = \dfrac{C}{2L}$
	Fixed/free	$f = \dfrac{C}{4L}$
	Simply supported, $h =$ thickness	$f = \dfrac{\pi C h / a^2}{4[3(1-\nu^2)]^{1/2}}\left[1 + \left(\dfrac{a}{b}\right)^2\right]$
	Fixed, $h =$ thickness	$f = \dfrac{3Ch}{\pi a^2}\left[\dfrac{7}{6(1-\nu^2)}\right]^{1/2}\left[\left(\dfrac{a}{b}\right)^4 + \dfrac{4}{7}\left(\dfrac{a}{b}\right)^2 + 1\right]^{1/2}$
	Simply supported plate	$f \cong \dfrac{0.3Ch}{a_0^2}$
	Fixed plate	$f \cong \dfrac{0.49Ch}{a_0^2}$
	Thin ring or cylinder	$f = \dfrac{C}{2\pi r}$
	Thin sphere	$f = \dfrac{0.27C}{r}$

time and in the right direction, increasing its amplitude of vibration and consequently its stresses. A simple check on the fundamental frequency of a component can tell us at a glance if the component should be redesigned to avoid the possibility of a resonance with the excitation frequency.

2. As explained in Chap. 11, knowledge of the natural period of vibration (reciprocal of fundamental frequency) should help us to select the method of dynamic analysis. This is accomplished by comparing the time interval of the external loading applied to the component with the natural period of vibration.
3. The design formulas given in Table 16.2 may be used in experimental work. For instance, observing the first fundamental mode and knowing the geometry and physical properties of a component, the corresponding sonic velocity C can be computed.

RESPONSE OF BURIED STRUCTURES

A number of interesting phenomena, which relate to the design formulas and principles outlined in this section, are found in the area of underground nuclear and conventional explosion experiments. It is often necessary, for example, to evaluate the nature of the dynamic response of a buried structure. In such situations the study of soil-structure interaction requires knowledge of the material properties and ground motion parameters. Near the point of detonation, the surrounding medium behaves as a fluid subjected to intense stress and velocity fields, and the structure may be experiencing rigid-body displacements. The extent of the initial ground shock can be estimated if the effective duration of the stress wave and the natural period of vibration of the particular structural component are known [96].

Although a spherical stress wave is generated from the point of the disturbance, it is often customary to consider a one-dimensional stress wave in the analysis, neglecting the influence of lateral inertia. With this assumption, one-directional frequency response, such as that indicated in Table 16.2, can be utilized to determine the level of analytical sophistication required in dealing with the problem. For instance, consider a ground shock enveloping a buried steel pipe of radius 24 in. Let the effective duration of explosion in a nuclear test be 30 msec. Using the formulas from Tables 16.1 and 16.2, the natural period of vibration of the pipe in the breathing mode becomes

$$T = 2\pi r \left[\frac{\rho(1-\nu^2)}{E}\right]^{1/2} \tag{16.12}$$

For $E = 30 \times 10^6$ psi, $\nu = 0.3$, and $\rho = (0.284/386.4)$ lb-sec^2/in.^{4}, Eq. (16.12) gives

$$T = 6.28 \times 24 \left(\frac{0.284 \times 0.91}{30 \times 10^6 \times 386.4}\right)^{1/2} = 0.71 \text{ msec}$$

According to the rule given in Chap. 11, the time of load application is certainly much longer than the natural period of a breathing mode, calculated for the pipe, with the required ratio higher than 3. Hence, with only a minor error, the loading in the explosion discussed above can be represented by a step pulse of infinite duration, so that the buried pipe can be analyzed statically.

From the physics point of view, isothermal deformation is expected under conditions of static loading. Under dynamic loading the process of deformation is considered to be nearly adiabatic. The deformation leading to the plastic flow at the level of impulsive loads, lasting only a few microseconds, experiences a time delay. The corresponding secant modulus of elasticity is known to increase with the rate of loading so that the particular structural member is not expected to yield if the time of response is less than the time delay to yielding. In practice, however, these effects are seldom accounted for, and static elastic constants can be used in the analysis.

STRESS PROPAGATION IN GRANULAR MEDIUM

The radial and tangential stresses in the surrounding medium, such as soil, rock, or stemming materials, must first be determined before a buried structure such as a pipe, canister, or underground room can be designed. The fundamentals of stress wave propagation for these media are essentially the same as those utilized in the study of the homogeneous materials such as metals. Unfortunately, the nature of granular, soil-like materials or rock formations represents a multitude of boundaries affecting the propagation and reflections of sonic stress waves. Consequently, only the approximate, gross values of ground motion parameters such as sonic velocity, particle velocity, and density can reasonably be estimated.

The mechanical model for predicting particle motions for ground motion due to underground explosions can be based on the following three assumptions [97, 98]:

1. Particle velocity decreases as $1/R^3$ in the inelastic region and as $1/R^2$ in the elastic region, where R is the scaled range, defined as $R_0/W^{1/3}$. In this relation R_0 denotes the distance from the point of explosion in feet and W is the yield in kilotons.
2. Peak particle velocity depends on the square of the sonic velocity of the media in which the explosion takes place.
3. Propagation velocity of the peak stress depends linearly on the sonic velocity.

Knowledge of wave initiation, propagation, and reflection is closely tied to seismic studies. Out of these three features the initiation process is probably least understood, despite a number of empirical observations and long field experience.

When a stress wave produces impulsive external pressure all around a cylindrical shell, complex vibration modes can be excited involving "breathing" and flexural response [97, 98]. The interaction between the purely extensional and flexural modes is found to precipitate permanent wrinkles. The subject of this response is too lengthy for inclusion in this brief review of wave propagation. It may be observed, however, that under dynamic conditions the buckling modes depend not only on the structure but also on the magnitude of the applied loads. The dynamic buckling often occurs where the impulsive loads are sufficiently high to cause an appreciable plastic flow.

APPLICATIONS TO MACHINERY

In many areas of industrial engineering the theoretical analysis of dynamic response may be too costly and time consuming. The alternative here is to follow an empirical approach which, however, requires the development of a specialized dynamic test rig and data reduction procedures. The most common area of interest relates to the effect of a moderately high velocity of impact on the resistance of a material or a machine element. The general problem in this case is concerned with the propagation of elastic and plastic waves in solids. Despite half a century of progress, engineers and designers are still faced with using oversimplified assumptions and models of dynamic behavior.

Many high-velocity-impact machines have been developed utilizing gravity and pneumatic forces. The speeds of impact vary generally between 100 and 200 ft/sec.

Mechanics of Stress Propagation

Some impact machines have capacities exceeding 15,000 ft-lb, intended for tests to destruction. Force-time and velocity relations can be obtained for the purpose of correlation with the classical theories. Also, critical impact velocity data for the various structural materials can be updated.

Calculation of instantaneous response of a machine member requires some knowledge of the velocity change and sonic velocity of the material, as indicated by Eq. (16.7). If sonic velocity data are not available, the elastic stress can be obtained from Eqs. (16.2) and (16.7) by concluding that $\sigma = V(\rho E)^{1/2}$. For instance, if a wire is being wound vertically onto a drum at a velocity of 15 ft/sec, the question may be asked: What dynamic stress is involved if the drum is suddenly stopped? Suppose that the modulus of elasticity is 15×10^6 psi and the weight density is 0.3 lb/in^3. Using a formula based on Eqs. (16.2) and (16.7) gives

$$\sigma = 15 \times 12 \left(\frac{0.3 \times 15 \times 10^6}{386.4} \right)^{1/2} = 19,425 \text{ psi}$$

It is seen that V is considered here as a basic parameter in the calculations. The same reasoning would apply if the wire rope were initially at rest and then attained a velocity of 15 ft/sec through a sudden turn of the drum. Hence, the net amount of the velocity change is important. If the sonic velocity for the rope were specified instead as $C = 11,583$ ft/sec, we would calculate the dynamic stress as follows:

$$\sigma = \frac{0.3 \times 11,583 \times 15 \times 12 \times 12}{386.4} = 19,425 \text{ psi}$$

This and previous sections dealing with the nonsteady load applications involved only the simplest theories of dynamics, vibration, and stress propagation. However, their use is generally sufficient for the purpose of the preliminary design.

SYMBOLS

A	Cross-sectional area, in.^{2} (mm^2)
a	Width of plate, in. (mm)
a_0	Radius of plate, in. (mm)
B	Constant of integration
b	Length of plate, in. (mm)
C	Sonic velocity, in./sec (mm/sec)
C_p	Velocity of plastic wave, in./sec (mm/sec)
E	Modulus of elasticity, psi (N/mm^2)
f	Fundamental frequency, Hz
h	Thickness of plate, in. (mm)
L	Length of beam, in. (mm)
M	Mass of rigid body, lb-sec^2/in. (N-sec^2/mm)
R	Scaled range, ft (m)
R_0	Distance from point of explosion, ft (m)
r	Radius of cylinder or sphere, in. (mm)
T	Natural period of vibration, sec.

t	Time, sec.
v	Particle velocity, in./sec (mm/sec)
V	Velocity of rigid body, in./sec (mm/sec)
V_0	Impact velocity, in./sec (mm/sec)
$V_{x,t}$	Arbitrary particle velocity at a given time and coordinate, in./sec (mm/sec)
V_{CR}	Critical impact velocity, in./sec (mm/sec)
W	Yield of explosion, kilotons
x	Arbitrary distance, in. (mm)
ν	Poisson's ratio
ϵ	Strain, in./in. (mm/mm)
σ	General symbol for stress
σ_0	Impact stress at boundary, psi (N/mm^2)
$\sigma_{x,t}$	Stress normal to wave front, psi (N/mm^2)
ρ	Mass density of medium, lb-sec^2/in.^{4} (N-sec^2/mm^4)

17
Thermal Stresses and Materials

NATURE OF THERMAL RESPONSE

The structural response of material under heating or cooling is governed by the expansion or contraction of its volume. When a structural member is subjected to a thermal gradient, it is necessary to recognize that there can be the following conditions:

1. External forces provide a constraint against a change in shape.
2. The form of the structural member is incompatible with the natural tendency of various parts of the member to contract or expand.

The thermal behavior under condition 1 may be represented, for example, by a uniform, straight bar held at the ends and subjected to a constant temperature gradient. The case corresponding to condition 2 can best be illustrated with reference to a cylinder having a temperature gradient across its wall. Although in this case no external forces of constraint are applied, thermal stresses are produced because the strains are incompatible with the free thermal deformation. The two cases outlined above can also be characterized as those structural systems which are governed by external or internal constraints.

BASIC STRESS FORMULA

The simplest expression for calculating the thermal stresses is based on the treatment of an elastic, uniformly heated or cooled bar, restrained firmly at the ends

$$\sigma = E\alpha\,\Delta T \tag{17.1}$$

When the temperature of the bar is decreased, the bar develops tension. Compressive stress, on the other hand, is caused by an increase in temperature. Since only the elastic stresses are considered, the total strain is the sum of the stress- and temperature-dependent strains [99].

In Eq. (17.1), α denotes the coefficient of linear, one-directional, thermal expansion, which for the majority of structural materials varies between 5 and 15×10^{-6} per °F. The corresponding volumetric strain due to the temperature change is

$$\left(\frac{\Delta V}{V}\right)_T = 3\alpha\,\Delta T \tag{17.2}$$

Some typical thermal constants that enter stress calculations are summarized in Table 17.1 [100, 101].

Table 17.1 Typical Thermal Constants at Moderate Temperatures

Material	α (°F^{-1} × 10^{-6})	C (Btu/lb °F)	k (Btu/hr ft °F)	λ (ft^2/hr)
Pure aluminum	14.0	0.22	128.0	3.50
Aluminum alloy	13.0	0.22	91.0	2.30
Pure copper	9.5	0.09	228.0	4.50
Brass (60/40)	10.5	0.09	54.0	1.10
Bronze (90/10)	10.0	0.09	24.0	0.49
Gold	7.8	0.03	180.0	5.10
Silver	11.0	0.06	240.0	6.10
Carbon steel	6.7	0.11	26.4	0.49
Alloy steel	6.7	0.11	13.2	0.25
Lead	16.0	0.03	20.4	0.94
Magnesium alloy	14.0	0.25	47.0	1.73
Pure nickel	7.2	0.11	33.6	0.56
Iron-nickel (64/36)	1.1	0.11	7.2	0.12
Platinum	5.0	0.03	41.0	1.00
Tin	15.0	0.05	37.0	1.60
Zinc	14.5	0.09	67.0	1.67
Ceramics	1.7	0.20	0.72	0.021
Concrete	6.7	0.21	0.60	0.020
Typical glass	4.5	0.18	0.48	0.017
Ice		0.50	0.13	0.045
Typical plastics	11.0	0.37	0.24	0.006
Sandstone	4.5	0.19	0.96	0.029
Granite	4.5	0.19	0.30	0.057

Note: These values are good up to 400°F. Symbols: α, thermal coefficient of linear expansion; C, specific heat; k, thermal conductivity; λ, thermal diffusion.

In planning the design of equipment for low-temperature use, one is confronted with the basic material decision influenced by various compilations of properties available in industrial literature. Where the thermal and mechanical properties are not readily available, their preliminary estimates can be made using the simplified analytical methods [100].

THERMAL EFFECT ON STRENGTH

It should be recalled that in the majority of cases of solid materials, which do not undergo transitions, their strength in tension, hardness, and resistance to fatigue increase with a decrease in service temperature. This rule applies without reservation to such metals as aluminum, copper, nickel alloys, and austenitic stainless steels. In the case of ordinary carbon steels, however, the advantages of improved properties are seriously compromised by the materials' tendency to become brittle.

To get some idea as to the variation of strength properties at low temperatures, Fig. 17.1 shows the dependence of ultimate tensile strength on temperature for a few selected metals. The curves represent the upper and lower boundaries corresponding to the cold-worked and annealed test specimens [102]. The properties are plotted against temperature in Kelvin degrees.

There is only a limited amount of data on the low-temperature properties of plastics. However, it is known that when a large piece of unreinforced plastic is suddenly cooled, the resulting thermal shock can cause cracking. Many plastics,

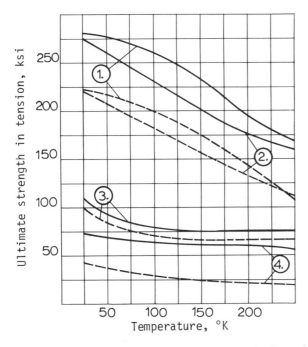

Fig. 17.1 Low-temperature ultimate strength of metals: (1) 304 stainless; (2) 347 stainless; (3) 2024 aluminum; (4) 6061 aluminum. Solid curves, cold-worked; dashed curves, annealed.

with the exception of Teflon, can become relatively brittle at lower temperatures, although these materials tend to resist thermal shock somewhat better than glass does.

It may be of interest to note that there is a certain similarity between the effects of low temperature and high strain rate. Experiments indicate that both effects tend to increase the yield and ultimate strength of steel while reducing the ductility. The observed influence is greater on the yield than on the ultimate strength, so that the relevant margin between the two properties is reduced. The usual stress-strain curve of the material under a high strain rate degenerates into a stress line parallel to the strain axis as if the material suddenly became completely plastic [40], with the ultimate strength markedly higher than that under static conditions.

MATERIALS FOR SPECIAL APPLICATIONS

In special applications involving nuclear propulsion systems, very high temperature strength is often required [103]. These may include structural components for fission reactors used as direct heat exchangers, nuclear propulsion systems, and aerospace reentry vehicles, to mention a few. Because of the extreme temperature environment, the choice of materials is very limited. In this class graphite, tungsten, and rhenium can be considered, although some of the carbides are also most valuable. Examples of the properties for refractory alloys are given in Table 17.2.

Some of the unusual properties for the selected carbides at room temperature are given in Table 17.3. Although these compounds are intended for highly specialized applications [104], it may be of interest to compare their mechanical properties with the typical data known for the structural materials in general use.

The effect of elevated temperatures on the percentage of retained strength for a variety of engineering materials is illustrated in Fig. 17.2. The diagram indicates the state-of-the-art knowledge in materials development for high-temperature applications [105].

Table 17.2 Selected Properties of Refractory Metals

Metal	Melting Point (°F)	Density (lb/in.^{3}) at 75°F	Modulus of Elasticity (psi)	NDT (°F)	Tensile Strength at 2200°F (psi)
Chromium	3450	0.76	42×10^6	625	8,000
Columbium (niobium)	4470	0.31	16×10^6	−185	10,000
Molybdenum	4730	0.37	47×10^6	85	22,000
Tantalum	5430	0.60	27×10^6	−320	15,000
Rhenium	5460	0.76	68×10^6	75	60,000
Tungsten	6170	0.70	58×10^6	645	32,000

Thermal Stresses and Materials

Table 17.3 Room-temperature Properties for Selected Carbides

Compound	Density (g/cm^3)	Coefficient of Thermal Expansion (10^{-6}/°C)	Modulus of Elasticity (10^6 psi)	Compressive Strength (10^3 psi)
Hafnium carbide (HfC)	12.7	6.0	61	—
Tantalum carbide (TaC)	14.5	6.5	55	—
Zirconium carbide (ZrC)	6.7	6.7	69	235
Niobium carbide (NbC)	7.8	6.5	49	—
Titanium carbide (TiC)	4.9	7.7	65	196
Tungsten carbide (WC)	15.8	5.0	102	900
Silicon carbide (SiC)	3.2	3.9	69	200
Boron carbide (B$_4$C)	2.5	4.5	42	420

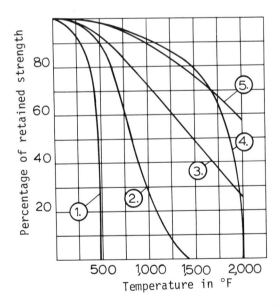

Fig. 17.2 Strength retention chart for special materials: (1) nylon; (2) "E" glass; (3) boron; (4) super alloys; (5) refractory metal alloys.

Table 17.4 Comparative Values of Thermal Stress Index for Typical Materials

Material	Ultimate Tensile Strength (psi)	Coefficient of Thermal Expansion (in./in. °F)	Thermal Conductivity (Btu/hr ft °F)	Modulus of Elasticity (psi)	Thermal Stress Index (Btu/ft hr)
Aluminum alloy, 24 ST	68,000	13×10^{-6}	75	10.4×10^6	37,800
Magnesium alloy	34,000	15×10^{-6}	45	6.5×10	15,300
Structural steel	60,000	7×10^{-6}	22	29×10^6	6,500
Nickel steel, A8	100,000	8×10^{-6}	15	30×10^6	6,250
Titanium	80,000	5.5×10^{-6}	10	16×10^6	9,100

THERMAL STRESS INDEX

In selecting the material for a particular application involving thermal effects, the *thermal stress index* can be employed. For the purpose of this discussion, the index will be denoted by TSI.

$$\text{TSI} = \frac{\sigma_u k}{\alpha E} \tag{17.3}$$

where σ_u is the ultimate tensile strength and E is the modulus of elasticity. The parameters k and α denote the thermal conductivity and the coefficient of linear expansion, respectively. Accordingly, the dimensions of the index must be Btu/ft hr.

The index may be defined as a qualitative measure of the ability of a given material to resist thermally induced stresses. It appears that the higher the index, the higher the ability of the material to resist thermal gradients. For instance, a few comparative values of TSI are given in Table 17.4.

THERMAL SHOCK

The index, given by Eq. (17.3) is particularly useful in correlating the material properties with reference to the resistance to fracture by thermal shock [105]. The maximum temperature that the material can withstand under the thermal shock conditions is dependent on the mechanical properties of the material but independent of thermal conductivity.

$$T_{\max} = \frac{\sigma_u(1-\nu)}{\alpha E} \tag{17.4}$$

Equations (17.3) and (17.4) are the necessary parameters for a complete determination of the thermal shock resistance of a given material. It is understood that

Thermal Stresses and Materials

in a quantitative determination of this particular property, thermal shock may be defined as the maximum sudden change in temperature that can be withstood by an infinite flat plate without fracture.

THERMAL CONDITIONS IN PIPING

In many practical cases of a long tube carrying a hot liquid and being cooled at the outer periphery, the temperature distribution can be established by evaluating heat balance across all the elemental rings of the cylinder.

When the amount of heat flow is specified, the temperature differential across the tube wall is

$$\Delta T = \frac{Q \ln(R_o/R_i)}{2\pi k} \qquad (17.5)$$

where R_o and R_i denote the outer and inner tube radii, respectively, and Q is the quantity of heat that must be conducted per unit length of tube. In terms of the stress parameters applicable to tubular members, the thermal stress index can be stated as follows:

$$\text{TSI} = \frac{Q}{4\pi(1-\nu)} \left(1 - \frac{2R_i^2}{R_o^2 - R_i^2} \ln \frac{R_o}{R_i}\right) \qquad (17.6)$$

This formula is useful in comparing brittle materials, and Eq. (17.6) shows that the material with the highest value of the thermal stress index will be able to withstand the highest amount of heat flow.

THERMAL STRESS FATIGUE

The brief discussion of the basic parameters given above applies to conventional thermal stresses and thermal shock conditions. The thermal stress fatigue, although similar to a conventional mechanical fatigue, involves a number of differences, which can be summarized as follows:

1. Plastic thermal strains tend to concentrate in the hottest regions of the body.
2. Accumulation of strain in thermal stress fatigue is localized.
3. Temperature cycling can have important effects on the material even without the mechanical fatigue loading.
4. Temperature and mechanical strains create superimposed effects and make it difficult to use low-cycle mechanical fatigue tests for the interpretation of thermal stress fatigue.
5. Tests indicate that for the same magnitude of total strain, thermal stress fatigue life can be markedly lower than the corresponding conventional life in mechanical fatigue.

PRELIMINARY THERMAL DESIGN

Since the solution of a thermal stress problem involves knowledge of temperature distribution in a structural member, thermal constants such as those given in Table 17.1 and the fundamentals of heat transfer in conduction, convection, and radiation [107] are required. In the usual engineering calculations, the modulus of elasticity and the coefficient of thermal expansion are assumed to be constant within a given temperature range. Some typical solutions to thermal stress problems due to the external and internal constraints are given in Tables 17.5 and 17.6 [108–110].

It is obvious from Tables 17.5 and 17.6 that only the simplest thermal stress problems have been addressed so far, and for any more complex cases the reader is referred to the specialized literature [2, 107]. Nevertheless, in many practical situations the formulas given in Tables 17.5 and 17.6 are sufficient for obtaining the results to bracket a particular design. It is of utmost importance to determine the bracketing values for a problem before embarking on any more complex and time-consuming investigation.

Table 17.5 Thermal Stresses Due to External Constraint

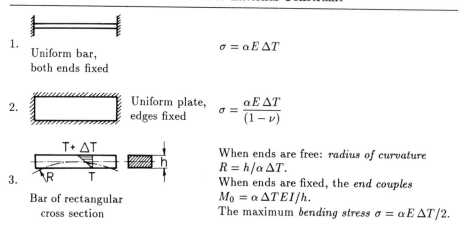

1. Uniform bar, both ends fixed
   $\sigma = \alpha E \Delta T$

2. Uniform plate, edges fixed
   $\sigma = \dfrac{\alpha E \Delta T}{(1 - \nu)}$

3. Bar of rectangular cross section
   When ends are free: *radius of curvature* $R = h/\alpha \Delta T$.
   When ends are fixed, the *end couples* $M_0 = \alpha \Delta T E I / h$.
   The maximum *bending stress* $\sigma = \alpha E \Delta T / 2$.

4. When instead of a bar of a rectangular cross section, a plate of thickness h is used, the radius of curvature is the same as that for case 3. The plate adopts a spherical curvature. The maximum bending stress $\sigma = \alpha E \Delta T / 2(1 - \nu)$.

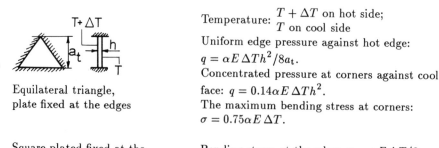

5. Equilateral triangle, plate fixed at the edges
   Temperature: $T + \Delta T$ on hot side; T on cool side
   Uniform edge pressure against hot edge: $q = \alpha E \Delta T h^2 / 8 a_t$.
   Concentrated pressure at corners against cool face: $q = 0.14 \alpha E \Delta T h^2$.
   The maximum bending stress at corners: $\sigma = 0.75 \alpha E \Delta T$.

6. Square plated fixed at the edges. Thermal gradient as above.
   Bending stress at the edge: $\sigma = \alpha E \Delta T / 2$ (approx.).

Thermal Stresses and Materials

Table 17.6 Thermal Stresses Due to Internal Constraint

1.

   Solid body of arbitrary shape

   Local gradient applied suddenly to a surface

   Compressive stress in surface layer:
   $\sigma = \alpha E \, \Delta T / (1 - \nu)$.

2.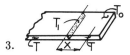

   Thin circular disk with heated central portion of radius a_0

   Maximum stress within heated zone:
   $\sigma = \alpha E \, \Delta T / 2$.
   Radial stress outside heated zone:
   $\sigma_r = \alpha E \, \Delta T \, a_0^2 / Sr^2$ (compression).
   Tangential stress outside heated zone:
   $\sigma_b = -\sigma_r$ (tension).
   Maximum shear stress at $r = a_0$:
   $\sigma_s = \alpha E \, \Delta T / 2$.

3.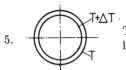

   Uniform heating across thickness and width at $x = 0$

   Tension along the edges at x:
   $\sigma_x = E\alpha(T - T_0)$.
   Maximum tensile stress: $\sigma = E\alpha(T_1 - T_0)$.
   Maximum compression stress at half-width of the plate: $\sigma_c = -E\alpha(T_1 - T_0)$.

4. The plate is heated as above, except that hotter surface has temp. T_2 and lower surface has temp. T_1. Min. temp. T_0.

   Maximum tensile stress at the edges, where $x = 0$:
   $$\sigma_x = \frac{\alpha E}{2}\left[T_1 + T_2 - 2T_0 + \frac{(1-\nu)}{(3+\nu)}(T_1 - T_2)\right]$$

5.

   Thin-walled tube linear gradient

   Maximum hoop stress: $\sigma = \alpha E \, \Delta T / 2(1 - \nu)$ (inner surface in compression, tension outside).
   Maximum longitudinal stress: $\sigma = \alpha E \, \Delta T / 2(1 - \nu)$ (compression inside, tension outside).

6. Hollow sphere

   Rate of surface temp. increase, m in deg/hour

   $$\sigma_r = \frac{\alpha m E}{15\lambda(1-\nu)}\left(\phi - \psi - r^2 - \frac{5a^3}{r}\right)$$

   $$\sigma_t = \frac{\alpha m E}{15\lambda(1-\nu)}\left(\phi + 0.5\psi - 2r^2 - \frac{5a^3}{2r}\right)$$

   $$\phi = \frac{b^5 + 5b^2 a^3 - 6a^5}{b^3 - a^3}$$

   $$\psi = \frac{b^5 a^3 - 6b^3 a^5 + 5b^2 a^6}{r^3(b^3 - a^3)}$$

SYMBOLS

a	Inner radius of sphere, in. (mm)
a_0	Radius of heated area, in. (mm)
a_t	Height of triangular plate, in. (mm)
b	Outer radius of sphere, in. (mm)
C	Specific heat, Btu/lb °F*
E	Modulus of elasticity, psi (N/mm^2)
h	Thickness of plate or bar, in. (mm)
I	Moment of inertia, in.^{4} (mm^4)
k	Thermal conductivity, Btu/hr ft °F
m	Temperature increase in °F/hr
M_0	Fixing couple, lb-in. (N-mm)
Q	Heat transfer per unit length of tube, Btu/ft
q	Edge pressure, lb/in (N/mm); corner load, lb (N)
R	Radius of curvature, in. (mm)
R_i	Inner radius of tube, in. (mm)
R_o	Outer radius of tube, in. (mm)
r	Arbitrary radius, in. (mm)
T_0, T_1, T_2	Temperatures, °F
T_{max}	Maximum temperature under thermal shock, °F
TSI	Thermal stress index, Btu/ft hr
ΔT	Temperature gradient, °F
V	Volume, in.^{3} (mm^3)
Δv	Volume change, in.^{3} (mm^3)
x	Arbitrary distance, in. (mm)
α	Linear coefficient of thermal expansion, °F^{-1}
λ	Thermal diffusivity, ft^2/hr
ν	Poisson's ratio
σ	General symbol for stress, psi (N/mm^2)
σ_c	Compressive stress, psi (N/mm^2)
σ_r	Radial stress, psi (N/mm^2)
σ_s	Shear stress, psi (N/mm^2)
σ_t	Tangential stress, psi (N/mm^2)
σ_u	Ultimate strength, psi (N/mm^2)
σ_x	Stress at any distance x, psi (N/mm^2)
ϕ	Auxiliary constant for sphere, in.^{2} (mm^2)
ψ	Auxiliary constant for sphere, in.^{2} (mm^2)

---

*The majority of industrial organizations in this country and elsewhere are still using the English system of heat units. However, the SI system defines the unit of thermal energy as the joule (J), where 1000 J is equivalent to 0.9478 Btu.

III
STRAIGHT MEMBERS

18
General Design Criteria

INTRODUCTION

Design of straight members can be based on the maximum stress criteria, stiffness characteristics, or stability, depending on the proportions of the members and the manner of loading. Straight members are by far the most frequently encountered in engineering calculations and a substantial amount of published data is available in the literature. The material selected for this chapter is governed by design office experience. Certain topics and types of problems appear to attract more attention than others when the student leaves the classroom environment and is confronted with developing practical answers in a design office. This chapter and other portions of the book are dedicated to reviewing those aspects of structural and mechanical design which have been found to be of predominant interest. The recommended formulas and techniques generally lead to conservative predictions as long as the criteria of elastic design are employed.

The basic elements of strength, stiffness, and stability have been briefly outlined in Part I. The material presented in this and the remaining portions of the book may be termed the application analysis. Only in special cases the theoretical developments leading to formula derivations are included where they are absolutely essential to a better understanding of the principles involved.

The definition of straight members refers here to all the standard beams, thick columns, shafts, and similar axially symmetric components having either constant or gradually changing cross-sectional areas which can be treated with the aid of a simple beam or column theory. The type of loading normally includes axial tension, axial compression, torsion, or bending. The resulting effects may be classified as stresses, elongation, twist, or transverse bending deflection. In the majority of the design cases, the engineer ends up specifying either the maximum stress or deflection criteria.

PRACTICAL RULES

The design of axially loaded members is often based on the normal stresses although the actual failure can occur on an inclined plane. At the same time the critical stress is expected at the minimum cross-sectional area unless an abrupt change in geometry or strength is involved.

The straight members subjected to torsion are likely to be designed according to the specified shear allowable. However, since the shearing stress can be interpreted in terms of the normal principal stress, the structural member should, at times, be checked for tension. For instance, a cast-iron part weaker in tension than shear can fail in tension despite the fact that the part is subjected to twist.

The design of a straight member in bending can be based on the concept of the section modulus alone provided that this member does not have to resist shear loading in addition to bending. When shear is present it may be necessary to use the theory of combined stresses.

Several practical rules in beam design may be summed up as follows:

1. Maximum and minimum bending moments occur where shear is zero.
2. No bending moment can be transmitted at a pin joint.
3. No deflection is possible at a hinged support.
4. Deflection and rotation are zero at fixed supports.
5. Complex deflections of beams can be obtained by superposition.
6. Truly fixed-end conditions as well as truly hinged supports are never found in practice.

ULTIMATE STRENGTH OF BEAMS

Although the maximum fiber stress in the elastic regime is generally recognized as the primary criterion of safety in beam design, the question is sometimes raised as to the ultimate strength and the reserve of load-carrying capacity beyond the elastic point. This question, as related to beam design, is particularly important because beam elements are often used in various structural models. Under such conditions we make the assumption that an individual beam element cannot deform indefinitely or collapse until the full plastic moment M_p is reached at some critical section. Once it is assumed that a given plastic moment acts at the critical point, the problem becomes fully determinate.

As the elastic limit of the material is attained, such as that, for instance, shown in Fig. 1.2 or 1.3, the shape of the stress-strain curve becomes nonlinear until the material finally reaches the ultimate stress condition corresponding to the maximum load-carrying capacity of the structure. The form of the true stress-strain curve can be approximated by a suitable mathematical expression involving the true stress at

General Design Criteria

the beginning of the plastic flow, true engineering strain, and the strain hardening coefficient.

The elementary concepts of stress and strain given in Chap. 1 refer to the engineering as well as true definitions of stress and strain. The designer is primarily interested in the original cross-sectional area needed to carry a certain load. However, when the structural member is stressed well beyond the elastic region, the relevant cross-sectional area may undergo certain changes in nominal dimensions. It was noted in Chap. 1 that a round bar in tension can experience necking so that the true stress is obtained by dividing the tensile force by the actual cross-sectional area. The true strain is based on the idea that a small increment of deflection should be divided by the actual rather than the original gage length. If required this strain can be found from Eq. (1.22).

The elementary theory of bending outlined in Chap. 3 shows that $M_y = IS_y/C$, where, according to Eq. (3.1), S_y denotes the yield strength of the material. If we now assume that with further increase in bending moment the yielding progresses at a constant stress, then for the condition of full plasticity, M_p will denote the plastic bending moment. The perfectly plastic material, often used in the calculations, shows no work hardening. The stress-strain diagram for this material shows two straight lines. The first line represents the elastic response, which is followed by the line parallel to the strain axis.

The analysis of beam collapse under transverse loading is beyond the scope of this book and is more apt to fall in the province of civil engineers concerned with rigorous as well as approximate methods of load calculations. However, it may be useful to characterize some of the typical beam cross sections in terms of a moment ratio M_p/M_y, where M_y corresponds to the yield strength of the material. Here, the material is assumed to be perfectly plastic, indicating yielding at a constant stress. Under ultimate load conditions, corresponding to the plastic moment, M_p, the plastic flow progresses toward the central axis of the cross section, and the corresponding stress distribution is rectangular instead of the customary triangular shape common to pure bending of beams in the elastic regime.

The magnitude of the moment ratio M_p/M_y signifies the reserve of strength of the particular beam in bending. For example, the theory of plasticity shows that for a rectangular cross section and perfectly plastic material, this ratio is equal to 1.5. Hence, to bring about structural collapse of the beam, the external bending moment should be 50% greater than that corresponding to the yield of the external fiber. Examples of the moment ratios for other cross-sectional geometry are given in Table 18.1.

In the case of a standard I beam, the calculations show that the moment ratio varies between 1.15 and 1.17. Therefore, if rectangular and I-beam sections are designed to the same factor of safety based on the onset of yield, the rectangular beam should have a larger reserve of strength than an I beam.

Table 18.1 Bending Moment Ratios for Typical Beam Cross Sections

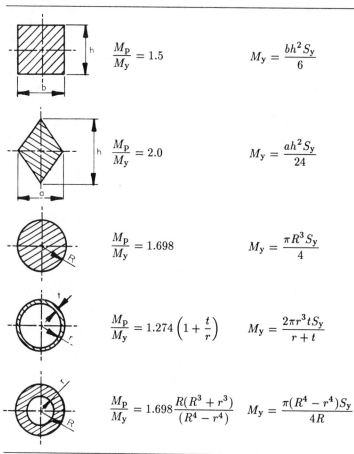

SYMBOLS

- a Width of diamond shape, in. (mm)
- b Width of rectangular cross section, in. (mm)
- C Distance from neutral axis to outer fiber, in. (mm)
- h Depth of beam section, in. (mm)
- I Moment of inertia, in.^{4} (mm^4)
- M_p Plastic bending moment, lb-in. (N-mm)
- M_y Bending moment at yield, lb-in. (N-mm)
- R Outer radius, in. (mm)
- r Inner radius, in. (mm)
- S_y Yield strength, psi (N/mm^2)
- t Wall thickness, in. (mm).

19
Axial Response of Straight and Tapered Bars

RESILIENCE OF SOLID BAR

The simplest component of a structural system is a straight bar of uniform cross section subjected to collinear loads coinciding with the longitudinal axis of the bar and inducing either tension or compression.

The elongation of a bar in tension, such as that shown in Fig. 19.1, may be given directly in terms of the bar length L, modulus of elasticity E, and resilience U, which defines the amount of elastic energy stored in a cubic inch of the material.

$$\Delta L = L \left(\frac{2U}{E} \right)^{1/2} \tag{19.1}$$

It is assumed here that the material is not strained beyond the elastic limit. The corresponding tensile stress is

$$S = (2UE)^{1/2} \tag{19.2}$$

The axial load is

$$W = 1.1 d^2 (UE)^{1/2} \tag{19.3}$$

The amount of energy U required in these calculations can be used for comparing the material properties provided that they have a finite elastic limit. For instance, U for a typical structural steel can be 30 times larger than that for a soft copper.

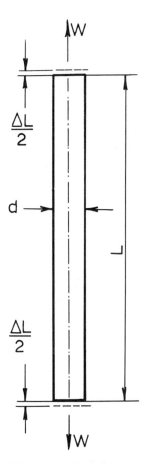

Fig. 19.1 Round bar in tension.

TAPERED AND STEPPED BARS

When a solid tapered bar of a circular cross section, shown in Fig. 19.2, is subjected to axial load W, the general formula for the total elongation, neglecting own weight of the bar, may be derived as follows.

Let $d(\Delta L)$ represent the elongation of the disklike element, having thickness dy. The mean radius for this element follows from the considerations of similar triangles CFH and ABH, so that

$$\frac{CF}{y} = \frac{AB}{L}$$

And since $CF = r - d/2$, and $AB = (D - d)/2$, solving the foregoing proportion for the mean radius r gives

$$r = \frac{d}{2} + \frac{y(D - d)}{2L}$$

Axial Response of Straight and Tapered Bars

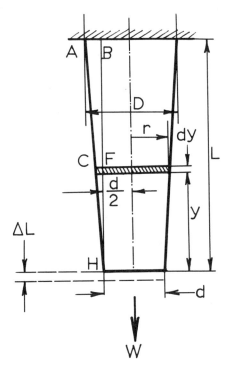

Fig. 19.2 Tapered bar under tensile end load.

For average cross-sectional area πr^2, the elongation of the single element becomes

$$d(\Delta L) = \frac{W\,dy}{\pi \left[\frac{d}{2} + \frac{y(D-d)}{2L}\right]^2 E}$$

Integrating this expression between the limits of zero and L with respect to the only variable, y, gives the general formula

$$\Delta L = \frac{4WL}{\pi D\,d E} \tag{19.4}$$

It is of interest to note that when $d = D$, Eq. (19.4) reduces to Eq. (19.1). However, the reader must be cautioned against extrapolating this formula to $D = 0$ or $d = 0$. This is one of the first practical considerations in stress analysis, indicating that certain formulas may have no meaning in the extreme cases, although our general aim is to derive such an expression for which the lower and upper values can be rationalized without incurring undue mathematical difficulties. In most cases practical formulas are good for specified ranges of variables and have some definite physical significance.

Design Problem 19.1

Determine the total amount of the elastic strain energy stored in a stepped bar subjected to a tensile load of 100,000 lb, as shown in Fig. 19.3. Assume the modulus of elasticity to be 30×10^6 psi. The weight of the bar is neglected.

Solution

Calculate stresses for the two portions separately as S_1 and S_2.

$$S_1 = 100{,}000 \Big/ \frac{\pi \times 3^2}{4} = 14{,}150 \text{ psi} \quad (97.6 \text{ N/mm}^2)$$

$$S_2 = 14{,}150 \times 9 = 127{,}350 \text{ psi} \quad (878.3 \text{ N/mm}^2)$$

The total strain energy is now obtained as the algebraic sum of the energies for the two separate portions. The corresponding areas are $A_1 = 9\pi/4$ and $A_2 = \pi/4$. Hence,

$$\begin{aligned} U_t &= \frac{S_1^2 L_1 A_1}{2E} + \frac{S_2^2 L_2 A_2}{2E} \\ &= \frac{14{,}150^2 \times 7 \times 9\pi + 127{,}350^2 \times 5 \times \pi}{4 \times 2 \times 30 \times 10^6} = 1227 \text{ lb-in.} \quad (138{,}632 \text{ N-mm}) \end{aligned}$$

It is seen that the step-down in diameter indicated in Fig. 19.3 causes a ninefold increase in the maximum tensile stress. If we made the bar 12 in. in length and 3 in. in diameter, the corresponding total strain energy would go down to about 282 lb.-in. The energy reduction here appears to have no relation to the ninefold change in stress. If we further

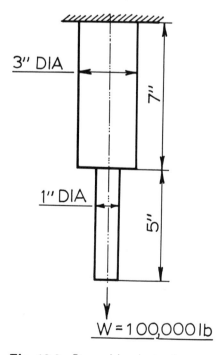

Fig. 19.3 Stepped bar in tension.

Axial Response of Straight and Tapered Bars

compare the maximum stresses on the basis of the equal amount of energy for the case of a stepped bar shown in Fig. 19.3 and a 12 in. bar of 1 in. diameter, the following stress relation is obtained:

$$U_t = \frac{S^2 AL}{2E} \quad \text{[see Eq. (6.2)]}$$

Then for $A = A_2 = \pi/4$ and $U_t = 1227$ lb-in., we get

$$S = \left(\frac{2EU_t}{A_2 L}\right)^{1/2} = \left(\frac{2 \times 30 \times 10^6 \times 1227 \times 4}{12\pi}\right)^{1/2} = 88{,}382 \text{ psi} \quad (609.5 \text{ N/mm}^2)$$

We have to compare this result with the 127,350 psi calculated for the stepped bar, which shows that for the case of the same energy, the maximum stress is increased for the stepped member by about 44%. It appears, then, that on the basis of energy the provision of the cross-sectional enlargement may actually be detrimental. The foregoing considerations do not include stress concentration. The assumption is made that a sufficiently mild transition can be machined for the purpose of stress mitigation. Other considerations of energy storage criteria are given in Chap. 6. ♦

TAPERED BAR UNDER OWN WEIGHT

To calculate the stress and elongation for a truncated solid cone under the action of its own weight, reference is made to Fig. 19.4. Utilizing the standard expression for the volume of a truncated cone, we get

$$Q_y = \frac{\pi \gamma y}{3} \left\{ \left[\frac{Ld + y(D-d)}{2L}\right]^2 + \frac{d}{2}\left[\frac{Ld + y(D-d)}{2L}\right] + \frac{d^2}{4} \right\}$$

Since the tensile stress at any section defined by y is

$$S = \frac{Q_y}{\pi r^2}$$

Putting $r = D/2$ and $y = L$ gives

$$S = \frac{\gamma L}{3D^2}(Dd + D^2 + d^2) \tag{19.5}$$

When $d = D$, that is, when the conical bar transforms into a bar of uniform cross section, we obtain the following useful stress formula:

$$S = \gamma L \tag{19.6}$$

When d becomes negligibly small, putting $d = 0$ in Eq. (19.5) yields

$$S = \frac{\gamma L}{3} \tag{19.7}$$

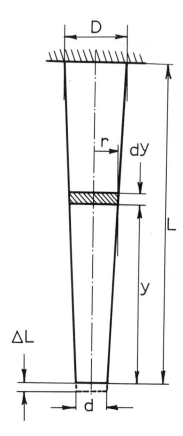

Fig. 19.4 Notation for a conical bar under its own weight.

Hence, we find another important practical result: the stress for the cone is equal to one-third of that in a circular bar, which is evident from Eqs. (19.6) and (19.7). The same comments also apply to a pyramid configuration.

If the dislike element shown in Fig. 19.4 has the thickness dy and average cross-sectional area of πr^2, the elongation of a single element becomes

$$d(\Delta L) = \frac{\gamma}{3} \left\{ \frac{3yL^2d^2 + 3y^2Ld(D-d) + y^3(D-d)^2}{[Ld + y(D-d)]^2 E} \right\} dy$$

To sum up all the elementary elongations given by the preceding equation, the integration may be carried out with the aid of the substitution

$$Z = Ld + y(D-d)$$

from which

$$dy = \frac{dZ}{D-d}$$

Axial Response of Straight and Tapered Bars

After these substitutions are made, the integration can be performed according to the standard rules of calculus, bearing in mind that the relevant integration limits of 0 and L must first be changed accordingly. When $y = 0$

$$Z = Ld$$

When $y = L$

$$Z = Ld + L(D - d) = LD$$

With the limits so altered, we obtain

$$\Delta L = \frac{\gamma L^2}{6D(D-d)^2 E}(D^3 + 2d^3 - 3Dd^2) \tag{19.8}$$

Checking the dimensions yields

$$\frac{\text{lb}}{\text{in.}^3} \times \frac{\text{in.}^2}{\text{in.} \times \text{in.}^2} \times \text{in.}^3 \times \frac{\text{in.}^2}{\text{lb}} = \text{in.}$$

DISCUSSION OF TAPERED BAR FORMULA

Additional check after such a derivation should involve reducing the more general expressions, such as Eq. (19.8), to any simpler and better known formulations [21]. For instance, making $d = 0$, Eq. (19.8) reduces to a standard formula for the total elongation of a sharp-pointed conical bar,

$$\Delta L = \frac{\gamma L^2}{6E} \tag{19.9}$$

However, further interpretation of Eq. (19.8), aimed at providing a simple formula for a bar of constant cross section, necessitates the condition $D = d$, which leads to an indeterminate form where both numerator and denominator vanish. This difficulty can be eliminated through the application of a powerful mathematical technique known as L'Hospital's rule [111]. The method usually involves one or more differentiations of the numerator and denominator individually until the form of the quotient becomes determinate. To illustrate this, consider Eq. (19.8). The first differentiation with respect to d gives

$$\frac{6d^2 - 6Dd}{6D(-2D + 2d)} = \frac{0}{0} \quad \text{for } d = D$$

Second differentiation results in

$$\frac{12d - 6D}{12D} = \frac{1}{2} \quad \text{for } d = D$$

Hence, the required formula for a bar of uniform cross section is

$$\Delta L = \frac{\gamma L^2}{2E} \tag{19.10}$$

This method of obtaining the formula was selected to merely check Eq. (19.8) by reducing it to a standard expression. Equation (19.10) could have been derived much more simply by starting with a bar of uniform cross section. The reader should follow the development of Eq. (19.8) as a matter of exercise in order to recall the potential mathematical complications that arise with closed-form solutions whenever a straight-line taper is involved. This is universally true in simpler as well as more advanced stress analysis problems, including, for instance, the analysis of a conical shell. Whenever possible, then, the taper should be represented by some approximate mathematical function, and considerable engineering judgment should be exercised in avoiding undue complications. There are many practical design situations under which the stress and elongation of structural members due to their own weight become of great importance.

Design Problem 19.2

Two long wires, made of aluminum and bronze, are suspended vertically. Calculate the length and the corresponding elongation of both wires for which the elastic limit of the material is not exceeded. The elastic limit and density of bronze are 8000 psi and 0.31 lb/in.^{3}, respectively. The corresponding values for wrought aluminum are assumed here as 7000 psi and 0.098 lb/in.^{3}. The relevant moduli of elasticity are 13×10^6 psi and 10×10^6 psi, respectively.

Solution

From Eq. (19.6),

$$L = \frac{S}{\gamma} = \frac{8000}{0.31} = 25{,}800 \text{ in.} = 2150 \text{ ft} \quad (655.32 \text{ m}) \quad \text{for bronze}$$

$$L = \frac{7000}{0.098} = 71{,}400 \text{ in.} = 5950 \text{ ft} \quad (1813.56 \text{ m}) \quad \text{for aluminum}$$

Equations (19.6) and (19.10) give

$$\Delta L = \frac{S^2}{2\gamma E} = \frac{8000 \times 8000}{2 \times 0.31 \times 13 \times 10^6} = 7.9 \text{ in.} \quad (200.66 \text{ mm}) \quad \text{for bronze}$$

and

$$\Delta L = \frac{7000 \times 7000}{2 \times 0.098 \times 10 \times 10^6} = 25 \text{ in.} \quad (635 \text{ mm}) \quad \text{for aluminum} \quad \blacklozenge$$

Design Problem 19.3

A string of steel piping intended for downhole emplacement consists of a 200-ft length of approximately 4.5 in. mean radius and 0.45 in. wall thickness with a 400-ft section of

Axial Response of Straight and Tapered Bars

5.7 in. mean radius and 0.60 in. wall. Calculate the total elongation and the maximum tensile strength when a string is suspended vertically, assuming a 100,000-lb weight at the bottom of the string. The modulus of elasticity and the density may be taken as 30×10^6 psi and 0.283 lb/in^3, respectively. The system is illustrated schematically in Fig. 19.5 using different scales for length and diameters.

The flanged joints of various pipe sections are not indicated and may be ignored in this calculation.

Solution

The formula for weight of piping of uniform cross section is

$$W = 2\pi RTL\gamma \tag{19.11}$$

where

R = mean radius of pipe, in.
T = wall thickness, in.
L = pipe length. in
γ = weight density, lb/in.^{3}

Using Eq. (19.11) for the upper portion of the string yields

$$W = 2 \times \pi \times 5.7 \times 0.6 \times 400 \times 12 \times 0.283 = 29{,}190 \text{ lb}$$

Similarly, for the lower portion

$$W = 2 \times \pi \times 4.5 \times 0.45 \times 200 \times 12 \times 0.283 = 8640 \text{ lb}$$

Hence, the total weight at the top is

$$W = 29{,}190 + 8640 + 100{,}000 = 137{,}830 \text{ lb}$$

The corresponding tensile stress at the top of the pipe is

$$S = \frac{137{,}830}{2 \times \pi \times 5.7 \times 0.6} = 6414 \text{ psi} \quad (44.22 \text{ N/mm}^2)$$

The stress at the top of the 200-ft section (Fig. 19.5) is

$$S = \frac{108{,}640}{2 \times \pi \times 4.5 \times 0.45} = 8538 \text{ psi} \quad (58.87 \text{ N/mm}^2)$$

Since the stress here is uniaxial and consists of the two component parts, one formula can be written for the whole length of the pipe using the method of superposition.

$$S = \frac{W}{A} + \gamma \left(y - L_1 + L_1 \frac{A_1}{A} \right)$$

Here y is the distance measured in inches from the bottom of the pipe string, as shown in Fig. 19.5, and A is the pipe cross-sectional area at the location considered. The cross-sectional areas and lengths are denoted by A_1, L_1 and A_2, L_2 for the lower and upper

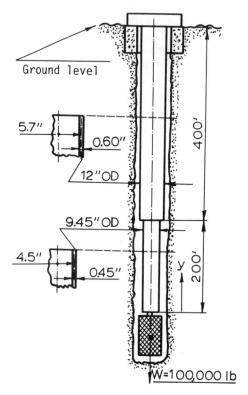

Fig. 19.5 Example of downhole pipe string.

portions of the pipe, respectively. For instance, when $y = L_1$ and $A = A_1$, the tensile stress is obtained at the junction of the two sections of the pipe. When $y = L_1 + L_2$ and $A = A_2$, we obtain the tensile stress at the top of the string. The total extension of the string can also be calculated with the aid of the method of superposition using the foregoing notation.

Let W_1 represent the weight of the lower portion of the pipe. The elongation of the lower portion is due to the pipe's own weight in this region and the weight of the package attached to it. Hence,

$$(\Delta L)_1 = \frac{W L_1}{A_1 E} + \frac{\gamma L_1^2}{2E}$$

The elongation of the upper portion of the pipe consists of three component values: (1) due to package weight, (2) due to weight of lower portion, and (3) due to its own weight. Hence,

$$(\Delta L)_2 = \frac{(W + W_1)L_2}{A_2 E} + \frac{\gamma L_2^2}{2E}$$

Axial Response of Straight and Tapered Bars

The total elongation at the bottom of the string, assuming the package to be infinitely rigid, is therefore

$$\Delta L = (\Delta L)_1 + (\Delta L)_2$$
$$= \frac{W}{E}\left(\frac{L_1}{A_1} + \frac{L_2}{A_2}\right) + \frac{W_1 L_2}{A_2 E} + \frac{\gamma}{2E}\left(L_1^2 + L_2^2\right)$$

The corresponding cross-sectional areas in this case are

$$A_1 = 2 \times 3.14 \times 4.5 \times 0.45 = 12.7 \text{ in.}^2$$
$$A_2 = 2 \times 3.14 \times 5.7 \times 0.60 = 21.4 \text{ in.}^2$$

Hence, substituting the relevant numerical data yields

$$\Delta L = \frac{100{,}000}{30 \times 10^6}\left(\frac{200 \times 12}{12.7} + \frac{400 \times 12}{21.4}\right) + \frac{8640 \times 400 \times 12}{21.4 \times 30 \times 10^6}$$
$$+ \frac{0.283 \times 144}{2 \times 30 \times 10^6}(200^2 + 400^2)$$
$$= 1.58 \text{ in.} \quad (40.13 \text{ mm}) \quad \blacklozenge$$

The foregoing method of analysis can be extended to any number of steps. Hence, in the more general case where we have to deal with a rod of variable cross section, for which a proper variation of the cross section with length is difficult to define, the rod shape may be approximated by a finite number of elements. The deflection can then be computed for all the individual elements and added directly to obtain the complete elongation.

Design Problem 19.4

An aluminum tube of 1 in. mean radius and 0.25 in. wall thickness is compressed between two rigid blocks by means of a long bolt of 0.75 in. outer diameter as shown in Fig. 19.6.

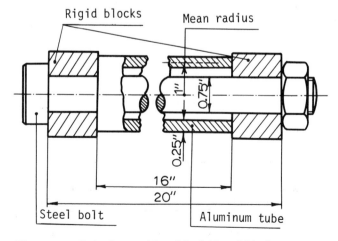

Fig. 19.6 Bolted assembly with rigid end blocks.

The tensile stress in the bolt on tightening of the nut is 10,000 psi. Calculate the tensile force in the bolt when a load of 1000 lb is applied to the rigid blocks, tending to pull them apart. Take the moduli of elasticity of 30×10^6 psi and 10×10^6 psi for steel and aluminum, respectively.

Solution

The bolt and tube cross-sectional areas are

$$A_b = \frac{\pi \times 0.75^2}{4} = 0.44 \text{ in.}^2$$

$$A_t = 2\pi \times 1 \times 0.25 = 1.57 \text{ in.}^2$$

The original bolt load is

$$W = SA_b = 10{,}000 \times 0.44 = 4400 \text{ lb}$$

The tube decreases in length under this load according to Hooke's law:

$$\Delta L = \frac{WL}{A_t E} = \frac{4400 \times 16}{1.57 \times 10 \times 10^6} = 0.0045 \text{ in.}$$

After the force is applied to the end blocks, let F represent the tensile load on the bolt, so that the additional extension of the bolt becomes

$$\delta = \frac{(F - 4400)20}{0.44 \times 30 \times 10^6}$$

The compressive force on the tube must then be decreased by the amount

$$\delta' = \frac{(F - 1000)16}{1.57 \times 10 \times 10^6}$$

Hence, the resultant compression of the tube is equal to the additional bolt extension

$$0.0045 - \frac{(F - 1000)16}{1.57 \times 10 \times 10^6} = \frac{(F - 4400)20}{0.44 \times 30 \times 10^6}$$

from which

$$F = 4800 \text{ lb} \quad (21{,}351.36 \text{ N})$$

In summary, then,

$$\Delta L - \delta' = \delta$$

The sign ahead of δ' is minus since the amount of axial compression of the tube must be reduced.

Another method of solving Design Problem 19.4 is used by most mechanical engineers concerned with fastener analysis. The method is recommended by the Industrial Fastener Institute, Cleveland, Ohio. The analytical approach [251, 252] is based on considering spring constants of the components bolted together as shown in Fig. 19.6. The derivation steps are as follows:

Stiffness constant:

$$k = \frac{F}{\delta} = \frac{AE}{l}$$

For bolt:

$$k_b = \frac{0.44(30 \times 10^6)}{20} = 6.6 \times 10^5$$

For tube:

$$k_t = \frac{1.57(10 \times 10^6)}{16} = 9.81 \times 10^5$$

$$\frac{k_t}{k_b} = \frac{9.81 \times 10^5}{6.6 \times 10^5} = 1.4864$$

$$k_t = 1.4864 k_b$$

$$F = \frac{k_b F'}{k_b + k_t} + F_i$$

Thus,

$$F = \frac{k_b F'}{k_b + 1.4864 k_b} + F_i = \frac{F'}{2.4864} + F_i$$

$$= \frac{1000}{2.4864} + 4400$$

$$= 4800 \text{ lb} \quad \blacklozenge$$

ANALYSIS OF COMPOSITE BARS

Design Problem 19.4 illustrates practical applications, where we have to take into account the relative rigidity of the various component members of the assembly. For instance, when a flanged joint is considered, the elastic response of the gaskets is included in the analysis, since any tension applied to the joint will be balanced by extra tension in the bolts and partial reduction of compression in the gaskets. The reverse would be true if an additional compressive force were applied to the joint, increasing gasket compression and at the same time decreasing bolt tension.

Another stress problem area, which utilizes the elementary version of Hooke's law, concerns the behavior of a composite bar subjected to mechanical loading. Consider here a system consisting of two concentric tubes, rigidly connected at the ends as shown in Fig. 19.7. The amount of longitudinal compression is the same for both tubes, hence according to Hooke's law

$$\frac{\Delta L}{L} = \frac{S_1}{E_1} = \frac{S_2}{E_2}$$

The total load on the system is

$$W = S_1 \times 2\pi RT + S_2 \times 2\pi rt$$

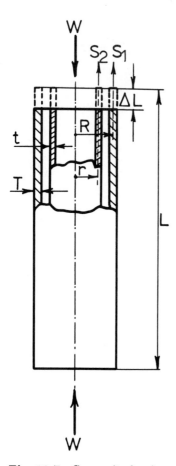

Fig. 19.7 Composite bar in compression.

Solving the foregoing relations for S_1 and S_2 yields

$$S_1 = \frac{WE_1}{2\pi(E_1 RT + E_2 rt)} \tag{19.12}$$

and

$$S_2 = \frac{WE_2}{2\pi(E_1 RT + E_2 rt)} \tag{19.13}$$

or in more general terms, putting $A_1 = 2\pi RT$ and $A_2 = 2\pi rt$, we get

$$S_1 = \frac{WE_1}{A_1 E_1 + A_2 E_2} \tag{19.14}$$

and

$$S_2 = \frac{WE_2}{A_1 E_1 + A_2 E_2} \tag{19.15}$$

Axial Response of Straight and Tapered Bars

When the materials are the same, that is, $E_1 = E_2$, the two stresses become equal, as evident from the above formulas. The application of this principle may be found in reinforced concrete design where steel rods and concrete are constrained to contract or extend equally. Another example concerns a rigid weight suspended by a number of wires held horizontally in a given position so that L remains the same for all the supporting wires.

KERN LIMIT

Various structural members discussed in this chapter so far were relatively short and rigid, so that any axial compressive loading was unlikely to produce buckling. When such a loading is applied eccentrically, the resulting stresses may be obtained by superposition of direct compression and flexural stresses.

In this section we briefly consider the concept of *kern*, sometimes referred to as "kernel" or "core of the cross section." Kern can be defined as the area in the plan of the section through which the line of action of the external force should pass to assure the same kind of normal stress at all points of the cross section. Consequently, the *kern limit* may be defined as a characteristic dimension of the central portion of the cross section, or the locus of points, within which the line of action of the external force should fall. The kern limit concept is useful in designing short prisms, columns, and piers subjected to eccentric thrust. Such a thrust causes a direct axial stress and a bending stress which can be superimposed as long as the response is purely elastic. As a rule, such stresses are not critical in the customary design of machine elements and steel members of various cross-sectional geometry, provided that local buckling is not a problem. However, if a structural member is made of a material that is good in compression but poor in tension, such as is the case with masonry columns, the analysis based on the concept of kern limit should be made [112].

The principle of kern limit design can be explained with reference to Fig. 19.8. Suppose that the external force P is applied at a distance a measured from the centroid of the prism having a symmetrical cross section, as indicated in Fig. 19.8. For simplicity of the derivation, the thrust loading is offset with respect to one axis of the cross section only. The uniform compressive stress is

$$S_c = \frac{P}{bh}$$

The tensile stress component due to offset bending is

$$S_t = \frac{6Pa}{bh^2}$$

In general, the actual stress distribution can be of the type shown in Fig. 19.8. It means that for a certain value of a, a portion of the cross section can be in a state of tension when the direct and bending stresses are superimposed. However, to assure that the combined stress is compressive at all points of the cross section, such as the one indicated in Fig. 19.8 as "desired stress distribution," it is necessary

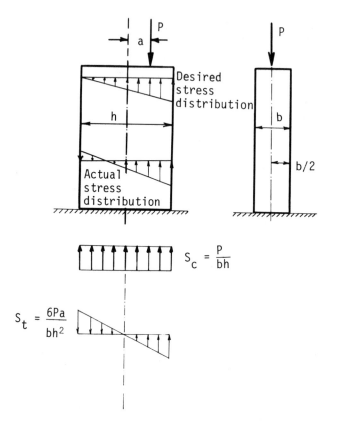

Fig. 19.8 Kern limit concept.

to satisfy the following condition:

$$S_t - S_c = 0$$

Hence, putting $a = e$ gives

$$\frac{6Pe}{bh^2} = \frac{P}{bh}$$

from which

$$e = \frac{h}{6}$$

This value is shown in Table 19.1, case 1, as one of the dimensions defining a diamond-shaped area characteristic of kern for a rectangular cross section of the structural member. If the line of thrust of the force P is offset with respect to both axes of symmetry of the cross section, the theoretical basis for the calculation follows directly from the foregoing derivation. The geometries selected for Table 19.1 cover a great many practical applications, giving relatively simple expressions for defining kern limits. For the case of a prism or a column of arbitrary cross section, the

Axial Response of Straight and Tapered Bars

Table 19.1 Kern Limits for Compression Members

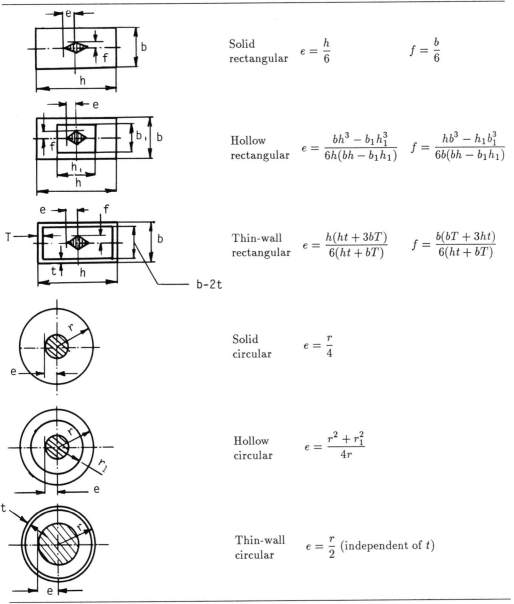

Shape	e	f
Solid rectangular	$e = \dfrac{h}{6}$	$f = \dfrac{b}{6}$
Hollow rectangular	$e = \dfrac{bh^3 - b_1 h_1^3}{6h(bh - b_1 h_1)}$	$f = \dfrac{hb^3 - h_1 b_1^3}{6b(bh - b_1 h_1)}$
Thin-wall rectangular	$e = \dfrac{h(ht + 3bT)}{6(ht + bT)}$	$f = \dfrac{b(bT + 3ht)}{6(ht + bT)}$
Solid circular	$e = \dfrac{r}{4}$	
Hollow circular	$e = \dfrac{r^2 + r_1^2}{4r}$	
Thin-wall circular	$e = \dfrac{r}{2}$ (independent of t)	

Note: Kern limit areas are defined by shaded portions of sections.

analysis of kern limits is much more involved and requires a semigraphical procedure with successive approximations.

SYMBOLS

A, A_1, A_2	Cross-sectional areas, in.² (mm²)
A_b	Bolt cross-sectional area, in.² (mm²)
A_t	Tube cross-sectional area, in.² (mm²)
a	Offset of end load, in. (mm)
b, b_1	Widths of rectangular cross sections, in. (mm)
D	Maximum diameter of bar, in. (mm)
d	Minimum diameter of bar, in. (mm)
E, E_1, E_2	Elastic moduli, psi (N/mm²)
e	Kern limit, in. (mm)
F	Bolt load, lb (N)
F_i	Initial bolt load, lb (N)
F'	Load on rigid block, lb (N)
f	Kern limit, in. (mm)
h, h_1	Depths of rectangular cross sections, in. (mm)
k	Spring constant of assembly, lb/in. (N/mm)
k_b	Spring constant of bolt, lb/in. (N/mm)
k_t	Spring constant of tube, lb/in. (N/mm)
L, L_1, L_2	Length dimensions, in. (mm)
ΔL	Extension, in. (mm)
P	Axial thrust, lb (N)
Q_y	Partial weight of conical bar, lb (N)
R	Mean radius of tube, in. (mm)
r, r_1	Tube radii, in. (mm)
S	General symbol for stress, psi (N/mm²)
S_1, S_2	Tensile stresses, psi (N/mm²)
S_c	Compressive stress, psi (N/mm²)
S_t	Tensile stress due to bending, psi (N/mm²)
T, t	Wall thicknesses, in. (mm)
U	Resilience, lb-in./in.³ (N-mm/mm³)
U_t	Total strain energy, lb-in. (N-mm)
W, W_1	Downward loads, lb (N)
y	Arbitrary distance, in. (mm)
Z	Auxiliary parameter, in.² (mm²)
γ	Specific weight, lb/in.³ (N/mm³)
δ, δ'	Deflections, in. (mm)

20
Uniform Cantilever Beams

APPLICATIONS AND IMPLICATIONS

When a beam is supported at only one end in such a way that its axis cannot rotate at that point, it is always referred to as a *cantilever beam*. The supported end is usually assumed to be totally restrained for the purpose of theoretical calculations, although it is well known that the total fixity can seldom be achieved. This may be an important consideration in the design of turbine blading and experimental work in the field of vibration. In practice, for instance, the degree of blade-root fixing depends on the design and the rotational speed of the turbine shaft. In testing it is necessary to simulate the clamping pressures at the root by running pilot experiments. The clamping has also some effect on the free length of the cantilever blade, and therefore on its frequency. The cantilever frequency is inversely proportional to the square of the length. The clamping pressure is usually increased gradually until the effect of the fixity on the amplitude of vibration fades out [113, 114]. In normal static applications, the problem of cantilever fixity at the support is almost never considered because of the various manufacturing and materials tolerances, which can prove to be more influential.

The selection of the method for developing the deflection formulas for cantilever beams of uniform cross section can be used as an illustration of the general philosophy of applying the principles of solid mechanics to this class of problems. The theory of Castigliano and the method of differential equation, coupled with the principle of superposition, appear to be the most popular approaches in this type of analysis. Although the mathematics here is generally straightforward, the choice of a particular model and its description may not be obvious and can lead to some complications. This can sometimes be avoided when sufficient experience in this category of problems is available. To illustrate some of the issues involved, let us consider the case of a cantilever beam carrying a concentrated load W at a distance a_0 measured from the free end of the beam, such as that shown in Fig. 20.1.

Suppose that by x we will denote an arbitrary distance from the free end of the beam. The bending moment equation applicable here is

$$M = W(x - a_0) \tag{20.1}$$

The method of the differential equation, or *double integration method*, Eq. (7.5) gives

$$EI\frac{d^2y}{dx^2} = M \tag{20.2}$$

Combining Eqs. (20.1) and (20.2) gives

$$EI\frac{d^2y}{dx^2} = W(x - a_0) \tag{20.3}$$

The statement of the problem implies that the flexural rigidity EI must be constant and that the only independent variable is x. Consequently, the first integration gives

$$EI\frac{dy}{dx} = W\left(\frac{x^2}{2} - a_0 x\right) + A \tag{20.4}$$

Since the relation (20.4) defines the slope of the beam at any distance, such as $a_0 \le x \le L$, the first integration constant A can be obtained from the fixed boundary $dy/dx = 0$, when $x = L$. This gives

$$A = \frac{WL}{2}(2a_0 - L) \tag{20.5}$$

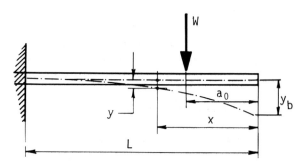

Fig. 20.1 Cantilever beam with intermediate load.

Uniform Cantilever Beams

Second integration of Eq. (20.3) yields

$$EIy = W\left(\frac{x^3}{6} - \frac{x^2 a_0}{2}\right) + Ax + B \tag{20.6}$$

The deflection must vanish at $x = L$. Hence, substituting Eq. (20.5) into Eq. (20.6) and making $y = 0$ at $x = L$ gives

$$B = \frac{WL^2}{6}(2L - 3a_0) \tag{20.7}$$

Hence, substituting Eqs. (20.5) and (20.7) into Eq. (20.6) leads to the following general expression:

$$y = \frac{W}{6EI}\left[x^2(x - 3a_0) + 3xL(2a_0 - L) + L^2(2L - 3a_0)\right] \tag{20.8}$$

When $x = L$, the deflection vanishes in line with the selected boundary condition. When $x = a_0$, the deflection under the applied load becomes

$$y = \frac{W}{3EI}\left[L^3 + 3a_0 L(a_0 - L) - a_0^3\right] \tag{20.9}$$

Also, when $a_0 = 0$, Eq. (20.9) reduces to a standard formula for maximum deflection of a cantilever beam when the load is applied at the free end.

Suppose we now assume that the distance x is measured from the fixed instead of the free end of the beam. Here the bending moment equation becomes

$$M = W(L - a_0 - x) \tag{20.10}$$

so that

$$EI\frac{d^2y}{dx^2} = W(L - a_0 - x) \tag{20.11}$$

Note that in some books on strength of materials, the differential equation is written as $EI(d^2y/dx^2) = -M$. The negative sign merely implies a specific convention, and the final absolute value of the deflection is not affected. In this derivation we continue to use $+M$ instead of $-M$. Hence, the two consecutive integrations of Eq. (20.11) gives

$$EI\frac{dy}{dx} = W\left(Lx - a_0 x - \frac{x^2}{2} + A\right) \tag{20.12}$$

and

$$EIy = W\left(\frac{Lx^2}{2} - \frac{a_0 x^2}{2} - \frac{x^3}{6}\right) + Ax + B \tag{20.13}$$

Observing that at $x = 0$, $y = dy/dx = 0$, Eqs. (20.12) and (20.13) immediately yield $A = B = 0$. This simplifies the derivation process considerably, so that Eq. (20.13) yields

$$y = \frac{W}{6EI}\left(3Lx^2 - 3a_0 x^2 - x^3\right) \tag{20.14}$$

when $x = 0$, $y = 0$ at the fixed point of the support. For $a_0 = 0$ and $x = L$, Eq. (20.14) reduces to a standard deflection formula. Also, when $x = L - a_0$, the expression for the deflection becomes identical with that given by Eq. (20.9), as should have been anticipated. However, it is of interest to note the implications of the assumed model. A simple change in the reference distance x can significantly affect the extent of the algebraic work in determining the constants of integration and the final form of the deflection formulas.

Suppose that we now wish to use the Castigliano principle for the derivation of Eq. (20.9). By reference to Eq. (6.10), the deflection under load W, applied at a distance a_0 from the free end, can be stated as

$$y = \frac{1}{EI}\int_{a_0}^{L} M \frac{\partial M}{\partial W}\, dx \tag{20.15}$$

From Eq. (20.1), we get

$$\frac{\partial M}{\partial W} = x - a_0 \tag{20.16}$$

Hence, substituting Eqs. (20.1) and (20.16) into Eq. (20.15) leads to these steps

$$y = \frac{W}{EI}\int_{a_0}^{L}(x - a_0)^2\, dx$$
$$= \frac{W}{EI}\left(\frac{x^3}{3} - a_0 x^2 + x a_0^2\right)_{a_0}^{L}$$
$$= \frac{W}{EI}\left(\frac{L^3}{3} - a_0 L^2 + L a_0^2 - \frac{a_0^3}{3} + a_0^3 - a_0^3\right)$$

On rearranging and simplifying various terms in the preceding expression we again obtain Eq. (20.9). In looking over the derivations using three modified approaches, it appears that the last theoretical model is the easiest to handle.

The deflection at the free end can be arrived at, for instance, with the aid of Eqs. (20.0) and (20.14) using the principle of superposition. The slope under load W can be found from Eq. (20.14) by differentiating.

$$\frac{dy}{dx} = \frac{W}{2EI}(2Lx - 2a_0 x - x^2) \tag{20.17}$$

Uniform Cantilever Beams

When $x = L - a_0$, Eq. (20.17) gives

$$\frac{dy}{dx} = \frac{W(L-a_0)^2}{2EI} \qquad (20.18)$$

Hence, employing Eq. (20.9) and (20.18), together with the principle of superposition, yields

$$y = \frac{W}{3EI}\left[L^3 + 3a_0L(a_0 - L) - a_0^3\right] + \frac{W(L-a_0)^2 a_0}{2EI}$$

which on simplifying becomes

$$y = \frac{W}{6EI}\left(2L^3 - 3a_0 L^2 + a_0^3\right)$$

Note that this is identical with Eq. (7.12), derived previously from the theorem of Castigliano without the help of the principle of superposition.

DESIGN CHARTS AND TABLES

With the help of simplified design charts [115], numerous cantilever beam problems can be solved as shown, for instance, in Fig. 20.2 and 20.3. Note that when $a_0 = 0$,

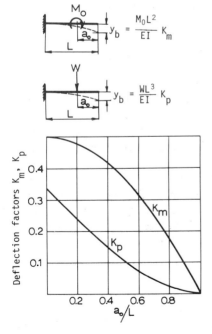

Fig. 20.2 Cantilever deflection factors for concentrated forces. E, elastic modulus (psi); I, moment of inertia (in.^{4}); all linear dimensions in inches.

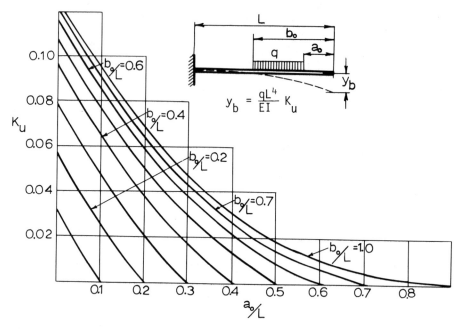

Fig. 20.3 Cantilever deflection factor for partially distributed load.

Fig. 20.2 gives standard deflection formulas $M_0L^2/2EI$ and $WL^3/3EI$. Similarly, extrapolating to $a_0/L = 0$ in Fig. 20.3 yields $qL^4/8EI$, which is consistent with standard solution [35].

In calculating the stresses and deflections, the rule of superposition is used extensively, provided that the stresses fall below the yield strength of the material. This rule is illustrated in Fig. 20.4.

The maximum stresses and deflections of cantilever beams are frequently estimated in the course of preliminary design. Several typical design formulas for this purpose are given in Table 20.1. This information is intended for typical structural members which are sufficiently slender. Under such conditions, it is only necessary to calculate the deflection and stresses due to bending. This deflection is denoted here by Y_b.

The first four cases given in Table 20.1 are encountered most frequently. However, when the positions of concentrated forces and partial uniform loading are significantly different from those given in the table, the designer may use Figs. 20.2 and 20.3. For instance, when the load W is placed at the midpoint of the cantilever, $a_0/L = 0.5$, and $K_p = 0.1$ is obtained from Fig. 20.2 (lower curve).

EFFECT OF SHEAR

In a simplified theory of bending of beams, it is customary to assume that plane sections remain plane during the process of bending. In reality, however, certain shearing distortions take place and successive elements of a beam tend to slide

Uniform Cantilever Beams

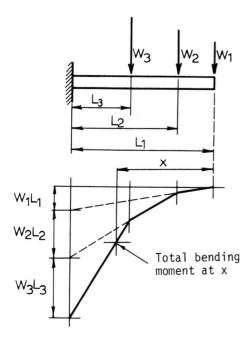

Fig. 20.4 Example of moment superposition.

under the transverse loading, producing additional deflection. It was also stated in Chap. 2 that the shear force is not distributed uniformly over the cross section.

Because of these effects, the question may arise as to the point in the analysis at which the designer can safely ignore the deformation due to shear [116, 117]. The general rule is sometimes advanced that for beam span/depth ratios greater than 10, there is no need to account for the shear deformation in the deflection studies. Unfortunately we seldom see a numerical illustration of this problem, so that the engineer or designer could gain a better appreciation of the complete subject. There also seems to be a certain amount of confusion as to the use of a shear distribution factor (Chap. 2), which deals with the shear stress, and the *form factor*, which enters the process of the derivation of the formulas for the deflection due to shear. To indicate the nature of the two factors involved, consider the following example.

Design Problem 20.1

Derive the value of the form factor for a cantilever beam of a rectangular cross section carrying a concentrated shear load W, shown in Fig. 20.5.

Solution

Since the variation of shear stress over the rectangular cross section is known to be parabolic [5], the expression for the shear stress at any point, such as that defined by x in Fig. 20.5, is

$$\tau = \frac{6W}{bh^3}\left(\frac{h^2}{4} - x^2\right) \tag{20.19a}$$

Table 20.1 Maximum Deflections and Moments for Cantilever Beams

Beam	Maximum Bending Moment	Maximum Deflection
Cantilever with point load W at tip, length L	$M = WL$	$Y_b = \dfrac{WL^2}{3EI}$
Cantilever with uniform load q over length L	$M = \dfrac{qL^2}{2}$	$Y_b = \dfrac{qL^4}{8EI}$
Cantilever with uniform load q from a to L	$M = \dfrac{q(L^2 - a^2)}{2}$	$Y_b = \dfrac{q(L-a)[6L(L+a)^2 - 3a(a^2 + 2L^2)]}{48EI}$
Cantilever with applied moment M_0 at tip	$M = M_0$	$Y_b = -\dfrac{M_0 L^2}{2EI}$
Cantilever with triangular load (max at a, zero at L)	$M = \dfrac{qe(L-a)}{2}$, $e = \dfrac{2a + L}{3}$	$Y_b = \dfrac{q(L-a)[14(L-a)^3 + 405Le^2 - 135e^3]}{1620EI}$
Cantilever with triangular load (max at wall, zero at tip)	$M = \dfrac{qL^2}{6}$	$Y_b = \dfrac{qL^4}{30EI}$
Cantilever with triangular load (zero at a, max at L)	$M = \dfrac{qe(L-a)}{2}$, $e = \dfrac{a + 2L}{3}$	$Y_b = \dfrac{q(L-a)[17(L-a)^3 + 90e^2(7L-a)]}{3240EI}$
Cantilever with triangular load (zero at wall, max at tip)	$M = \dfrac{qL^2}{3}$	$Y_b = \dfrac{11qL^4}{120EI}$

The strain energy due to shear for an elementary volume ($bL\,dx$) parallel to the neutral axis, is expressed in terms of Eq. (20.19a)

$$dU_S = \left[\dfrac{6W}{bh^3}\left(\dfrac{h^2}{4} - x^2\right)\right]^2 \dfrac{bL\,dx}{2G} \qquad (20.19\text{b})$$

Uniform Cantilever Beams

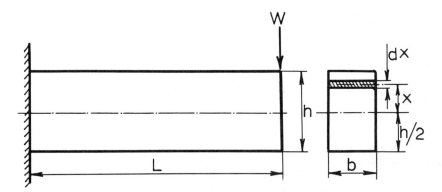

Fig. 20.5 Short cantilever notation for deriving form factor.

Integrating this energy over the entire beam, we get

$$U_S = 2 \int_0^{h/2} \frac{18W^2 L}{Gbh^6} \left(\frac{h^4}{16} - \frac{h^2 x^2}{2} + x^4 \right) dx \qquad (20.19c)$$
$$= \frac{3W^2 L}{5GBh}$$

The work done by the load W in translating the end of the cantilever beam by the amount Y_s is equal to

$$U_W = \frac{W Y_s}{2} \qquad (20.19d)$$

Equating the work done, Eq. (20.19d), and the total shear energy stored, Eq. (20.19c), gives

$$Y_s = \frac{6WL}{5Gbh} \qquad (20.19e)$$

The form factor in Eq. (20.19e) is 6/5, which means that the actual shear deflection is 6/5 times that obtained if the shear stress were assumed to be uniform. The shear distribution factor for the rectangular section, however, is 3/2, as shown by Eq. (20.19a) when $x = 0$. ◆

A brief summary of shear deflection formulas intended for more common cases of cantilever loading is given in Table 20.2. The table also includes the ratios of shear to bending deflections, indicating that basically the limiting nondimensional ratios such as $(h/L)^2$ and $(r/L)^2$ define the effect of shear deformation on the total beam deflection. For example, assuming that shear deflection should not exceed, say, 10% of the total deflection for case 1 in Table 20.2, the limiting ratio of beam span/depth calculates to be 2.6. Hence, for any contribution due to shear to be less than 10%, the beam span/depth ratio must be higher than 2.6. Although the 10% magnitude was selected rather arbitrarily, it is not unreasonable because the basic mechanical properties and manufacturing tolerances are seldom known with the accuracy better than 5 to 10%.

Table 20.2 Shear Deflection Criteria for Cantilever Beams

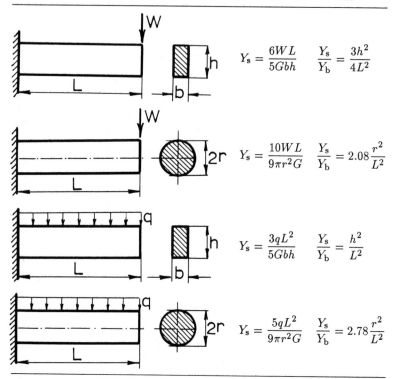

SYMBOLS

a	Distance to uniform or variable load from the support, in. (mm)
a_0	Distance to external load from free end, in. (mm)
b	Width of rectangular cross section, in. (mm)
b_0	Distance to end of uniform load, in. (mm)
E	Modulus of elasticity, psi (N/mm^2)
e	Distance to center of gravity of triangular loading, in. (mm)
G	Modulus of rigidity, psi (N/mm^2)
h	Depth of rectangular cross section, in. (mm)
I	Moment of inertia of beam cross section, in.^{4} (mm^4)
K_m	Deflection factor for concentrated moment
K_p	Deflection factor for concentrated load
K_u	Deflection factor for uniform load
L, L_1, L_2, L_3	Arbitrary length, in. (mm)
M	Maximum bending moment, lb-in. (N-mm)
M_0	Concentrated moment, lb-in. (N-mm)
q	Unit load, lb/in. (N/mm)
r	Radius of circular beam, in. (mm)
U_S	Strain energy due to shear, lb-in. (N-mm)

Uniform Cantilever Beams

U_W	Elastic work done, lb-in. (N-mm)
x	Arbitrary distance to point of interest, in. (mm)
W, W_1, W_2, W_3	Concentrated loads, lb (N)
y	General symbol for deflection, in. (mm)
Y_b	Deflection due to bending, in. (mm)
Y_s	Deflection due to shear, in. (mm)
τ	Shear stress, psi (N/mm^2)

21
Variable Section Cantilever Beams

BASIC CONSIDERATIONS

The expression for the vertical displacement of any beam of uniform cross section under a given concentrated load takes the following general form [37]:

$$\Delta = (\text{constant})\frac{WL^3}{EI} \qquad (21.1)$$

The load in this equation can be concentrated, uniformly distributed, or variable, as long as each is treated separately and added according to the method of superposition. In a similar fashion the deflection formula can be expressed in terms of a unit bending stress S and depth of beam.

$$\Delta = (\text{constant})\frac{SL^2}{hE} \qquad (21.2)$$

The rules expressed by Eqs. (21.1) and (21.2) are easy to observe when the beam has uniform cross section, because the moment of inertia is constant throughout the length of the beam. In many design situations the beam problems involve only the variation of the bending moment over the span. However, in some cases it may be more efficient to employ the tapered or constant stress beams by suitably varying the cross-sectional dimensions over the total beam length. Provided that no rapid cross-sectional changes are involved, the usual engineering theory of bending can be employed. It is well to note, however, that in the general case of a tapered beam,

Variable Section Cantilever Beams

the maximum bending stress may not occur at the point of the maximum bending moment.

CONSTANT PARAMETER CRITERIA

The principle defining a beam section for a constant bending stress follows directly from the elementary engineering formula $MC/I = M/Z$. For a given value of the maximum design stress σ_m, the section modulus must vary in direct proportion to the bending moment

$$Z = \frac{M}{\sigma_m} \tag{21.3}$$

For a simple cantilever beam of rectangular cross section loaded at the free end, the following three alternatives are available for maintaining the constant bending stress [29]:

Constant depth criterion

$$b = \left(\frac{6Wx}{\sigma_m h^2}\right) \tag{21.4}$$

Constant width criterion

$$h = \left(\frac{6W}{b\sigma_m}\right)^{1/2} x^{1/2} \tag{21.5}$$

Constant proportions criterion

$$h = \left(\frac{6W}{K\sigma_m}\right)^{1/3} x^{1/3} \tag{21.6}$$

In Eq. (21.6), $K = b/h$, the parameter that is kept unchanged in setting out the variation in beam depth. The only real independent variable in the three cases above is an arbitrary distance x, measured from the loaded point. The corresponding design details are shown in Fig. 21.1. In the constant proportions design the section modulus varies as the cube of a dimension.

ANALYSIS OF TAPERED CANTILEVERS

Rigorous analysis of end-loaded tapered cantilever beams [5] indicates that a small error is involved in calculating the maximum fiber stresses with the help of the engineering formula MC/I. The location of the maximum shear stresses is also affected by the taper, although this stress is seldom found to be critical. The factors of safety employed in design more than compensate for such discrepancies.

The calculation of the deflection in beams where the moment of inertia varies from one section to another can be a tedious procedure. The two general equations

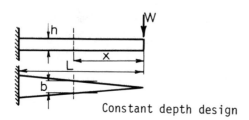
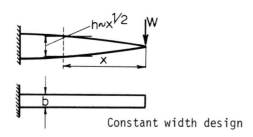
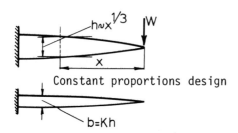

Fig. 21.1 Design of variable section cantilevers for constant bending stress.

that are taken in this analysis can be solved by numerical or graphical integration

$$\frac{dy}{dx} = A_1 - \frac{1}{E} \int \frac{M\,dx}{I} \tag{21.7}$$

and

$$y = A_1 x + B_1 - \frac{1}{E} \iint \frac{M}{I}\,dx\,dx \tag{21.8}$$

The closed-form solution is possible only when M/I functions are integrable. Here A_1 and B_1 denote the integration constants to be determined from the relevant boundary conditions.

Design Problem 21.1

Develop the deflection formula for a tapered cantilever beam of circular cross section illustrated in Fig. 21.2, assuming that the diameter at the fixed end is three times that at the free end.

Variable Section Cantilever Beams

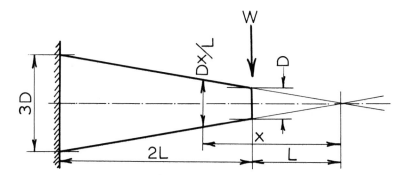

Fig. 21.2 Example of tapered cantilever beam.

Solution

The moment of inertia at any point along the beam length is

$$I = \frac{\pi x^4 D^4}{64 L^4} \qquad (21.9a)$$

The bending moment is

$$M = W(x - L) \qquad (21.9b)$$

Hence, using Eqs. (21.9a) and (21.9b) gives

$$\frac{M}{EI} = \frac{d^2 y}{dx^2}$$
$$E\frac{d^2 y}{dx^2} = \frac{M}{I} = W(x-L)\frac{64 L^4}{\pi x^4 D^4} = k(x^{-3} - L x^{-4}) \qquad (21.9c)$$

where $k = 64WL^4/\pi D^4$. On the first integration of Eq. (21.9c), the slope equation is obtained:

$$E\frac{dy}{dx} = k\left(-\frac{x^{-2}}{2} + \frac{Lx^{-3}}{3}\right) + C_1 \qquad (21.9d)$$

The slope is zero at the built-in end when $x = 3L$. Hence, $C_1 = 7k/162L^2$, so that Eq. (21.9d) becomes

$$E\frac{dy}{dx} = k\left(\frac{7}{162L^2} + \frac{Lx^{-3}}{3} - \frac{x^{-2}}{2}\right) \qquad (21.9e)$$

Subsequent integration of Eq. (21.9e) gives

$$Ey = k\left(\frac{x^{-1}}{2} - \frac{Lx^{-2}}{6} + \frac{7x}{162L}\right) + C_2 \qquad (21.9f)$$

When $x = 3L$, $y = 0$, so that $C_2 = -5k/18L$. Hence, the final expression for the deflection becomes

$$y = \frac{k}{E}\left(\frac{7x}{162L^2} + \frac{1}{2x} - \frac{L}{6x^2} - \frac{5}{18L}\right) \tag{21.9g}$$

when $x = 3L$, $y = 0$. When $x = L$, the expression for the maximum downward displacement is obtained

$$y = \frac{512WL^3}{81\pi ED^4} \quad \blacklozenge \tag{21.9h}$$

Design Problem 21.2

Find the location of the maximum bending stress for the cantilever beam shown in Fig. 21.2.

Solution

Since the diameter of the section varies according to the law xD/L, the section modulus in bending at any point along the beam is

$$Z = \frac{\pi x^3 D^3}{32L^3} \tag{21.10a}$$

The bending moment is

$$M = W(x - L) \tag{21.10b}$$

Hence, the general expression for the bending stress becomes

$$\sigma = \frac{32WL^3}{\pi D^3}\left(\frac{x-L}{x^3}\right) \tag{21.10c}$$

The position of the maximum bending stress is now found from Eq. (21.10c) with the aid of the condition $d\sigma/dx = 0$. This gives

$$x^3 - 3x^2(x - L) = 0$$

from which

$$x = 1.5L \quad \blacklozenge$$

The foregoing two examples indicate the analytical methods applicable to beams of variable cross sections. Detailed information on minimum-weight optimum design of variable section members is available [39]. At this stage of the preliminary design, however, the data given in this section should be sufficient for most practical purposes.

TAPERED FLAT SPRINGS

The general area of variable section cantilever beams is closely related to the flat, mechanical springs of special interest to designers. The three most important configurations are shown in Fig. 21.3. Out of these the simplest case is that of a triangular spring, Fig.21.3a. The bending stress at the built-in edge is

$$S = \frac{6PL}{Bh^2} \tag{21.11}$$

The corresponding maximum deflection under load P is

$$y = \frac{6PL^3}{EBh^3} \tag{12.12}$$

The deflection to cause yield stress follows directly from Eqs. (21.11) and (21.12)

$$y = \frac{L^2 S_y}{hE} \tag{21.13}$$

In the case of a trapezoidal type of a mechanical spring, Fig. 21.3b, the analysis is somewhat more complicated and requires the introduction of a taper factor β_T, plotted in Fig. 21.4. The relevant deflection at the free end of the spring is then

$$y = \frac{PL^3 \beta_T}{EBh^3} \tag{21.14}$$

Note that for $b = 0$, Fig. 21.4 yields $\beta_T = 6$, so that Eq. (21.14) reduces to Eq. (21.12). On the other hand, when $b = B$, Eq. (21.14) reduces to a standard deflection formula for a cantilever beam with uniform rectangular cross section. The formula for the bending stress at the fixed end is the same as that for Fig. 21.3a and c. Hence, the deflection corresponding to the yield stress, for the case shown in Fig. 21.3b, becomes

$$y = \frac{L^2 S_y \beta_T}{6Eh} \tag{21.15}$$

Finally, for the case of a cantilever spring of uniform width and a parabolically varying thickness, Fig. 21.3c, the maximum deflection is

$$y = \frac{8PL^3}{EBh^3} \tag{21.16}$$

In terms of the yield stress developed at the built-in edge, the deflection becomes

$$y = \frac{4L^2 S_y}{3Eh} \tag{21.17}$$

In designing the spring configurations shown in Fig. 21.3, normal procedure should involve selecting the working stress at the built-in edge before deciding on the

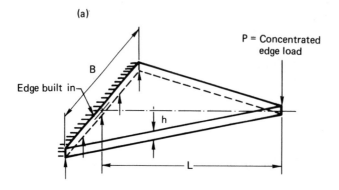
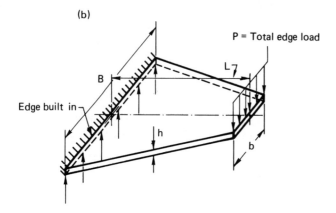
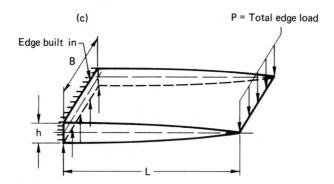

Fig. 21.3 Tapered cantilever springs.

Variable Section Cantilever Beams

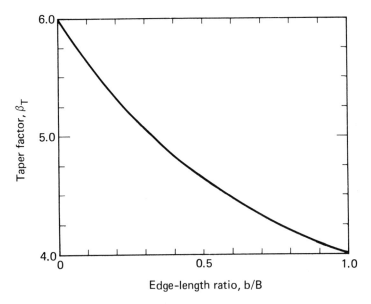

Fig. 21.4 Taper factor.

thickness and the deflection criteria. It will be assumed here that the applied load, modulus of elasticity, and the overall dimensions of the spring are specified at the start of the calculations.

DESIGN OF MULTIPLE-LEAF SPRINGS

Standard references in machine design seldom include engineering data on multiple-leaf springs. Normally, these structural members consist of several leafs of rectangular cross section and are sometimes known as laminated or carriage springs. These have been widely used on automobiles and sometimes on locomotives. The applications of multiple-leaf springs can still be found, and therefore this section includes a brief summary of the working formulas for stresses and deflections.

According to the standards of the Society of Automotive Engineers, the design stresses for the multiple-leaf springs in industry range between 90,000 and 110,000 psi for passenger cars and about 50,000 to 80,000 psi for trucks [145]. To improve the fatigue life of these springs, the leaf surfaces on the tension side should be shot peened.

The simplest case of a multiple-leaf spring is shown in Fig. 21.5. Here the design formulas can be stated as

$$y = \frac{6PL^3}{Ebnh^3} \tag{21.18}$$

$$S = \frac{6PL}{bnh^2} \tag{21.19}$$

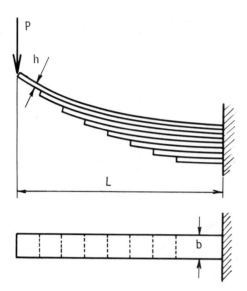

Fig. 21.5 Multiple-leaf cantilever spring.

and the deflection in terms of stress becomes

$$y = \frac{SL^2}{Eh} \qquad (21.20)$$

For the case of a multiple-leaf spring centrally loaded as a beam on simple supports, Fig. 21.6, the formulas are

$$y = \frac{3PL^3}{8Ebnh^3} \qquad (21.21)$$

$$S = \frac{3PL}{2bnh^2} \qquad (21.22)$$

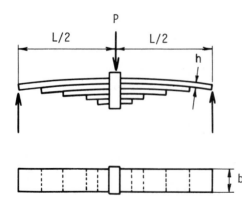

Fig. 21.6 Multiple-leaf symmetrical spring.

Variable Section Cantilever Beams

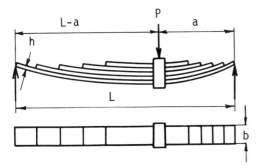

Fig. 21.7 Multiple-leaf unsymmetrical spring.

and

$$y = \frac{SL^2}{4Eh} \tag{21.23}$$

Finally, for an unsymmetric multiple-leaf spring shown in Fig. 21.7, we have

$$y = \frac{6P(L-a)^2 a^2}{EbnLh^3} \tag{21.24}$$

$$S = \frac{6P(L-a)a}{bnLh^2} \tag{21.25}$$

and the maximum deflection in terms of the design stress becomes

$$y = \frac{S(L-a)a}{Eh} \tag{21.26}$$

Note that when $a = L/2$, Eqs. (21.24) to (21.26) reduce to Eqs. (21.21) to (21.23). The foregoing formulas are essentially based on the premise of a beam of uniform bending strength. In practice, interleaf friction, variations in leaf thickness, use of shackles, and other effects may cause significant deviations from the calculated stress and deflection values. For this reason, the above formulas should only be used in the preliminary work. The analysis of stresses and deflections in conventional mechanical springs is given in Chap. 41.

SYMBOLS

A_1	Constant of integration
a	Arbitrary distance, in. (mm)
B	Width of cross section, in. (mm)
B_1	Constant of integration
b	Width of rectangular section, in. (mm)
C	Distance to stressed fiber, in. (mm)
C_1, C_2	Constants of integration
D	Diameter of cantilever, in. (mm)

E	Modulus of elasticity, psi (N/mm^2)
h	Depth of rectangular cross section, in. (mm)
I	Moment of inertia, in.^{4} (mm^4)
K	Side ratio
k	Auxiliary parameter, lb (N)
L	Length of beam, in. (mm)
M	General symbol for moment, lb-in. (N-mm)
n	Number of leafs
P	Edge load on springs, lb (N)
S	General symbol for stress, psi (N/mm^2)
S_y	Yield strength, psi (N/mm^2)
W	Concentrated load, lb (N)
x	Arbitrary distance, in. (mm)
y	Deflection, in. (mm)
Z	Section modulus, in.^{3} (mm^3)
β_T	Taper factor
Δ	General symbol for displacement, in. (mm)
σ_m	Maximum bending stress, psi (N/mm^2)

22
Beams on Simple Supports

BASIC ASSUMPTIONS

A considerable number of engineering problems in everyday structural and mechanical design can be reduced to a conservative model of a simply supported beam. A simple support is that which is designed to transmit the concentrated forces but it is not able to exert any reactive moments. The simple support, then, offers no restraint to the angular rotation of the ends of the beam as it deflects under loads.

It is also proper to assume under these conditions that at least one support is able to undergo some horizontal movement, so that no axial beam forces are generated. Problems that involve a combination of the transverse and axial loading are considerably more involved and are seldom solved in the design office. This chapter therefore is limited to beams that are subject to transverse loading only.

Overhanging beams may be included in the category of beams on simple supports. Such load-carrying members are freely supported at two or more points and have one or two ends extending beyond the supports.

The general rules that govern straight beam criteria are based on the elastic response and are subject to the usual engineering assumptions. These are pertinent to all the chapters in this book dealing with straight but relatively slender members. The rules can be summed up as follows:

1. The modulus of elasticity is the same in tension and compression.
2. The beam cross section is uniform.
3. The beam is relatively long (span/depth=8 or more).
4. The beam is relatively narrow.

5. The maximum working stress does not exceed the elastic limit of the material.
6. Slight curvature of beam can be tolerated.
7. Plane sections remain plane during bending.

The analysis is performed for the distributed or concentrated loads, while the strength and rigidity are the two basic requirements in design. It is well to note that a concentrated load implies in reality an infinitely high bearing pressure exerted on the beam. In practice, all loads must be considered as distributed, thereby resulting in a finite, acceptable bearing pressure.

MATHEMATICAL CONCEPTS

When discussing various beam problems, it is helpful to keep in mind the basic mathematical concepts that govern relations among the shearing loads, loading intensity, bending moments, and beam curvature. A summary of these concepts is given in Table 22.1.

For a short length of beam such as that shown in Fig. 22.1, an element of length dx carries a uniformly distributed load q. Let the shear force change over this element from Q to $Q + dQ$, and the corresponding bending moment from M to $M + dM$. From the resolution of vertical forces

$$Q + q\,dx = Q + dQ$$

and

$$q = \frac{dQ}{dx}$$

Taking moments about the right-hand side of the element gives

$$M + Q\,dx + q\,dx\,\frac{dx}{2} = M + dM$$

Table 22.1 Summary of Mathematical Concepts for Beam Analysis

Beam curvature	(1/in.)	$\dfrac{d^2y}{dx^2} \cong \dfrac{1}{R}$
Bending Moment	(lb-in.)	$M = EI\dfrac{d^2y}{dx^2}$
Shearing force	(lb)	$Q = \dfrac{dM}{dx} = EI\dfrac{d^3y}{dx^3}$
Load intensity	(lb/in.)	$q = \dfrac{dQ}{dx} = EI\dfrac{d^4y}{dx^4} = \dfrac{d^2M}{dx^2}$

Beams on Simple Supports

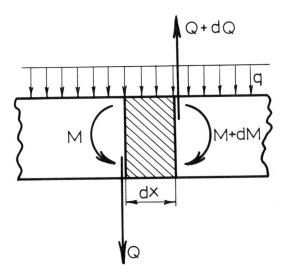

Fig. 22.1 Equilibrium of beam element.

and ignoring the small quantity of the second order yields

$$Q = \frac{dM}{dx}$$

This is an important relation because it shows that the position of the maximum bending moment corresponds to the zero shearing force. Furthermore, combining the foregoing relations, it is found that load intensity is

$$q = \frac{d^2 M}{dx^2}$$

It should be noted, however, that in some cases the location of the maximum bending moment can be better found by inspection than by a mathematical deduction.

DESIGN TABLES FOR SIMPLE BEAMS

As indicated in Chap. 8, beams on two simple supports are statically determinate, as shown by the summary of the more common formulas in Table 22.2. Specific loading conditions selected for this and other tables are purposely based on varied sources [118–123] and practical design office experience. These loading cases appear to be most in demand, and design situations seldom require more detailed stress analysis. The formulas for shear deflection are included as a matter of general illustration of this effect on the total beam deformation as a function of beam proportions, type of loading, and support (Tables 20.2 and 22.3).

Somewhat more specialized, but still of interest are the formulas in Table 22.4 pertaining to cases where the deflection may be one of the basic criteria. This is found applicable to structural support systems in delicate equipment and instrumentation based on line-of-sight tolerances and similar considerations. For instance,

Table 22.2 Maximum Moments and Deflections for Simply Supported Beams

	Maximum Bending Moments	Maximum Deflection
(W at L/2)	$M = \dfrac{WL}{4}$	$Y_b = \dfrac{WL^3}{48EI}$
(W at a, b)	$M = \dfrac{Wab}{L}$	$Y_b = \dfrac{Wab(a+2b)[3a(a+2b)]^{1/2}}{27EIL}$ at $x = 0.58(a^2 + 2ab)^{1/2}$ when $a > b$
(uniform q over L)	$M = \dfrac{qL^2}{8}$	$Y_b = \dfrac{5qL^4}{384EI}$
(triangular q)	$M = 0.064qL^2$ at $x = 0.58L$	$Y_b = \dfrac{0.0065qL^4}{EI}$ $x = 0.52L$
(triangular peak at L/2)	$M = \dfrac{qL^2}{12}$	$Y_b = \dfrac{qL^4}{120EI}$
(V-shaped load)	$M = \dfrac{qL^2}{24}$	$Y_b = \dfrac{3qL^4}{640EI}$
(M_0 at end)	$M = M_0$	$Y_b = \dfrac{0.064M_0L^2}{EI}$ at $x = 0.42L$
(M_0 at a,b)	$0 < x < a$ $M = -\dfrac{M_0 x}{L}$ $a < x < L$ $M = M_0 - \dfrac{M_0 x}{L}$	$Y_b = \dfrac{M_0 x(3b^2 - L^2 + x^2)}{6EIL}$ $Y_b = -\dfrac{M_0(L-x)(3a^2 - 2Lx + x^2)}{6EIL}$

a beam designed for an optical device, corresponding to case 1, Table 22.4, may have deflection requirements such that $\Delta_A = \Delta_C$, which gives

$$48k^4 + 96k^3 + 24k^2 - 16k - 5 = 0$$

Since a cut-and-try method is often lengthy in the solution of this equation the following semigraphical approach may be found useful. Let $F_1 = F_2$, where $F_1 = 48k^4 + 96k^3$ and $F_2 = 5 + 16k - 24k^2$. By plotting both functions against $k = a/L$,

Beams on Simple Supports

Table 22.3 Some Shear Deflection Criteria for Simply Supported Beams

Beam		
(center point load W, span 2L)	$Y_s = \dfrac{3WL}{10GBH}$	$\dfrac{Y_s}{Y_b} = 3\left(\dfrac{H}{L}\right)^2$
(point load W at distances a, b; span L)	$Y_s = \dfrac{6Wab}{5GBHL}$	$\dfrac{Y_s}{Y_b} = \dfrac{3H^2}{4ab}$
(uniform load q over span L)	$Y_s = \dfrac{3qL^2}{20GBH}$	$\dfrac{Y_s}{Y_b} = \dfrac{12}{5}\left(\dfrac{H}{L}\right)^2$

Table 22.4 Maximum Moments and Deflections for Simply Supported Beams with Overhangs

Beam	Moments	Deflections
uniform load, overhangs a both sides	$M_B = \dfrac{qa^2}{2}$ $M_C = \dfrac{q(L^2 - 4a^2)}{8}$	$\Delta_A = \dfrac{qa(3a^3 + 6a^2L - L^3)}{24EI}$ $\Delta_C = \dfrac{qL^2(5L^2 - 24a^2)}{384EI}$
uniform load on overhang a	$M_B = \dfrac{qa^2}{2}$	$\Delta_A = \dfrac{qa^2L(4L + 3a)}{24EI}$ $\Delta_C = -\dfrac{qaL^3}{31EI}$ at $x = 0.58L$
point loads W at both overhang ends	$M_B = Wa$	$\Delta_A = \dfrac{Wa^2(3L + 2a)}{6EI}$ $\Delta_C = -\dfrac{WaL^2}{8EI}$
point load W at overhang end	$M_B = Wa$	$\Delta_A = \dfrac{Wa^2(L + a)}{3EI}$ $\Delta_C = -\dfrac{0.064WaL^2}{EI}$ at $x = 0.58L$

$+\Delta$ Positive deflection
$-\Delta$ Negative deflection

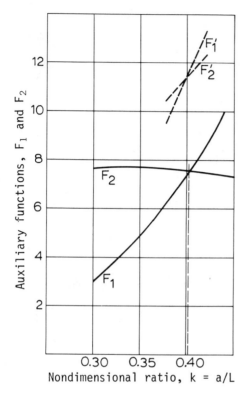

Fig. 22.2 Example of graphical solution for $\Delta_A = \Delta_C$, case 1, Table 22.4.

the solution is easily obtained, as shown in Fig. 22.2. The choice of F_1 and F_2 is, of course, largely arbitrary as long as $F_1 - F_2 = 0$. For example, $F_1 = 48k^4 - 5$ and $F_2 = 96k^3 + 24k^2 - 16k = 4k(24k^2 + 6k - 4)$ should give essentially the same result within the reading accuracy of the graph. The character of the intersecting curves may be different from those shown in Fig. 22.2. To illustrate this point further, take the functions

$$F_1' = 24k^2(2k^2 + 4k + 1)$$
$$F_2' = 16k + 5$$

so that

$$F_1' - F_2' = 0$$

Substituting for k from Fig. 22.2 into the straight-line relationship, we get

$$F_2' = 16 \times 0.402 + 5 = 11.43$$

Similarly, utilizing the expression for F_1' gives

$$24 \times 0.402 \times 0.402(2 \times 0.402 \times 0.402 + 4 \times 0.402 + 1) = 11.43$$

Beams on Simple Supports

It is seen from Fig. 22.2 that the new point of intersection is still on the same vertical line of about $k = 0.4$, which is sufficiently accurate for most practical purposes.

Design case 1, Table 22.4, therefore, contains similar features as far as the deflection criteria are concerned. These can be summed up as follows:

$$\Delta_A = 0 \quad \text{when } a/L = 0.374$$
$$\Delta_C = 0 \quad \text{when } a/L = 0.457$$
$$\Delta_A = \Delta_C \text{when } a/L = 0.402$$

Standard cases of maximum deflection in simply supported beams are given in Table 22.2. It is of interest to note that the position of the maximum deflection is not too sensitive to the manner of loading. Hence, in many design calculations, central deflection is often computed instead of the maximum deflection, which would require more sophisticated approach to the problem.

SYMBOLS

a	Arbitrary distance, in. (mm)
B	Width of cross section, in. (mm)
b	Arbitrary distance, in. (mm)
E	Modulus of elasticity, psi (N/mm^2)
F_1, F_2, F_1', F_2'	Auxiliary functions
G	Modulus of rigidity, psi (N/mm^2)
H	Depth of cross section, in. (mm)
I	Moment of inertia, in.^{4} (mm^4)
$k = a/L$	Length ratio
L	Length of beam, in. (mm)
M, M_B, M_C	Bending moments, lb-in. (N-mm)
M_0	Concentrated moment, lb-in. (N-mm)
Q	Shear force, lb (N)
q	Uniform load, lb/in. (N/mm)
R	Radius of curvature, in. (mm)
W	Concentrated load, lb (N)
x	Arbitrary distance, in. (mm)
y	Transverse displacement, in. (mm)
Y_b	Deflection due to bending, in. (mm)
Y_s	Deflection due to shear, in. (mm)
$\Delta, \Delta_A, \Delta_C$	Deflection in beams with overhangs, in. (mm)

23
Beams with Constraint

GENERAL CONSIDERATIONS

The general rules common to cantilever and simply supported beams loaded transversely also apply to beams with constraint. This category of straight members implies, however, that a constraint is present at the supports in the form of end-fixing moments or statically indeterminate reactions. For instance, four types of beams, illustrated in Fig. 23.1, have some form of a constraint that will require additional boundary conditions in evaluating the reactions and bending moments. The constraint at the supports, however, does not prevent a horizontal strain, so that the beams do not become laterally loaded tie bars.

In the first case of Fig. 23.1, we recognize a statically indeterminate problem such as that previously shown in Fig. 8.1. In that particular event the reaction at the propped end was considered as an unknown quantity and the basic equations of equilibrium were supplemented with the expression for elastic energy in bending. The method of approach to the problem, outlined briefly in Chap. 8, can be used here to obtain working formulas for various propped cantilevers. A summary of the more common formulas found in this category is given in Table 23.1 [27, 37].

BEAM WITH SINKING SUPPORT

An interesting example of a beam with constraint is shown in Fig. 23.1 as case 2. If a load W is applied at one end in such a manner that the support at the right,

Beams with Constraint

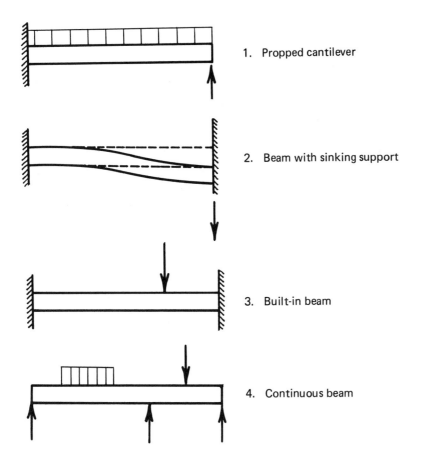

Fig. 23.1 Examples of beams with constraint.

say, is lowered to a different level through a distance Y, it may be observed that the beam behaves, in effect, as a double cantilever. Cutting the beam in half, we obtain the equilibrium such as that shown in Fig. 23.2. Each of the two cantilevers is then subjected to free deflection at the end equal to

$$\frac{Y}{2} = \frac{W}{3EI}\left(\frac{L}{2}\right)^3$$

which gives

$$Y = \frac{WL^3}{12EI} \tag{23.1}$$

The end-fixing moment is

$$M_0 = \frac{WL}{2} \tag{23.2}$$

Table 23.1 Forces and Displacements for Propped Cantilevers

	End Reaction	Maximum Bending Moments	Maximum Deflection Due to Bending
	$R = \dfrac{5W}{16}$ $(W - R)$ at A	$M = \dfrac{3WL}{16}$ at A	$Y_B = \dfrac{WL^3}{107EI}$ at $x = 0.45L$
	$R = \dfrac{Wb^2(a+2L)}{2L^3}$	when $a \leq 0.41L$ $M = \dfrac{Wab^2(a+2L)}{2L^3}$ when $a \geq 0.41L$ $M = \dfrac{Wab(b-2L)}{2L^2}$	when $a \leq 0.41L$ and $x = \dfrac{L(L^2 + a^2)}{3L^2 - a^2}$ $Y_B = \dfrac{Wa(L^2 - a^2)^3}{3EI(3L^2 - a^2)^2}$ when $a \geq 0.41L$ and $x = L\left(\dfrac{a}{2L+a}\right)^{1/2}$ $Y_B = \dfrac{Wab^2}{6EI}\left(\dfrac{a}{2L+a}\right)^{1/2}$
	$R = \dfrac{3qL}{8}$ $(qL - R)$ at A	$M = \dfrac{qL^2}{8}$ at A	$Y_B = \dfrac{qL^4}{185EI}$ at $x = 0.42L$
	$R = \dfrac{qL}{10}$ $\left(\dfrac{qL}{2} - R\right)$ at A	$M = \dfrac{qL^2}{15}$ at A	$Y_B = \dfrac{qL^4}{420EI}$ at $x = 0.45L$
	$R = \dfrac{11qL}{40}$ $\left(\dfrac{qL}{2} - R\right)$ at A	$M = \dfrac{7qL^2}{120}$ at A	$Y_B = \dfrac{qL^4}{328EI}$ at $x = 0.40L$

CLEVIS DESIGN

The case illustrated in Fig. 23.2 can be interpreted either as a double cantilever or as a beam with sinking supports. It should be noted here that despite its obvious simplicity, the double cantilever concept has numerous practical applications. For instance, consider a single-pin clevis of a rectangular geometry subjected to a pull-up load in such a way that the originally parallel surfaces are maintained essentially parallel under load, as shown in Fig. 23.3. This can be accomplished by bolting the

Beams with Constraint

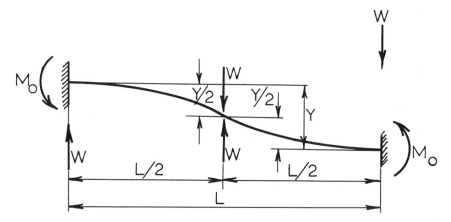

Fig. 23.2 Double cantilever or a beam with sinking support.

clevis arms against a rigid insert. Each arm then behaves as a double cantilever beam, illustrated in Fig. 23.2.

The maximum bending stress can be obtained directly with the aid of Eq. (23.2)

$$S_b = \frac{3WL}{BH^2} \tag{23.3}$$

The corresponding pull-up deflection follows from Eq. (23.1)

$$Y = \frac{WL^3}{EBH^3} \tag{23.4}$$

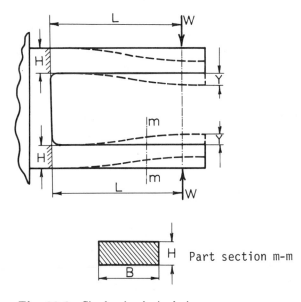

Fig. 23.3 Single-pin clevis design.

Making, say, $S_b = S_y$, and eliminating W between the two preceding formulas, yields

$$Y = \frac{S_y L^2}{3HE} \qquad (23.5)$$

This is the fundamental design formula for a single-pin clevis, which determines the allowable pull-up deflection for a specified value of yield strength S_y, or a fraction thereof, if a factor of safety is put on stress.

FRAME UNDER LATERAL LOAD

In some instances involving building structures subjected to horizontal disturbances due to earthquake ground motion, an explosion blast, or a gust of wind, it is customary to employ a simplified method of analysis based on the concept of a rectangular frame acted upon by a horizontal, shear load. The girder and floor system in a building, for instance, is considerably more rigid in shear than the vertical support columns. This kind of a system is often referred to as a shear building [12, 13]. The mode of response of the system in a natural period of free lateral vibration is illustrated in Fig. 23.4. The deflection required for the estimate of the natural period follows from Eq. (23.1)

$$Y = \frac{WL^3}{24EI} \qquad (23.6)$$

Equation (23.6) can now be used in conjunction with Eq. (11.1) in determining the natural period of vibration in one degree of freedom. Note that half of the total load is assigned to each of the two support columns, so that the numerical factor 24 appears in Eq. (23.6).

The third category of beam constraint, according to Fig. 23.1, is due to built-in supports. A brief summary of more frequently used formulas is given in Table 23.2.

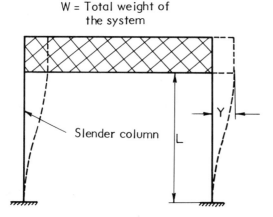

Fig. 23.4 Model of lateral response for frames and buildings.

Beams with Constraint

Table 23.2 Maximum Moments and Deflections for Built-In Beams

	Maximum Bending Moment	Maximum Deflection Due to Bending
	For $a < b$ $M_A = \dfrac{Wab^2}{L^2}$ For $a > b$ $M_B = \dfrac{Wa^2b}{L^2}$	For $a < b$ and $x = \dfrac{2Lb}{L+2b}$ $Y_B = \dfrac{2Wa^2b^3}{3EI(L+2b)^2}$ For $a > b$ and $x = \dfrac{2La}{L+2a}$ $Y_B = \dfrac{2Wa^3b^2}{3EI(L+2a)^2}$
	$M = \dfrac{WL}{8}$	$Y_B = \dfrac{WL^3}{192EI}$
	$R_1 = \dfrac{qb}{4L^3}$ $\times [4e^2(L+2d) - b^2(c-a)]$ $M_1 = \dfrac{qb}{24L^2}$ $\times \{b^2[L+3(c-a)] - 24e^2d\}$	For $0 < x < a$ $Y_b = \dfrac{3M_1x^2 + R_1x^3}{6EI}$ For $a < x < (a+b)$ $Y_B = \dfrac{12M_1x^2 + 4R_1x^3 - q(x-a)^4}{24EI}$
	$M = -\dfrac{qL^2}{12}$	$Y_B = \dfrac{qL^4}{384EI}$
	$R_1 = -\dfrac{6M_0ab}{L^3}$ $M_1 = -\dfrac{M_0b(L-3a)}{L^2}$ $M_2 = -\dfrac{M_0a(2L-3a)}{L^2}$ $R_2 = -R_1$	For $x = -\dfrac{2M_1}{R_1}$ and $a > \dfrac{L}{3}$ $Y_B = \dfrac{2M_1^3}{3EIR_1^2}$ For $x = L + \dfrac{2M_2}{R_2}$ and $a < 2\dfrac{L}{3}$ $Y_B = \dfrac{2M_2^3}{3EIR_2^2}$
	$R_1 = \dfrac{3qL}{20}$ $R_2 = \dfrac{7qL}{20}$ $M_B = \dfrac{qL^2}{20}$ (maximum)	$Y_B = \dfrac{qL^4}{764EI}$ (maximum) at $x = 0.525L$

When the concentrated load is not at center or where the uniform load extends over a portion of the beam only, the derivation of deflection formulas becomes considerably more involved, particularly where the position and magnitude of the deflection are required. Similar comments apply to the case of variable intensity of loading such as the triangular distribution shown in Table 23.2 [37].

PARTIAL UNIFORM LOAD

When simply supported and built-in beams carry partial uniform loading, the design procedures for estimating reactions, as well as the magnitude and locations of maximum deflections, can be made simpler by developing design charts [124]. Although these charts are too numerous to bring out here in detail, certain observations on the relative location of the points of the maximum deflection should be

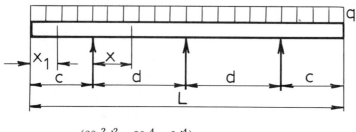

$$M_{\max} = \frac{q(28c^2d^2 - 36c^4 - 9d^4)}{128d^2}$$

at

$$x = \frac{6c^2 + 3d^2}{8d}$$

$$Y_{\max} = \frac{0.0054qL^4}{16EI}$$

at

$$x = 0.422d \quad \text{and} \quad c = 0$$

Deflection between supports:

$$Y = \frac{qL^4}{16EI(k+1)^4}\left\{\frac{m^4}{24} - \frac{m^3}{16} + \frac{m}{48} - \frac{k^2m^3}{8} + \frac{k^2m^2}{4} - \frac{k^2m}{8}\right\}$$

where $k = c/d$ and $m = x/d$

$$Y = \frac{qL^4}{16EI(k+1)^4}\left\{\frac{n-k}{48} + \frac{n^4}{24} + \frac{k^4 + k^3 - k^2n}{8} - \frac{k^3n}{6}\right\}$$

where $k = c/d$ and $n = x_1/c$

Fig. 23.5 Example of beam on three supports.

Beams with Constraint

Table 23.3 Nonsymmetrical Loading on Three Supports

(diagram: uniform load q over left span; R_1 at left, R_2 between, R_3 at right; spans L and L; x measured from R_1)	$R_1 = 0.438qL$ $R_2 = 0.625qL$ $R_3 = -0.063qL$ $M_{\max} = 0.096qL^2$ at $x = \dfrac{7L}{16}$ $Y_{\max} = \dfrac{0.0092qL^4}{EI}$ at $x = 0.472L$
(diagram: point load W at midspan of left portion; R_1, R_2, R_3 supports; spans $L/2$, $L/2$, L)	$R_1 = 0.406W$ $R_2 = 0.688W$ $R_3 = -0.094W$ $M_{\max} = 0.203WL$ (under load W) $Y_{\max} = \dfrac{0.015WL^3}{EI}$ at $x = 0.480L$

made. The detailed calculations for simply supported beams, for instance, show that the positions of the maximum deflections for various lengths of uniform partial loading vary between, say, $0.50L$ and $0.57L$, or $0.50L$ and $0.43L$, depending on the relative position of the loading. The variations in the positions of the maximum deflections is somewhat wider for built-in beams, ranging from $0.37L$ to $0.63L$. Fortunately, for the designer of mechanical and structural hardware, the exact position of the maximum deflection is seldom required.

SPECIAL DESIGN PROBLEMS

The last category of a common beam variety in design is shown in Fig. 23.1, as the fourth case. The problem here is generally made more complex by the requirements of finding the best location for supports in minimizing bending moments and deflections [125, 126]. An example of a solution for a beam on three supports is given in Fig. 23.5. In addition, a couple of cases of unsymmetrical loading for beams with three supports are shown in Table 23.3 [120].

In reviewing the strength and rigidity of beams fixed at the supports, it would immediately appear that such beams should always be stronger than the corresponding beams resting on simple supports. The engineering practice, however, shows that built-in beams are often purposely avoided. One of the reasons for this is that slight misalignment of the supports during mechanical assembly or unavoidable changes in temperature can cause additional stresses of appreciable magnitude. Furthermore, any form of fatigue loading on a built-in structure, such as that which may be found in a bridge, would tend to alter the original degree of fixity. One of the methods of obviating such difficulties is to design using the principle of a double-cantilever beam, such as that shown in Fig. 23.2. For instance, by placing the hinge at the point of beam deflection, there is no bending moment there to resist and the beam will appear to be simply supported by the free ends of the two cantilevers.

SYMBOLS

a, b, c, d, e	Arbitrary lengths, in. (mm)
B	Width of cross section, in. (mm)
E	Modulus of elasticity, psi (N/mm^2)
H	Depth of beam cross section, in. (mm)
I	Moment of inertia, in.^{4} (mm^4)
k, m, n	Nondimensional lengths
L	Length of beam, in. (mm)
M	General symbol for bending moment, lb-in. (N-mm)
M_1, M_2, M_A, M_B	Bending moments at specific points, lb-in. (N-mm)
M_0	Concentrated moment, lb-in. (N-mm)
q	Distributed load, lb/in. (N/mm)
R, R_1, R_2, R_3	Reactions at specific supports, lb (N)
S_b	Bending stress, psi (N/mm^2)
S_y	Yield strength of material, psi (N/mm^2)
M_{max}	Maximum bending moment, lb-in. (N-mm)
W	Concentrated load, lb (N)
x, x_1	Distances from support, in. (mm)
Y	General symbol for deflection, in. (mm)
Y_B	Deflection due to bending, in. (mm)
Y_{max}	Maximum deflection, in. (mm)

24
Special Beam Problems

INTRODUCTION

The majority of problems involving straight beams which are solved in the design office have been outlined in Chaps. 18 to 23. These involve estimates of the maximum bending stresses or deflections, and the loads are usually applied in the direction transverse to the beam axis. These beams are assumed to have relatively small cross-sectional dimensions and the transverse loads produce pure bending only.

It was previously shown that for beams of small span/depth ratio, the deflection due to shear may not be negligible. The corresponding deflection ratios involving shear and bending for several typical cases were shown in Table 20.2. The elementary theory of bending gives then sufficiently accurate prediction of fiber stresses in beams for spans greater than 3 to 4 times the depth. When concentrated loads are involved, high localized compressive and shear stresses develop which can cause reduction in fiber stresses. For these conditions to exist, however, the span-to-depth ratio must be rather small. In most cases this value is found to be close to unity or less.

DEEP SECTION BEAMS

The strength of a conventional beam is nearly always governed by the fiber stress, provided that buckling due to bending is not a criterion. Some cases of buckling

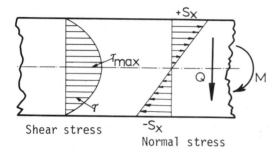

Fig. 24.1 Stress distribution in a transversely loaded beam.

due to bending were given in Table 10.6. Whether a combined or the maximum fiber stress governs the design for a beam of a relatively great depth, it is convenient to keep in mind the basic stress distribution, such as that illustrated in Fig. 24.1. It is clear from the sketch that for a given bending moment M and a shear load Q the maximum normal and shear stresses do not occur at the same point. When a question arises, then, as to which stress governs the design, reference should be made to one of the basic formulas, such as Eq. (4.4), from which the maximum principal stresses can be calculated. For instance, making $S_y = 0$, $S_x = S$, and $\tau_{xy} = \tau$, Eq. (4.4) gives

$$S_{\max} = \frac{S}{2} + \frac{1}{2}(S^2 + 4\tau^2)^{1/2} \qquad (24.1)$$

Equation (24.1) applies at any point of a beam cross section. In the preliminary engineering analysis, we assume that the maximum bending stress at the outer fiber is also the maximum normal stress. As we proceed from the outer fiber of the beam toward the neutral axis, the bending stress drops off linearly while the shear stress increases parabolically, as shown, for example, in Fig. 24.1. Hence, at an arbitrary distance x from the neutral axis, we have two finite stress values, S and τ, which can be combined according to Eq. (24.1) to give the maximum principal stress, which may or may not be smaller than the maximum bending stress at the extreme fiber. This aspect of stress analysis may be of importance in knuckle joint and pin design, where the ratio of span to depth can be rather small. Application of this principle is described further in Chap. 40.

WIDE BEAMS

In the case of relatively wide beams, such as thin metallic strips, the deflections computed from simple beam formulas are somewhat high. This effect can be allowed for by substituting $E/(1 - \nu^2)$ for E.

When the beams are wide and short, the deflections and stresses are not uniform. To circumvent this design difficulty, it is customary to utilize the concept of effective width together with the standard beam formulas [127–129]. The case of special interest is that of a very wide cantilever shown in Table 24.1 [130]. Although the formula in case 4 is strictly applicable to a plate of infinite width, fairly good

Special Beam Problems

Table 24.1 Wide and Short Beams and Slabs (Formulas for Effective Width)

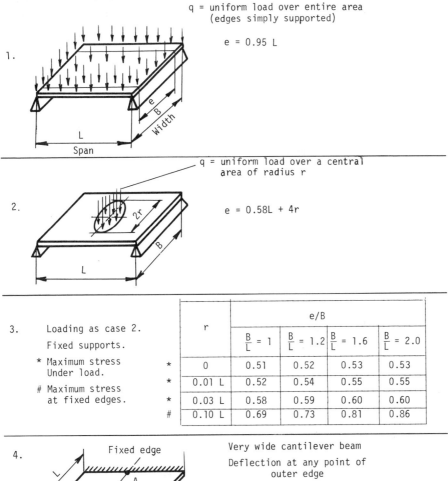

1. q = uniform load over entire area (edges simply supported)
   $e = 0.95 L$

2. q = uniform load over a central area of radius r
   $e = 0.58L + 4r$

3. Loading as case 2. Fixed supports.
   * Maximum stress Under load.
   # Maximum stress at fixed edges.

r	e/B			
	$\frac{B}{L} = 1$	$\frac{B}{L} = 1.2$	$\frac{B}{L} = 1.6$	$\frac{B}{L} = 2.0$
* 0	0.51	0.52	0.53	0.53
* 0.01 L	0.52	0.54	0.55	0.55
* 0.03 L	0.58	0.59	0.60	0.60
# 0.10 L	0.69	0.73	0.81	0.86

4. Very wide cantilever beam
   Deflection at any point of outer edge
   $Y = KWL^2(1-\nu^2)/Et^3$

x/L	0	0.25	0.50	1.00	2.00	2.50
K	2.01	1.80	1.45	0.81	0.19	0.15

Maximum bending stress at A
$S_{max} = 3.05\, W/t^2$

agreement can be expected with the slabs having widths as small as four times the span L.

Design Problem 24.1

Wide cantilever plate is rigidly fixed at the long edge and carries a concentrated load W at the midpoint of the unsupported long edge. Assuming that the maximum bending stress in the middle of the fixed edge is limited to $S_{max} = 30{,}000$ psi, calculate the maximum deflection under load W. The material is structural steel and the extent of overhang L is 24 in. Assume that the plate is 0.5 in. thick. Find the approximate deflection, assuming that standard deflection formula for a simple cantilever beam is applicable when the width of the plate is equal to $2L$.

Solution

From case 4, Table 24.1, we have

$$Y = \frac{KWL^2(1-\nu^2)}{Et^3}$$

and

$$S_{\max} = \frac{3.05W}{t^2}$$

From the foregoing stress formula

$$W = \frac{t^2 S_{\max}}{3.05}$$

Substituting in the expression for the deflection and taking Poisson's ratio $\nu = 0.3$ gives

$$Y = \frac{KL^2(1-0.3^2)}{Et^3} \frac{t^2 S_{\max}}{3.05}$$
$$= 0.2984 \frac{KL^2 S_{\max}}{Et}$$

For $x/L = 0$, case 4 in Table 24.1 yields $K = 2.01$. Hence,

$$Y = \frac{0.6L^2 S_{\max}}{Et} \tag{24.2a}$$

For the case of a standard cantilever beam having a width b and length L, we have

$$Y = \frac{WL^3}{3EI}$$

and

$$S_{\max} = \frac{6WL}{bt^2}$$

Since

$$I = \frac{bt^3}{12}$$

then

$$Y = \frac{4WL^3}{Ebt^3}$$

Special Beam Problems

From the cantilever stress formula

$$W = \frac{bt^2 S_{\max}}{6L}$$

Hence substituting for W in the deflection formula yields

$$Y = \frac{4L^3}{Ebt^3} \frac{bt^2 S_{\max}}{6L}$$
$$= \frac{2L^2 S_{\max}}{3Et} \tag{24.2b}$$

Taking $E = 30 \times 10^6$ psi, Eq. (24.2a) gives

$$Y = \frac{0.6 \times 24 \times 24 \times 30{,}000}{0.5 \times 30 \times 10^6} = 0.691 \text{ in.} \quad (17.56 \text{ mm})$$

Using the standard cantilever formula, Eq. (24.2b), gives $Y = 0.768$ in. This value is about 11% too high, because additional constraint of a wide beam is not included in Eq. (24.2b). The term $1/(1 - \nu^2)$ suggested as a correction amounts in this case to about 10%. ◆

All the beam problems discussed so far were based on the assumption that the loads were applied in the plane of symmetry; that is, the analysis involved cross sections for which the center of gravity and the center of twist coincided. When these two centers do not coincide, the transverse load is usually resolved into a direct shear force and a twisting couple equal to the product of the shear force and the distance between the two centers. Hence, the stresses are obtained directly by the method of superposition in the case of shear, Chaps. 2 and 3, or by the methods described in Chap. 5 when the resultant effects of shear and flexure are required.

COMPOSITE BEAMS

Some beams in engineering applications consist of two or more materials which are attached rigidly together throughout their entire length. Typical examples of composite sections are shown in Fig. 24.2. If by M we denote the total bending moment resisted by the section made of two materials 1 and 2, the first condition to be fulfilled is

$$M = M_1 + M_2 \tag{24.3}$$

Since the radius of curvature must be the same for the two parts, we get

$$\frac{M_1}{M_2} = \frac{E_1 I_1}{E_2 I_2} \tag{24.4}$$

Hence

$$M_1 = M \frac{E_1 I_1}{E_1 I_1 + E_2 I_2} \tag{24.5}$$

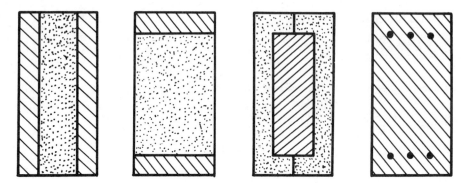

Fig. 24.2 Examples of composite beam cross sections.

and

$$M_2 = M \frac{E_2 I_2}{E_1 I_1 + E_2 I_2} \tag{24.6}$$

The corresponding bending stresses are then

$$S_1 = \frac{M_1}{Z_1} \tag{24.7}$$

and

$$S_2 = \frac{M_2}{Z_2} \tag{24.8}$$

In calculating the deflections, we note that the beam must behave as though the total bending stiffness EI were equal to the sum of the component stiffnesses

$$EI = E_1 I_1 + E_2 I_2 \tag{24.9}$$

BEAM COLUMNS

Certain structural members and machine parts can be simultaneously subjected to direct compressive forces and lateral loads. Such members are classified as beam columns because they contain characteristics of columns and straight beams. It may be instructive here to briefly review some of the solutions to a beam column problem shown in Fig. 24.3. Suppose that the beam is bent under the combined loading P and W in such a way that a central deflection Δ is produced. The maximum bending stress can be defined, for example, as

$$S_{\max} = \frac{P}{A} + \frac{WLC}{4I} \tag{24.10}$$

Special Beam Problems

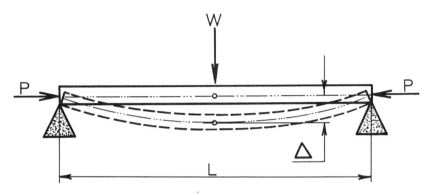

Fig. 24.3 Example of a beam-column.

Here I/C denotes the section modulus for the beam cross section, and Eq. (24.10) may be defined as the first approximation to the solution. Equation (24.10) ignores the bending moment due to the eccentric application of axial forces P as the beam deflects. Hence, the more correct approximation may be described as

$$S_{\max} = \frac{P}{A} + \left(\frac{WL}{4} + P\Delta\right)\frac{C}{I} \tag{24.11}$$

It is noted that the solution of Eq. (24.11) requires several steps of approximation and the deflection becomes rather sensitive to small changes in the eccentricity as the axial load P approaches the critical buckling value. The exact solution to this problem should be based on the bending moment relation

$$M_{\max} = \frac{W \tan \frac{L}{2}\left(\frac{P}{EI}\right)^{1/2}}{2\left(\frac{P}{EI}\right)^{1/2}} \tag{24.12}$$

Hence, the maximum true stress must be

$$S_{\max} = \frac{P}{A} + \frac{M_{\max} C}{I} \tag{24.13}$$

Numerous exact solutions to the beam column problems are available throughout the literature, including a comprehensive collection of detailed working charts [33, 34]. Many design applications and formulas have also been worked out which have been incorporated in codes of standard practice in the field of steel construction [120]. Selected theoretical formulas for the maximum bending moments and deflections are given in Table 24.2. When it is necessary to provide a measure of safety for a particular design, separate factors of safety should be applied to both the transverse and axial loads before the corresponding maximum bending moment is determined from Eq. (24.12) and the true maximum stress from Eq. (24.13). This total stress should not exceed either the yield strength of the material or the critical buckling stress in compression, whichever is smaller.

Table 24.2 Maximum Moments and Deflections for Beam Columns

$$M_{\max} = W\sqrt{\frac{EI}{P}} \tan L\sqrt{\frac{P}{EI}} \qquad Y_{\max} = \frac{W}{P}\left(\sqrt{\frac{EI}{P}} \tan L\sqrt{\frac{P}{EI}} - L\right)$$
at A $\qquad\qquad\qquad\qquad\qquad$ at free end

$$M_{\max} = q\sqrt{\frac{EI}{P}}\left[\sqrt{\frac{EI}{P}}\left(1 - \sec L\sqrt{\frac{P}{EI}}\right) + L \tan L\sqrt{\frac{P}{EI}}\right] \quad \text{at } A$$

$$Y_{\max} = \frac{q\sqrt{\frac{EI}{P}}}{P}\left[\sqrt{\frac{EI}{P}}\left(1 + \frac{PL^2}{2EI} - \sec L\sqrt{\frac{P}{EI}}\right)\right.$$
$$\left. + L\left(\tan L\sqrt{\frac{P}{EI}} - L\sqrt{\frac{P}{EI}}\right)\right]$$
at free end

$$M_{\max} = -\frac{W}{2}\sqrt{\frac{EI}{P}}\tan\frac{L}{2}\sqrt{\frac{P}{EI}} \qquad\qquad \text{at } x = L/2$$

$$Y_{\max} = \frac{W}{2P}\sqrt{\frac{EI}{P}}\left(\tan\frac{L}{2}\sqrt{\frac{P}{EI}} - \frac{L}{2}\sqrt{\frac{P}{EI}}\right) \qquad \text{at } x = L/2$$

$$M_{\max} = -\frac{qEI}{P}\left(\sec\frac{L}{2}\sqrt{\frac{P}{EI}} - 1\right) \qquad\qquad \text{at center}$$

$$Y_{\max} = \frac{qEI}{P^2}\left(\sec\frac{L}{2}\sqrt{\frac{P}{EI}} - 1 - \frac{PL^2}{8EI}\right) \qquad \text{at center}$$

$$M_A = M_C = -\frac{W}{2}\sqrt{\frac{EI}{P}}\left(\frac{1-\cos\frac{L}{2}\sqrt{\frac{P}{EI}}}{\sin\frac{L}{2}\sqrt{\frac{P}{EI}}}\right)$$

$$M_B = -\frac{W}{2}\sqrt{\frac{EI}{P}}\left(\tan\frac{L}{2}\sqrt{\frac{P}{EI}} - \frac{1-\cos\frac{L}{2}\sqrt{\frac{P}{EI}}}{\sin\frac{L}{2}\sqrt{\frac{P}{EI}}\cos\frac{L}{2}\sqrt{\frac{P}{EI}}}\right)$$

$$Y_{\max} = \frac{W}{2P}\sqrt{\frac{EI}{P}}\left[\tan\frac{L}{2}\sqrt{\frac{P}{EI}} - \frac{L}{2}\sqrt{\frac{P}{EI}} - \frac{\left(1-\cos\frac{L}{2}\sqrt{\frac{P}{EI}}\right)^2}{\sin\frac{L}{2}\sqrt{\frac{P}{EI}}\cos\frac{L}{2}\sqrt{\frac{P}{EI}}}\right]$$

$$M_A = M_C = -\frac{qEI}{P}\left(1 - \frac{\frac{L}{2}\sqrt{\frac{P}{EI}}}{\tan\frac{L}{2}\sqrt{\frac{P}{EI}}}\right)$$

$$M_B = \frac{qEI}{P}\left(\frac{\frac{L}{2}\sqrt{\frac{P}{EI}}}{\sin\frac{L}{2}\sqrt{\frac{P}{EI}}} - 1\right)$$

$$Y_{\max} = \frac{qEI}{P^2}\left[\sec\frac{L}{2}\sqrt{\frac{P}{EI}} - \frac{PL^2}{8EI} - 1 - \left(1 - \frac{\frac{L}{2}\sqrt{\frac{P}{EI}}}{\tan\frac{L}{2}\sqrt{\frac{P}{EI}}}\right)\right.$$
$$\left.\times\left(\frac{1-\cos\frac{L}{2}\sqrt{\frac{P}{EI}}}{\cos\frac{L}{2}\sqrt{\frac{P}{EI}}}\right)\right]$$

Special Beam Problems

Design Problem 24.2

A uniform beam of length $L = 120$ in. is simply supported at both ends and carries a central transverse load $W = 200$ lb. If in addition to this load, an axial compressive load P equal to 100 lb is applied at the ends of the beam, calculate the maximum stress and the deflection under load W. Assume the material to be structural steel with $E = 30 \times 10^6$ psi. The beam has a tubular cross section with mean radius 0.875 in. and wall thickness 0.125 in.

Solution

The required moment of inertia of the tube cross section is

$$I = \pi R^3 T = \pi \times 0.875^3 \times 0.125 = 0.2631 \text{ in.}^4$$

Hence, the term needed for the calculation of the tangent function according to Table 24.2 is

$$\frac{120}{2}\left(\frac{100}{30 \times 10^6 \times 0.2631}\right)^{1/2} = 0.2136 \text{ rad}$$

$$0.2136 \times 57.3 = 12.24 \text{ deg}$$

so that

$$\tan \frac{L}{2}\sqrt{\frac{P}{EI}} = \tan(12.24) = 0.2169$$

Also

$$\sqrt{\frac{EI}{P}} = \left(\frac{30 \times 10^6 \times 0.2631}{100}\right)^{1/2} = 280.94 \text{ in.}$$

Hence, the maximum bending moment from Table 24.2 is

$$M_{\max} = -\frac{W}{2}\left(\frac{EI}{P}\right)^{1/2} \tan\left[\frac{L}{2}\left(\frac{P}{EI}\right)^{1/2}\right]$$

$$= -\frac{200}{2} \times 280.94 \times 0.2169 = -6093.59 \text{ lb-in.}$$

The negative sign for the bending moment can be ignored if we add the two compressive stress components according to Eq. (24.13):

$$S_{\max} = \frac{P}{A} + \frac{M_{\max}}{Z}$$

where

$$A = 2\pi RT = 2\pi \times 0.874 \times 0.125 = 0.6872 \text{ in.}^2$$

and

$$Z = \pi R^2 T = \pi \times 0.875^2 \times 0.125 = 0.3007 \text{ in.}^3$$

The maximum stress is then

$$S_{\max} = \frac{100}{0.6872} + \frac{6093.59}{0.3007} = 20{,}410 \text{ psi} \quad (140.8 \text{ N/mm}^2)$$

Again using Table 24.2, the maximum deflection becomes

$$Y_{\max} = \frac{W}{2P}\left(\frac{EI}{P}\right)^{1/2}\left[\tan\frac{L}{2}\left(\frac{P}{EI}\right)^{1/2} - \frac{L}{2}\left(\frac{P}{EI}\right)^{1/2}\right]$$

$$= \frac{200}{2 \times 100} \times 280.94(0.2169 - 0.2136) = 0.927 \text{ in.} \quad (23.55 \text{ mm}) \quad \blacklozenge$$

The material given in Table 24.2 treats a number of relatively conservative cases because the axial load P, shown as compressive, tends to increase the deflection and stresses of the beam columns under consideration.

The decision whether to use Table 24.2 or to rely on a simplified approach, ignoring the effect of the deflection, can be made on the basis of the following ratio:

$$\frac{\text{Actual end load}}{\text{Euler load}}$$

The error introduced by ignoring the deflection is approximately equal to the magnitude of the above ratio. For example, using the classical Euler formula, Eq. (10.1), in conjunction with Design Problem 24.2 gives

$$P_{\text{CR}} = \frac{\pi^2 EI}{L^2} = \frac{\pi^2 \times 30 \times 10^6 \times 0.2631}{120 \times 120} = 5409.8 \text{ lb}$$

Hence, the numerical ratio is

$$\frac{100}{5409.8} = 0.0185$$

This is less than 2%, which can also be verified using a standard deflection formula for a simply supported beam subjected to a central, concentrated load. Substituting the relevant data from Design Problem 24.2 yields

$$Y = \frac{WL^3}{48EI} = \frac{200 \times 120^3}{48 \times 30 \times 10^6 \times 0.2631} = 0.912 \text{ in.}$$

Hence, the ratio based on the deflections is

$$\frac{0.927 - 0.912}{0.912} = 0.0164$$

For all practical purposes, the foregoing result is quite close to 0.0185. Similarly, on the basis of a standard beam calculation, we have

$$S = \frac{M}{Z} = \frac{WL}{4\pi R^2 T}$$

$$= \frac{200 \times 120}{4\pi \times 0.875^2 \times 0.125} = 19{,}956 \text{ psi}$$

Special Beam Problems

Hence, the stress ratio gives

$$\frac{20,410 - 19,956}{19,956} = 0.023$$

The foregoing considerations indicate that although the material from Table 24.2 and similar sources can be used without undue difficulty, it may be advisable to perform a simple Euler load calculation before evaluating detailed interactions of the transverse and axial loads.

Finally, it should be added that when the axial force P is tensile instead of compressive, it produces a restoring moment that makes the deflections smaller and acts as a stabilizing factor. In effect, the bending moment changes its sign and the deflection formulas, such as those given in Table 24.2, should be rederived using $-P$ instead of $+P$.

BEAMS ON ELASTIC FOUNDATION

The original interest in the analysis of railroad tracks supported by cross ties and soil formation led to the early application of the theory of beams on elastic foundation [131]. Detailed developments of this theory, however, were accomplished much later [132]. The elastic resistance of a foundation is defined as the amount of pressure necessary to produce a unit deflection. This resistance, known as the *modulus of foundation*, is denoted here as k_0. Dimensionally, the parameter is expressed as the force per length to the third power and represents the intensity of the distributed elastic reactions along the axis of the deflected beam. With reference to Fig. 24.4, the pertinent relations may be defined as

$$k = k_0 B \left(\frac{\text{lb}}{\text{in.}^3}\right) \times \text{in.} \tag{24.14}$$

and

$$P_r = ky \left(\frac{\text{lb}}{\text{in.}^2}\right) \times \text{in.} \tag{24.15}$$

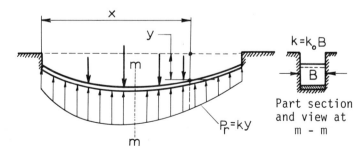

Fig. 24.4 Notation for theory of beams on an elastic foundation.

where k and P_r are the elastic foundation constant and foundation reaction, respectively. The relevant equation for a beam on elastic foundation is written in the following way:

$$EI\frac{d^4y}{dx^4} = -ky \tag{24.16}$$

The general solution of Eq. (24.16) is

$$y = e^{\beta x}(C_1 \cos \beta x + C_2 \sin \beta x) + e^{-\beta x}(C_3 \cos \beta x + C_4 \sin \beta x) \tag{24.17}$$

Here

$$\beta = \left(\frac{k}{EI}\right)^{1/4} \tag{24.18}$$

and C_1 through C_4 represent arbitrary constants of integration to be evaluated from the particular boundary conditions. The parameter β can be defined as the damping factor since it enters the functions $\cos \beta x$ and $\sin \beta x$, which determine the characteristic wave form of diminishing amplitude as x is increased. The response of a very long beam on elastic foundation can be, for example, illustrated in Fig. 24.5, if we assume a single, concentrated load W, producing local deflection y against the resistance of the foundation. The deflection, slope, bending moment, and shear are shown to vary rather drastically as the distance x is varied. For the case of the uniform load q and local moment M_0, Table 24.3 can be used.

There are a number of practical problems where the foregoing theory can be applied to get an idea of structural behavior in such systems as railways, roadbeds, buried utility piping, piles, and other. When instead of a very long beam, the theory is applied to a semi-infinite beam on elastic foundation, the areas of application can be extended to pressure vessels, containers, support skirts, and shell structures involving geometrical transitions, junctures, and local attachments where secondary stresses develop. These types of stresses are generally local, self-limiting, and difficult to evaluate, even under conditions of simple geometry and manner of loading. In this particular instance, the following working equations may be of interest

$$y = \frac{2W^*\beta}{k}e^{-\beta x}\cos \beta x - \frac{2M_0^*\beta^2}{k}e^{-\beta x}(\cos \beta x - \sin \beta x) \tag{24.19}$$

$$\theta = -\frac{2W^*\beta^2}{k}e^{-\beta x}(\cos \beta x + \sin \beta x) + \frac{4M_0^*\beta^3}{k}e^{-\beta x}\cos \beta x \tag{24.20}$$

$$M^* = -\frac{W^*}{\beta}e^{-\beta x}\sin \beta x + M_0^*e^{-\beta x}(\cos \beta x + \sin \beta x) \tag{24.21}$$

$$Q^* = W^*e^{-\beta x}(\cos \beta x - \sin \beta x) - 2M_0^*\beta e^{-\beta x}\sin \beta x \tag{24.22}$$

The corresponding notation is shown in Fig. 24.6. At $x = 0$ the deflection and slope can be calculated as follows:

$$y_{\max} = \frac{2\beta}{k}(W^* - \beta M_0^*) \tag{24.23}$$

Special Beam Problems

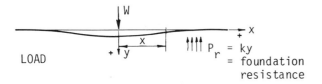
LOAD

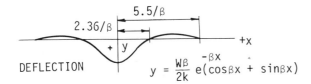
DEFLECTION

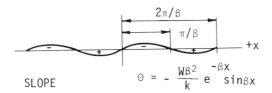
SLOPE

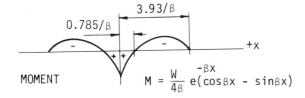
MOMENT

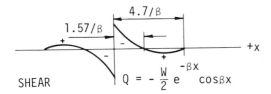
SHEAR

Fig. 24.5 Formulas and notation for long beams on an elastic foundation.

$$\theta_{\max} = -\frac{2\beta^2}{k}(W^* - 2\beta M_0^*) \tag{24.24}$$

To clarify some of the basic concepts and definitions of the theory of beams on elastic foundation in relation to the design analysis of shell-like components, let us first consider a portion of a cylindrical vessel subjected to edge loading as shown in Fig. 24.7. An elemental, longitudinal strip is cut out of the cylinder wall with

Table 24.3 Formulas for Uniform Load and Local Bending Couple on Beams on Elastic Foundation

$$y_{max} = \frac{q}{2k}(2 - e^{-\beta a}\cos\beta a - e^{-\beta b}\cos\beta b)$$

$$\theta_{max} = \frac{q\beta}{2k}\left[e^{-\beta a}(\cos\beta a + \sin\beta a) - e^{-\beta b}(\cos\beta b + \sin\beta b)\right]$$

$$M_{max} = \frac{q}{4\beta^2}(e^{-\beta a}\sin\beta a + e^{-\beta b}\sin\beta b)$$

$$Q_{max} = \frac{q}{4\beta}\left[e^{-\beta a}(\cos\beta a - \sin\beta a) - e^{-\beta b}(\cos\beta b - \sin\beta b)\right]$$

$$y = \frac{M_0\beta^2}{k}e^{-\beta x}\sin\beta x$$

$$\theta = \frac{M_0\beta^3}{k}e^{-\beta x}(\cos\beta x - \sin\beta x)$$

$$M = \frac{M_0}{2}e^{-\beta x}\cos\beta x$$

$$Q = -\frac{M_0\beta}{2}e^{-\beta x}(\cos\beta x + \sin\beta x)$$

the x and y coordinates coinciding with the longitudinal and radial directions, respectively. The elemental strip is loaded by the shear force W^* and the bending moment M_0^*, at $x = 0$. Both components of loading act in the radial plane of the cylinder. Although only one element is shown in Fig. 24.7, it is postulated that the shell consists of a finite number of identical longitudinal elements, each carrying

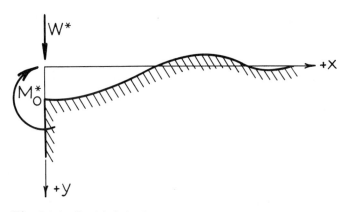

Fig. 24.6 Semi-infinite beam on an elastic foundation.

Special Beam Problems

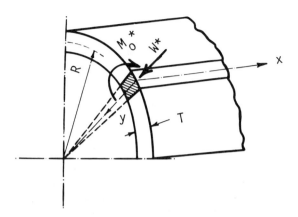

Fig. 24.7 Longitudinal strip of cylindrical wall modeled as a beam on an elastic foundation.

the same edge loads. Hence, the edge of the cylinder as a whole carries uniformly distributed shear and bending loading W^* and M_0^*, having the dimensions of lb/in. and lb-in./in., respectively. Each element can then be considered to be of unit width measured along the mean circumference of the cylinder. With these assumptions, each elemental strip can be compared to a beam on elastic foundation, carrying end shear load W^* and bending moment M_0^*, as long as the overall loading on the cylinder remains to be rotationally symmetric.

Considering next a relatively simple case of a pressure vessel having flat end closures, as shown in Fig. 24.8, the problem of local stresses can be reviewed with

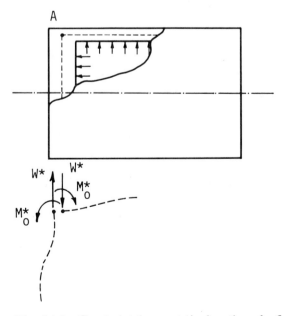

Fig. 24.8 Constraint forces at the junction of a flat closure and cylindrical wall.

the help of the basic concepts of beams on elastic foundation. The end closure behaves as a circular plate supported at the edge and uniformly loaded by the internal pressure. When free to deform under this load, the closure tends to bend to an approximately spherical shape. This process results in a finite slope at the edge of the closure. At the same time, the cylindrical part of the vessel experiences an axisymmetric radial growth due to the internal pressure. However, this growth must be restricted at the juncture, resulting in a shear force and a bending moment at each point of the circumference.

In the lower portion of Fig. 24.8, we have a sketch of forces at point A of the juncture, where W^* and M_0^* are the shear force and the bending moment per inch of circumference required to keep the joint in static equilibrium. These local forces are created because the internal pressure tends to separate the end closures from the cylindrical portion of the vessel. Once the magnitudes and directions of W^* and M_0^* are known, the relevant displacements and stresses can be calculated from the formulas derived for beams on elastic foundation.

Note that such formulas are given as functions of x, representing damped-out waves of rapidly decreasing amplitudes. These particular characteristics of beams on elastic foundation are very well suited to modeling the discontinuity effects, such as secondary stresses.

It is recalled that secondary stresses are developed due to the constraints of the adjacent parts. As shown in Fig. 24.8, for example, the end closure provides a constraint because of the differential amounts of growth and dilation of the connected parts under internal pressure. In contrast to such secondary stresses, the vessel is subjected to primary stresses. In this category, for example, we have the well-known membrane stresses, which will be discussed later in conjunction with piping and pressure vessel design.

Equations (24.23) and (24.24) are particularly useful in studying discontinuity stresses in pressure vessels. It is also to be noted that in solving various problems of beams on elastic foundation involving several loads acting simultaneously, the method of superposition and reciprocity apply. The latter theorem dates back more than 100 years [133, 134]. It means that when load W moves, say, from point A to B on a beam, the deflection at B with load at A will be equal to deflection at A with the same load applied at B.

Essentially, then, the theory of beams on elastic foundation can be applied to two classes of problems. The first one is concerned with the local loads, such as that shown in Fig. 24.5, where the effect of the displacements and forces is damped out rather rapidly on both sides of the applied load. The other case may be defined by Fig. 24.6 and the design formulas given by Eqs. (24.19) to (24.24). Here we assume that the effect is also local and the beam is infinitely long, starting at $x = 0$. This case can then represent the effect of forces W^* and M_0^* when they are applied to a rotationally symmetric edge such as that of a circular cylinder, sphere, or a conical shell.

The problem becomes very complicated when the beam or shell cannot be regarded as relatively long. In other words, structural transitions such as vessel heads, circumferential stiffeners, or other abrupt changes in the structure are so close to each other that the disturbances in the deformation and stress fields interfere with one another. Under these conditions it becomes necessary to evaluate all four constants, such as C_1 through C_4 in Eq. (24.17). Such solutions have been worked in

Special Beam Problems

great detail by Hetenyi [132]. However, they are often considered to be beyond the scope of traditional courses in the area of mechanical design.

One of the practical difficulties in applying Eq. (24.17), or one of its derivatives, is to find the correct numerical value for the parameter β. Although the term of flexural rigidity denoted by EI in Eq. (24.18) is easy to calculate if we know the cross-sectional geometry of the beam and its material, the elastic foundation constant k and modulus of foundation k_0 are not readily found in the literature. The only exception to this case is perhaps in the area of footing and foundation design for building walls, columns, and similar structures, where numerical data on safe bearing capacity of soils are required by the building codes. However, even in this case of a well-established design practice, the allowable bearing capacity of soils can vary over a rather wide spectrum. For example, the allowable bearing pressure on a granite substratum can be on the order of 100 tons/ft^2, whereas the corresponding value for soft clay is only 1 ton/ft^2. Nevertheless, these and some intermediate values can be used on the basis of Eq. (24.15) in order to determine the foundation constant k. For soft clay this procedure should yield $2000/144 = 13.9$ psi.

Design Problem 24.3

Calculate the maximum deflection of a long I beam resting on compact soil, assuming a concentrated load W equal to 5000 lb. The estimated soil resistance is 150 lb/in. when the flange portion of the I beam is depressed 0.5 in. into the substratum. The beam has a moment of inertia in the stronger direction equal to 31.7 in.^{4} and a modulus of elasticity of 29×10^6 psi. The load is applied in the manner indicated in Fig. 24.5.

Solution

According to Eq. (24.15), the foundation constant is

$$k = \frac{P_r}{y} = \frac{150}{0.5} = 300 \text{ psi}$$

The parameter β follows from Eq. (24.18)

$$\beta = \left(\frac{k}{EI}\right)^{1/4} = \left(\frac{300}{29 \times 10^6 \times 31.7}\right)^{1/4} = 0.0239 \text{ in.}^{-1}$$

From Fig. 24.5, the general expression for the deflection is

$$y = \frac{W\beta}{2k} e^{-\beta x} (\cos \beta x + \sin \beta x)$$

Here $y = y_{\max}$ when $x = 0$, so that

$$y_{\max} = \frac{W\beta}{2k}$$

Hence

$$y_{\max} = \frac{5000 \times 0.0239}{2 \times 300} = 0.2 \text{ in.} \quad (5.06 \text{ mm}) \quad \blacklozenge$$

When dealing with cylindrical vessels, piping and similar configurations, all longitudinal elements of the vessel wall, loaded symmetrically with respect to the axis of the vessel, behave as beams on elastic foundation. The elastic foundation is provided by the adjacent elements of the vessel. Any radial displacement of the longitudinal element creates a strain in the hoop direction because the sides of an individual element are not free to rotate and are not able to accommodate any lateral deformation due to the Poisson's ratio effect. This set of constraint conditions can be represented by a reactive force on the element which opposes any deflection (Fig. 24.9). In mathematical terms, this model can be defined in relation to cylinder dimensions such as radius R and wall thickness T. According to the theory of elasticity, the strain in the hoop direction is equal to y/R if the radial displacement is denoted by y. Consequently, by Hooke's law the corresponding hoop stress must be equal to Ey/R, and hoop force per unit length of the element becomes $N = ETy/R$. Since the angle subtended by the element of unit width is equal to $1/R$, it can be shown that the relevant reactive force F is

$$F = \frac{EyT}{R^2} \tag{24.25}$$

Since the modulus of foundation can be looked upon as a spring constant equal to F/y, we get

$$k_0 = \frac{ET}{R^2} \tag{24.26}$$

The flexural rigidity EI for a flat plate element can be defined as

$$EI = \frac{ET^3}{12(1 - \nu^2)} \tag{24.27}$$

Hence, substituting Eqs. (24.26) and (24.27) into Eq. (24.18) and putting $\nu = 0.3$ yields

$$\beta = \frac{1.285}{(RT)^{1/2}} \tag{24.28}$$

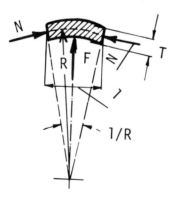

Fig. 24.9 Equilibrium diagram for a wall element.

Special Beam Problems

Equation (24.28) is very important because it allows us to use the theory of beams on elastic foundation in the analysis of shells, vessels, and similar configurations without any further recourse to the concept and specific numerical values of the modulus of foundation.

Design Problem 24.4

A cylindrical vessel of mean radius $R = 10$ in. and wall thickness $T = 0.25$ in. carries a uniformly distributed edge moment $M_0^* = 150$ lb-in. per inch of circumference. Calculate the maximum bending stress and the stress at a distance $x = 10$ in.

Solution

From Eq. (24.28),

$$\beta = \frac{1.285}{(10 \times 0.25)^{1/2}} = 0.813$$

From Eq.(24.21), when W^* is equal to zero, we find

$$M^* = M_0^* e^{-\beta x} (\cos \beta x + \sin \beta x)$$

The maximum bending stress is found at $x = 0$. Hence the above formula becomes

$$M^* = M_0^*$$

The parameter $\beta x = 10 \times 0.813 = 8.13$ rad. This gives 465.8 deg. Hence, at a distance $x = 10$ in., we get

$$M^* = 150 e^{-8.13} [\cos(465.8) + \sin(465.8)] = 0.03 \text{ lb-in./in.}$$

The corresponding bending stresses then become as follows. At $x = 0$

$$S = \frac{6 M_0^*}{T^2} = \frac{6 \times 150}{0.25^2} = 14,400 \text{ psi} \quad (99.3 \text{ N/mm}^2)$$

At $x = 10$ in.,

$$S = \frac{6 \times 0.03}{0.25^2} = 2.88 \text{ psi} \quad (0.02 \text{ N/mm}^2) \quad \blacklozenge$$

The foregoing calculations indicate how rapidly local bending stresses can be attenuated. However, local bending stresses in a pressure vessel can be very high, for example, at the junction of the shell and the adjoining stiffer part, such as its closure head or a circumferential stiffener. For more details of this nature, the reader is directed to the specialized literature on this subject [178, 192, 198, 200, 211, 212, 215].

SYMBOLS

A	Cross-sectional area, in.^{2} (mm^2)
a	Arbitrary distance, in. (mm)

B	Arbitrary length, in. (mm)
b	Arbitrary distance or width, in. (mm)
C	Distance to extreme fiber, in. (mm)
C_1, C_2, C_3, C_4	Integration constants
e	Effective width, in. (mm)
E, E_1, E_2, E_3	Moduli of elasticity, psi (N/mm^2)
F	Reactive force, lb (N)
I, I_1, I_2	Moments of inertia, in.^{4} (mm^4)
K	Wide beam factor
k	Elastic foundation constant, lb/in.^{2} (N/mm^2)
k_0	Modulus of foundation, lb/in.^{3} (N/mm^3)
L	Length of beam, in. (mm)
M	Bending moment in general, lb-in. (N-mm)
M^*	Bending moment per inch of circumference, lb-in./in. (N-mm/mm)
M_1, M_2	Bending moments in composite sections, lb-in. (N-mm)
M_0	Localized bending couple, lb-in. (N-mm)
M_0^*	Edge moment per unit length, lb-in./in. (N-mm/mm)
M_{max}	Maximum bending moment, lb-in. (N-mm)
N	Hoop force per unit length, lb/in. (N/mm)
P	Axial load in beam columns, lb (N)
P_r	Foundation reaction, lb/in. (N/mm)
Q	Shear load, lb (N)
Q^*	Shear load per inch of circumference, lb/in. (N/mm)
q	Uniform load, lb/in. or lb/in.^{2} (N/mm) or (N/mm^2)
R	Mean radius of cylinder, in. (mm)
r	Radius of loaded area, in. (mm)
S	General symbol for stress, psi (N/mm^2)
S_1, S_2	Stress in composite beams, psi (N/mm^2)
S_x	Stress in x direction, psi (N/mm^2)
S_y	Stress in y direction, psi (N/mm^2)
S_{max}	Maximum normal stress, psi (N/mm^2)
T	Thickness of plate or shell, in. (mm)
t	Thickness of slab, in. (mm)
W	Concentrated or total beam load, lb (N)
W^*	Edge shear per unit length, lb/in. (N/mm)
x	Coordinate, in. (mm)
Y	General symbol for deflection, in. (mm)
y_{max}	Maximum transverse deflection, in. (mm)
y	Deflection of beam on elastic foundation, in. (mm)
Z_1, Z_2	Section moduli in composite beams, in.^{3} (mm^3)
β	Parameter in beams on elastic foundation, in.$^{-1}$ (mm^{-1})
Δ	Deflection of beam column, in. (mm)
θ	Slope, rad
ν	Poisson's ratio
τ	General symbol for shear stress, psi (N/mm^2)
τ_{xy}	Two-dimensional shear, psi (N/mm^2)
τ_{max}	Maximum shear stress, psi (N/mm^2)

IV
CURVED MEMBERS

25
Curved Cantilevers

RELATED CONSIDERATIONS

The background material for the design analysis of elastic stresses and deflections in curved structural members is available in a number of classical works [136–139].

As long as the particular curved member is relatively thin, there is no need to consider the effects of the position of the neutral axis. In this case the cross section of the curved member and its response to the external bending moment can be estimated in the same manner as that for a straight beam. In other words, the bending stress follows directly from the customary engineering formulas $S_b = M/Z$, where S_b is the bending stress and M and Z are the bending moment and section modulus, respectively.

The derivation of the design formulas for the deflection of thin curved members, however, requires a different methodology and is best accomplished by the use of Castigliano's theory. Several ramifications of this principle have been discussed in Chaps. 6 and 7. The material that follows represents an application analysis of Castigliano's theory to practical engineering problems. Furthermore, in all cases considered, the bending moment equations are given, which can be used directly in the calculation of the bending stresses.

The general area of application of thin, curved members is rather extensive despite the fact that only a limited number of engineering texts are devoted to the subject. One type of a curved structure can be termed arched cantilever beam [135]. Although its response under load is statically determinate, the relevant load-deflection characteristics can also be applied to problems involving static indeterminacy. For instance, engineers frequently deal with redundant frame members formed in a circular arc, subtending any angle and carrying an in-plane load. In this type of analysis, one end of the cantilever is usually fixed while the other may be displaced by the application of forces and moments. From the relation between loading and deflection at the free end, the magnitude and direction of reactions can be found. When all the forces including reactions are known, stresses at any point of a curved cantilever can be estimated. Deflection analysis is considerably simplified by assuming the member to be relatively thin so that the contribution of shear and direct stresses to the total deflection can be neglected. In addition, the following simplified assumptions are made:

The cross section of a curved member is compact and uniform.
The deflections are relatively small.
The strains are elastic.
The loads are applied gradually
Flanged and thin-walled, open sections are excluded.
Hollow sections are permitted only when the customary moment-curvature relationship is deemed to apply.

The foregoing assumptions are not concerned with any ramifications of the neutral axis, because these curved cantilevers are taken to be thin. In other situations, however, where such curved members may be relatively thick, the reader is referred to Chap. 28.

ARCHED CANTILEVER UNDER VERTICAL LOAD

The more frequently required design formulas deal with a concentrated end load, a bending couple, or uniformly distributed load on arched cantilevers. A simple case of such a structural member under a concentrated loading is shown in Fig. 7.2. Design Problem 7.2 illustrates the basic procedure for using the Castigliano principle in deriving a specific working deflection formula. In the more general case of an arched cantilever shown in Fig. 25.1, the applicable bending moment equation is

$$M = PR(\cos\theta - \cos\beta) \tag{25.1}$$

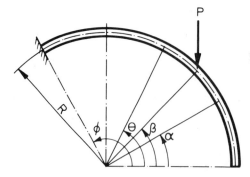

Fig. 25.1 General case of arched cantilever under vertical load.

Curved Cantilevers

Because of the discontinuity at $\theta = \beta$, the deflection equations should be derived separately for the two regions. The vertical and horizontal deflections Y and X occurring at the angle α (Fig. 25.1) can be found by the following equations for the conditions $\alpha < \beta$ and $\alpha > \beta$, where β is the angle at which the load is applied. When $\alpha < \beta$:

$$Y = \frac{PR^3}{EI}[(\phi - \beta)\cos\alpha\cos\beta - (\cos\alpha + \cos\beta)(\sin\phi - \sin\beta) \\ + 0.5(\phi - \beta) + 0.25(\sin 2\phi - \sin 2\beta)] \qquad (25.2)$$

$$X = \frac{PR^3}{EI}[\sin\alpha(\sin\phi - \sin\beta) - (\phi - \beta)\sin\alpha\cos\beta \\ + \cos\beta(\cos\beta - \cos\phi) + 0.25(\cos 2\phi - \cos 2\beta)] \qquad (25.3)$$

When $\alpha > \beta$:

$$Y = \frac{PR^3}{EI}[(\phi - \alpha)\cos\alpha\cos\beta - (\cos\alpha + \cos\beta)(\sin\phi - \sin\alpha) \\ + 0.25(2\phi + \sin 2\phi - 2\alpha - \sin 2\alpha)] \qquad (25.4)$$

$$X = \frac{PR^3}{EI}[\sin\alpha\sin\phi + \cos\beta(\cos\alpha - \cos\phi) \\ - (\phi - \alpha)\sin\alpha\cos\beta - \sin^2\alpha + 0.25(\cos 2\phi - \cos 2\alpha)] \qquad (25.5)$$

When $\alpha = \beta = 0$, we obtain the vertical and horizontal displacements of the free end of the curved cantilever under a concentrated vertical load. The relevant design formulas can be stated as

$$Y = \frac{PR^3 K_1}{EI} \qquad (25.6)$$

and

$$X = \frac{PR^3 K_2}{EI} \qquad (25.7)$$

where K_1 and K_2 are dimensionless factors depending only on angle ϕ. They can be obtained from Figs. 25.3. and 25.4.

ARCHED CANTILEVER UNDER HORIZONTAL LOAD

Employing the same methodology, the case of an arched cantilever subjected to horizontal load (Fig. 25.2) can be analyzed. For the symbols defined in Fig. 25.2, the bending moment becomes

$$M = HR(\sin\beta - \sin\phi) \qquad (25.8)$$

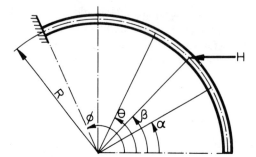

Fig. 25.2 General case of arched cantilever under horizontal load.

The corresponding deflection formulas for a general case of horizontal loading are then, when $\alpha < \beta$

$$Y = \frac{HR^3}{EI}[0.25(\cos 2\phi - \cos 2\beta) + \cos\alpha(\cos\beta - \cos\phi)$$
$$- (\phi - \beta)\sin\beta\cos\alpha + \sin\phi\sin\beta - \sin^2\beta] \qquad (25.9)$$

$$X = \frac{HR^3}{EI}[0.5(\phi - \beta) + 0.25(\sin 2\beta - \sin 2\phi)$$
$$- (\cos\beta - \cos\phi)(\sin\alpha + \sin\beta) + (\phi - \beta)\sin\alpha\sin\beta] \qquad (25.10)$$

When $\alpha > \beta$:

$$Y = \frac{HR^3}{EI}[0.25(\cos 2\phi - \cos 2\alpha) - (\phi - \alpha)\sin\beta\cos\alpha$$
$$+ \sin\beta(\sin\phi - \sin\alpha) + \cos^2\alpha - \cos\phi\cos\alpha] \qquad (25.11)$$

$$X = \frac{HR^3}{EI}[0.5(\phi - \alpha) + 0.25(\sin 2\alpha - \sin 2\phi)$$
$$+ (\phi - \alpha)\sin\alpha\sin\beta - (\cos\alpha - \cos\phi)(\sin\alpha + \sin\beta)] \qquad (25.12)$$

When $\alpha = \beta = 0$, Eqs. (25.9) to (25.12) yield simplified design formulas for the vertical and horizontal displacements, which can be expressed as follows:

$$Y = \frac{HR^3 K_2}{EI} \qquad (25.13)$$

and

$$X = \frac{HR^3 K_3}{EI} \qquad (25.14)$$

It is well to observe that the vertical displacement resulting from a horizontal load is numerically equal to the horizontal displacement due to a vertical load, provided that $P = H$. It is also of interest to recall that this correspondence of deflections agrees with the Maxwell's theorem of reciprocity for elastic structures [133, 134].

Curved Cantilevers

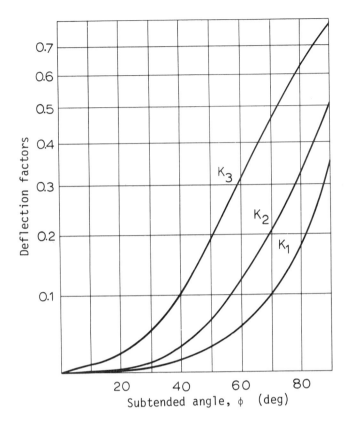

Fig. 25.3 Deflection factors for curved cantilevers with small angles.

The dimensionless deflection factors for the most common cases of curved cantilevers discussed above are shown in Figs. 25.3 and 25.4. These nomograms have been largely developed for practical purposes [140]. The analysis normally called for in predicting the load-deflection characteristics of such members is sometimes avoided by having a sample built in the shop and then subjecting it to tests. The time and cost involved in such a development can be minimized, if a simplified procedure, such as that given by Eqs. (25.6), (25.7), (25.13), and (25.14), and Figs. 25.3 and 25.4, is available. Utilizing the equations and charts given in this chapter, many theoretical and practical problems can be solved with the aid of the principle of superposition provided that the stresses are elastic. This can be illustrated as follows.

Design Problem 25.1

A curved cantilever beam of uniform tubular cross section, made of structural steel, subtends a circular arc of 150° and serves as a support structure for the external load W acting along the line inclined at 30° to the horizontal plane as shown in Fig. 25.5. The mean radii of curvature and the cross sections are 24 and 2 in., respectively. The yield strength of the material is 50,000 psi and its modulus is 30×10^6 psi. Determine the

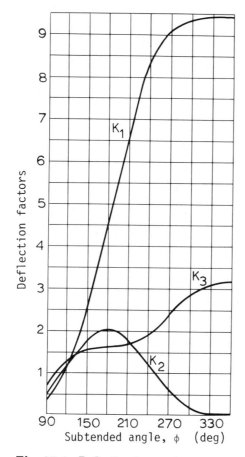

Fig. 25.4 Deflection factors for curved cantilevers with large angles.

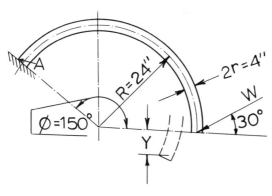

Fig. 25.5 Arched cantilever under inclined in-plane load.

Curved Cantilevers

maximum vertical displacement of the free end of the curved member corresponding to the onset of yield stress in bending.

Solution

Resolving load W into vertical and horizontal components gives

$$P = W \sin 30° = 0.5000W$$
$$H = W \cos 30° = 0.8660W$$

From Eqs. (25.1) and (25.8), the resultant bending moment is

$$M = R[P(\cos \phi - 1) - H \sin \phi]$$

This bending moment equation is the result of the addition of the bending moments used by the vertical and horizontal components of the load, as expressed by Eqs. (25.1) and (25.8) when $\theta = \phi = 150°$ and $\beta = 0°$. With the relevant substitutions, we get

$$M = 24[0.5W(-0.866 - 1) - 0.866W(0.5)] = -32.784W$$

and

$$S_y = -\frac{32.784W}{Z}$$

Here Z denotes the section modulus while the negative sign indicates that the outer surface at corner A (Fig. 25.5) is in tension according to the adopted sign convention for this numerical example. In further calculations this negative sign can be neglected. Substituting yield strength of the material, solving for W, and combining the result with the expressions for load components yields

$$P = 762.57Z$$

and

$$H = 1320.76Z$$

The total vertical displacement can be obtained by superposition of Eqs. (25.6) and (25.13)

$$Y = \frac{R^3}{EI}(PK_1 + HK_2)$$

For a tubular cross section, $I = rZ$. Finding next the value of the deflection factors from Fig. 25.4 gives

$$K_1 = 2.70$$
$$K_2 = 1.75$$

Substituting these values together with the expressions for P and H into the foregoing formula for the deflection, we find that the term Z cancels out and the required deflection becomes

$$Y = 1.01 \text{ in.} \quad (25.7 \text{ mm}) \quad \blacklozenge$$

ARCHED CANTILEVER UNDER UNIFORM LOAD

A rather important type of a thin, circular cantilever with uniform cross section loaded in plane can also be analyzed by the methods indicated in Chap. 7. When a curved member such as that shown in Fig. 25.6 is subject to the uniform in-plane loading due to the external pressure or inertia effects, it is first necessary to define the bending moment caused by the distributed load q. The elementary bending moment about a section defined by θ is

$$dM_q = qR^2(\cos\epsilon - \cos\theta)\,d\epsilon \tag{25.15}$$

Integrating Eq. (25.15) between the limits of 0 and θ gives

$$M_q = qR^2(\sin\theta - \theta\cos\theta) \tag{25.16}$$

The process of finding the vertical and horizontal displacements of the free end of the beam requires the introduction of fictitious quantities $\bar{P}$ and $\bar{H}$ applied in the direction of the desired deflections as shown in Fig. 25.6. Accordingly, the total bending moment at a section defined by θ is

$$M = qR^2(\theta\cos\theta - \sin\theta) + \bar{P}R(\cos\theta - 1) - \bar{H}R\sin\theta \tag{25.17}$$

It is well to recall here that Castigliano's principle, described in Chap. 6 by Eq. (6.11), states that the deflection under and in the direction of load P is equal to the partial derivative of the elastic strain energy U stored with respect to P, written

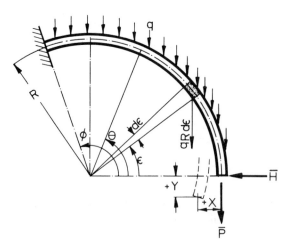

Fig. 25.6 Uniformly loaded arched cantilever.

Curved Cantilevers

as $\partial U/\partial P$. If instead of a real load P, we take a fictitious load $\bar{P}$, the process of derivation will still be the same. In our particular case here, we will have

$$Y = \frac{\partial U}{\partial \bar{P}} = \frac{R}{EI}\int_0^\phi M\frac{\partial M}{\partial \bar{P}}\,d\theta$$

Similarly, for the horizontal deflection, we get

$$X = \frac{\partial U}{\partial \bar{H}} = \frac{R}{EI}\int_0^\phi M\frac{\partial M}{\partial \bar{H}}\,d\theta$$

Note that in both cases the applicable bending moment equation is given by Eq. (25.17), which includes both real and fictitious quantities. When both $\bar{P}$ and $\bar{H}$ are made equal to zero, the bending moment becomes

$$M = qR^2(\theta\cos\theta - \sin\theta)$$

This expression, together with the appropriate partial derivatives, enables us to calculate the displacements Y and X through the process of substitution and integration.

The partial derivative terms needed for the application of the principle expressed by Eq. (6.11) are found directly from Eq. (25.17)

$$\frac{\partial M}{\partial \bar{P}} = R(\cos\theta - 1) \tag{25.18}$$

and

$$\frac{\partial M}{\partial \bar{H}} = -R\sin\theta \tag{25.19}$$

Bearing in mind that the fictitious terms have been made equal to zero, the vertical and horizontal deflections can be defined as follows:

$$Y = \frac{qR^4}{EI}\int_0^\phi (\theta\cos\theta - \sin\theta)(\cos\theta - 1)\,d\theta \tag{25.20}$$

and

$$X = \frac{qR^4}{EI}\int_0^\phi (\sin\theta - \theta\cos\theta)\sin\theta\,d\theta \tag{25.21}$$

Integrating and substituting the relevant integration limits yields

$$Y = \frac{qR^4}{8EI}(2\phi\sin 2\phi - 8\phi\sin\phi + 3\cos 2\phi - 16\cos\phi + 2\phi^2 + 13) \tag{25.22}$$

$$X = \frac{qR^4}{8EI}(4\phi + 2\phi\cos 2\phi - 3\sin 2\phi) \tag{25.23}$$

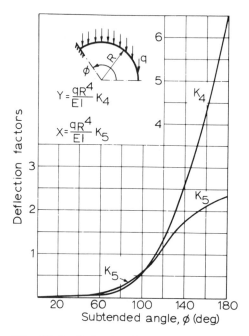

Fig. 25.7 Deflection factors for arched cantilever under uniform load.

The expressions for the vertical and horizontal displacements of the free end given above can be simplified as shown below.

$$Y = \frac{qR^4}{EI} K_4 \tag{25.24}$$

and

$$X = \frac{qR^4}{EI} K_5 \tag{25.25}$$

These functions are illustrated in Fig. 25.7 and are suitable for direct application to analysis and design.

Design Problem 25.2

The open ring shown in Fig. 25.8 is made of flat steel strip 2 in. × 0.1 in. and is shaped to form a circle with a gap Δ equal to 0.1 in. Calculate the external load q necessary to close the gap and the corresponding maximum bending stress in the ring. Take $E = 30 \times 10^6$ psi.

Solution

When $\phi = 180°$, Eq.(25.22) gives

$$Y = \left(\frac{\pi^2 + 16}{4}\right) \frac{qR^4}{EI}$$

Curved Cantilevers

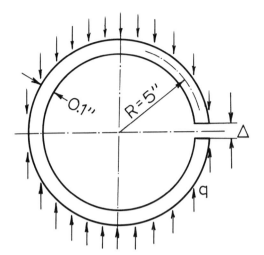

Fig. 25.8 Example of open ring under uniform external load.

Since $\Delta = 2Y$, then

$$q = \frac{2\Delta EI}{(\pi^2 + 16)R^4}$$

where $I = (2 \times 0.1^3)/12$, $\Delta = 0.1$ in., $E = 30 \times 10^6$ psi, and $R = 5$ in. Hence, substitution gives

$$q = \frac{2 \times 0.1 \times 30 \times 10^6}{(\pi^2 + 16)6 \times 5^4 \times 10^3} = 0.06 \text{ lb/in.} \quad (0.0105 \text{ N/mm})$$

From Eq. (25.17) for $\bar{P} = \bar{H} = 0$ and $\theta = \pi$,

$$M = \pi q R^2 = \pi \times 0.06 \times 25 = 4.72 \text{ lb-in.}$$

Since

$$Z = \frac{2 \times 0.1^2}{6} = 3.33 \times 10^{-3} \text{ in.}^3$$

the corresponding bending stress is

$$S_b = \frac{4.72}{3.33} \times 10^3 = 1417 \text{ psi} \quad (9.77 \text{ N/mm}^2)$$

Although the maximum stress in this case is rather negligible, the deflection is on the order of magnitude of the material thickness. This corresponds to the condition for which the small-deflection theory is still appropriate. The example indicates that this type of open ring, when unrestrained at the cut ends, represents a highly flexible structural member. In fact, it can be shown that the weight of the ring itself should close the major portion of the specified gap Δ. This is, therefore, a rather striking example of the deflection-governed design. ◆

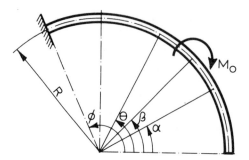

Fig. 25.9 Arched cantilever under in-plane couple.

ARCHED CANTILEVER UNDER MOMENT

When an arched cantilever is loaded eccentrically in the plane of curvature, as shown in Fig. 25.9, the bending moment is, of course, equal to M_0 for all values of θ between β and ϕ. The relevant deflection expressions become, when $\alpha < \beta$

$$Y = \frac{M_0 R^2}{EI}[(\phi - \beta)\cos\alpha - \sin\phi + \sin\beta] \tag{25.26}$$

$$X = \frac{M_0 R^2}{EI}[\cos\beta - \cos\phi - (\phi - \beta)\sin\alpha] \tag{25.27}$$

When $\alpha > \beta$

$$Y = \frac{M_0 R^2}{EI}[(\phi - \alpha)\cos\alpha - \sin\phi + \sin\alpha] \tag{25.28}$$

$$X = \frac{M_0 R^2}{EI}[\cos\alpha - \cos\phi - (\phi - \alpha)\sin\alpha] \tag{25.29}$$

When $\alpha = \beta = 0$, the foregoing formulas simplify considerably and can be defined as follows:

$$Y = \frac{M_0 R^2}{EI} K_6 \tag{25.30}$$

and

$$X = \frac{M_0 R^2}{EI} K_7 \tag{25.31}$$

For convenience of design, Eqs. (25.30) and (25.31) have been plotted in Fig. 25.10.

COMPLEX-SHAPE CANTILEVER

Another configuration of a curved cantilever is shown in Fig. 25.11. The structure consists of a circular portion and a straight beam extension. It is acted upon by a system of end forces, which cause two components of deflection and slope, with

Curved Cantilevers

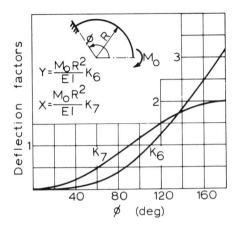

Fig. 25.10 Deflection factors for arched cantilever under in-plane couple.

a positive sign convention as indicated. This configuration may be of interest in the study of redundant frames, piping, or engine lines subjected to the movement of one support of a line, for example, with respect to the other support because of the temperature gradients or assembly misalignments [141, 142]. If such a frame or line is restricted against any movement of supports by the adjacent parts of a machine or structure under consideration, damaging stresses can be induced either in the line or the support.

In calculating the deflection of such members as shown in Fig. 25.11, the designer sometimes employs the concept of a developed length in conjunction with the elementary formula for a straight cantilever beam, $Y = PL^3/3EI$. It can be shown by a rational analysis that neglecting the true effect of the curvature can lead to significant errors for certain proportions of curved-end cantilevers [11].

Because of the change in shape of the cantilever as we progress from the unsupported end, it is necessary to define two separate expressions for the bending moments. Along the straight portion AB, we have

$$M_1 = -Px - M_0 \tag{25.32}$$

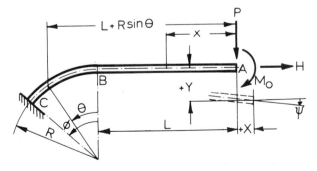

Fig. 25.11 General case of curved cantilever.

Along the curved portion BC, the bending moment is

$$M_2 = -PR(k + \sin\theta) - HR(1 - \cos\theta) - M_0 \qquad (25.33)$$

Note that k denotes a convenient dimensionless ratio L/R.

In deriving individual deflection formulas employing the theorem of Castigliano [8, 9], the relevant terms can be taken from Eqs. (25.32) and (25.33). The system of forces indicated in Fig. 25.11 may be interpreted as real or imaginary, and it is not necessary to carry superscript vertical bars such as those shown in Fig. 25.6. This procedure may be illustrated briefly as follows.

Suppose we wish to calculate the horizontal deflection X for a curved cantilever given in Fig. 25.11, which carries a vertical load P. Because of the change in the bending moment equations, the integration should be performed over the two separate regions. In line with Eqs. (6.10) and (6.11), and neglecting superscript notation for the imaginary forces, the general expression for calculating the horizontal displacement under vertical load P is

$$X = \frac{1}{EI}\int_0^L M_1 \frac{\partial M_1}{\partial H} dx + \frac{1}{EI}\int_0^\phi M_2 \frac{\partial M_2}{\partial H} R\, d\theta \qquad (25.34)$$

The applicable partial derivatives from Eqs. (25.32) and (25.33) are

$$\frac{\partial M_1}{\partial H} = 0 \qquad (25.35)$$

$$\frac{\partial M_2}{\partial H} = -R(1 - \cos\theta) \qquad (25.36)$$

The relevant bending moments representing real forces are

$$M_1 = -Px \qquad (25.37)$$

and

$$M_2 = -PR(k + \sin\theta) \qquad (25.38)$$

The procedure is then to substitute Eqs. (25.35) to (25.38) into Eqs. (25.34) and to carry out the integration in accordance with the indicated limits. Utilizing this approach, various deflection formulas can be derived, and useful charts developed for design of special mechanical springs and similar components [143–145]. Some of the general formulas applicable to this type of a problem can be summarized as follows. For the case of a transverse load P, as shown in Fig. 25.11, we have

$$Y = \frac{PR^3}{EI}(0.33k^3 + k^2\phi + 2k - 2k\cos\phi + 0.5\phi - 0.25\sin 2\phi) \qquad (25.39)$$

$$X = \frac{PR^3}{EI}(k\phi - k\sin\phi - \cos\phi + 0.25\cos 2\phi + 0.75) \qquad (25.40)$$

$$\psi = \frac{PR^2}{EI}(0.5k^2 + k\phi - \cos\phi + 1) \qquad (25.41)$$

Curved Cantilevers

When only horizontal load H is involved, the relevant expressions become

$$Y = \frac{HR^3}{EI}(k\phi - \cos\phi - k\sin\phi + 0.25\cos 2\phi + 0.75) \qquad (25.42)$$

$$X = \frac{HR^3}{EI}(1.5\phi - 2\sin\phi + 0.25\sin 2\phi) \qquad (25.43)$$

$$\psi = \frac{HR^2}{EI}(\phi - \sin\phi) \qquad (25.44)$$

Finally, the displacements due to the moment M_0 applied at the free end are

$$Y = \frac{M_0 R^2}{EI}(0.5k^2 + k\phi - \cos\phi + 1) \qquad (25.45)$$

$$X = \frac{M_0 R^2}{EI}(\phi - \sin\phi) \qquad (25.46)$$

$$\psi = \frac{M_0 R}{EI}(k + \phi) \qquad (25.47)$$

The above equations can be used with the aid of the principle of superposition for any combined loading conditions, such as those involving simultaneous action of P, H, and M_0. These equations may be employed in the study of statically determinate and indeterminate structures [146]. For special cases involving $\phi = \pi/2$ and $\phi = \pi$, simple design charts can be developed such as those given in Figs. 25.12 and 25.13.

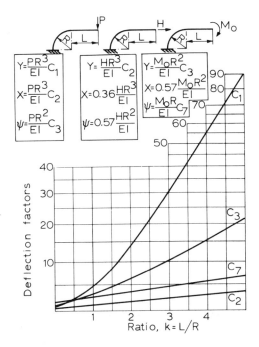

Fig. 25.12 Chart for quarter-circle curved cantilevers.

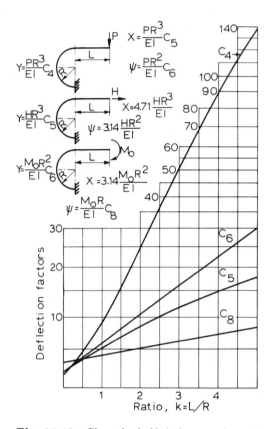

Fig. 25.13 Chart for half-circle curved cantilevers.

Design Problem 25.3

Curved cantilever with a quarter-circle bend carries a concentrated load W, inclined $45°$ to the straight portion of the member and applied to the free end of the member. The general case is similar to that shown in Fig. 25.11 if we resolve the external load W into vertical and horizontal components, defined as P and H. Assuming that the length ratio $k = L/R = 1$ and denoting by E and I the relevant modulus of elasticity and the moment of inertia, respectively, derive the design formula for the deflection component normal to the straight portion of the member. This deflection component is denoted by Y in Fig. 25.11. Use the equations derived for the general case of a curved cantilever and the notation given in Fig. 25.11.

Solution

From the statement of the problem, we have

$$P = H = \frac{W\sqrt{2}}{2} \quad \text{since } \sin 45° = \cos 45° = \frac{\sqrt{2}}{2}$$

Let Y_1 and Y_2 denote the vertical deflections due to P and H, respectively.

Curved Cantilevers

Putting $k = 1$ and $\phi = 90°$, Eq. (25.39) reduces to

$$Y_1 = \frac{PR^3}{EI}\left(\frac{28 + 9\pi}{12}\right)$$

Also for the same k and ϕ values, Eq. (25.42) gives

$$Y_2 = \frac{HR^3}{EI}\left(\frac{\pi - 1}{2}\right)$$

According to the principle of superposition, the two component deflections can be added directly as long as the corresponding stresses are elastic. Hence, we get $Y = Y_1 + Y_2$, or

$$Y = \frac{WR^3(15\pi + 22)\sqrt{2}}{24EI} \qquad \blacklozenge$$

CURVED-END CANTILEVER

When a structural member is curved at the free rather than at the built-in end, the general theoretical development is essentially the same as that for other curved members. For example, if our design configuration is similar to that shown in Fig. 25.14a, and if the external concentrated force is applied at the free end in either the Y or the X direction, we can start by writing down the bending moment expression separately for the curved and straight portions. We assume that P acts along y while H is directed along the x coordinate. In terms of the orientation of the curved-end cantilevers shown in Fig. 25.14, we will denote the vertical and horizontal loads by P and H, respectively.

Certainly, the more frequently encountered configurations involve quarter- and half-circle bends, which simplify the derivation of the design formulas considerably. The problems become more complex, however, when the curved portions subtend angles other than 90 or 180°.

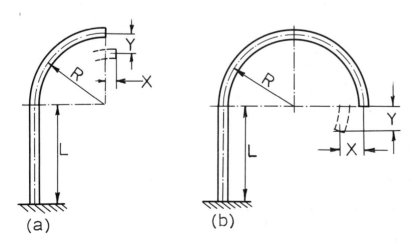

Fig. 25.14 Example of curved-end cantilevers.

Table 25.1 Displacement Factors for Curved-End Cantilevers

Type	Deflection	End Load (Vertical)	End Load (Horizontal)
Quarter circle: Fig. 25.14a	Vertical Horizontal	$k + 0.7854$ $0.5k^2 + k + 0.5$	$0.5k^2 + k + 0.5$ $0.3333k^3 + k^2 + k + 0.3562$
Half circle: Fig. 25.14b	Vertical Horizontal	$4k + 4.7124$ $2 - k^2$	$2 - k^2$ $0.3333k^3 + 1.5708$

When the external end load is applied at an arbitrary angle, the solution can be obtained by first resolving the end load into vertical and horizontal components and then by using the method of superposition. This general approach has been illustrated in practical terms in conjunction with Design Problem 25.1 and 25.3.

Analytical treatment of complex shape members, involving straight and curved portions simultaneously, was discussed in some detail in the preceding section. This account will be limited to providing a working summary of the design formulas for the most frequently needed practical cases. Furthermore, since the estimate of the bending stress follows directly from the knowledge of the bending moment at a particular point, the design summary will include the dimensionless factors such as those shown in Table 25.1. The factors should be used in conjunction with Fig. 25.14, which describes the general condition of loading and support with the appropriate references to the vertical and horizontal directions. In general, then, the deflection can be defined as follows:

Deflection = dimensional parameter × dimensionless factor

Here the typical dimensional parameter consists of PR^3/EI or HR^3/EI, while the dimensionless quantity may be equal to one of the displacement factors listed in Table 25.1 [11, 144].

For instance, if the deflection X is required for a half-circle cantilever under a horizontal load H acting in the direction of X, the relevant formula is $(HR^3/EI)(0.3333k^3 + 1.5708)$. In all cases pertaining to Fig. 25.14 and Table 25.1, the loads are presumed to be applied at the free end. A specific application example, pertinent to Table 25.1, is given in Design Problem 25.4. Note that in this case the dimensional parameter is PR^3/EI, while all the dimensionless factors are functions of the length ratio k. Because of this feature, Table 25.1 is applicable to a variety of proportions of curved-end cantilevers.

Design Problem 25.4

A tubular support column made in the form of a curved-end cantilever carries a vertical load of $P = 200$ lb, as shown in Fig. 25.15. The material is structural steel with the modulus of elasticity $E = 28 \times 10^6$ psi. The relevant dimensions are indicated in the figure. Assuming that the effect of tube flattening due to bending is ignored, calculate the amount of the displacement of the free end and the corresponding maximum bending stress.

Curved Cantilevers

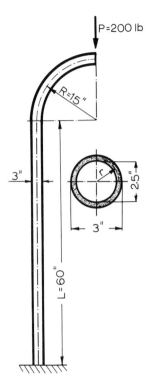

Fig. 25.15 Tubular curved-end cantilever.

Solution

The displacement equations for the horizontal and vertical directions follow from Table 25.1, corresponding to the cantilever with a quarter-circle bend:

$$X = \frac{PR^3}{EI}(0.5k^2 + k + 0.5)$$
$$Y = \frac{PR^3}{EI}(k + 0.7854)$$

The straight length-to-radius ratio is $k = 60/15 = 4$. Also, for mean radius of tubular cross section $r = 1.375$ in. and wall thickness $t = 0.25$ in., the moment of inertia is

$$I = \pi r^3 t$$
$$= \pi \times 1.375^3 \times 0.25 = 2.0417 \text{ in.}^4$$

Hence, using the expressions above we get

$$X = \frac{200 \times 15^3 (0.5 \times 16 + 4 + 0.5)}{28 \times 10^6 \times 2.0417} = 0.148 \text{ in.}$$

and

$$Y = \frac{200 \times 15^3 (4 + 0.7854)}{28 \times 10^6 \times 2.0417} = 0.057 \text{ in.}$$

The resultant displacement is then

$$u = (0.148^2 + 0.057^2)^{1/2} = 0.159 \text{ in.} \quad (4.03 \text{ mm})$$

Since the section modulus is

$$Z = \pi \times 1.375^2 \times 0.25 = 1.485 \text{ in.}^3$$

The maximum bending stress at any location along the straight portion becomes

$$S_b = \frac{PR}{Z}$$
$$= \frac{200 \times 15}{1.485} = 2020 \text{ psi} \quad (13.93 \text{ N/mm}^2) \quad \blacklozenge$$

SYMBOLS

C_1 through C_8	Deflection factors for curved cantilevers
E	Modulus of elasticity, psi, (N/mm^2)
H	Horizontal load, lb (N)
$\bar{H}$	Fictitious horizontal load, lb (N)
I	Moment of inertia, in.^{4} (mm^4)
K_1 through K_7	Deflection factors for arched cantilevers
$k = L/R$	Straight length-to-radius ratio
L	Length of straight portion, in. (mm)
M	Bending moment, lb-in. (N-mm)
M_0	Externally applied bending couple, lb-in. (N-mm)
M_q	Bending moment due to uniform load, lb-in. (N-mm)
M_1, M_2	Bending moments for curved cantilevers, lb-in. (N-mm)
P	Vertical load, lb (N)
$\bar{P}$	Fictitious vertical load, lb (N)
q	Uniform load, lb/in. (N/mm)
R	Mean radius of curvature, in. (mm)
r	Mean radius of tube, in. (mm)
S_b	Bending stress, psi (N/mm^2)
S_y	Yield stress, psi (N/mm^2)
t	Wall thickness, in. (mm)
u	Resultant deflection, in. (mm)
W	Skew load, lb (N)
X	Horizontal deflection, in. (mm)
x	Arbitrary distance, in. (mm)
Y	Vertical deflection, in. (mm)
Z	Section modulus, in.^{3} (mm^3)
α	Angle at which deflection is required, rad
β	Angle at which load is applied, rad
Δ	Gap in split ring, in. (mm)
ϵ	Auxiliary angle, rad

Curved Cantilevers

θ	Angle at which forces are considered, rad
ϕ	Angle subtended by curved portion, rad
ψ	Slope, rad

26
Complex-Shape Springs

DEFINITIONS AND ASSUMPTIONS

The methods of analysis employed in the derivation of formulas for curved cantilevers apply directly to a large number of shapes involving a combination of straight and circular contours. The popular term "flat springs" is, of course, a misnomer and only implies that complex-shape springs are made of flat stock before they are formed into specific bent configurations. Because of the great variety of shapes which it is theoretically possible to design, a complete discussion of this subject is certainly beyond the scope of this chapter. The treatment is limited, therefore, to a number of typical shapes, the design theory of which may be helpful in further studies of similar components [147]. A small collection of typical complex spring forms is shown in Fig. 26.1. The analysis of conventional mechanical springs is included in Chap. 41.

Predicting deflections and stresses of complex springs has obvious advantages, since with the help of the calculations, fewer test samples can be built in the shop before any final design is firmed up and mass production established.

In developing various analytical expressions, it is assumed that the external forces applied to a spring are delivered without a shock. The cross-sectional areas are constant and only small deflections will be considered so that the principle of superposition remains valid.

Complex-Shape Springs

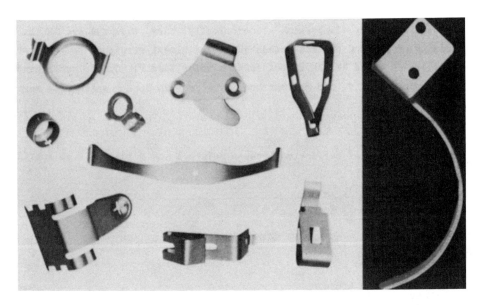

Fig. 26.1 Typical complex spring shapes. (Courtesy of Associated Spring Barnes Group, Inc.).

SNAP RING

One of the simpler configurations which may be described as a snap-ring spring made of flat stock of width b and thickness h is shown in Fig. 26.2. Because of symmetry, it is sufficient to analyze one-half of the spring, leading to the following bending moment expression at point B:

$$M = PR(\cos\alpha - \cos\theta) \tag{26.1}$$

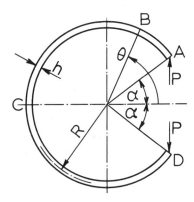

Fig. 26.2 Snap-ring spring.

For a specified value of α, the maximum bending stress becomes

$$S_b = \frac{6PR(1+\cos\alpha)}{bh^2} \tag{26.2}$$

When α increases, the portion of the arc AC decreases until $\cos\alpha = -1$ and the bending effect disappears. Theoretically, then, when the arc in the vicinity of point C becomes rather short and flattens out, the only stress in existence can be the direct tension under load P. On the other hand, when α approaches 0, the maximum bending stress in such a split ring must be

$$S_b = \frac{12PR}{bh^2} \tag{26.3}$$

The total amount of deflection between points A and D follows from application of Eq. (6.11)

$$Y = \frac{2}{EI}\int_\alpha^\pi M\frac{\partial M}{\partial P} R\, d\theta \tag{26.4}$$

From Eq. (26.1)

$$\frac{\partial M}{\partial P} = R(\cos\alpha - \cos\theta) \tag{26.5}$$

Hence, substituting Eqs. (26.1) and (26.5) into Eq. (26.4) and integrating between the indicated limits yields

$$Y = \frac{PR^3}{EI}\left[(\pi-\alpha)(1+2\cos^2\alpha)+1.5\sin 2\alpha\right] \tag{26.6}$$

Equation (26.6) also applies when the direction of loading is reversed, as is the case with all other formulas when the response of the structure is elastic. When $\alpha = 0$, that is, for a snap-ring spring with a minute gap between points A and D, the deflection becomes

$$Y = \frac{3\pi PR^3}{EI} \tag{26.7}$$

The approximate numerical check in this case can also be obtained with the help of the deflection factor K_1 from Fig. 25.4.

The split ring shown in Fig. 26.3 is subjected to pull along the vertical diameter AB. This configuration applies to the design of piston rings, and the gap at the split end is regarded to be rather small compared with the radius of curvature R. The design equations can be developed on the assumption that the stresses are essentially those for a conventional straight member and the deflection is caused by the bending stresses alone.

The increase in opening at D is equal to twice the displacement of the free end D relative to C, as shown by the arched cantilever model in Fig. 26.4. Since there is

Complex-Shape Springs

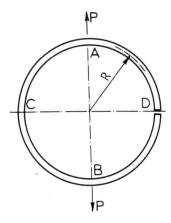

Fig. 26.3 Snap-ring spring under diametral load.

no force at point D, a fictitious force $\bar{P}$ is applied at the point and in the direction of the required deflection. The bending moment for the portion AD is then

$$M_1 = \bar{P}R(1 - \cos\theta) \tag{26.8}$$

Similarly, for the second quadrant

$$M_2 = PR\cos(\pi - \theta) + \bar{P}R[1 + \cos(\pi - \theta)] \tag{26.9}$$

The general expression for the deflection is

$$Y = \frac{2}{EI}\int_0^{\pi/2} M_1 \frac{\partial M_1}{\partial \bar{P}} R\,d\theta + \frac{2}{EI}\int_{\pi/2}^{\pi} M_2 \frac{\partial M_2}{\partial \bar{P}} R\,d\theta \tag{26.10}$$

where

$$\frac{\partial M_1}{\partial \bar{P}} = R(1 - \cos\theta) \tag{26.11}$$

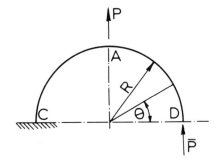

Fig. 26.4 Arched cantilever model for piston ring analysis.

and

$$\frac{\partial M_2}{\partial \bar{P}} = R[1 + \cos(\pi - \theta)] \qquad (26.12)$$

Substituting Eqs. (26.8), (26.9), (26.11), and (26.12) into Eq. (26.10), integrating, and making $\bar{P} = 0$, we obtain

$$Y = \frac{PR^3(4 + \pi)}{2EI} \qquad (26.13)$$

PRECURVED CANTILEVER

In some applications, a precurved type of a cantilever spring can be employed as a loading or a support member as shown in Fig. 26.5. One end of the spring can be considered fixed while the loaded end can be either free or constrained. For the case of unconstrained end A, the bending moment at any section is

$$M = PR \sin \theta \qquad (26.14)$$

The maximum bending stress for this spring, then, is

$$S_b = \frac{6PR}{bh^2} \qquad (26.15)$$

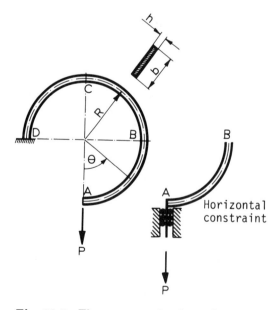

Fig. 26.5 Three-quarter circular spring.

Complex-Shape Springs

Using Eqs. (6.11) and (26.14), the displacement of the free end becomes

$$Y = \frac{3\pi P R^3}{4EI} \tag{26.16}$$

If the unsupported end AB of the spring is now constrained by guides to move in a vertical direction only as shown in the figure, the bending moment equation, Eq. (26.14), must be modified by introducing a constraining couple M_f as a redundant and unknown quantity

$$M = PR\sin\theta - M_\mathrm{f} \tag{26.17}$$

Since, owing to the constraint, the angle of rotation of the end A must be equal to zero, the following expression can be used for calculating M_f:

$$\int_0^{3\pi/2} M \frac{\partial M}{\partial M_\mathrm{f}} \, d\theta = 0 \tag{26.18}$$

Solution of Eq. (26.18) with the aid of the bending moment relation, Eq. (26.17), gives $M_\mathrm{f} = 2PR/3\pi$, and the corresponding bending moment formula becomes

$$M = PR\left(\sin\theta - \frac{2}{3\pi}\right) \tag{26.19}$$

Again using Eq. (6.11), (26.19), and the partial derivative from Eq. (26.19) with respect to P gives

$$Y = \frac{PR^3}{EI}\left(\frac{9\pi^2 - 8}{12\pi}\right) \tag{26.20}$$

For the particular case considered, the constraint reduces downward deflection by about 9%. However, this effect is not constant and depends on the angle subtending the spring arc. It may be of interest to note that the maximum bending stress in the spring is still given by Eq. (26.16). The deflections can also be expressed as a function of stress for the free-end and the guided-end design conditions, giving

$$Y = \frac{3\pi R^2 S}{2Eh} \quad \text{(free end)} \tag{26.21}$$

and

$$Y = \frac{(9\pi^2 - 8)R^2 S}{6\pi Eh} \quad \text{(guided end)} \tag{26.22}$$

Design Problem 26.1

A flat S spring, shown in Fig. 26.6, has width of cross section b and depth h. If the maximum bending stress S_b and the total deflection under load P are specified, determine a formula for the required minimum depth of cross section h.

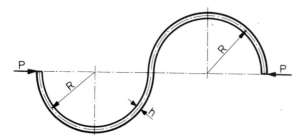

Fig. 26.6 Half-circle S spring.

Solution

Since the total deflection is equal to the sum of the deflections of four arched cantilevers subtending 90° each, Eq. (25.10) can be utilized. Let $H = P$ and in line with Fig. 25.2, $\alpha = \beta = 0$ and $\phi = \pi/2$. Hence, Eq. (25.10) gives $X = \pi H R^3/4EI$. Putting $Y = 4X$, $I = bh^3/12$, and $H = P$ yields

$$Y = \frac{12\pi P R^3}{bEh^3}$$

The maximum bending stress is

$$S_b = \frac{6PR}{bh^2}$$

Hence, solving the above two equations gives

$$h = \frac{2\pi R^2 S_b}{YE} \quad \blacklozenge$$

THREE-QUARTER WAVE SPRING

With increasing complexity of configurations consider next a three-quarter circular wave shown in Fig. 26.7. Portions AC and CF can be analyzed separately by writing the bending moment equations for points B and D, respectively. For point B, Eq. (26.14) describes the effect of P. For point D, however, the bending moment equation should be modified to give

$$M = PR(2 - \cos\theta) \tag{26.23}$$

It is obvious that at $\theta = 0$ for the CDF portion of the spring, $M = PR$; and at $\theta = \pi$, $M = 3PR$. The maximum stress for this type of a spring is due to bending, equal to

$$S_b = \frac{18PR}{bh^2} \tag{26.24}$$

Complex-Shape Springs

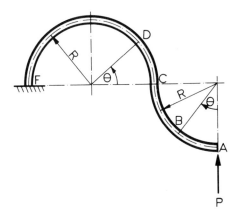

Fig. 26.7 Three-quarter circular wave spring.

The total resilience of the spring is obtained from the algebraic sum of the two component values corresponding to the portions ABC and CDF, and therefore the total deflection can be described as follows:

$$Y = \frac{PR^3}{EI} \int_0^{\pi/2} \sin^2\theta \, d\theta + \frac{PR^3}{EI} \int_0^{\pi} (2 - \cos\theta)^2 \, d\theta \qquad (26.25)$$

Integrating Eq. (26.25) gives the required formula for the deflection under load P

$$Y = \frac{19\pi PR^3}{4EI} \qquad (26.26)$$

The term "resilience" implies the elastic strain energy stored in the spring which can be totally recovered upon the release of load P.

If a straight portion of length L is now added as shown in Fig. 26.8, the maximum bending stress remains unchanged because the extension FG acts as a cantilever beam, built in at G and loaded by a couple $(3PR)$ at F, provided that the

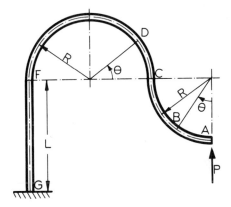

Fig. 26.8 Three-quarter circular wave spring with extension.

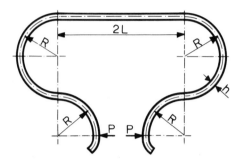

Fig. 26.9 Clip spring.

direct stresses are ignored. The total resilience of the spring, however, is increased by the effect of the end couple on the cantilever, so that the modified deflection formula becomes

$$Y = \frac{PR^3}{EI}\left(9k + \frac{19\pi}{4}\right) \tag{26.27}$$

Here $k = L/R$, as before, so that for $k = 0$, Eq. (26.27) reduces to Eq. (26.26). The direct extension of the portion FG under load P is ignored as being small relative to the deflection at point A.

CLIP SPRING

Armed with the result given by Eq. (26.27), it is now a simple matter to calculate the amount of spread of a clip spring under loads P as shown in Fig. 26.9.

$$Y = \frac{2PR^3}{EI}\left(9k + \frac{19\pi}{4}\right) \tag{26.28}$$

Combining Eqs. (26.24) and (26.28) gives the maximum bending stress in the spring in terms of the deflection and spring dimensions.

$$S_b = \frac{hYE}{R^2(12k + 19.9)} \tag{26.29}$$

GENERAL U SPRING

In developing the design formulas for the general type of U spring shown in Fig. 26.10, the theory of a curved cantilever, such as that given previously for Fig. 25.11, can be used. For instance, the component force, P_1, acting perpendicular to the straight leg L, can be defined as

$$P_1 = P\cos\nu \tag{26.30}$$

Complex-Shape Springs

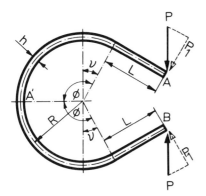

Fig. 26.10 General U spring.

If Y_1 is the displacement of loaded end A in the direction of P_1 and Y, the total change in the distance AB due to P, then

$$Y_1 = 0.5Y \cos \nu \tag{26.31}$$

From the previous results such as Eq. (25.39), we get

$$Y_1 = \frac{P_1 R^3}{EI} F(k, \phi) \tag{26.32}$$

where

$$F(k, \phi) = 0.33k^3 + k^2\phi + 2k(1 - \cos \phi) + 0.5\phi - 0.25 \sin 2\phi \tag{26.33}$$

Hence

$$Y = \frac{2PR^3}{EI} F(k, \phi) \tag{26.34}$$

The corresponding maximum bending stress at point A', Fig. 26.10, can be shown to be

$$S_b = \frac{6PR}{bh^2} \left(1 + \sin \nu + \frac{k}{\cos \nu}\right) \tag{26.35}$$

For the validity of Eq. (26.35), $\phi = \nu + \pi/2$. When, on the other hand, $\phi \leq \pi/2$, the foregoing stress formula becomes

$$S_b = \frac{6PR}{bh^2}(1 - \cos \phi + k \cos \nu) \tag{26.36}$$

For the special case of $\nu = 0$ and $\phi = \pi/2$, Eq. (26.36) reduces to

$$S_b = \frac{6PR(1 + k)}{bh^2} \tag{26.37}$$

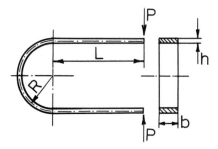

Fig. 26.11 Flat U beam without constraint.

INSTRUMENT-TYPE U SPRING

In many machine design applications and especially in the areas of instruments and precision equipment, U-shaped springs are used successfully. They can be manufactured with the aid of a relatively simple machinery and can be easily assembled. These springs are frequently loaded in plane and can be used both as tension and compression devices. There are essentially three types of U springs commonly found in hardware design as shown in Figs. 26.11–26.13. The maximum bending moment and deflection formulas for these configurations can be summarized in Table 26.1. The relevant design factors needed for the formulas are given in Figs. 26.14–26.16 [148, 149].

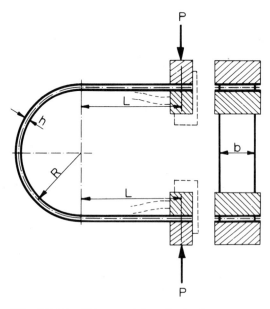

Fig. 26.12 U beam with ends fixed as to slope.

Complex-Shape Springs

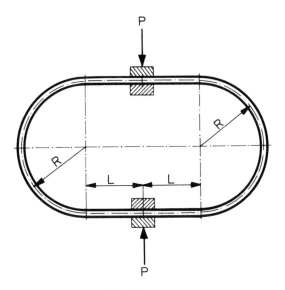

Fig. 26.13 Double U beam.

Design Problem 26.2

A flat U spring is made of aluminum and has the dimensions shown in Fig. 26.17. If the modulus of elasticity is $E = 10 \times 10^6$ psi, calculate the decrease in the distance between the free ends under the concentrated loads P. Also, estimate the corresponding bending stress.

Solution

From Fig. 26.17, $k = L/R = 3$. Since $\phi = \pi/2$, Eq. (26.33) gives

$$F(k,\phi) = \frac{k^3}{3} + \frac{\pi k^2}{2} + 2k + \frac{\pi}{4}$$

Hence, putting $k = 3$ and $I = bh^3/12$ into Eq. (26.34) yields

$$Y = \frac{6(60 + 19\pi)PR^3}{Ebh^3}$$

Table 26.1 Moment and Deflection Formulas for U Springs

Type of Spring	Moment	Deflection
Single U spring without constraint	$M = PRC_3$	$Y = \frac{PR^3}{EI}C_1$
Single U beam with constraint	$M = PRC_4$	$Y = \frac{PR^3}{EI}C_2$
Double U beam	$M = 0.5PRC_4$	$Y = \frac{PR^3}{2EI}C_2$

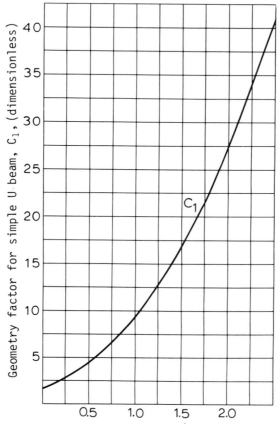

Fig. 26.14 Deflection chart for U beam without constraint.

Substituting the numerical data from Fig. 26.17 gives $Y = 0.368$ in. Finally, using Eq. (26.37), we obtain the maximum bending stress of $S_b = 15,360$ psi. ◆

Design Problem 26.3

Develop the general expression for the bending moment at the restrained end of a U spring, with the ends fixed as to slope, as shown in Fig. 26.12.

Solution

For one-half of the spring, the bending moment equations for the straight and curved portions are

$$M_1 = M_f - Px$$
$$M_2 = M_f - PR(k + \sin\theta)$$

Complex-Shape Springs

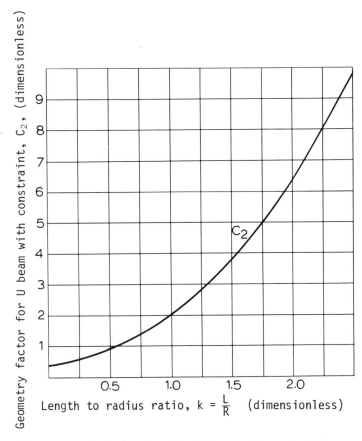

Fig. 26.15 Deflection chart for U beams with constraint.

Here x and θ define the particular point on the straight and curved portions of the spring and M_f is a statically indeterminate moment at the fixed end. In line with the general concepts of treating statically indeterminate quantities, discussed in Chap. 8, the relevant boundary condition can be stated as follows:

$$\int_0^L M_1 \frac{\partial M_1}{\partial M_f} \, dx + \int_0^{\pi/2} M_2 \frac{\partial M_2}{\partial M_f} R \, d\theta = 0$$

Since $\partial M_1/\partial M_f = \partial M_2/\partial M_f = 1$, utilizing this value together with the appropriate bending moment expressions, integrating, and solving for M_f yields

$$M_f = PR \left(\frac{k^2 + \pi k + 2}{\pi + 2k} \right)$$

Here $k = L/R$, as before. By reference to Table 26.1, then $M = PRC_4$ where

$$C_4 = \frac{k^2 + \pi k + 2}{\pi + 2k}$$

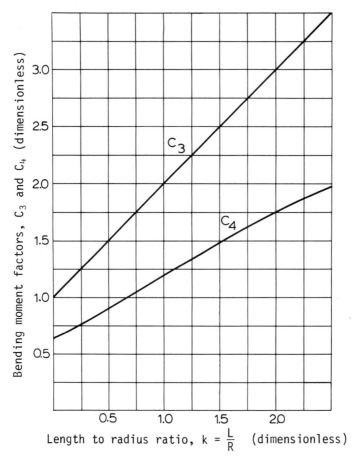

Fig. 26.16 Bending moment chart for U beams.

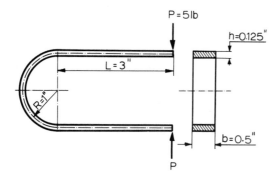

Fig. 26.17 Flat U spring (Design Problem 26.2).

Complex-Shape Springs

For instance, when $k = 2$, $C_4 = 1.72$, which is quite close to the value indicated in Fig. 26.16. ◆

SYMMETRICAL WAVE SPRING

When a symmetrical wave spring, shown in Fig. 26.18, is centrally loaded, the analysis can be performed for a cantilever portion ABC, fixed at C and loaded by $P/2$ at the support. Assuming that contact friction between the supporting surface and the end of the spring at A is negligible, the bending moment equations can be stated in customary terms

$$M_1 = \frac{PR \sin \theta}{2} \tag{26.38}$$

and

$$M_2 = \frac{PR}{2}(2 \sin \phi - \sin \theta) \tag{26.39}$$

The displacement of point A relative to C can now be found from the Castigliano theorem.

$$Y = \int_0^\phi M_1 \frac{\partial M_1}{\partial P} R \, d\theta + \int_0^\phi M_2 \frac{\partial M_2}{\partial P} R \, d\theta \tag{26.40}$$

From the bending moment expressions, the partial derivatives are

$$\frac{\partial M_1}{\partial P} = \frac{R \sin \theta}{2} \tag{26.41}$$

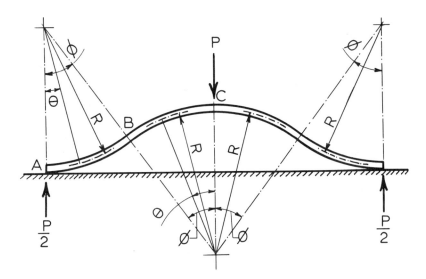

Fig. 26.18 Symmetrical wave spring.

and

$$\frac{\partial M_2}{\partial P} = \frac{R(2\sin\phi - \sin\theta)}{2} \tag{26.42}$$

Introducing Eqs. (26.38), (26.39), (26.41), and (26.42) into Eq. (26.40) and integrating gives

$$Y = \frac{PR^3}{EI}(0.25\phi + \phi\sin^2\phi - \sin\phi + 0.375\sin 2\phi) \tag{26.43}$$

When $\phi = 0$, the deflection must vanish in accordance with Eq. (26.43). By varying ϕ, a number of interesting spring configurations are possible for which the relevant spring constants can be calculated from Eq. (26.43).

FRAME SPRING VERSUS WAVE CONFIGURATION

The wave configuration shown in Fig. 26.18 appears to have certain similar characteristics to that of the frame-type spring shown in Fig. 26.19.

The maximum deflection for the frame-type spring can be obtained from the expression

$$Y = \frac{P[(a+d)^3 + 3a^2c]}{6EI} \tag{26.44}$$

Suppose that $a = d = R\sin\phi$ and $c = 2R(1 - \cos\phi)$ so that the overall spans and depths of the wave and frame-type springs are the same. Hence, Eq. (26.44) becomes

$$Y = \frac{PR^3[8\sin^3\phi + 6\sin^2\phi(1 - \cos\phi)]}{6EI} \tag{26.45}$$

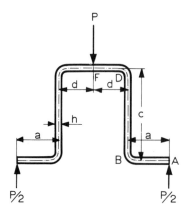

Fig. 26.19 Frame-type spring.

Complex-Shape Springs

It may now be of interest to compare the flexibility of the two springs if we divide Eq. (26.45) by Eq. (26.43). This gives

$$\lambda = \frac{(3 + 4\sin\phi - 3\cos\phi)\sin^2\phi}{1.125\sin 2\phi + 0.75\phi + 3\sin\phi(\phi\sin\phi - 1)} \tag{26.46}$$

Examination of Eq. (26.46) indicates that for all values of ϕ between 0 and $\pi/2$, the ratio λ varies roughly between 2 and 2.4. It is assumed here, of course, that the question of elastic stability of the vertical sides of the spring shown in Fig. 26.19 does not arise and that the corner radii are so small that they have no effect on the overall flexibility. On the other hand, it is also well to point out that sharp corner radii can cause stress concentration. In the case of a ductile material under static loading, the effects of stress concentration can be substantially modified by the presence of plastic yielding. Under fatigue conditions, however, ductility of the material does not provide any measure of immunity to stress concentration.

SYMBOLS

a	Length, in. (mm)
b	Width of rectangular section, in. (mm)
c	Length, in. (mm)
C_1, C_2, C_3, C_4	Moment and deflection factors for U springs
d	Length, in. (mm)
E	Modulus of elasticity, psi (N/mm^2)
$F(k, \phi)$	Function of k and ϕ, Eq. (26.33)
H	Horizontal load, lb (N)
h	Depth of cross section, in. (mm)
I	Moment of inertia, in.^{4} (mm^4)
K_1	Deflection factor for arched cantilever
$k = L/R$	Straight length-to-radius ratio
L	Straight portion of spring, in. (mm)
M, M_1, M_2	Bending moments, lb-in. (N-mm)
M_f	Fixing moment, lb-in. (N-mm)
P	Vertical load, lb. (N)
$\bar{P}$	Fictitious vertical load, lb (N)
P_1	Component of vertical load, lb (N)
R	Mean radius of curvature, in. (mm)
S_b	Bending stress, psi (N/mm^2)
x	Arbitrary distance, in. (mm)
X	Horizontal deflection, in. (mm)
Y	Vertical deflection, in. (mm)
Y_1	Component of vertical deflection, in. (mm)
α	Half-angle of snap spring, rad
β	Angle at which load is applied, rad
θ	Angle at which forces are considered, rad
λ	Deflection ratio, Eq. (26.46)
ν	Auxiliary angle in U springs, rad
ϕ	Angle subtended by curved portion, rad

27
Thin Rings and Arches

ASSUMPTIONS

Various problems of circular rings under in-plane and out-of-plane loading have been solved in classical works [5, 137–139]. Several books [11, 35, 150] have compiled and reviewed practical ring formulas and examples of their use [152]. Here the usual assumptions are made which state that one of the principal axes of the moment of inertia of the ring cross section is parallel to the axis of revolution. Equations are also available for coupled modes of in-plane and out-of-plane deformation [151]. The latter types of problems, however, are seldom needed in practice. For this reason, the data that follow are devoted to solutions where the two basic modes of deformation can be considered as uncoupled.

THIN ELASTIC RING

The classical case of a thin ring loaded in the plane of curvature represents a statically indeterminate problem. This seems somewhat surprising considering the regularity of the geometrical shape and structural response. An example of a simple case of a plain, thin ring subjected to diametral loading is illustrated in Fig. 27.1. The equilibrium of forces for one quadrant of the ring is essentially the same for both tension and compression and the problem is simplified considerably when the symmetry of loading and support is observed.

Thin Rings and Arches

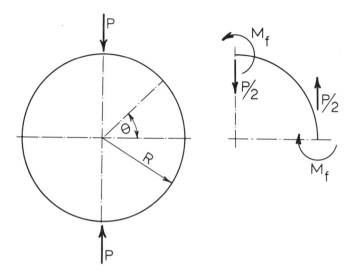

Fig. 27.1 Equilibrium of forces for thin ring.

In line with the notation given in Fig. 27.1, the bending moment at any point of the ring is equal to

$$M = \frac{PR}{2\pi}(2 - \pi \cos \theta) \tag{27.1}$$

The decrease in the vertical diameter obtained by the method described in Chap. 6 is

$$Y = \frac{PR^3(\pi^2 - 8)}{4\pi EI} \tag{27.2}$$

The corresponding increase of the horizontal diameter is given by

$$X = \frac{PR^3(4 - \pi)}{2\pi EI} \tag{27.3}$$

Equating the moment given by Eq. (27.1) to zero yields angle $\theta = 50.5°$. At this point the bending moment changes its sign. It is also of interest to note the ratio of the vertical to horizontal deflection, which is equal to 1.089. This value is applicable to thin rings, that is, when only the first bending term of Eq. (7.4) is used in the calculation. The error introduced by ignoring the effects of shear and direct stresses is in most cases rather small. More exact formulas for deformation of a circular ring of rectangular cross section may be given as follows:

$$Y = \frac{P\chi}{bE}(1.7856\chi^2 + 0.7854 + 2.0453) \tag{27.4}$$

$$X = \frac{P\chi}{bE}(1.6392\chi^2 - 0.5000 + 1.3020) \tag{27.5}$$

Table 27.1 Percentage Stress Contribution to Maximum Deflection for Thin Rings

	Bending Stress	Direct Stress	Shear Stress
Vertical deflection	+98.44	+0.43	+1.13
Horizontal deflection	+99.51	−0.30	+0.79

where χ is the ratio of radius of curvature to depth of section.

Assuming that a typical thin ring can be characterized by $\chi = 10 = R/h$, where h denotes the depth of a rectangular cross section, the percentage contribution of the various stresses can be computed from Eqs. (27.4) and (27.5). The relevant results are summed up in Table 27.1.

DESIGN CHARTS FOR CIRCULAR RINGS

It is evident from Table 27.1 that the simplified analysis of rings, involving bending stresses alone, will be acceptable in most design situations. A considerable amount of mathematical work associated with the derivation of formulas including the effects of direct and shear stresses can, therefore, be avoided.

The parameters most frequently required in the analysis involve radial deflection, slope, and bending moment as a function of angle θ, defining the particular location along the circular contour. Numerous closed-form solutions have been worked out and can be found in design literature [11, 35]. A summary of the more common loading configurations is given in Table 27.2. The relevant design factors are denoted by K_u, K_ψ, and K_M for the deflection, slope, and moment, respectively.

ESTIMATE BY SUPERPOSITION

Several important cases summarized in Table 27.2 can be used to answer the majority of practical design questions related to the response of closed, thin rings. Should a question regarding a more complex type of ring loading arise that is not covered specifically by Table 27.2, the designer may wish to solve the problem using the principle of superposition. It will be recalled that this principle allows us to add algebraically the stresses or deflections at a point of a structure caused by two or more independent loads. This principle is applicable as long as the resultant stresses and strains remain elastic.

To illustrate some basic steps in applying the principle of superposition to ring design, consider the response of a thin, elastic ring subjected to a four-way tension, as shown at the left in Fig. 27.2. Symbolically, the principle of superposition is illustrated in Fig. 27.2 in terms of a diagrammatic summary of the two effects. The basic component in this study is the ring subjected to diametral tension, as shown at the top of Table 27.2. In its vertical orientation it represents a two-way tension along the vertical axis indicated in Fig. 27.2 as the first component. The same ring

Thin Rings and Arches

Table 27.2 Circular Rings Loaded in Plane

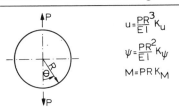

$u = \dfrac{PR^3}{EI} K_u$

$\psi = \dfrac{PR^2}{EI} K_\psi$

$M = PR\, K_M$

Symbol	Function	Range of Application
K_u	$0.2500 \sin\theta + (0.3927 - 0.2500\,\theta)\cos\theta - 0.3183$	$0 - \pi$
K_ψ	$(0.2500\,\theta - 0.3927) \sin\theta$	$0 - \pi$
K_M	$0.5000 \sin\theta - 0.3183$	$0 - \pi$

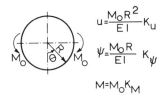

$u = \dfrac{M_0 R^2}{EI} K_u$

$\psi = \dfrac{M_0 R}{EI} K_\psi$

$M = M_0 K_M$

Symbol	Function	Range of Application
K_u	$0.5000 - 0.3183\,\theta \sin\theta - 0.4775 \cos\theta$	$0 - \pi/2$
K_ψ	$0.1592 \sin\theta - 0.3183\,\theta \cos\theta$	$0 - \pi/2$
K_M	$0.5000 - 0.6366 \cos\theta$	$0 - \pi/2$
	$-0.5000 - 0.6366 \cos\theta$	$\pi/2 - \pi$

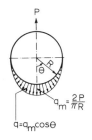

$q = q_m \cos\theta$

$q_m = \dfrac{2P}{\pi R}$

$u = \dfrac{PR^3}{EI} K_u \quad \psi = \dfrac{PR^2}{EI} K_\psi \quad M = PR\, K_M$

Symbol	Function	Range of Application
K_u	$0.1989\,\theta \sin\theta + (0.4081 - 0.0796\,\theta^2)\cos\theta - 0.3618$	$0 - \pi/2$
	$(0.3750 - 0.0398\,\theta)\sin\theta + (0.3658 - 0.2500\,\theta)\cos\theta - 0.3618$	$\pi/2 - \pi$
K_ψ	$(0.0796\,\theta^2 - 0.2092)\sin\theta + 0.0398\,\theta \cos\theta$	$0 - \pi/2$
	$(0.2500\,\theta - 0.4055)\sin\theta + (0.1250 - 0.0398\,\theta)\cos\theta$	$\pi/2 - \pi$
K_M	$0.3183\,\theta \sin\theta + 0.2387 \cos\theta - 0.3618$	$0 - \pi/2$
	$0.5000 \sin\theta - 0.0796 \cos\theta - 0.3618$	$\pi/2 - \pi$

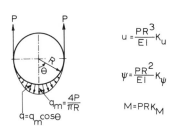

$q = q_m \cos\theta \qquad q_m = \dfrac{4P}{\pi R}$

$u = \dfrac{PR^3}{EI} K_u$

$\psi = \dfrac{PR^2}{EI} K_\psi$

$M = PR\, K_M$

Symbol	Function	Range of Application
K_u	$0.5570\,\theta \sin\theta + (0.9382 - 0.1592\,\theta^2)\cos\theta - 0.9053$	$0 - \pi/2$
	$(0.0796\,\theta - 0.2500)\sin\theta + 0.0681 \cos\theta + 0.0947$	$\pi/2 - \pi$
K_ψ	$(0.1592\,\theta^2 - 0.3812)\sin\theta + 0.2387 \cos\theta$	$0 - \pi/2$
	$0.0115 \sin\theta + (0.0796\,\theta - 0.2500)\cos\theta$	$\pi/2 - \pi$
K_M	$0.6366\,\theta \sin\theta + 0.7958 \cos\theta - 0.9053$	$0 - \pi/2$
	$0.1592 \cos\theta + 0.0947$	$\pi/2 - \pi$

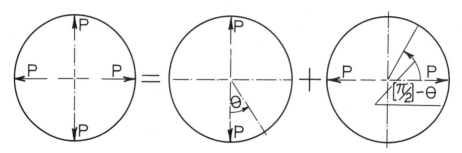

Fig. 27.2 Symbolic representation of superposition of ring deflection.

oriented horizontally represents the effect of a two-way tension along the horizontal axis, as shown symbolically in Fig. 27.2.

It is assumed that in a vertical orientation the angle θ is measured from the vertical axis counterclockwise. For the case of $\theta = 0$, $\sin \theta = 0$, and $\cos \theta = 1$, so that the deflection factor from Table 27.2 gives

$$K_u = 0.3927 - 0.3183 = 0.0744$$

The effect of the vertically oriented loading on the horizontal displacement is obtained when $\theta = \pi/2$. Since $\sin(\pi/2) = 1$ and $\cos(\pi/2) = 0$, we get

$$K_u = 0.2500 - 0.3183 = -0.0683$$

The sign is negative here because the ring attains an oval shape, with the vertical diameter increasing and the horizontal diameter decreasing.

If next we consider the basic ring to be oriented horizontally, as shown in Fig. 27.2, and if we still use the same expression for K_u, the question arises as to the appropriate angle convention. Clearly, using the same convention for the two orientations will not be satisfactory if we wish to derive a general formula. Let us see what kind of a substitution we would have to make in order to have the two K_u factors consistent with the two orientations of loading. Suppose that in a vertical orientation, angle θ defines a point on the ring at which the displacement is considered. The same point on the ring, when viewed from the horizontal line of loading, can be reached at an angle that is complementary to 90°, resulting in $(\pi/2) - \theta$. This assumption implies, for instance, that with $\theta = \pi/2$, the vertically oriented ring load (center of Fig. 27.2) produces a horizontal displacement. For the horizontally oriented load, $\theta = \pi/2$ gives the angle equal to zero and the displacement along the horizontal line of action. This statement follows from the symbolic diagram in Fig. 27.2.

For vertically oriented ring loading, Table 27.2 gives

$$K_u = 0.2500 \sin \theta + (0.3927 - 0.2500\theta) \cos \theta - 0.3183$$

Thin Rings and Arches

For a horizontally oriented pair of forces P, angle θ must be replaced by $(\pi/2) - \theta$, which gives

$$K_u = 0.2500 \cos\theta + \left[0.3927 - 0.2500\left(\frac{\pi}{2} - \theta\right)\right] \sin\theta - 0.3183$$

Adding the preceding two expressions yields

$$K'_u = 0.2500(1 + \theta)\sin\theta + (0.6427 - 0.2500\theta)\cos\theta - 0.6366 \qquad (27.6)$$

so that the radial deflection for a four-way tension becomes

$$u = \frac{PR^3}{EI} K'_u \qquad (27.7)$$

Note that for $\theta = 0$ and $\theta = \pi/2$, the expression for the deflection factor K'_u in a four-way tension gives two identical results, indicating that the deformation pattern of this ring is radially symmetric. In other words, radial displacement under each of the four loads is

$$u = 0.0061 \frac{PR^3}{EI}$$

The same result can be obtained if the deflection factor K'_u is calculated using previously quoted numerical values for $\theta = 0$ and $\theta = \pi/2$

$$K'_u = 0.0744 - 0.0683 = 0.0061$$

RING WITH CONSTRAINT

As stated previously, a circular ring loaded in a simple manner, such as that shown in Fig. 27.1, represents a statically indeterminate case. The complexity of the problem increases rather rapidly when additional constraints are provided. For example, consider a thin ring loaded and constrained as shown in Fig. 27.3. The equilibrium of one quadrant of the ring can be maintained if we add two statically indeterminate forces, H and M_f. Because of the symmetry of loading and support, all quadrants of the ring must deform in an identical manner. The bending moment at an arbitrary section defined by θ is

$$M = HR\sin\theta - \frac{PR}{2}(1 - \cos\theta) - M_f \qquad (27.8)$$

The deflection for this case is identical with that for a centrally loaded arch with fixed supports given later in this chapter.

Since the horizontal displacement and slope at $\theta = 0$ must be equal to zero, because of the rigid connection between the ring and the bar, the following conditions

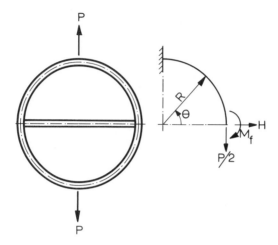

Fig. 27.3 Thin ring with rigid restraining bar.

apply.

$$\int_0^{\pi/2} M \frac{\partial M}{\partial H} d\theta = 0 \tag{27.9}$$

and

$$\int_0^{\pi/2} M \frac{\partial M}{\partial M_\mathrm{f}} d\theta = 0 \tag{27.10}$$

Substituting Eq. (27.8) and its derivatives into Eqs. (27.9) and (27.10) and integrating gives two simultaneous equations in terms of P, R, H, and M_f, which can be solved to give

$$H = P\left(\frac{4-\pi}{\pi^2 - 8}\right) \tag{27.11}$$

and

$$M_\mathrm{f} = PR\left(\frac{4 + 2\pi - \pi^2}{2(\pi^2 - 8)}\right) \tag{27.12}$$

Hence, the original bending moment expression given by Eq. (27.8) becomes

$$M = PR(0.4591 \sin\theta + 0.5000 \cos\theta - 0.6106) \tag{27.13}$$

It may be instructive to observe the effect of the diametral constraint illustrated in Fig. 27.3. When $\theta = \pi/2$, the absolute value of the bending moment under load P found from Eq. (27.13) is $M = 0.1515PR$, which is equal to about 48% of that obtained from Eq. (27.1). When $\theta = 0$, Eq. (27.13) yields $M = 0.1106PR$, which represents some 61% of the moment for a diametrically loaded ring without any

Thin Rings and Arches

horizontal restraint. Hence, as far as the preliminary design is concerned, we have established the important bracketing values of the bending moment. Assuming, then, that the horizontal bar shown in Fig. 27.3 is not perfectly rigid, the values of the bending moment for $\theta = 0$ and $\theta = \pi/2$ should increase. The maximum bending stresses are found at the two points of load application. Therefore when a certain amount of horizontal constraint is present, the stresses at the critical locations are reduced.

ANALYSIS OF PISTON RINGS

Many types of piston, retaining, and snap rings encountered in machine design have one common feature. Structurally, they all represent split, circular rings of uniform or variable cross section, depending on the particular loading and stress requirements. In the literature they are sometimes referred to as concentric or eccentric rings, corresponding to uniform or variable thickness. In the case of retaining or snap rings, they can be produced with rectangular, square, or circular cross sections. The split rings used with the pistons of steam engines and gasoline motors are known under the general heading of "piston rings" and are produced with rectangular and rectangular element cross sections. All the split rings described in this section can be basically of two types, external or internal. In the case of the external types, the ring is designed to fit into a circumferential groove on a shaftlike member. With the internal type, the groove is cut into the side of a cylindrical housing.

The analysis of piston, retaining, and snap-type rings involves the theory of curved structural members, which is beyond the scope of the elementary texts on machine design. Commercially produced split rings are available as off-the-shelf items because they lend themselves to a mass production. For these and similar reasons, the great majority of handbooks seldom treat this important topic of machine design, which is of special interest to those engaged in the field of internal combustion engines, steam engines, compressors, and pumping machinery. Since the basic problem of a piston ring involves several facets of solid mechanics, materials science, and practical engineering, this section is devoted to a more complete presentation of the subject material in a manner directly applicable to engine design.

The piston of a gasoline motor or a steam engine must slide freely in a cylinder and it is only slightly smaller than the diameter of the cylinder. To ensure a completely tight joint between a piston and the cylindrical surface packing rings are used to take up the clearance between the piston wall and the cylinder. Such rings vary in design and, throughout the years, efforts have been made to manufacture a ring that would give an ideally tight joint. Although an ideal design has not been found, many known configurations serve the purpose and are generally inexpensive to manufacture. Several patented designs are available for gasoline engines, and a number of standard configurations can be found in design specifications in this country and elsewhere. An interesting example of this type of a specification is found, for instance, in Germany [27].

It should be pointed out that the basic reason for employing a varying ring thickness is to assure a relatively uniform pressure against the cylinder wall all around the circumference. In practice it is difficult to conform to the exact theoretical shape. However, any reasonable approximation to this design may be better than an ordinary ring of uniform cross section. In the final analysis the choice will depend on a particular design application and the quality of manufacture. The main objection to the tapered or variable section design is the additional clearance found on the thin side of the ring if one assumes that the depth of the groove in the piston is constant around the circumference. Such an additional clearance creates an easier passage for gas or steam.

The method of assembly of the piston ring indicates that there must be two separate conditions for which the ring should be designed. In the first place the ring must be sprung open enough to allow it to go over the piston. This action causes strains within the ring. Hence, the ring must be proportioned in such a way as not to overstress the material. After the ring is inserted into the piston groove, the ring undergoes compression to the cylinder size on assembly, and corresponding maximum bending stress should be about the same as that experienced during the process of springing the ring open. In both cases, of course, the stress may be greatest at a point opposite the joint. The margin of safety will depend on the amount of expanding and compressing of the ring and on the quality of the ring material. Certain rules of thumb indicate that a piston ring of uniform cross section made of good cast iron can have thickness on the order of 1/32 of the cylinder diameter in order not to break during the spring-open phase of the assembly. The maximum thickness of the variable cross-section ring, using the same criteria, can be equal to 1/28 of the cylinder diameter. The ratios of ring width to maximum thickness appear to differ for various manufacturers, with the upper value seldom exceeding 1.5. For reasons of decreasing the contact pressure, the ratio of width to thickness can be made higher. However, this and other rough rules of thumb serve only as the first approximation and are no substitute for detailed design analysis. This is particularly true in the case of the amount of cutout of the ring.

The choice of piston ring materials depends primarily on the required resistance to wear. This property is influenced by such factors as sliding velocity, temperature, mechanical load, and frequency of engine use. Other chronic causes of trouble include airborne particles of dirt, corrosion, and insufficient quality of lubrication. The most common material for piston rings in the United States is gray cast iron in the range 40,000 to 50,000 psi. Occasionally, steels of the type 1070 and 52100 could be used, on the basis of their resistance to abrasive wear. Iron and steel rings are sometimes chromium plated for the purpose of reducing the rate of wear in such applications as heavy-duty diesel engines, natural gas engines, and in some cases passenger cars. The disadvantage here is that chromium plating can reduce the fatigue limit of a particular part, leading to a requirement for higher-strength varieties of steel or cast iron. In some instances bronze piston rings are used in steam locomotive design. However, this material is unsuitable for internal combustion engines, where higher temperatures are encountered. Whatever the choice of the material, it is well to keep in mind the rule that the piston ring should not be appreciably harder than the cylinder wall. Otherwise, the more expensive item, such as the engine cylinder, will wear more rapidly than the inexpensive piston ring. The rate of wear is known to decrease linearly with increase in surface hardness. In some cases of piston ring manufacture for reciprocating internal combustion engines, a surface layer is formed by flame spraying a mixture of molybdenum

Thin Rings and Arches

and ceramic materials. Such a layer, according to open trade literature, has good lubricant-retention qualities and superior resistance to wear.

Operational characteristics of piston rings are, at times, subject to the effect of residual stresses. These stresses may be caused by manufacturing processes, such as casting, machining, or grinding. Assuming that piston ring castings have a uniform thickness, the residual stresses should not be severe, provided that cooling rate is uniform. Residual stresses developed during metal cutting are generally compressive in nature and their magnitude depends on the degree of sharpness of the cutting tool. On the other hand, residual stresses caused by grinding at moderate and higher grinding-wheel speeds are normally tensile. It should be pointed out, however, that not all piston rings are subjected to grinding. Although in the majority of practical applications residual stress effects may be relatively small, it is well to keep in mind this aspect of piston ring fabrication.

The first theoretical problem of piston ring design may be concerned with the mechanics of expanding the ring over the piston in order to insert it into the groove. The analysis of a snap ring given in Chap. 26 provides the first insight into the steps necessary in the preliminary design, bearing in mind that the cast iron is inherently weaker in tension than in compression. The design formula for the maximum allowable opening of the ring at the junction is found from Eq. (26.6) by assuming that $\alpha = 0$ in Fig. 26.2 and by eliminating the load term P. This procedure leads to the following result.

$$Y = \frac{3\pi R^2 S_b}{Eh} \tag{27.14}$$

If Y denotes the total opening of the ring, the corresponding configuration and loading can be illustrated in Fig. 27.4.

It appears that for a typical rectangular or a square cross section of the piston ring, the allowable opening is independent of the width of the ring and it is directly proportional to the critical stress. This stress in the case of cast iron is $S_b = S_t$, where S_t denotes the maximum tensile stress at the inner surface of the ring being expanded over the piston. We assume that the modulus of elasticity E, depth of the cross section h, and mean radius R are kept constant.

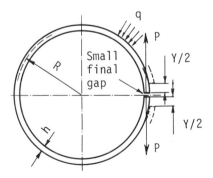

Fig. 27.4 Uniform-thickness (concentric) piston ring.

For a specified width of the cross section b, allowable tensile stress S_t, mean radius R, and depth h, the corresponding ring-opening load P is found from Eq. (26.3)

$$P = \frac{bh^2 S_t}{12R} \tag{27.15}$$

When the extent of the maximum opening along line AB is required, Fig. 26.3 and Eq. (25.4) can be used. Note that in this particular application load P is applied at the split end D while the displacement is calculated along the diameter AB. Using the notation of Fig. 25.1 and putting $\beta = 0$, $\alpha = \pi/2$, and $\phi = \pi$ in Eq. (25.4) gives one-half of the required deflection. Hence, the total deflection becomes

$$Y = \frac{6(\pi + 4)PR^3}{Ebh^3} \tag{27.16}$$

It is well to observe that for $I = bh^3/12$, Eq. (27.16) is identical with Eq. (26.13). This is consistent with the Maxwell's principle of reciprocal deflections, which, for the case of Fig. 26.3 may be stated as follows: Vertical displacement at D due to load P at A is equal to the vertical displacement at A when load P is applied at D.

Design Problem 27.1

A class 40 ($S_t = 40{,}000$ psi) gray cast-iron ring has an outside diameter of 3 in., a width of 0.188 in., and a depth of 0.125 in. Calculate the maximum opening of the ring at the junction of the ends that can be obtained before it will break. What is the expected maximum opening along AB if the load is applied at D, as shown in Fig. 26.3? Assume the elastic modulus to be equal to 18.3×10^6 psi.

Solution

The mean radius is

$$R = \frac{3}{2} - \frac{0.125}{2} = 1.4375 \text{ in.}$$

Hence, using Eq. (27.14) gives the maximum opening

$$Y_D = \frac{3\pi R^2 S_t}{Eh} = \frac{3\pi \times 1.4375^2 \times 40{,}000}{18.3 \times 10^6 \times 0.125} = 0.34 \text{ in.} \quad (8.7 \text{ mm})$$

The corresponding ring-opening load applied at point D, Fig. 26.3, follows from Eq. (27.15).

$$P = \frac{bh^2 S_t}{12R} = \frac{0.188 \times 0.125^2 \times 40{,}000}{12 \times 1.4375} = 6.81 \text{ lb} \quad (30.3 \text{ N})$$

The maximum opening along AB is found from Eq. (27.16)

$$Y_{AB} = \frac{6(\pi + 4)PR^3}{Ebh^3} = \frac{6(\pi + 4) \times 6.81 \times 1.4375^3}{18.3 \times 10^6 \times 0.188 \times 0.125^3} = 0.13 \text{ in.} \quad (3.3 \text{ mm}) \quad \blacklozenge$$

Thin Rings and Arches

In the design of a standard piston ring subjected to uniform radial loading, as shown in Fig. 27.5, the basic analytical problem is to find the value of uniform, radial load q needed to close the circumferential gap Y. It is convenient here to employ the concept of a fictitious load discussed in Chap. 7.

The bending moment due to the radial unit load q is

$$M_q = qR^2 \int_0^\theta \sin(\theta - \epsilon)\, d\epsilon \qquad (27.17)$$

The total bending moment involving the real and fictitious loading can now be obtained after integrating Eq. (27.17)

$$M = qR^2(1 - \cos\theta) + PR(1 - \cos\theta) \qquad (27.18)$$

Hence, by analogy to Eq. (6.11), the deflection can be given as follows.

$$Y = \frac{2R}{EI} \int_0^\pi M \frac{\partial M}{\partial P}\, d\theta \qquad (27.19)$$

Integrating Eq. (27.19) for $P = 0$ and solving for the required external loading q in a piston ring with a rectangular cross section yields

$$q = \frac{YEbh^3}{36\pi R^4} \qquad (27.20)$$

The maximum bending stress is found at $\theta = \pi$ and is equal to

$$S_b = \frac{YEh}{3\pi R^2} \qquad (27.21)$$

Extension of the theory permits selection of ring thickness h as a function of θ for which the external pressure and the bending stress can be kept constant. This will

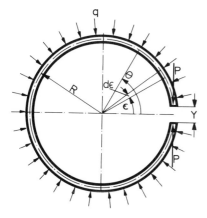

Fig. 27.5 Piston ring of uniform cross section.

lead to the following design equation.

$$h = 2.45R \left[\frac{q(1-\cos\theta)}{bS_b} \right]^{1/2} \tag{27.22}$$

In this type of a design, the piston ring is expected to exert a uniform radial pressure against the cylinder wall.

Based on Fig. 27.5 and Eq. (27.22), the radial depth of the piston ring can, theoretically, vary from 0 to some maximum value h, depending on the magnitudes R, q, b, and S_b. The maximum bending stress is found at $\theta = \pi$ equal to

$$S_b = \frac{12qR^2}{bh^2} \tag{27.23}$$

In this case h is taken as the maximum radial thickness. The formula given by Eq. (27.23) is applicable to the case of a piston ring with uniform radial thickness h. The expression for the initial gap in a piston ring with continuously varying radial thickness between 0 and h is given by the formula [5]

$$Y = \frac{24qR^4}{bEh^3} \tag{27.24}$$

An example of a variable section piston ring is shown in Fig. 27.6. The minimum thickness is found to be small but not equal to zero.

The analysis of piston rings outlined above treats essentially two types of geometry. The more elementary case is concerned with a ring of uniform rectangular cross section, Eqs. (27.14) to (27.16). The approximate analysis for a piston ring with the continuously varying radial depth is represented by Eqs. (27.20) to (27.22). For practical reasons, ring depth cannot vary from a maximum to zero. However, the error introduced by providing a small but finite radial depth at the gap end of the ring ($\theta = 0$) is expected to be rather small.

It should be added that when the pressure is specified, instead of the uniform load q the equivalent term is q/b. Furthermore, in some instances the radial devi-

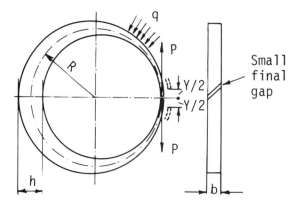

Fig. 27.6 Variable-thickness (eccentric) piston ring.

Thin Rings and Arches

ation ΔR is given as a variable instead of the gap Y. Under these circumstances R denotes the radius of the inner wall of the piston while $R + \Delta R$ corresponds to the radius of curvature of the piston ring measured in the unstressed state at the outer surface of the ring. For a relatively small gap Y, the relation between the two variables can be established as follows.

$$2\pi(R + \Delta R) - 2\pi R = Y$$

from which

$$Y = 2\pi(\Delta R) \tag{27.25}$$

Design Problem 27.2

It is proposed to use a piston ring of uniform rectangular cross section having width $b = 0.5$ in. and mean radius of curvature $R = 5$ in. The material selected for this purpose is cast iron with a strength at yield of 45,000 psi and a modulus of elasticity of 22×10^6 psi. The design factor of safety on yield is given as 3, and contact pressure between the piston wall and the ring is assumed to be 1 psi. Determine the required radial depth h and the initial gap Y not to exceed the maximum design value of the bending stress.

Solution

The allowable bending stress $S_b = 45,000/3 = 15,000$ psi. The radial unit load $q = 0.5$ lb/in. From Eq.(27.22) at $\theta = \pi$, or Eq. (27.23), we obtain

$$h = 2R\left(\frac{3q}{bS_b}\right)^{1/2} = 2 \times 5 \left(\frac{3 \times 0.5}{0.5 \times 15,000}\right)^{1/2} = 0.14 \text{ in.}$$

From Eq. (27.21),

$$Y = \frac{3\pi S_b R^2}{Eh} = \frac{3\pi \times 15,000 \times 25}{22 \times 10^6 \times 0.14} = 1.15 \text{ in.} \quad (29.21 \text{ mm}) \quad \blacklozenge$$

Design Problem 27.3

Estimate the magnitude of the initial separation between the free ends of a piston ring assuming a constant width of rectangular cross section and continuously varying radial depth from zero to $h = 0.25$ in. The ring has mean radius of curvature $R = 5$ in., and the material is gray cast iron. The corresponding yield strength of the material is 40,000 psi and the modulus of elasticity is 20×10^6 psi. The design factor of safety on yield is 3.

Solution

Eliminating q and b from Eqs. (27.23) and (27.24), we obtain

$$Y = \frac{2R^2 S_b}{Eh}$$

The allowable design stress is

$$S_b = \frac{40{,}000}{3} = 13{,}333 \text{ psi}$$

Hence, substitution gives

$$Y = \frac{2 \times 25 \times 13{,}333}{20 \times 10^6 \times 0.25} = 0.133 \text{ in.} \quad (3.39 \text{ mm})$$

This numerical example does not represent a realistic condition when zero thickness is assumed at the joint. In practice a very thin edge may be subject to frictional and inertia effects which could precipitate ring breakage. The design problem is intended only as an illustration of the theoretical approach to piston ring design where relatively constant bending stress and constant pressure are required. ♦

Design Problem 27.4

Develop the design curve for the variation of radial depth of piston ring with the angle measured from the center of the slot in the ring. Assume that the maximum bending stress S_b, uniform contact load q, radius of curvature R, and width of the ring b are combined to yield a dimensionless parameter.

Solution

From Eq. (27.22), the required dimensionless parameter is

$$\frac{h}{R}\left(\frac{bS_b}{6q}\right)^{1/2} = (1 - \cos\theta)^{1/2}$$

Varying angle θ from 0 to π gives the relation illustrated in Fig. 27.7. ♦

One of the practical problems of piston ring design is concerned with determining the right amount of cutout in the ring to allow compressing it to the cylinder

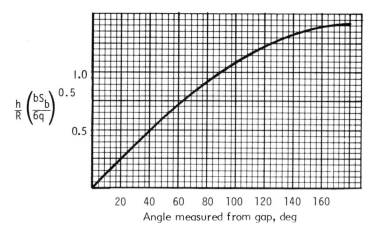

Fig. 27.7 Variation of radial depth of piston ring.

Thin Rings and Arches

diameter. Essentially, this numerical value is equal to the circumference of the ring prior to splitting, less the circumference of the ring in the compressed state, plus a small amount for thermal expansion and grinding. The following allowances can be used as a general guide for design:

Thermal expansion   0.004 × cylinder diameter
Grinding   0.008 × cylinder diameter

The actual joint of the ring can be of the transverse lap type with a step or plane diagonal-cut type. The angle between the diagonal cut and the side of the ring can vary between 30 and 45°. The most commonly used joint angle is 30°.

It should be noted that one of the most important manufacturing requirements is to make a piston ring truly cylindrical after radial compression to the size of the cylinder bore. For this purpose the split rings are turned and ground in a compressed configuration conforming to the cylinder diameter.

To relate the amount of cutout to the required ring size before the cut, Eq. (27.25) can be utilized. If the total amount of the allowances related to the mean radius R can be taken as $0.024R$, then the mean diameter of the ring before the cutout can be calculated as

$$d = 2.048R + 0.318Y \qquad (27.26)$$

Design Problem 27.5

A piston ring made of gray cast iron is required to work in a cylinder at a contact pressure of 20 psi. Assuming the inner diameter of the cylinder to be 3 in., the width of the ring cross section 0.15 in., and the depth of the cross section 0.10 in., calculate the required gap and the mean diameter of the ring before it is split. The modulus of elasticity of the ring material is 18.3×10^6 psi.

Solution

The compressive load per inch of the ring circumference is

$$q = 20 \times 0.15 = 3 \text{ lb/in.}$$

The mean radius of the ring after assembly is

$$R = \frac{3 - 0.10}{2} = 1.45 \text{ in.}$$

Hence, the theoretical gap using Eq. (27.20) is

$$Y = \frac{36\pi R^4 q}{Ebh^3} = \frac{36\pi \times 1.45^4 \times 3}{18.3 \times 10^6 \times 0.15 \times 0.10^3} = 0.546 \text{ in.} \quad (13.88 \text{ mm})$$

The mean diameter of the ring before the cut follows from Eq. (27.26):

$$d = 2.048R + 0.318Y$$
$$= 2.048 \times 1.45 + 0.318 \times 0.546 = 3.143 \text{ in.} \quad (79.84 \text{ mm}) \quad \blacklozenge$$

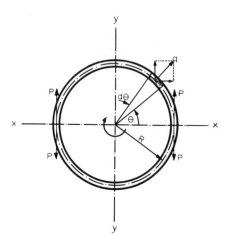

Fig. 27.8 Rotating ring in plane of curvature.

ANALYSIS OF SPECIAL RINGS

In yet another case of a more important type of in-plane loading, the ring is assumed to rotate in its plane of curvature simulating a rim of a flywheel in a load-leveling or similar energy storage device. Denoting the uniform loading due to the centrifugal action by q and the corresponding internal forces by P, the equation of equilibrium for the model shown in Fig. 27.8 becomes

$$P = \int_0^{\pi/2} qR \sin\theta \, d\theta \tag{27.27}$$

The corresponding tensile stress expressed in terms of the peripheral velocity v and weight density of the material γ can be shown to be

$$S = \frac{\gamma v^2}{g} \tag{27.28}$$

When v is expressed in in./sec, γ in lb/in.^{3}, and $g = 386.4$ in./sec^2, Eq. (27.28) should give yield stress in psi. The theory assumes that the tensile stress given by Eq. (27.28) is distributed uniformly over the ring cross section. This approximation is sufficiently accurate for many practical applications involving relatively thin rotating rims.

Design Problem 27.6

A circular ring of mean radius R is made from a thin bar of uniform cross section with two ends at C connected through a pin joint, as shown in Fig. 27.9. The ring is in equilibrium under three radial forces which cause a horizontal reaction F at the joint. Develop a parametric equation for F in terms of P and α.

Thin Rings and Arches

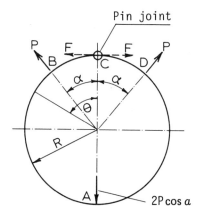

Fig. 27.9 Ring model for Design Problem 27.6.

Solution

Because of symmetry with respect to the AC diameter, it is sufficient to analyze one-half of the ring, such as ABC. Angle θ can be smaller or greater than α, as shown in the sketch. For the portion of ring $BC(\theta < \alpha)$,

$$M_1 = FR(1 - \cos\theta)$$

For the portion $AB(\theta > \alpha)$,

$$M_2 = FR(1 - \cos\theta) - PR\sin(\theta - \alpha)$$

The condition for finding the redundant quantity F is given, as before, by the following expression.

$$\int_0^\alpha M_1 \frac{\partial M_1}{\partial F} R\, d\theta + \int_\alpha^\pi M_2 \frac{\partial M_2}{\partial F} R\, d\theta = 0$$

where

$$\frac{\partial M_1}{\partial F} = \frac{\partial M_2}{\partial F} = R(1 - \cos\theta)$$

On integration and substitution of the foregoing limits, the equation is solved for F to give

$$F = \frac{P}{3\pi}[2(1 + \cos\alpha) + (\pi - \alpha)\sin\alpha]$$

It should be noted that this equation is applicable only to α values varying between zero and $\pi/2$, since at $\alpha = \pi/2$, vertical reaction at the fixed end must vanish for $\cos(\pi/2) = 0$. At the same time, the value of the pin reaction is $F = (4 + \pi)P/6\pi$, or approximately $F \cong 3P/8$. When, on the other hand, α becomes rather small, vertical reaction tends to the value of $2P$, and the force on the pin approaches $F = 4P/3\pi$. The maximum force on the pin can be found from solving $dF/d\alpha = 0$, which gives $\tan\alpha = (\pi - \alpha)/3$. This should correspond to $F = 0.54P$. Surprisingly, however, the force on the pin indicates only small variations for a range of α between zero and $90°$. ♦

SIMPLY SUPPORTED ARCH

Increase in the calculational complexity is found in dealing with structures and machine parts made in the form of circular frames and arches [137, 153, 154]. Most frequently, however, the design engineer wants to know the maximum stresses and deflections for centrally loaded circular arches. Again, it is convenient to use the concept of elastic strain energy and the theorems of Castigliano, which apply equally well to statically determinate and indeterminate structures. It will also be assumed that the arched members considered here have uniform cross sections and large radii of curvature in comparison with the radial thicknesses. The deflections are taken to be relatively small and the strain energy due to bending alone is used in the analysis.

The easiest case is that of a simply supported circular arch under central load P as shown in Fig. 27.10. If we postulate that there is no friction at the supports and no constraint of any kind, the reactions can be obtained from a simple equation of statics. The bending moment at a section defined by θ is written as follows.

$$M = VR(\cos\alpha - \cos\theta) \tag{27.29}$$

The deflection of each support, if the section at $\theta = \pi/2$ is fixed, can be calculated directly

$$Y = \frac{VR^3}{EI} \int_\alpha^{\pi/2} (\cos\alpha - \cos\theta)^2 \, d\theta \tag{27.30}$$

Integrating and substituting the limits yields

$$Y = \frac{PR^3}{EI} G_1 \tag{27.31}$$

where

$$G_1 = 0.125[(\pi - 2\alpha)(1 + 2\cos^2\alpha) - 8\cos\alpha + 3\sin 2\alpha] \tag{27.32}$$

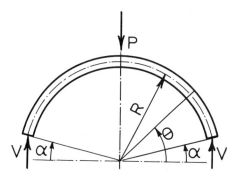

Fig. 27.10 Simply supported circular arch under central load.

Thin Rings and Arches

Note here that in developing Eq. (27.32) we take $V = P/2$, after partial differentiation is completed. Hence, the integration is accomplished with the aid of $\partial M/\partial V = R(\cos\alpha - \cos\theta)$. In dealing with the reactions, it may be advisable to employ a different symbol as long as it can be treated as an independent quantity. In this particular case, one-half of the arch behaves as a curved cantilever under the action of end load V. In performing a quick check on the derivation, it is helpful to examine the two bracketing conditions. For instance, when $\alpha = \pi/2$, $G_1 = 0$, and for $\alpha = 0$, $G_1 = 0.1781$, which is consistent with a standard deflection formula for a semicircular arch and with the value of K_1 given in Fig. 25.3. Note that G_1 corresponds to $P/2$ loading, so that $K_1 = 2G_1 = 0.3562$ at $\phi = \pi/2$. To facilitate the calculations, the deflection factor G_1 is plotted in Fig. 27.11.

PIN-JOINTED ARCH

Consider next a pin-jointed arch with a central, concentrated load as shown in Fig. 27.12. As indicated in Chap. 8, the conditions of static equilibrium are here insufficient for the determination of a horizontal reaction H_p. Static equilibrium tells us only that the horizontal reactions must be equal and opposite in direction.

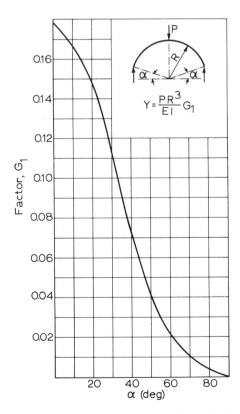

Fig. 27.11 Deflection factor for a simply supported arch.

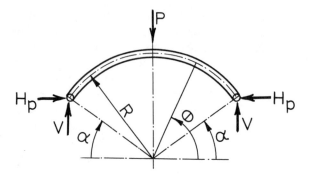

Fig. 27.12 Pin-jointed circular arch under central load.

The bending moment at a section defined by θ is then

$$M = VR(\cos\alpha - \cos\theta) - H_p R(\sin\theta - \sin\alpha) \tag{27.33}$$

Since the displacement of the arch support is assumed to be zero, the condition for determining the redundant quantity H_p can be stated as follows:

$$\int_\alpha^{\pi/2} M \frac{\partial M}{\partial H_p} d\theta = 0 \tag{27.34}$$

From Eq. (27.33) the relevant partial derivative is

$$\frac{\partial M}{\partial H_p} = -R(\sin\theta - \sin\alpha) \tag{27.35}$$

Introducing Eqs. (27.33) and (27.35) into (27.36) and integrating gives

$$H_p = PG_2 \tag{27.36}$$

where

$$G_2 = \frac{4\sin\alpha + 3\cos 2\alpha - (\pi - 2\alpha)\sin 2\alpha - 1}{2(\pi - 2\alpha)(1 + 2\sin^2\alpha) - 6\sin 2\alpha} \tag{27.37}$$

When $\alpha = 0$, Eq. (27.37) gives a standard handbook value of 0.318, as shown in Chap. 8. Equation (27.37) indicates, however, that thrust coefficient G_2 increases at a high rate with an increase in α until the arch begins to yield in compression. It is well to keep in mind, then, that relatively flat arches may be subject to appreciable horizontal forces if the supports are kept apart at a fixed distance.

The maximum bending moment now follows from Eqs. (27.33) and (27.36) when $\theta = \pi/2$. This yields

$$M = PR\left[\frac{\cos\alpha}{2} - (1 - \sin\alpha)G_2\right] \tag{27.38}$$

Thin Rings and Arches

The deflection under load P can be found by the Castigliano theorem, considering one-half of the arch as an arched cantilever subjected to a vertical force V and horizontal thrust H_p acting as the two statistically independent forces. The concept of static independence is justified here since the external work done by H_p in the direction of V is zero. Following the usual procedure, we get

$$Y = \frac{PR^3}{EI} G_3 \tag{27.39}$$

where

$$G_3 = G_1 - \frac{[4\sin\alpha + 3\cos 2\alpha - (\pi - 2\alpha)\sin 2\alpha - 1]^2}{8(\pi - 2\alpha)(1 + 2\sin^2\alpha) - 24\sin 2\alpha} \tag{27.40}$$

For $\alpha = 0$, Eq. (27.39) reduces to the well-known formula for a semicircular arch with horizontal constraint

$$Y = \frac{0.0189 PR^3}{EI} \tag{27.41}$$

The functions G_2 and G_3 are illustrated in Fig. 27.13.

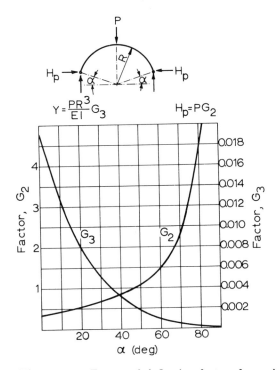

Fig. 27.13 Force and deflection factors for a pin-jointed arch.

BUILT-IN ARCH

The analysis of a circular arch built in at the supports is much more involved, even for the case of a concentrated central load, as shown in Fig. 27.14. The reason for this is that this arch represents a doubly redundant structure with regard to the horizontal thrust H_b and the fixing couple M_f. In terms of these two unknown quantities and the vertical reaction V, which follows from statics, the bending moment at any section defined by θ can be stated as

$$M = VR(\cos\alpha - \cos\theta) - H_b R(\sin\theta - \sin\alpha) - M_f \qquad (27.42)$$

The boundary conditions for the calculation of H_b and M_f can be expressed with the aid of the theorem of Castigliano

$$\int_\alpha^{\pi/2} M \frac{\partial M}{\partial H_b} d\theta = 0 \qquad (27.43)$$

and

$$\int_\alpha^{\pi/2} M \frac{\partial M}{\partial M_f} d\theta = 0 \qquad (27.44)$$

Integrating Eqs. (27.43) and (27.44) gives two simultaneous equations for finding the unknown redundants H_b and M_f.

$$2H_b R A_2 - PR A_1 - 8M_f A_3 = 0 \qquad (27.45)$$

and

$$2H_b R A_3 - PR A_4 - (\pi - 2\alpha)M_f = 0 \qquad (27.46)$$

Solving Eqs. (27.45) and (27.46) yields

$$H_b = \frac{P[(\pi - 2\alpha)A_1 - 8A_3 A_4]}{2(\pi - 2\alpha)A_2 - 16A_3^2} \qquad (27.47)$$

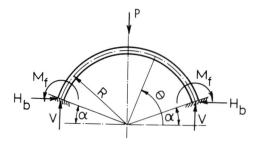

Fig. 27.14 Built-in circular arch under central load.

Thin Rings and Arches

and

$$M_f = \frac{PR(A_1 A_3 - A_2 A_4)}{(\pi - 2\alpha)A_2 - 8A_3^2} \tag{27.48}$$

where

$$A_1 = 4\sin\alpha + 3\cos 2\alpha - (\pi - 2\alpha)\sin 2\alpha - 1 \tag{27.49}$$
$$A_2 = (\pi - 2\alpha)(1 + 2\sin^2\alpha) - 3\sin 2\alpha \tag{27.50}$$
$$A_3 = 0.5(\pi - 2\alpha)\sin\alpha - \cos\alpha \tag{27.51}$$
$$A_4 = 1 - 0.5(\pi - 2\alpha)\cos\alpha - \sin\alpha \tag{27.52}$$

The derivation of the design formula for the central deflection follows the same rules as those used in conjunction with other arches, although the amount of algebraic work here is substantially increased. The final result in this case is

$$Y = \frac{PR^3}{EI}\left\{G_1 - \frac{4A_2 A_4^2 - 8A_1 A_3 A_4 + 0.5(\pi - 2\alpha)A_1^2}{4[(\pi - 2\alpha)A_2 - 8A_3^2]}\right\} \tag{27.53}$$

All the results for the fixed-end arch can now be simplified by introducing additional symbols for the combined trigonometric functions

$$M_f = PRG_4 \tag{27.54}$$
$$H_b = PG_5 \tag{27.55}$$
$$Y = \frac{PR^3}{EI}G_6 \tag{27.56}$$

To simplify rather tedious calculations further, factors G_4, G_5, and G_6 are plotted in Fig. 27.15. The general equation for bending moment under concentrated load is

$$M = PR[0.5\cos\alpha - G_5(1 - \sin\alpha) - G_4] \tag{27.57}$$

It may also be of interest to reduce all the general formulas for the fixed arch by introducing $\alpha = 0$. Because of the similarities in loading and support conditions, the following equations are also applicable to Fig. 27.3.

$$H_b = \frac{P(4 - \pi)}{\pi^2 - 8} \tag{27.58}$$
$$M_f = \frac{PR(\pi^2 - 2\pi - 4)}{2(\pi^2 - 8)} \tag{27.59}$$
$$M = \frac{PR(2\pi - 6)}{\pi^2 - 8} \tag{27.60}$$
$$Y = \frac{PR^3}{EI}\left[\frac{\pi^3 - 20\pi + 32}{8(\pi^2 - 8)}\right] \tag{27.61}$$

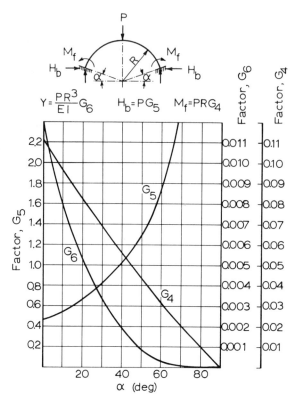

Fig. 27.15 Force and deflection factors for a built-in circular arch.

The moment given by Eq. (27.60) is found under the concentrated load P in Fig. 27.14 when $\alpha = 0$.

PINNED ARCH UNDER UNIFORM LOAD

In certain design applications it is necessary to know the deformation of a pin-jointed circular arch subjected to uniform vertical loading as shown in Fig. 27.16. This type of loading and support is feasible, for instance, in large cylindrical containers or the casing of a compressor of a jet engine subjected to inertia load under a sudden change in the direction of flight. Another possible application is that of a buried, thin-walled cylinder responding to a seismic ground motion or a soil compression wave caused by an underground explosion. Since a long pipe or a cylindrical container can be treated as a number of rings connected together, the model of a pin-jointed arch can be used in the analysis. In the case of a compressor casing, we can have a longitudinal joint holding the two casing halves together, so that the model illustrated in Fig. 27.16 is particularly appropriate. When an analysis of the deformation pertains to any point on the arch, such as that defined by θ, the derivation can be rather involved. The relevant statically indeterminate quantity is $qR/2$, where q denotes weight of the arch per inch of circumference.

Thin Rings and Arches

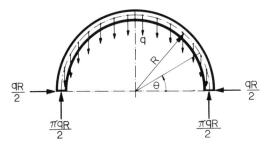

Fig. 27.16 Load equilibrium for a pin-jointed arch under its own weight.

Accordingly, the bending moment at any section θ is

$$M = 0.5qR^2(\pi - \pi\cos\theta - 3\sin\theta + 2\theta\cos\theta) \qquad (27.62)$$

or

$$M = qR^2 B \qquad (27.63)$$

where B is a bending moment factor given in Fig. 27.17. The maximum deflection is

$$Y = \frac{0.0135qR^4}{EI} \qquad (27.64)$$

For small angles of θ, on the order of 30°, vertical deflection can undergo the change in sign. At these values, however, the deflection is relatively small.

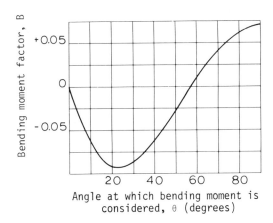

Fig. 27.17 Bending moment chart for an arch under its own weight.

SYMBOLS

A_1, A_2, A_3, A_4	Load factors for built-in arch
B	Bending moment factor
b	Width of rectangular section, in. (mm)
d	Mean diameter of piston ring before cut, in. (mm)
E	Modulus of elasticity, psi (N/mm^2)
F	Pin-joint reaction, lb (N)
G_1 through G_6	Arch factors
g	Acceleration due to gravity, in./sec^2 (mm/sec^2)
H	Horizontal load, lb (N)
H_b	Horizontal thrust in built-in arch, lb (N)
H_p	Horizontal thrust in pin-jointed arch, lb (N)
h	Depth of cross section, in. (mm)
I	Moment of inertia, in.^{4} (mm^4)
K_1	Arched cantilever factor
K_M	Factor for bending moment
K_u, K'_u	Factors for radial deflection
K_ψ	Factor for slope
M	Bending moment, lb-in. (N-mm)
M_f	Fixing moment, lb-in. (N-mm)
M_0	External bending couple, lb-in. (N-mm)
M_q	Bending moment under uniform load, lb-in. (N-mm)
M_1, M_2	Bending moments for various portions, lb-in. (N-mm)
P	Concentrated load, lb (N)
$\bar{P}$	Fictitious load (Fig. 27.4), lb (N)
q	Uniform load, lb/in. (N/mm)
q_m	Maximum load per unit length, lb/in. (N/mm)
R	Mean radius of curvature, in. (mm)
S	Stress, psi (N/mm^2)
S_b	Bending stress, psi (N/mm^2)
u	Radial displacement, in. (mm)
v	Peripheral velocity, in./sec (mm/sec)
V	Vertical reaction, lb (N)
X	Horizontal deflection, in. (mm)
Y	Vertical deflection, in. (mm)
α	Ring or arch angle, deg
γ	Specific weight, lb/in.^{3} (N/mm^3)
ϵ	Auxiliary angle, rad
θ	Angle at which forces are considered, rad
ϕ	Angle subtended by arched cantilever, rad
$\chi = R/h$	Ratio of radius of curvature to depth of section
ψ	Slope, rad

28
Curved Beams and Hooks

EARLY DEVELOPMENTS

The analysis and design of curved members such as hooks, chain links, bearing housings, proving rings, and structural frames for machines involves application of the well-known principles of elastic response of curved beams. In this type of a machine element, the radius of curvature is usually small in comparison with the overall dimensions of the beam, and the neutral axis is said to be displaced toward the center of curvature.

Early developments of curved-beam theory date back at least to the middle of the nineteenth century and are due to Winkler [155]. However, experimental verification of the original theory and its subsequent modifications received rather scant attention from early investigators in the field of elasticity. Part of the reason for this poor state of knowledge at the time was the lack of suitable apparatus for experimental work. It was not until 1906 that a rational test program got under way at the University of Illinois [156] with the specific purpose of providing design data for chain links. The test of Winkler theory, however, was not feasible because of the inherent doubt regarding the distribution of pressure between adjacent links. It was then necessary to go to experiments on circular rings with rectangular cross sections, for which a more true knife-edge bearing was possible. These ring tests seemed to confirm the theoretical analysis of deflection, and it was concluded that the fundamental curved beam equation used at the time could also predict the true stresses. The Winkler formula (later to be known as Winkler-Bach) was then expressed as follows.

$$S = \frac{M}{AR}\left[1 + \frac{c}{\lambda(R+c)}\right] \tag{28.1}$$

In this equation, S denotes the circumferential stress at a distance c measured from the central axis of the transverse cross section of the beam at which the bending moment is M. The distance from the centroidal axis to the center of curvature of the unstressed beam is R. Finally, A denotes the area of the cross section and λ is a geometrical property of the cross section, which is defined as

$$\lambda = \frac{1}{A}\int \frac{c}{R+c}\,dA \tag{28.2}$$

The Winkler-Bach equations take into account the effect of beam curvature on circumferential stresses, but they require prior determination of the section parameter given by Eq. (28.2). Some typical analytical expressions for λ are found in Table 28.1. The exact analysis of the stress distribution should involve simultaneous consideration of the direct stresses in radial and circumferential directions, together with the superimposed effect of shear. Since, however, the maximum stresses in the circumferential direction considerably exceed the maximum radial stresses the approximate Winkler-Bach theory, based on circumferential stresses only, has a good deal of practical justification. Until about the year 1914, English and American practice was to estimate the stresses in hooks and curved bars by the rules applicable to straight beams and to ignore the effects of curvature. At that time, however, Morley published a discussion of the engineering approach to curved-beam design [157] and gave support to the Winkler-Bach theory on practical grounds. The adequacy of this theory was also demonstrated by the early tests [158], while fundamental curved-beam equations were derived from the first principles using the strain energy approach [159].

Although Winkler-Bach theory has certainly stood the test of time, it is well to recognize that the general limitations that apply to the formula $S_b = M/Z$ for straight beams are also valid for the curved-beam equations [139]. Additional restrictions must also be imposed on curved members having H, I, T, and similar cross-sections. Specifically, flanged sections should not be excessively wide or thin in order to avoid local buckling or unusual stretching. This is particularly important in the case of relatively brittle materials, which can easily fail due to a local tensile fracture. Some of these failures can, of course, be prevented by adding either welded or riveted stiffeners, provided that the local stress concentration at the heat-affected zone does not significantly alter the load-carrying capacity of the entire structure.

CORRECTION FOR NEUTRAL AXIS

When a curved beam is subjected to the combination of loading, superposition of bending and direct elastic stresses gives the general expression

$$S = \frac{P}{A} + \frac{M}{AR}\left[1 + \frac{c}{\lambda(R+c)}\right] \tag{28.3}$$

Curved Beams and Hooks

Table 28.1 Analytical Expressions for λ

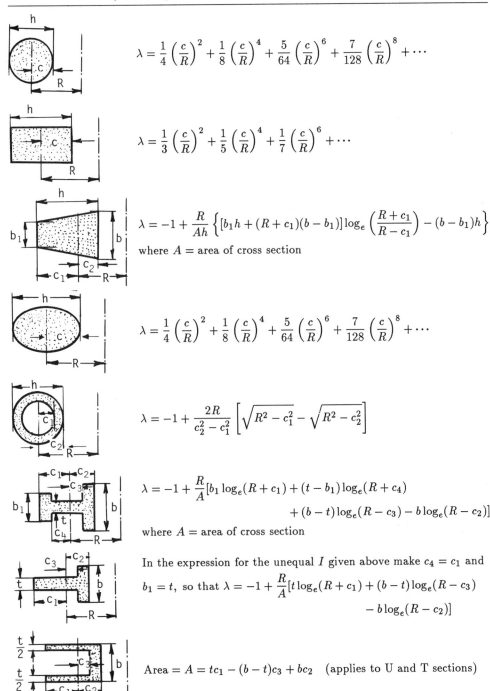

$$\lambda = \frac{1}{4}\left(\frac{c}{R}\right)^2 + \frac{1}{8}\left(\frac{c}{R}\right)^4 + \frac{5}{64}\left(\frac{c}{R}\right)^6 + \frac{7}{128}\left(\frac{c}{R}\right)^8 + \cdots$$

$$\lambda = \frac{1}{3}\left(\frac{c}{R}\right)^2 + \frac{1}{5}\left(\frac{c}{R}\right)^4 + \frac{1}{7}\left(\frac{c}{R}\right)^6 + \cdots$$

$$\lambda = -1 + \frac{R}{Ah}\left\{[b_1 h + (R+c_1)(b-b_1)]\log_e\left(\frac{R+c_1}{R-c_1}\right) - (b-b_1)h\right\}$$

where A = area of cross section

$$\lambda = \frac{1}{4}\left(\frac{c}{R}\right)^2 + \frac{1}{8}\left(\frac{c}{R}\right)^4 + \frac{5}{64}\left(\frac{c}{R}\right)^6 + \frac{7}{128}\left(\frac{c}{R}\right)^8 + \cdots$$

$$\lambda = -1 + \frac{2R}{c_2^2 - c_1^2}\left[\sqrt{R^2 - c_1^2} - \sqrt{R^2 - c_2^2}\right]$$

$$\lambda = -1 + \frac{R}{A}[b_1\log_e(R+c_1) + (t-b_1)\log_e(R+c_4)$$
$$+ (b-t)\log_e(R-c_3) - b\log_e(R-c_2)]$$

where A = area of cross section

In the expression for the unequal I given above make $c_4 = c_1$ and $b_1 = t$, so that $\lambda = -1 + \frac{R}{A}[t\log_e(R+c_1) + (b-t)\log_e(R-c_3)$
$$- b\log_e(R-c_2)]$$

Area = $A = tc_1 - (b-t)c_3 + bc_2$ (applies to U and T sections)

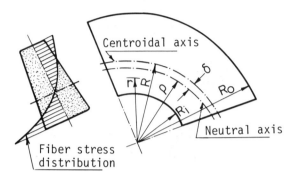

Fig. 28.1 Curved-beam notation.

The parameter λ can be calculated for a number of typical sectional configurations, as shown for example by Table 28.1. The form of Eq. (28.3) is sometimes rearranged in terms of the parameter δ defining the amount of the radial displacement of the neutral axis

$$S = \frac{P}{A} + \frac{M(R - R_i - \delta)}{AR_i\delta} \tag{28.4}$$

If ρ denotes the radius that locates the position of the neutral axis and r is an arbitrary distance from the center of curvature to a point in the cross section shown in Fig. 28.1, the displacement of the neutral axis is

$$\delta = R - \rho \tag{28.5}$$

The term ρ can be defined as

$$\rho = \frac{A}{\int \frac{dA}{r}} \tag{28.6}$$

Again, this parameter can be estimated using a closed-form approach for a number of regular, compact cross sections, and various published tables are available for this purpose. The main disadvantage in using Eq. (28.4) is, however, that the result is sensitive to relatively small changes in the value of δ.

EXPERIMENTAL FACTORS IN DESIGN

To simplify the design procedure for curved beams, Wilson and Quereau conducted an extensive series of tests some 60 years ago [160]. Based on these tests and on the theory of neutral axis, simple correction factors have been developed for the use with the original Winkler-Bach formula. Denoting the relevant correction factor by ϕ_0, the expression for the maximum stress can be stated as

$$S = \phi_0 \left(\frac{P}{A} + \frac{Mc}{I} \right) \tag{28.7}$$

Curved Beams and Hooks

Table 28.2 Wilson and Quereau Factors for Stresses in Curved Beams

	R/c	ϕ_0 Inner Face	ϕ_0 Outer Face	δ/R
	1.2	3.41	0.54	0.224
	1.4	2.40	0.60	0.151
	1.6	1.96	0.65	0.108
	1.8	1.75	0.68	0.084
	2.0	1.62	0.71	0.069
	3.0	1.33	0.79	0.030
	4.0	1.23	0.84	0.016
	6.0	1.14	0.89	0.0070
	8.0	1.10	0.91	0.0039
	10.0	1.08	0.93	0.0025
	1.2	2.89	0.57	0.305
	1.4	2.13	0.63	0.204
	1.6	1.79	0.67	0.149
	1.8	1.63	0.70	0.112
	2.0	1.52	0.73	0.090
	3.0	1.30	0.81	0.041
	4.0	1.20	0.85	0.021
	6.0	1.12	0.90	0.0093
	8.0	1.09	0.92	0.0052
	10.0	1.07	0.94	0.0033
	1.2	3.01	0.54	0.336
	1.4	2.18	0.60	0.229
	1.6	1.87	0.65	0.168
	1.8	1.69	0.68	0.128
	2.0	1.58	0.71	0.102
	3.0	1.33	0.80	0.046
	4.0	1.23	0.84	0.024
	6.0	1.13	0.88	0.011
	8.0	1.10	0.91	0.0060
	10.0	1.08	0.93	0.0039
	1.2	3.09	0.56	0.336
	1.4	2.25	0.62	0.229
	1.6	1.91	0.66	0.168
	1.8	1.73	0.70	0.128
	2.0	1.61	0.73	0.102
	3.0	1.37	0.81	0.046
	4.0	1.26	0.86	0.024
	6.0	1.17	0.91	0.011
	8.0	1.13	0.94	0.0060
	10.0	1.11	0.95	0.0039

Examples of the numerical values of ϕ_0 for the inner and outer faces of the curved members for various cross sections are given in Table 28.2, which continues on pp. 342 and 343.

APPROXIMATION OF STRESS FACTORS

Examination of these values appears to indicate that despite rather striking differences in the cross-sectional geometry, the stress correction factors do not differ excessively over a wide range of R/c or $R/(R - R_i)$ ratios. For example, plotting

Table 28.2 (cont.)

Shape	R/c	ϕ_0 Inner Face	ϕ_0 Outer Face	δ/R
5b / 4b / b / c / R (trapezoidal)	1.2	3.14	0.52	0.352
	1.4	2.29	0.54	0.243
	1.6	1.93	0.62	0.179
	1.8	1.74	0.65	0.138
	2.0	1.61	0.68	0.110
	3.0	1.34	0.76	0.050
	4.0	1.24	0.82	0.028
	6.0	1.15	0.87	0.012
	8.0	1.12	0.91	0.0060
	10.0	1.10	0.93	0.0039
3b/5 / b / c / R (triangular-trapezoid)	1.2	3.26	0.44	0.361
	1.4	2.39	0.50	0.251
	1.6	1.99	0.54	0.186
	1.8	1.78	0.57	0.144
	2.0	1.66	0.60	0.116
	3.0	1.37	0.70	0.052
	4.0	1.27	0.75	0.029
	6.0	1.16	0.82	0.013
	8.0	1.12	0.86	0.0060
	10.0	1.09	0.88	0.0039
9t/2 / 3t/2 / t / 4t / c / R (T-section)	1.2	3.63	0.58	0.418
	1.4	2.54	0.63	0.299
	1.6	2.14	0.67	0.229
	1.8	1.89	0.70	0.183
	2.0	1.73	0.72	0.149
	3.0	1.41	0.79	0.069
	4.0	1.29	0.83	0.040
	6.0	1.18	0.88	0.018
	8.0	1.13	0.91	0.010
	10.0	1.10	0.92	0.0065
t / 3t / 2t / 4t / 6t / c / R (I-section)	1.2	3.55	0.67	0.409
	1.4	2.48	0.72	0.292
	1.6	2.07	0.76	0.224
	1.8	1.83	0.78	0.178
	2.0	1.69	0.80	0.144
	3.0	1.38	0.86	0.067
	4.0	1.26	0.89	0.038
	6.0	1.15	0.92	0.018
	8.0	1.10	0.94	0.010
	10.0	1.08	0.95	0.0065

the relevant values for the inner surfaces of the beams, where the critical stresses are usually found, a wide range of the factors may be represented by a single design curve such as that shown in Fig. 28.2. Since according to the customary procedure in curved-beam and hook design generous factors of safety are recommended, ballpark estimates such as those based on Fig. 28.2 may be acceptable [161] during the preliminary stages of hardware sizing. This approximation increases in accuracy as the cross sections of the curved members become more compact and have fewer abrupt changes of contour.

Curved Beams and Hooks

Table 28.2 (cont.)

R/c	ϕ_0 Inner Face	ϕ_0 Outer Face	δ/R
1.2	2.52	0.67	0.408
1.4	1.90	0.71	0.285
1.6	1.63	0.75	0.208
1.8	1.50	0.77	0.160
2.0	1.41	0.79	0.127
3.0	1.23	0.86	0.058
4.0	1.16	0.89	0.030
6.0	1.10	0.92	0.013
8.0	1.07	0.94	0.0076
10.0	1.05	0.95	0.0048
1.2	3.28	0.58	0.269
1.4	2.31	0.64	0.182
1.6	1.89	0.68	0.134
1.8	1.70	0.71	0.104
2.0	1.57	0.73	0.083
3.0	1.31	0.81	0.038
4.0	1.21	0.85	0.020
6.0	1.13	0.90	0.0087
8.0	1.10	0.92	0.0049
10.0	1.07	0.93	0.0031
1.2	2.63	0.68	0.399
1.4	1.97	0.73	0.280
1.6	1.66	0.76	0.205
1.8	1.51	0.78	0.159
2.0	1.43	0.80	0.127
3.0	1.23	0.86	0.058
4.0	1.15	0.89	0.031
6.0	1.09	0.92	0.014
8.0	1.07	0.94	0.0076
10.0	1.06	0.95	0.0048

STRESSES IN HOOKS

In the common type of a lifting hook represented by the curved-beam model shown in Fig. 28.3, the load is applied as a direct shear. The actual shear stresses, however, are rather small, and the maximum stress is found at or close to the section $A_1 A_2$ of the hook perpendicular to the direction of the load at point A_1. It is clear that both the greatest bending moment and direct load are developed on the section $A_1 A_2$, and the corresponding circumferential stresses for the tensile and the compressive regions of the hook can be expressed as follows [159].

$$S_t = \frac{P}{A} \left[\left(\cos\theta - \frac{H_0}{R} \right) \frac{c}{\lambda(R-c)} + \frac{H_0}{R} \right] \tag{28.8}$$

and

$$S_c = -\frac{P}{A} \left[\left(\cos\theta - \frac{H_0}{R} \right) \frac{d-c}{\lambda(R+d-c)} - \frac{H_0}{R} \right] \tag{28.9}$$

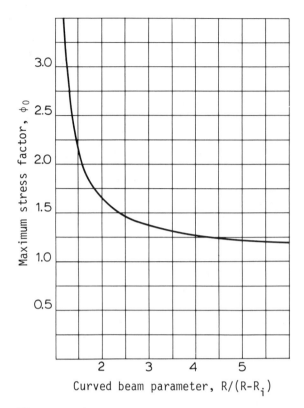

Fig. 28.2 Approximate stress factor for curved beams.

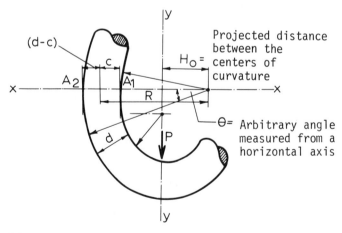

Fig. 28.3 Notation for a working portion of a machine hook.

Curved Beams and Hooks

In many cases distance H_0 for the working portion of the hook is zero, so that the relevant equations for circumferential and compressive stresses reduce to

$$S_t = \frac{Pc\cos\theta}{A\lambda(R-c)} \qquad (28.10)$$

and

$$S_c = -\frac{P(d-c)\cos\theta}{A\lambda(R+d-c)} \qquad (28.11)$$

When the hook cross section is constant and $\theta = 0$, the maximum stresses become

$$S_t = \frac{Pc}{A\lambda(R-c)} \qquad (28.12)$$

and

$$S_c = -\frac{P(d-c)}{A\lambda(R+d-c)} \qquad (28.13)$$

It is noted that c in the foregoing equations is measured from the inner surface of the critical section to the center of gravity of the cross section. The total depth of the hook cross section in this case is denoted by d, as shown in Fig. 28.3.

Design Problem 28.1

A machine support bracket has a uniform T cross section and subtends 180°. Estimate the magnitude of the maximum stress at point B, assuming the dimensions given in Fig. 28.4 and a maximum downward load $P = 10,000$ lb acting through the center of gravity of the section.

Solution

From the sectional geometry and dimensions, we get

$A = 2.625$ in.^{2}
$c = 1.0$ in.
$R = 6.0$ in.
$I = 2$ in.^{4}

From Fig. 28.2 $\phi_0 = 1.19$. Hence, using Eq. (28.7) the stress becomes

$$S = 1.19\left(\frac{10,000}{2.625} + \frac{1 \times 10,000 \times 12}{2}\right) = 76,000 \text{ psi} \quad (524 \text{ N/mm}^2) \quad \blacklozenge$$

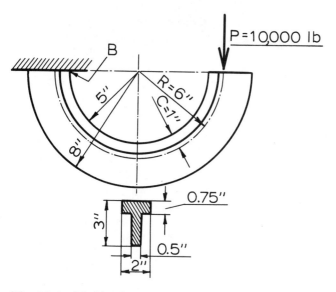

Fig. 28.4 Machine bracket geometry for Design Problem 28.1.

Design Problem 28.2

A machine clamp having a form of a half circle has the same cross-sectional geometry as that analyzed in Design Problem 28.1. Assuming that the load $P = 12{,}000$ lb acts in the direction shown in Fig. 28.5, calculate the maximum hook-type stress in the clamp, using the exact and approximate methods of the analysis given in this section.

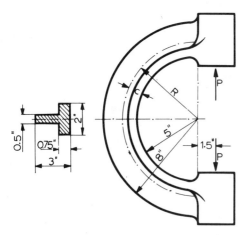

Fig. 28.5 Half-circle clamp.

Curved Beams and Hooks

Solution

For $A = 2.625$ in.^{2} and $R = 6.0$ in., Table 28.1 gives $\lambda = 0.0163$. Hence, using Eq. (28.12), we obtain

$$S_t = \frac{12{,}000 \times 1.0}{2.625 \times 0.0163(6.0 - 1.0)} = 56{,}091 \text{ psi} \quad (387 \text{ N/mm}^2)$$

Since the bending moment for the clamping device given in Fig. 28.5 is somewhat higher than that for a simple hook because of the additional offset equal to 1.5 in., the total stress can be corrected as follows.

$$\frac{P}{A} + \frac{7.5}{6}\left(S_t - \frac{P}{A}\right)$$
$$= 4571 + \frac{7.5}{6}(56{,}091 - 4571) = 68{,}971 \text{ psi} \quad (476 \text{ N/mm}^2)$$

From Eq. (28.7), representing the approximate method of design, one gets

$$S = 1.19\left(\frac{12{,}000}{2.625} + \frac{7.5 \times 12{,}000 \times 1.0}{2}\right) = 58{,}990 \text{ psi} \quad (407 \text{ N/mm}^2) \quad \blacklozenge$$

Hence, the result obtained by the approximate is off by about 14%. Other cases may indicate somewhat higher differences, but in the majority of practical designs, such differences should be well within the customary factors of safety.

Since the numerical values of the circumferential stresses in a typical crane hook are quite different for the inner and outer surfaces, the choice of the cross-sectional geometry is seldom economical. Hence, the thrust of some of the major investigations in the past was directed toward the development of methods of selection of optimized cross-sectional geometry. It soon became obvious that the greatest economy of the material could be effected by making the two extreme circumferential stresses essentially equal. This theoretical criterion is

$$\frac{R_i}{2R_o} = \frac{R - R_i}{R_o - R_i} \tag{28.14}$$

Although strictly theoretical condition has always implied the possibility of utilizing a triangular section, the limitations of the manufacturing process have influenced the development of a bull-head type of a hook cross section, with I and T sections, however, being efficient and practicable alternatives. Once the basic cross-sectional shape was finally determined, the next parameter of interest became the ratio of the depth of the cross section to the inner radius of curvature for a given specified stress. The most useful practical relations in hook design involved ultimate load, stress, radius of curvature, and the area of cross section.

According to one of the very valuable papers dealing with the scientific and practical aspects of lifting gear components [159], the selection of the depth of the cross section can be accomplished in a simplified manner for the two most common geometries.

Circular section

$$R_o - R_i = 0.023(P)^{1/2} + 0.18R_i \tag{28.15}$$

Trapezoidal or bull-head section

$$R_o - R_i = 0.026(P)^{1/2} + 0.20R_i \tag{28.16}$$

Over the years the industry has developed an almost infinite variety of lifting components. A few typical hook shapes found in recent applications are shown in Fig. 28.6. Trapezoidal, bull-head, and circular hook cross sections appear to comprise at least 90% of all the configurations involved for which the line of action of the load passes through the center of curvature of the horizontal section.

When the line of the hook-load does not pass through the center of curvature of the horizontal section, the stresses should also be checked for the hook area where the large radius runs into the smaller radius of curvature below the horizontal plane. Although in this location the external forces on the hook may be smaller than those found along a typical horizontal section, the effect of a greater local curvature may sometimes be sufficient to make the local stress higher. In some standard hook designs, it is customary to make identical cross sections for the horizontal and vertical planes, with the intermediate sections made slightly greater in depth. Theoretically, the stress along the inner surface of the hook could be maintained constant by reducing each intermediate section in proportion to $\cos\theta$, where θ is the angle measured from the horizontal section as shown in Fig. 28.3. The lower limit imposed on hook depth at the vertical section y–y, however, must depend on the shear load. Although this cosine theory of sizing is theoretically sound, the mechanics of possible hook deformations under load suggests a potential problem area. For instance, as the load causes yield at the horizontal section, the hook begins to open up and makes the subsequent weaker sections more nearly perpendicular to the load. Such sections then yield, in turn, promoting further opening of the hook and further yielding of other sections, leading to complete failure. To prevent

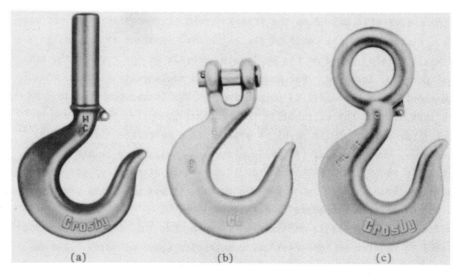

Fig. 28.6 Typical hook shapes from industry; (a) shank hook; (b) clevis slip hook; (c) eye hook. (Courtesy of the Crosby Group.)

this type of progressive opening and yielding, standard hook configurations contain greater cross-sectional areas between the horizontal and vertical sections, denoted by the x–x and y–y axes in Fig. 28.3. As far as the upper sections of the hook are concerned, it is only necessary to assure that the stress in them nowhere exceeds that found at the horizontal section. Usually, some reduction in area is found above the horizontal section, permitting a gradual transition to the shank or eye portion of the hook. The design of the eye can be developed on the basis of ring theory, but it is often influenced by experience and convention. Several aspects of the design theory of eye-bar connections that bear on this problem are discussed in the next section. The design of the hook point is mainly conventional, provided that any other dangerous possibilities, such as load jumping off the hook, are minimized.

The development of hooks and similar lifting components is generally governed by elastic theory. This is a sound and straightforward approach for all the cases involving working loads only. However, when the same analysis is used for estimating the maximum stresses under the proof load conditions, equivalent normally to about twice the working load, local plastic stresses can develop. The redistribution of such stresses due to the plastic action will depend on the stress-strain characteristics of the materials. Furthermore, the ratio of the calculated apparent stress to the actual yield stress, based on the Mc/I rule, depends on the type of the sectional geometry involved. For instance, for a rectangular section, this ratio should give 1.7 and 2.0, respectively. The latter number also indicates that despite the variability of sectional geometry, the maximum apparent fiber stress should never be greater than twice the yield stress. Furthermore, the values of the maximum stresses, estimated on the basis of the elastic theory, are such that the actual yield strength of the material is never exceeded. Also, when failure occurs, it can always be postulated that the material's breakdown is due to tension.

The effect of plastic deformation is such that the outside fibers overstressed in tension go into residual compression upon unloading. The reverse is also true about the fibers stressed originally in compression. Here the argument is sometimes advanced that the theory of elasticity must be erroneous. Although the calculated elastic stresses often exceed the tensile strength of the material by a substantial margin, the component not only fails to fracture at the first load application but also shows very little permanent deformation. Consequently, one might assume that elastic theory has no real bearing on the problem and the component should withstand unlimited repetitions of the load. Such a statement, of course, is dangerous because it ignores the fundamental principles of material behavior in fatigue discussed in Chap. 13.

As the arguments for and against the use of the classical theory of elasticity in lifting gear design progressed, some tests on hooks have at times been made. Unfortunately, the preponderance of the test data is not readily available because of the proprietary nature of the information obtained by the competitive branches of industry. Some manufacturers rely simply on the industrial standards without questioning the basic theory or conducting research. Others develop their own improved configurations, review existing design techniques, and conduct verification tests without, however, divulging any such data. Recent surveys and repeated contacts with the industry have confirmed the existence of such restrictive practices.

Observations on the results of hook testing given in this chapter are based on a selected reference [159] and private communications concerned with the hooks designed according to the British standards. Early chain links and hooks were forged in either wrought iron or mild steel and the tests involved monitoring deflection, permanent set, and typical modes of failure. In all the examined cases structural failures occurred at the calculated stresses above the elastic limit of the material, accompanied by a negligible amount of permanent set. In addition to the standard, low-strain-rate tests on these components some data were obtained under fatigue conditions. In several instances factors of safety as low as 1.2 were found. Although such values appeared to be unduly low, in actual service the hooks were not designed to withstand an indefinitely great number of load applications. Nevertheless, an important lesson should be derived from such experiments, since the presence of surface discontinuities by design or by ill usage decrease the number of load cycles to failure. It was also noted that the effect of heavier hook sections between the vertical and horizontal planes increased the fatigue factor of safety. The observed failures appeared to have the origins of cracks at the points of maximum stresses accompanied by a limited degree of overall distortion. Of the two early materials considered in the tests, a slight superiority for mild steel over wrought iron was noted. As far as the impact strength was concerned, prior repeated load applications produced a finite decrease of impact resistance values.

DESIGN OF CURVED BEAMS

In the common problems of curved beams, the analyst is concerned with the calculation of maximum stresses. In fully analyzing the deflection of thick curved bars, such as those shown in Fig. 28.7 and 28.8, use can be made of Eq. (7.4), where account is taken of the elastic strain energy due to bending, normal, and shear stresses. From the resolution of forces, the relevant expressions for the bending moment M, normal force N, and shear force Q are obtained. For instance, for the case of a horizontal load H (Fig. 28.7), the relevant moment and forces are

$$M = -HR\sin\theta \qquad (28.17)$$
$$N = H\sin\theta \qquad (28.18)$$
$$Q = H\cos\theta \qquad (28.19)$$

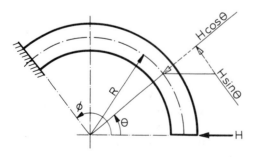

Fig. 28.7 Curved bean under horizontal end load.

Curved Beams and Hooks

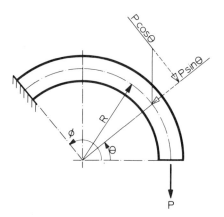

Fig. 28.8 Curved beam under vertical end load.

The partial derivatives necessary for the solution of Eq. (7.4) follow directly from Eqs. (28.17) to (28.19). Hence, integrating Eq. (7.4) within the limits of 0 and ϕ gives the following general formula for the horizontal deflection under horizontal load

$$X = \frac{HR}{4A}\left[\frac{(2\phi - \sin 2\phi)(R + 3\delta)}{\delta E} + \frac{\xi(2\phi + \sin 2\phi)}{G}\right] \quad (28.20)$$

When $\phi = \pi/2$, $G = 0.4E$, and $\xi = 3/2$, Eq. (28.20) reduces to

$$X = \frac{\pi H R(4R + 27\delta)}{16AE\delta} \quad (28.21)$$

For this equation, the value ξ corresponds to that for a rectangular cross section. The rigorous solution of the same problem based on the theory of elasticity [2] gives

$$X = \frac{\pi H}{bE}\left[\frac{1 + 4\chi^2}{(1 + 4\chi^2)\log_e\left(\frac{2\chi+1}{2\chi-1}\right) - 4\chi}\right] \quad (28.22)$$

where $\chi = R/h$. For instance, taking $b = 1$ in., $h = 2$ in., and $R = 3$ in., Eqs. (28.21) and (28.22) give $11.2\pi H/E$ and $10.7\pi H/E$, respectively. In other words, the elastic strain energy solution differs only by about 4.7% from the more rigorous solution. It is recalled here that in Eq. (7.4), the moment of inertia was replaced by the term $AR\delta$. When δ is very small, that is, where the central and neutral axes coincide, the term involving δ in the numerator of Eq. (28.21) may be ignored. Putting $\delta = I/AR$ into Eq. (28.21) gives the design formula for a thin curved member, subtending 90° and carrying a horizontal, concentrated load.

When a curved beam such as that shown in Fig. 28.8 is considered, the required moment and force equations are rather straightforward.

$$M = -PR(1 - \cos\theta) \quad (28.23)$$

$$N = -P\cos\theta \tag{28.24}$$
$$Q = P\sin\theta \tag{28.25}$$

Utilizing the strain energy approach, the deflection under load P becomes

$$Y = \frac{PR}{4A}\left[\frac{2\phi(3R-\delta)}{\delta E} + \frac{(\sin 2\phi - 8\sin\phi)(R-\delta)}{\delta E} + \frac{\xi(2\phi - \sin 2\phi)}{G}\right] \tag{28.26}$$

For $\phi = \pi/2$, Eq. (28.26) simplifies to

$$Y = \frac{PR}{4A}\left[\frac{R(3\pi - 8) + \delta(8 - \pi)}{\delta E} + \frac{\pi\xi}{G}\right] \tag{28.27}$$

For $\delta \ll R$ and $I = AR\delta$, Eq. (28.27) reduces to a standard design formula for a quarter-circle arched cantilever under vertical load.

Design Problem 28.3

The short pipe element shown in Fig. 28.9 is rigidly fixed at one end and deflects 0.01 in. under a horizontal, concentrated load H. If the material is steel with $E = 30 \times 10^6$ psi, calculate the approximate bending stress, assuming that $G = 0.4E$, $\delta = 0.4$ in., and $\xi = 1$.

Solution

Substituting $\phi = \pi/2$, $G = 0.4E$, and $\xi = 1$ into Eq. (28.20) gives

$$X = \frac{HR}{AE\delta}(0.7854R + 4.3197\delta)$$

Since the bending moment $M = HR$, the foregoing expression yields

$$M = \frac{XAE\delta}{0.7854R + 4.3197\delta}$$

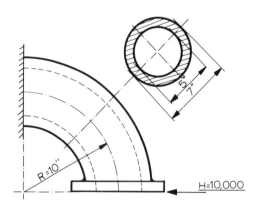

Fig. 28.9 Quarter-circle pipe bend.

Curved Beams and Hooks

Hence, the bending stress, using Eq. (28.7), is

$$S = M\phi_0 \left(\frac{1}{AR} + \frac{1}{Z}\right)$$

Since $R/(R - R_i) = 10/3.5 = 2.86$, from Fig. 28.2

$$\phi_0 = 1.35$$

Also from Fig. 28.9

$$A = 18.85 \text{ in.}^2$$
$$Z = 28.27 \text{ in}^3$$

and

$$M = \frac{0.01 \times 18.85 \times 30.10^6 \times 0.4}{0.7854 \times 10 + 0.4 \times 4.3197} = 236{,}070 \text{ lb-in.}$$

so that

$$S = 1.35 \times 236{,}070 \left(\frac{1}{188.5} + \frac{1}{28.27}\right) = 12{,}964 \text{ psi} \quad (89.4 \text{ N/mm}^2) \quad \blacklozenge$$

CURVED BEAMS WITH VARIABLE CROSS SECTIONS

In designing machines for riveting, stitching, drilling, and welding, as well as press and jigsaw work, curved beams of variable cross section are often encountered. This design feature complicates the analysis very significantly and often requires the help of various approximations. The techniques for analyzing deflections of curved members of variable cross sections are seldom considered in engineering handbooks. Lofgren's semigraphical and mathematical procedures [162] should be noted for clarity and practicality of the approach to calculating the deflections. In general terms, the deflection of a curved member with a variable cross section can be stated as

$$Y = \frac{1}{E}\int_0^L \frac{M}{I_x}\frac{\partial M}{\partial P} ds \qquad (28.28)$$

Here I_x is the moment of inertia at an arbitrary point along the neutral axis and L is the "stretched-out" length of the member. The method is best outlined by reference to Fig. 28.10, which shows a curved frame with 12 subdivision points, making certain that the relevant sections are perpendicular to the neutral axis. Since prior knowledge of the position of neutral axis is required, reference to Table 28.2 should be of some help. For the particular case illustrated in Fig. 28.10, Eq. (28.28) becomes

$$Y = \frac{P}{E}\int_0^s \frac{a_x^2}{I_x} ds \qquad (28.29)$$

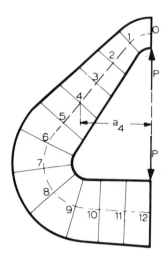

Fig. 28.10 Curved member with varying cross section.

Expression (28.29) reflects the principle of Castigliano applied to a semigraphical procedure of finding deflections, and it can be explained in the following way. The term a_x represents a moment arm measured from the line of action of load P to an arbitrary point along the neutral axis. At station 4, , for instance, the relevant bending moment is $M = Pa_4$ and its partial derivative needed for Eq. (28.28) is $\partial M/\partial P = a_4$. The term $M\,\partial M/\partial P$ gives the square of the moment arm at each point, as shown by Eq. (28.29). The area under the curve is the required integral as shown in Fig. 28.11, which can now be obtained directly by means of a planimeter or other suitable method. The deflection is then found by multiplying this area by the constant term P/E. The quantity under the integral sign is inversely proportional to length. Some examples of application of this method can be found in the literature [11, 162].

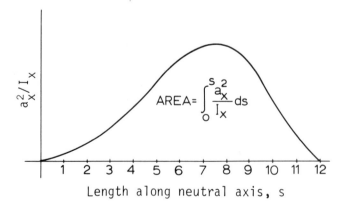

Fig. 28.11 Graphical integral.

Curved Beams and Hooks

SYMBOLS

A	Area of cross section, in.^{2} (mm^2)
a_x	Moment arm, in. (mm)
b	Width of section, in. (mm)
c	Distance from inner surface to central axis, in. (mm)
d	Round bar diameter, in. (mm)
E	Modulus of elasticity, psi (N/mm^2)
G	Modulus of rigidity, psi (N/mm^2)
H	Horizontal load, lb (N)
h	Depth of cross section, in. (mm)
H_0	Distance between centers of curvature, in. (mm)
I, I_x	Moments of inertia, in.^{4} (mm^4)
L	Length of straight portion, in. (mm)
M	Bending moment, lb-in. (N-mm)
N	Normal force, lb (N)
P	Vertical load, lb (N)
Q	Transverse shearing force, lb (N)
R	Radius to center of gravity, in. (mm)
R_i	Inner radius, in. (mm)
R_o	Outer radius, in. (mm)
r	Arbitrary radius, in. (mm)
S	Stress, psi (N/mm^2)
S_c	Compressive stress, psi (N/mm^2)
S_t	Tensile stress, psi (N/mm^2)
s	Distance along curved member, in. (mm)
X	Horizontal deflection, in. (mm)
Y	Vertical deflection, in. (mm)
Z	Section modulus, in.^{3} (mm^3)
δ	Distance from neutral to central axis, in. (mm)
θ	Angle at which forces are considered, rad
λ	Winkler's parameter
ξ	Shear distribution factor
ρ	Cross-sectional area factor
ϕ	Angle subtended by curved member, rad
ϕ_0	Curved-beam stress factor
$\chi = R/h$	Ratio of radius of curvature to depth of section

29
Links and Eyebars

INTRODUCTION

Structural rings, coupling links, chain components, eyebars, and similar machine parts represent multiply connected members that have to be analyzed as statically indeterminate structures. Despite their relative geometrical simplicity, their analysis can be quite tedious including the application of strain energy methods. Whenever possible, the relevant analysis is simplified by neglecting the effects of direct and shear stresses in comparison with those due to bending.

Since proving rings fall in the same general category as the conventional machine parts it may be of interest to briefly review the particular response of a thick ring under diametral loading. This analysis should be applicable not only to chain links but also to such machine elements as bearing rings and the rims of heavy gears. The derivation that follows is based on the expressions for the elastic strain energy [11, 163–165], although a more rigorous approach to this problem has also been considered [2, 166].

THICK-RING THEORY

For a typical thick ring such as that shown in Fig. 29.1, the equilibrium of forces for one quarter gives the following relations in terms of the real and fictitious forces.

$$M = \frac{PR(1-\cos\theta)}{2} + \frac{HR\sin\theta}{2} - M_{\mathrm{f}} \qquad (29.1)$$

Links and Eyebars

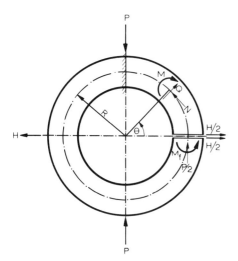

Fig. 29.1 Thick ring in diametral compression.

$$N = \frac{H \sin \theta}{2} - \frac{P \cos \theta}{2} \tag{29.2}$$

$$Q = \frac{P \sin \theta}{2} + \frac{H \cos \theta}{2} \tag{29.3}$$

In these equations, the symbol H without a bar is regarded as a fictitious quantity introduced solely for the purpose of calculating the change in the horizontal diameter due to load P. Since this is a classical and well-known case, only the principal equations will be given. To find the statically indeterminate fixing couple $M_{\rm f}$, the following boundary condition is obtained from Eq. (7.4).

$$\int_0^{\pi/2} (M - N\delta) \frac{\partial M}{\partial M_{\rm f}} \, d\theta = 0 \tag{29.4}$$

Utilizing Eqs. (29.1) and (29.2), integrating Eq. (29.4), and solving for $M_{\rm f}$ gives

$$M_{\rm f} = \frac{P(\pi R - 2R + 2\delta)}{2\pi} \tag{29.5}$$

Also, for $H = 0$, Eq. (29.1) yields

$$M = \frac{P(2R - 2\delta - \pi R \cos \theta)}{2\pi} \tag{29.6}$$

Again using the complete general expression, Eq. (7.4), the vertical and horizontal deflections become

$$Y = \frac{P}{4AE} \left[\frac{\pi^2 R(R - 2\delta) - 8(R - \delta)^2}{\pi \delta} + \pi R(1 + 2.5\xi) \right] \tag{29.7}$$

and

$$X = \frac{P}{2AE}\left[\frac{(4R - \pi R - 4\delta)(R - \delta)}{\pi\delta} + 2.5R\xi\right] \tag{29.8}$$

THICK-RING EXPERIMENT

Over the years, various calculations of thick-ring deflections have been made using either the Castigliano principle or the theory of elasticity. However, despite the relative importance of the application of thick-ring elements to design, experimental verifications of the various theories have been rather few [11]. One of the more interesting aspects of this analysis concerns the comparison between the measured and calculated deflections at the inner surfaces of the ring when the load is applied to the outer surface, such as that shown in Fig. 29.1. In this sketch H is considered to be extremely small and is carried here only as a mathematical convenience. Extensive tests made at London University [164] indicate that curved-beam theory, expressed by Eqs. (29.7) and (29.8), overestimates the amount of diametral change of the ring in both the vertical and horizontal directions. The agreement between the theory and measurements in a horizontal sense, however, is better than that in the line of load P, suggesting that the effect of radial strain due to the direct compression of the rim may be responsible for the greatest share of the difference. This effect is not accounted for in the customary theory of curved beams. On the basis of the London tests, simple correction factors were proposed for use with Eqs. (29.7) and (29.8). These factors are shown in Fig. 29.2 as K_y and K_x for the vertical and horizontal deflections, respectively. In order, then, to estimate the real diametral change in the direction of load P, for instance, calculate Y from Eq. (29.7) and multiply the result by K_y from Fig. 29.2 for the actual ratio of outer to inner ring diameter. This should give the displacement of the inner surface of the ring in line with the action of load P.

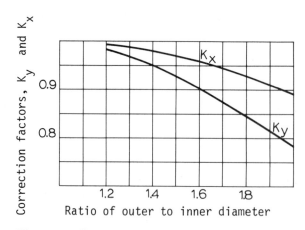

Fig. 29.2 Empirical factors for deflection of a thick ring in diametral compression.

Links and Eyebars

THEORY OF CHAIN LINKS

The theoretical model of a symmetrical chain link is shown in Fig. 29.3. Since the usual interest is the maximum stress, it is first necessary to calculate the bending moment under load W. Since the link is a statically indeterminate structure find the moment M_0 on the assumption that there is no rotation at $\theta - 0$ because of symmetry, as shown in Fig. 29.3. This condition is normally stated in mathematical terms.

$$\int_0^{\pi/2} M_1 \frac{\partial M_1}{\partial M_0} R\, d\theta + \int_0^{L/2} M_2 \frac{\partial M_2}{\partial M_0}\, dx = 0 \tag{29.9a}$$

For the model shown,

$$M_1 = M_0 - PR \sin \theta \tag{29.9b}$$

and

$$M_2 = M_0 - PR \tag{29.9c}$$

and since $\partial M_1/\partial M_0 = \partial M_2/\partial M_0 = 1$, substituting this result together with Eqs. (29.9b) and (29.9c) into Eq. (29.9a) and integrating yields

$$\left(\frac{M_0 \pi}{2} - PR\right) R + (M_0 - PR)\frac{L}{2} = 0 \tag{29.9d}$$

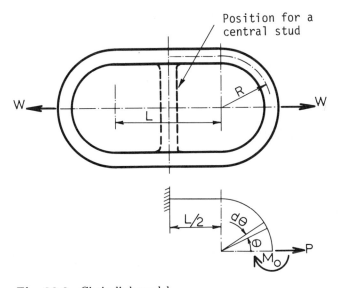

Fig. 29.3 Chain link model.

Solving next for M_0, Eq. (29.9d) gives

$$M_0 = \frac{WR(2R+L)}{2(\pi R + L)} \qquad (29.10)$$

Examination of Eqs. (29.9b) and (29.9c) with the help of Eq. (29.10) shows that M_0 is actually the maximum bending moment in this type of a chain link. When the two ends of the link have different radii, the general procedure for finding the redundant moments is the same, but the actual expression becomes too unwieldy for a general solution because in addition to the redundant moments, direct redundant forces are involved. Solution of the two simultaneous equations is then best obtained by the introduction of the actual link dimensions [159]. Whereas in the usual cases of curved beams and hooks the maximum stress appears always at the inner surface of the component, the analysis of links with even and uneven ends can also indicate the maximum stresses at the outer surfaces. This relation changes further with the change in length of the straight portion of the link. As the link becomes more and more like a circular ring, the stresses at the inner surfaces again become important. Extensive charts are available for detailed examination of the relevant stresses. These were originally used as a basis for developing British Standard Specifications [159]. If by r we denote the radius of a circular cross section and if R is the radius of curvature of a link with even ends, the majority of r/R ratios in all links in practice fall between 0.2 and 0.5. The ratios of L/R are found to vary between 0 and 4. For example, many railway car coupling links are characterized by $L/R = 4$. Note that the condition of $k = 0$ corresponds to the response of a plane circular ring.

LINK REINFORCEMENT

When the strength of a typical open link, such as that shown in Fig. 29.3, is found to be insufficient, the insertion of a central stud, shown by dashed lines, is known to increase its strength quite significantly. The structural analysis for this special case can be made with reference to the equilibrium diagram shown in Fig. 29.4. There are two redundant quantities, M_0 and H, in terms of which the following bending moment expressions can be set up.

$$M_1 = M_0 - \frac{WR}{2}\sin\theta + HR(1 - \cos\theta) \qquad (29.11a)$$

and

$$M_2 = M_0 - \frac{WR}{2} + H(R + x) \qquad (29.11b)$$

The relevant boundary conditions are based on the idea that because of symmetry there is no change in slope between the ends of the quadrants and that there is virtually no deflection in the direction of H. The first of these conditions has already been satisfied by Eq. (29.9a). The condition of zero displacement can be

Links and Eyebars

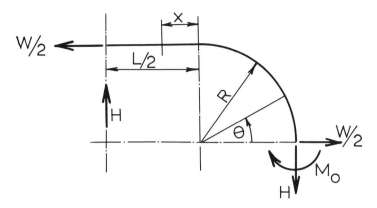

Fig. 29.4 Equilibrium model of a quarter of a studded link.

stated in a similar manner

$$\int_0^{\pi/2} M_1 \frac{\partial M_1}{\partial H} R\, d\theta + \int_0^{L/2} M_2 \frac{\partial M_2}{\partial H}\, dx = 0 \tag{29.11c}$$

Here, again, using Eqs. (29.11a) and (29.11b), we get

$$\frac{\partial M_1}{\partial M_0} = \frac{\partial M_2}{\partial M_0} = 1 \tag{29.11d}$$

$$\frac{\partial M_1}{\partial H} = R(1 - \cos\theta) \tag{29.11e}$$

and

$$\frac{\partial M_2}{\partial H} = R + x \tag{29.11f}$$

Hence, introducing Eqs. (29.11a), (29.11b), and (29.11d) to (29.11f) into Eqs. (29.9a) and (29.11c), integrating, and solving the two simultaneous equations, yields

$$M_0 = \frac{WRC_1}{2} \tag{29.12}$$

and

$$H = \frac{WC_2}{2} \tag{29.13}$$

where

$$C_1 = \frac{(k+2)[k^3 + 6k^2 + 12k(4-\pi) + 48(\pi-3)]}{k^4 + 4\pi k^3 + 48k^2 + 24\pi k + 24(\pi^2 - 8)} \tag{29.14}$$

and

$$C_2 = \frac{12(k+2)[(\pi-2)k + 2(4-\pi)]}{k^4 + 4\pi k^3 + 48k^2 + 24\pi k + 24(\pi^2 - 8)} \tag{29.15}$$

In the foregoing equations, $k = L/R$. It can be shown that for $k = 0$, the relevant redundant quantities are essentially the same as those obtained previously as Eqs. (27.9) and (27.10) for a circular ring with a restraining bar. Past investigations indicate [156, 159] that the provision of a link stud could decrease the maximum tensile stress by about 20%. The relevant maximum compressive stress can be reduced by as much as 50%, although this type of stress is generally less important in link design.

PROOF RING FORMULAS

Whenever standard rings of circular cross section were used as lifting components in the past, it was customary to employ the simplified expression for calculating the proof load [159]

$$W = \frac{33,000 d^3}{D_i + 0.3d} \tag{29.16}$$

In this formula the maximum strength of the material was taken to be on the order of 54,000 psi. When it is required to extend Eq. (29.16) to other material allowables, consider the following criteria.

$$S_y = \frac{M}{Z} \tag{29.17}$$

where

$$M = \frac{W(R - \delta)}{\pi} \tag{29.18}$$

and

$$Z = \frac{\pi d^3}{32} \tag{29.19}$$

Combining Eqs. (29.17) to (29.19) gives

$$W = \frac{0.62 S_y d^3}{D_i + d - 4\delta} \tag{29.20}$$

According to this derivation Eq. (29.20) should give a smaller proof load W than that obtained from Eq. (29.16).

The analysis of circular links provides an introduction to the study of eyebars and similar mechanical joints [167, 168], which, despite their apparent simplicity,

Links and Eyebars

often become a point of contention as to the design criteria and the potential modes of failure. There is surprisingly little information on the effect of basic variables on the critical stresses in eyebars. In particular, local areas of the maximum tensile stresses may be of concern because of the modern requirements of fracture safe design.

KNUCKLE JOINT

In the case of a typical knuckle joint, such as that illustrated in Fig. 29.5, the basic strength of a pin is of importance. However the remaining two components of the connection behave essentially as eyebars which, theoretically, can fail in various ways. Examples of such eyebar failure modes, encountered in machine design, are illustrated in Fig. 29.6. The three most likely modes include local compression due to the pin contact, primary tension, and tear-out shear. The average compressive stress for the eyebar given in stress calculations can be obtained from the projected area. This procedure is probably satisfactory provided the pin fits the eyebar with zero clearance. However, even under these assembly conditions, the pressure around the pin is expected to vary according to a definite pattern.

One of the plausible approaches to calculating the compressive stresses may be based on the cosine load distribution indicated in Fig. 29.7. From the equilibrium of forces, the horizontal components of q are in balance with each other. The vertical components can be related to the external load W as follows

$$W = 2r \int_0^{\pi/2} q \cos\theta \, d\theta \qquad (29.21)$$

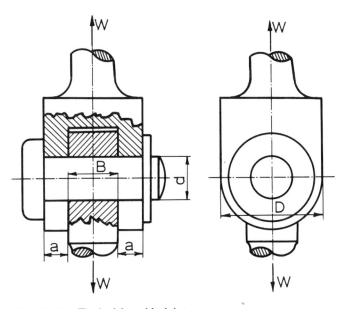

Fig. 29.5 Typical knuckle joint.

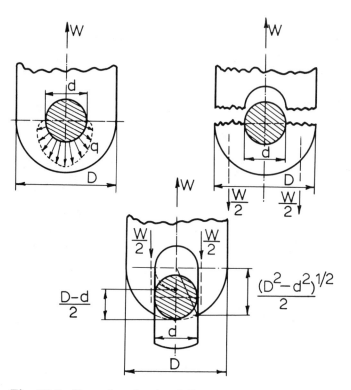

Fig. 29.6 Examples of eyebar failure.

Since by definition $q = q_{max} \cos \theta$, substituting this term in Eq. (29.21) and integrating gives

$$q_{max} = \frac{2W}{\pi r} \tag{29.22}$$

In Eq. (29.22), q_{max} defines the unit load in lb/in. and r is the inner radius of the eyebar. Normally, this radius is assumed to be nearly equal to the pin radius. Dividing q_{max} by the width of the eyebar cross section gives the compressive stress based on the above model.

EYEBAR WITH ZERO CLEARANCE

The approach described so far provides a quick answer to the question of contact stresses if the clearance between the pin and the eyebar is not excessive. However, the magnitude of the acceptable clearance in a pin joint appears to be rather poorly defined in engineering literature because of the theoretical complexity of the problem where the elastic theory breaks down. All that can be stated for certain is that the cosine load distribution is probably a reasonable approximation to the manner of loading under zero clearance, while a concentrated load model should correspond to a relatively large clearance between the pin and the eyebar.

Links and Eyebars

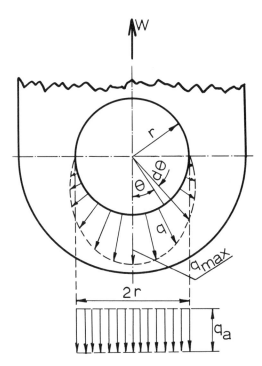

Fig. 29.7 Cosine loading on eyebar.

When the clearance is found to be relatively small and the pin can be assumed to be rigid, the analysis of the maximum stresses in the eyebar portion of the joint can be performed with the aid of the following simple expression [21].

$$S = \frac{W\phi}{BR} \tag{29.23}$$

In Eq. (29.23), ϕ denotes a design factor which depends on the angular position θ and the nominal radii R and r, as shown in Fig. 29.8. The numerical values of ϕ are given in Fig. 29.9 as a function of θ and R/r. According to this model, the maximum tensile stress is found at the inner surface of the eyebar, where $\theta = \pi/2$. The chart in Fig. 29.9 is intended for the usual range of R/r and shows that the effect of this ratio on the maximum stress in the eye is relatively small in the critical regions of $\theta = 0$ and $\theta = \pi/2$.

As far as the theory of fracture-safe criteria is concerned, only the maximum tensile stress is of an immediate interest to the designer. Accordingly, in line with the eyebar theory discussed above, the tensile stress becomes

$$S = \frac{3.52W}{BR} \tag{29.24}$$

In this context Eq. (29.24) should be applicable to a typical eyebar geometry for all ratios of R/r between 2 and 4.

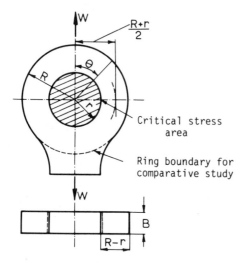

Fig. 29.8 Eyebar geometry after Faupel [21].

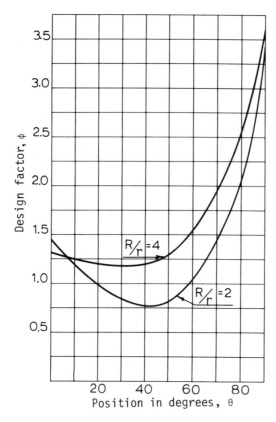

Fig. 29.9 Design factor based on Faupel [21].

Links and Eyebars

The primary tensile stress, corresponding to the simplified mode of failure depicted in Fig. 29.6, is

$$S_t = \frac{W}{2B(R-r)} \tag{29.25}$$

In comparing the two models based on Eqs. (29.24) and (29.25) a convenient nondimensional ratio $\lambda = R/r$ can be used. This procedure then yields directly

$$\frac{S}{S_t} = \frac{7.04(\lambda - 1)}{\lambda} \tag{29.26}$$

In the conventional practice of machine design, $\lambda = 2$, for which the stress ratio based on Eq. (29.26) yields a value of 3.52. The nominal tensile stress of the type given by Eq. (29.25) is the most damaging from the point of view of fracture propagation because of the relatively large amount of elastic energy stored and available to drive the crack. Furthermore, as the crack develops, the net area is being progressively lost, thereby increasing the nominal stress. Since the elastic energy stored is proportional to the square of the stress per unit volume of the stressed material it is easy to see the role of the nominal stress. Hence, the local tensile stress of the type described by Eq. (29.24) is most likely to be responsible for crack initiation. By the application of relatively high factors of safety, the nominal stresses can be kept at a low level and the corresponding calculations should not present undue difficulties. Unfortunately, the more rigorous analysis of a local stress concentration can be very complicated and requires advanced knowledge of materials behavior supported by a well-conceived experimental program.

The characteristics of eyebar geometry, reported by Faupel [21] can also be investigated on the basis of a thick-ring model. The hypothetical boundary for such a ring is illustrated in Fig. 29.8.

THICK-RING METHOD OF EYEBAR DESIGN

In general, when performing the analysis of a curved member of a relatively sharp curvature, it is customary to assume that plane sections remain plane during bending while the neutral axis is displaced toward the center of curvature, which in this particular case coincides with the center of the eyebar. For a circular ring of a rectangular cross section, such a displacement can be obtained in a closed-form solution and then approximated using the theorem of Maclaurin. Denoting the relevant displacement of the neutral axis by δ, the following simplified relation may be obtained:

$$\delta = rF(\lambda) \tag{29.27}$$

where

$$F(\lambda) = \frac{(\lambda - 1)^2(\lambda + 1)}{8(\lambda^2 + \lambda + 1)} \tag{29.28}$$

The geometry factor $F(\lambda)$ given by Eq. (29.28) can be compared with the more rigorous solution involving a logarithmic term [11]. Such a comparison indicates numerical differences between the approximate and more exact solutions on the order of 15 to 20% for $\lambda = 2$ and $\lambda = 4$, respectively. However, since the parameter δ has only a limited effect on the magnitude of the bending moment, the use of the approximate Eq. (29.28) may be justified within the range of λ considered.

It can be shown with reference to Fig. 29.8 that the bending moment varies with the subtended angle θ as

$$M = \frac{W}{2\pi}\left[R + r - 2\delta - \frac{\pi(R+r)}{2}\sin\theta\right] \qquad (29.29)$$

Since the maximum tension develops at the inner radius where $\theta = \pi/2$, putting $\lambda = R/r$ and $\delta = (R-r)^2(R+r)/8(R^2 + Rr + r^2)$, Eq. (29.29) gives

$$M = WrG(\lambda) \qquad (29.30)$$

where

$$G(\lambda) = \frac{\lambda+1}{4\pi}\left[\pi - 2 + \frac{(\lambda-1)^2}{2(\lambda^2 + \lambda + 1)}\right] \qquad (29.31)$$

The geometry factor $F(\lambda)$ and the bending moment factor $G(\lambda)$ are illustrated in Fig. 29.10 for reference. Since the section modulus for the critical cross-sectional area of the eyebar depicted in Fig. 29.8 is now $Br^2(\lambda-1)^2/6$, the corresponding bending stress is

$$S_b = \frac{6WG(\lambda)}{Br(\lambda-1)^2} \qquad (29.32)$$

The total stress at the extreme fiber on the inner eyebar diameter is the sum of bending and tension. According to the original Winkler-Bach theory, the general expression for the maximum stress may be expressed as

$$S_{max} = \phi_0(S_t + S_b) \qquad (29.33)$$

In Eq. (29.33), ϕ_0 denotes a correction factor, as before, allowing for a hyperbolic instead of linear distribution of the resultant normal stress over the depth of the cross section. Consequently, the stress developed at the inner face of the curved beam may be assumed to be substantially higher than that predicted by the theory of straight members.

So far, the calculations of maximum stresses using the curved-beam model were made for a single concentrated load. This discussion has been included here because many engineering estimates are based on a single-load assumption, which often proves to be highly conservative. On the other hand the expression reported by Faupel, Eq. (29.23), considers a rigid pin with zero clearance, involving a distribution loading similar but not necessarily identical with that shown in Fig. 29.7.

Links and Eyebars

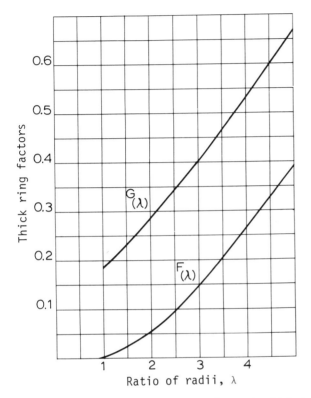

Fig. 29.10 Moment and geometry factors in thick-ring theory.

Design Problem 29.1

A large swivel eye designed for the traveling block of a commercially available crane is shown in Fig. 29.11. The material specified for this part by the manufacturer indicates a yield strength of 100,000 psi. Estimate the critical tensile stresses in the eye portion of the swivel assuming a maximum working load of 800,000 lb and utilizing the rigid-pin and thick-ring methods of analysis. Figure 29.11 shows some major nominal dimensions based on the actual commercial drawing.

Solution

The eye portion of the swivel and the relevant notation are shown in Fig. 29.12. According to the rigid-pin method, Eq. (29.23)

$$\frac{R}{r} = \frac{8.25}{4.03} = 2.05$$

From Fig. 29.9, at $\theta = \pi/2$, $\phi = 3.5$. Hence,

$$S = \frac{800,000 \times 3.5}{8.25 \times 17} = 19,960 \text{ psi}$$

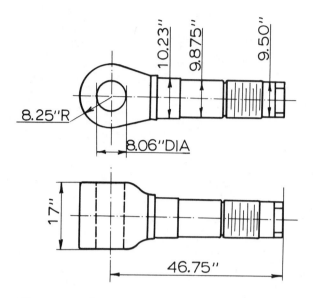

Fig. 29.11 Travelling block swivel eye.

The primary tensile stress based on Eq. (29.25) is

$$S_t = \frac{800,000}{2 \times 17(8.25 - 4.03)} = 5,575 \text{ psi}$$

From Fig. 29.10, $G(\lambda) = 0.29$. The approximate correction factor from Fig. 28.2 is obtained for curved-beam parameter equal to 2.91, This gives $\phi_0 = 1.37$. Then, from

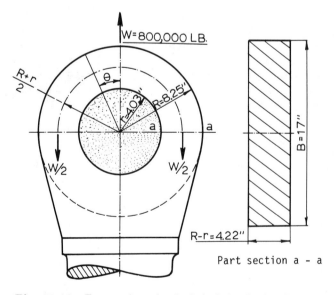

Fig. 29.12 Eye portion of swivel shaft for Design Problem 29.1.

Links and Eyebars

Eqs. (29.32) and (29.33), we get

$$S_b = \frac{6 \times 800,000 \times 0.29}{17 \times 4.03(2.05-1)^2} = 18,430 \text{ psi}$$

and

$$S_{max} = 1.37(5575 + 18,430) = 32,890 \text{ psi} \quad (227 \text{ N/mm}^2)$$

Based on the indicated yield strength of the material, the corresponding factors of safety are then 5.01 and 3.07, respectively. ◆

API STANDARD

The spread between the critical stresses obtained from Design Problem 29.1 is expected because of the difference in the assumption of load distribution on the swivel eye in the rigid-pin and thick-ring methods of analysis. The factors of safety found in this design problem appear to be reasonable on the basis of accepted engineering standards. For example, a specification of the American Petroleum Institute for hoisting equipment [169] gives a chart for the safety factor based on the considerations of yield strength of the material and crane capacity in tons, as shown in Fig. 29.13. According to this information, the required factor of safety on yield should not be less than 2.4, as indicated by the dashed line in the diagram. Unfortunately, the proposed specification does not discuss the stress analysis criteria that should be used with recommended factors. This is of particular concern because the question of primary and secondary stresses often comes up during the design and must certainly affect the choice of the factors of safety. The importance of this

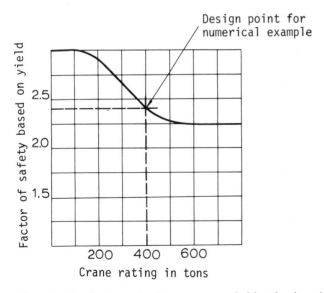

Fig. 29.13 Factors of safety recommended by the American Petroleum Institute.

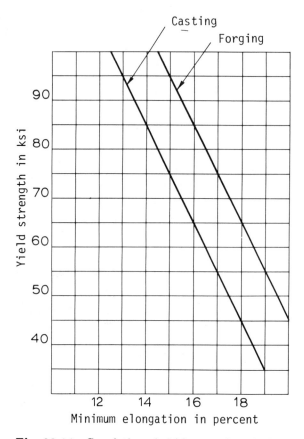

Fig. 29.14 Correlation of yield strength and minimum elongation.

aspect of design cannot be overemphasized, especially in the area of fracture-safe design. The API standard makes some general rules to avoid any undue brittleness of the materials used. These rules may be illustrated by the two curves of strength versus the minimum elongation in Fig. 29.14. Although it is essential to employ material with relatively high elongation for the purpose of mitigating local stress concentrations, the choice of the correct strength and strain parameters does not constitute assurance that the requirements of fracture-safe design are automatically met.

EYEBAR WITH FINITE CLEARANCE

The problem of pin clearance and the manner of load distribution must affect the absolute levels of the calculated critical stresses on which the factors of safety are based. Yet, as mentioned previously, very limited information is available on this subject. For a typical eyebar geometry for which $\lambda \cong 2$, some interpolation between the results of rigid-pin type and thick-ring type can be made if we assume that the two cases examined probably represent the two extreme loading conditions for the most practical purposes. The choice of $\lambda = 2$ may be supported by numerous

Links and Eyebars

examples of practical machine design applications which evolved over a long span of time. The only practical guide of the effect of clearance on the stress correction of the type given in Eq. (29.23) has been reported by Korkut [170], where the clearance is defined as 1/64 in. per inch of pin diameter and where the eyebar is constructed according to $\lambda = 2.58$. The effect of clearance varies with the type of design, illustrated in Fig. 29.15.

The clearance correction factor C, illustrated in Fig. 29.15, is obtained by dividing the maximum stress with clearance into the maximum stress without the clearance represented by Eq. (29.23). The information contained in Fig. 29.15 was originally developed in Germany. It indicates that the effect of design is only moderate, and for the purpose of our discussion it will suffice to assume the average value of the correction factor equal to $C = 1.38$ when the clearance between the pin and the eye is about 1.5% of the diameter. On this basis the stress from Eq. (29.23) computed for Design Problem 29.1, using the rigid-pin method, should be about 27,540 psi, which is only some 15% lower than that obtained by the thick-ring method if the effect of λ between 2.05 and 2.58 is disregarded. The last assumption seems to be quite reasonable when reference is made to Fig. 29.9, in which the effect of λ between the values of 2 and 4 is found to be negligible for stress at $\theta = \pi/2$.

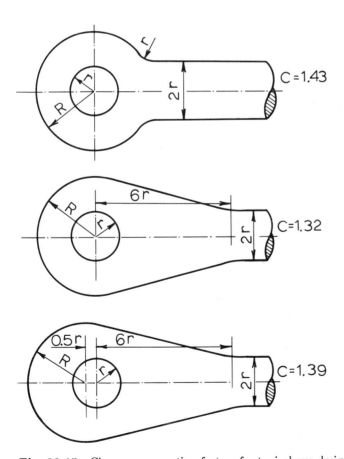

Fig. 29.15 Clearance correction factors for typical eye designs.

It is recalled here that the critical tensile stress was assumed to occur at $\theta = \pi/2$, on the inner surface of the eye.

The information at hand appears to be rather limited for establishing a general relation between the parameter C from Fig. 29.15 and the percentage of clearance. However, a useful functional relationship may be developed using the two data points and an asymptotic behavior of the model if we can estimate the general shape of the curve.

By analogy to a direct bearing problem, the pin and eye surfaces may be simulated by a cylinder pressing against a circular groove with the axes remaining parallel, as indicated in Fig. 29.16. According to the Hertz theory of elastic contact [38] in the case of long cylinders interacting along a generator, the width of contact can be expressed in terms of load per linear inch, material properties, and radii of curvature. When the cylinders becomes rather short, the problem reduces to that of a plane stress, although the general stress distribution remains quite similar to that for long cylinders.

Putting the numerical values of elastic constants into the standard formula and introducing the notation compatible with Fig. 29.16 yields

$$B' = 3 \left(\frac{W r r_0}{BE(r - r_0)} \right)^{1/2} \tag{29.34}$$

When the loading and dimensional data are substituted for the case of a large swivel eye from the numerical example discussed above, we obtain

$$\tan \beta = 0.03 \left(\frac{1 - \eta}{\eta} \right)^{1/2} \tag{29.35}$$

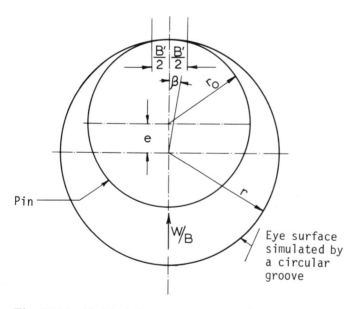

Fig. 29.16 Model of pin-to-eye contact surface.

Links and Eyebars

where η denotes the ratio of the radial clearance to the inner radius of the eyebar. Finally, when the total angle of contact 2β is plotted against the clearance ratio η, the curve in Fig. 29.17 is obtained. The shape of the curve derived in this manner has certain practical significance for establishing a guide for the selection of a rational clearance correction factor for the purpose of the preliminary design of eyebar joints. For instance, when $\eta = 0$, the stress correction factor ϕ is equal to about 3.5, as indicated by Eq. (29.23). This condition corresponds to the very steep portion of the curve plotted in Fig. 29.17. As the angle of contact becomes gradually smaller, the curve flattens out rather rapidly, tending asymptotically to zero for a large value of η. For most practical reasons the contact angle of 10° is considered here to be consistent with the assumption of a concentrated load. This is particularly convenient during the analysis when the critical stress is located along the horizontal axis, that is, at $\theta = \pi/2$, as shown in Fig. 29.8. The condition of a concentrated load gives a correction factor $C = 5.01/3.07 = 1.63$ based on the calculations in Design Problem 29.1.

The intermediate point is taken from Fig. 29.15 according to German practice with the average value of $C = 1.38$ when $\eta = 0.015$. This clearance ratio is obtained from the assumption of 1/64 in. per inch of pin diameter [170]. The stress factor curve in Fig. 29.18 is then constructed from three points. At $\eta = 0$, Faupel solution [21] gives $\psi = 3.5$, the intermediate point at $\eta = 0.015$ [170] corre-

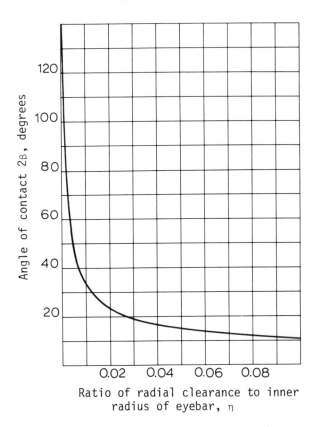

Fig. 29.17 Effect of clearance on contact angle.

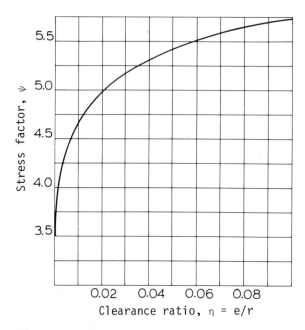

Fig. 29.18 Stress factor for typical eyebar design.

sponds to $\psi = 1.38 \times 3.5 = 4.83$, and the conventional thick-ring analysis results in $\psi = 1.63 \times 3.5 = 5.71$ at $\eta = 0.10$.

Based on the foregoing analysis, the maximum tensile stress in the eyebar can be computed from the following simplified expression.

$$S_{tmax} = \frac{W\psi}{\lambda r B} \tag{29.36}$$

where ψ denotes a new stress factor, given as a function of the clearance ratio η in Fig. 29.18.

It may also be convenient to express the maximum stress from Eq. (29.36) in terms of the nominal tension indicated in Fig. 29.6 and defined by Eq. (29.25). The calculation is based on the primary tensile stress on the net cross-sectional area. This yields the critical stress at the inner boundary of the eyebar

$$\frac{S_{tmax}}{S_t} = \frac{2\psi(\lambda - 1)}{\lambda} \tag{29.37}$$

For instance, assuming a 2% clearance ratio for the swivel eye geometry given in Design Problem 29.1, the primary tensile stress is 5580 psi, $\lambda = 2.05$, and $\psi = 4.98$, so that $S_{tmax} = 28,470$ psi, or an equivalent stress concentration of 5.1 according to Eq. (29.37). By reducing the clearance to zero, the maximum stress becomes 20,120 psi. How closely this zero-clearance feature can be maintained in practice is obviously not a matter of theoretical prediction but the accuracy of a fabrication process.

Links and Eyebars

EYEBAR EXPERIMENTS

The degree of conservatism in the design and the rationale for selecting the factors of safety in a particular application must be viewed with due caution in order not to overdesign the eyebar connections, many of which have been shown to work successfully despite rather high apparent elastic stresses. Equations (29.36) and (29.37) have been found to be conservative in recent photoelastic experiments designed to assess the approximate effect of radial clearance and geometry [253] on the maximum tensile stress in the eye of the "bullnose" type of a joint. The approximate shape of the eyebar used in this study is illustrated by Fig. 29.19. The revised design factors are plotted in Fig. 29.20. However, the results of this work show relative insensitivity of the eyebar stresses to the clearance effect when the values of λ exceed 2.0.

Although Figs. 29.19 and 29.20 pertain to a special eyebar design, the inscribed radius R is Fig. 29.19 suggests a clear boundary for a standard geometry. Since the region of the critical stress remains at the inner boundary intersecting the horizontal axis, the material located outside the R contour is expected to have only a minor effect on the empirical and theoretical results. For all practical purposes therefore the design curves given in Fig. 29.20 should be applicable to majority of eyebar configurations with λ values between 1.6 and 3.0. This range is common to industrial practice.

In one of the latest applications [266] the $2\frac{1}{2}$-in. wire rope socket components were tested and analyzed to find a correlation between the measured and calculated values of the maximum stresses. The test procedures included the photoelastic-plastic coating technique and the conventional strain gases. The plastic coating provided fringes and isoclinic patterns for the interpretation of experimental results. Since the geometry of the socket components conformed rather closely to a typical eyebar configuration, the analysis of the critical stresses was made on the basis of Eq. (29.36). The fringe patterns confirmed rather effectively that the highly

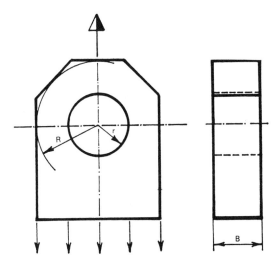

Fig. 29.19 "Bullnose" eyebar configuration.

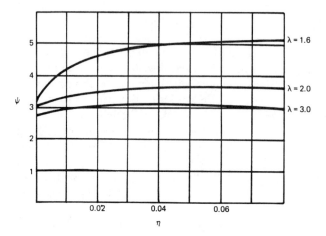

Fig. 29.20 Photoelastic factors for "bullnose" eyebars.

stressed area was at the inner radius and at an angle of 90° to the line of force. This is consistent with Figs. 29.8 and 29.9, as well as the assumptions on which Eq. (29.36) is based.

The socket-eyebar nominal dimensions used in the study included $R = 4.375$ in., $r = 2.16$ in., and $B = 1.875$ in. The resulting value of $\lambda = 2$ appears to be typical for industry. When Eq. (29.36) and Fig. 29.20 are invoked the ratio of calculated to measured stresses is on the order of 1.35 on the premise of the maximum pin clearance allowed in Fig. 29.20. For the minimum clearance consistent with this type of hardware the stress ratio may be reduced to about 1.20.

The experimental work has confirmed that for all practical purposes the design model represented by Eq. (29.36) remains conservative. Note that only one stress factor symbol is used with the various models and experiments under study.

OTHER MODES OF EYEBAR FAILURE

In reviewing other possible modes of failure of a typical eyebar connection indicated in Fig. 29.6, the question of a tear-out mode can be resolved rather simply. Since this failure requires a double-shear equilibrium, and since the shear areas are assumed to be rectangular, the theoretical value of the maximum shear stress becomes

$$S_{s\max} = \frac{3W}{4Br(\lambda^2 - 1)^{1/2}} \tag{29.38}$$

For the purpose of the calculations involving shear, the ratio of the allowable shear strength to yield strength in tension should be taken as $1/\sqrt{3}$. Hence, the factor of safety based on shear is

$$F_{ss} = \frac{S_y}{\sqrt{3} S_{s\max}} \tag{29.39}$$

Links and Eyebars

For the purpose of correlation with tension, using Eq. (29.36) the corresponding factor of safety may be defined as

$$F_{st} = \frac{S_y}{S_{tmax}} \tag{29.40}$$

Hence, utilizing Eqs. (29.36) and (29.38) to (29.40) yields

$$\frac{F_{st}}{F_{ss}} = \frac{3\sqrt{3}\lambda}{4\psi(\lambda^2 - 1)^{1/2}} \tag{29.41}$$

For the case of zero clearance and the practical range of values of λ equal between 2 and 4, Eq. (29.41) gives safety factor ratios of 0.426 and 0.331, respectively. Therefore, one of the primary considerations in eyebar joint design is to evaluate the factor of safety related to the critical tensile stress criteria rather than shear. This is particularly important in fracture-safe design.

CONTROL OF SLING LOADS

Although it appears reasonable to design the eye portion of the eyebar on the basis of axial loading alone, it is well to realize that in the majority of slinging operations, the line of loading may be inclined to the axis of the shank. This type of a load produces bending in the eye, collar, and shank, and the analysis of such a redundant system, even in the plane of the eye, is rather involved [159]. In industry, the shape and proportions of the manufactured eyebolts and similar components are established by specifications. Despite this control, however, there is a great diversity of existing eyebar configurations, some of which may be inadequate for the intended application. To help with checking the performance of a given unit subjected to inclined sling loading, the equivalent shear stress for the shank can be computed as

$$S_s = \frac{W \cos \alpha}{2A_c} \left[(1 + G \tan \alpha)^2 + 4 \tan^2 \alpha\right]^{1/2} \tag{29.42}$$

where

$$G = 2D_s \left(\frac{16nD_s + 8d_s}{16n^2 D_s^2 + d_s^2}\right) \tag{29.43}$$

The relevant notation for this analysis is shown in Fig. 29.21. It should also be noted that adequate proportioning of the collar is a very real consideration, since under the inclined loading, partial failure of the collar is likely to occur through local crushing of the edge. The core area of the shank is denoted here by A_c and n defines the location on the collar diameter at which the maximum crushing load might develop.

The experience with eyebolts in lifting loads inclined to the axis of the bolt shows that the shank is subjected to tensile, shearing, and bending actions due to

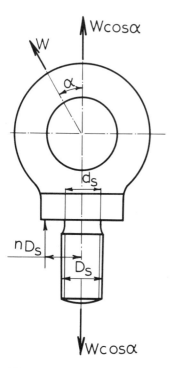

Fig. 29.21 Eyebar under inclined loading.

the components $W \cos \alpha$, $W \sin \alpha$ as well as to a bending moment consisting of the term $W \sin \alpha$ multiplied by the distance from the collar surface to the center of the eyebolt. Based on the original British investigations [159], this distance can be taken to be equal to $2D_s$. Similarly, the recommended diameter of the collar can be on the order of $2.25D_s$, making this a well-sized eyebolt configuration. Of course such variables as the extent of friction of a collar screwed into the block, internal stresses, and surface conditions are more difficult to define and can only be judged by empirical findings.

The use of the recommended formulas, Eqs. (29.42) and (29.43), can be made easy if we assume that the vertical reaction shown in Fig. 29.21 is situated at a given distance of nD_s where $n = 17/16$. The second assumption may be based on the average value of d_s/D_s, equal to 0.84, which should not be much in error for the entire range of shank diameters between 0.5 and 3 in. With these stipulations the variable G can be replaced by a constant, leading to the criterion

$$S_s = \frac{W \cos \alpha}{2A_c}[(1 + 2.53 \tan \alpha)^2 + 4 \tan^2 \alpha]^{1/2} \qquad (29.44)$$

The relative value of the load lifted vertically by the eyebolt can now be evaluated as a function of the angle of lift α measured in degrees. The results are given in Table 29.1.

Hence in order to determine the numerical value of the lifted load ($W \cos \alpha$) it is only necessary to select an appropriate value of S_s representing the allowable

Links and Eyebars

Table 29.1 Sling Load Versus Angle of Lift

Lifted Load, percent	100	67	48.5	36.5	28	24.5
Angle of lift, degrees	0	10	20	30	40	45

shear stress of the shank. The effect of the angle of lift given in Table 29.1 also suggests that in a well controlled lifting operation the angle of the sling should not be allowed to exceed 45 degrees.

SYMBOLS

A	Area of bar cross-section, in.² (mm²)
A_c	Core area of shank, in.² (mm²)
a	Width of knuckle joint, in. (mm)
B	Width of eyebar, in. (mm)
B'	Width of contact area, in. (mm)
C	Correction factor for eyebars
C_1, C_2	Force factors in studded link
D	Maximum eyebar diameter, in. (mm)
D_i	Inner diameter of link, in. (mm)
D_s	Nominal diameter of shank, in. (mm)
d_s	Core diameter of shank, in. (mm)
d	Bar diameter, in. (mm)
E	Modulus of elasticity, psi (N/mm²)
e	Clearance, in. (mm)
$F(\lambda)$	Auxiliary factor
F_{ss}	Factor of safety for shear
F_{st}	Factor of safety for tension
G	Shank diameter factor
$G(\lambda)$	Auxiliary factor
H	Fictitious ring load, lb (N)
K_y, K_x	Ring deflection factors
$K = L/R$	Length-to-radius ratio
L	Length of straight portion, in. (mm)
M, M_1, M_2	Total bending moments, lb-in. (N-mm)
M_f	Redundant moment in thick ring, lb-in. (N-mm)
M_0	Redundant moment in link, lb-in. (N-mm)
N	Direct force, lb (N)
n	Proportionality constant
P	Concentrated load, lb (N)
Q	Shear load, lb (N)
q	Unit load, lb/in. (N/mm)
q_a	Average unit load, lb/in. (N/mm)

q_{max}	Maximum unit load, lb/in. (N/mm)
R	Mean radius; major radius of eyebar, in. (mm)
r	Radius of bar; radius of eye, in. (mm)
r_0	Radius of pin, in. (mm)
S	General symbol for stress, psi (N/mm^2)
S_b	Bending stress, psi (N/mm^2)
S_s	Shear stress, psi (N/mm^2)
S_t	Tensile stress, psi (N/mm^2)
S_{max}	Maximum stress, psi (N/mm^2)
S_{smax}	Maximum shear stress, psi (N/mm^2)
S_{tmax}	Maximum tensile stress, psi (N/mm^2)
S_y	Yield stress, psi (N/mm^2)
W	Link or eyebar load, lb (N)
X	Horizontal deflection, in. (mm)
x	Length along straight portion, in. (mm)
Y	Vertical deflection, in. (mm)
Z	Section modulus, in.^{3} (mm^3)
α	Angle of inclined load, rad
β	Angle subtending contact arc, rad
δ	Displacement of neutral axis, in. (mm)
$\eta = e/r$	Clearance ratio
θ	Angle at which forces are considered, rad
$\lambda = R/r$	Eyebar ratio
ξ	Shear distribution factor
ϕ	Eyebar design factor
ϕ_0	Curved-beam stress factor
ψ	Pin clearance correction factor

30
Out-of-Plane Response

TORSIONAL STRENGTH FACTORS

The calculation of strength and rigidity of machine elements and structures involves the concepts of section modulus and moment of inertia related to the structural response in bending. Such section properties are commonly available in engineering literature and can often be derived by elementary techniques. In some cases the response of curved members involves twisting and out-of-plane bending requiring additional section properties. For instance, in dealing with the torsional rigidity it is necessary to utilize the concept of a torsional shape factor K having the same dimensions as those of the moment of inertia but different magnitudes. Some of the applications of the K factor have already been discussed in Part I. The term GK, denoting the product of the modulus of rigidity and the torsional shape factor, is referred to as the torsional rigidity of a section, by analogy to the flexural rigidity denoted by the term EI. The convenient ratio of the two rigidities can be defined as $\lambda = EI/GK$. This parameter enters the formulas for displacements of curved members loaded out of the plane of curvature.

To calculate the stress due to torsion in a member, the following stress formula can be used:

$$\tau = \frac{T}{K_s} \qquad (30.1)$$

or

$$\tau = \frac{TC}{K} \tag{30.2}$$

In Eq. (30.1), K_s denotes a sectional property defined previously as a section modulus for torsion, by analogy to the section modulus in pure bending, referred to normally as Z. Also by analogy to the theory of bending of beams, Eq. (30.1) could be replaced by Eq. (30.2). Unfortunately, the value of C is difficult to determine for complex shapes. Only in the case of a circular cross section can C be taken to be equal to the distance of the extreme fiber from the neutral axis.

A brief summary of the typical sectional properties is given in Table 30.1. In addition, the values of K and K_s are plotted in Fig. 30.1 for use with solid rectangular cross sections. These parameters are also given in Table 2.4. For more complex cross-sectional configurations of open type (for instance channels, I, T, and similar sections, but not hollow rectangles, etc.), the sum of K or K_s values of individual rectangular elements is equal to the total torsional property for the entire cross section.

BASIC EQUATIONS

The analysis of curved members loaded out of plane can be approached either on the basis of the Castigliano principle or through the use of a differential equation describing the elastic line. In the first case the transverse displacement under load P due to the combined effect of bending and twisting moments is

$$Y = \int_0^\theta \frac{M}{EI}\left(\frac{\partial M}{\partial P}\right) R\, d\theta + \int_0^\theta \frac{T}{GK}\left(\frac{\partial T}{\partial P}\right) R\, d\theta \tag{30.3}$$

The corresponding slope measured in the plane perpendicular to the plane of curvature is then

$$\psi = \frac{1}{R}\left(\frac{dY}{d\theta}\right) \tag{30.4}$$

The angle of twist at the point of application of the externally applied twisting couple T_0, treated as an independent variable, is

$$\eta = \int_0^\theta \frac{M}{EI}\left(\frac{\partial M}{\partial T_0}\right) R\, d\theta + \int_0^\theta \frac{T}{GK}\left(\frac{\partial T}{\partial T_0}\right) R\, d\theta \tag{30.5}$$

When the twisting couple T_0 is a known function of P, the deflection becomes

$$Y = \frac{\partial U}{\partial P} + \left(\frac{\partial U}{\partial T_0}\right)\left(\frac{\partial T_0}{\partial P}\right) \tag{30.6}$$

Out-of-Plane Response

Table 30.1 Typical Sectional Properties

Section	Section Area A	Moment of Inertia About x-x Axis I_x	Section Modulus About x-x Axis Z_x	Torsional Shape Factor K	Section Modulus For Torsion K_s
	$\dfrac{\pi d^2}{4}$	$\dfrac{\pi d^4}{64}$	$\dfrac{\pi d^3}{32}$	$\dfrac{\pi d^4}{32}$	$\dfrac{\pi d^3}{16}$
	$\pi(R_o^2 - R_i^2)$	$\dfrac{\pi}{4}(R_o^4 - R_i^4)$	$\dfrac{\pi(R_o^4 - R_i^4)}{4R_o}$	$\dfrac{\pi}{2}(R_o^4 - R_i^4)$	$\dfrac{\pi(R_o^4 - R_i^4)}{2R_o}$
	$2\pi r t$	$\pi r^3 t$	$\pi r^2 t$	$2\pi r^3 t$	$2\pi r^2 t$
	bh	$\dfrac{bh^3}{12}$	$\dfrac{bh^2}{6}$	For K/bh^3 See Fig. 30.1	For K_s/bh^2 See Fig. 30.1
	$bh - b_o h_o$	$\dfrac{bh^3 - b_o h_o^3}{12}$	$\dfrac{bh^3 - b_o h_o^3}{6h}$	$\dfrac{(h - h_o)(2b - h + h_o)^2(h + h_o)^2}{16(b + h_o)}$	$\dfrac{(h^2 - h_o^2)(2b - h + h_o)}{4}$

Table 30.1 (Continued)

Shape	Area				
Rectangle (b × t)	bt	$\dfrac{bt^3}{12}$	$\dfrac{bt^2}{6}$	$\dfrac{bt^3}{3}$	$\dfrac{bt^2}{3}$
Ellipse (2a × 2b)	πab	$\dfrac{\pi ab^3}{4}$	$\dfrac{\pi ab^2}{4}$	$\dfrac{\pi a^3 b^3}{a^2 + b^2}$	$\dfrac{\pi ab^2}{2}$
Hollow ellipse	$\pi (ab - a_o b_o)$	$\dfrac{\pi (ab^3 - a_o b_o^3)}{4}$	$\dfrac{\pi (ab^3 - a_o b_o^3)}{4b}$	$\dfrac{\pi a^3 b^3}{a^2 + b^2}\left[1 - \left(\dfrac{a_o}{a}\right)^4\right]$	$\dfrac{\pi ab^2}{2}\left[1 - \left(\dfrac{a_o}{a}\right)^4\right]$
Thin ellipse	$\pi t (a + b)$	$\dfrac{\pi b^2 t (3a + b)}{4}$	$\dfrac{\pi bt (3a + b)}{4}$	$\dfrac{4\pi^2 abt (ab - at - bt)}{\Theta}$ $\Theta = \pi (a + b - t)\left[1 + 0.27 \dfrac{(a-b)^2}{(a+b)^2}\right]$	$\pi t (2ab - bt - at)$
Equilateral triangle (60°)	$0.433 a^2$	$0.018 a^4$	max $a^3/16$ min $a^3/32$	$\dfrac{\sqrt{3}}{80} a^4$	$a^3/20$

Out-of-Plane Response

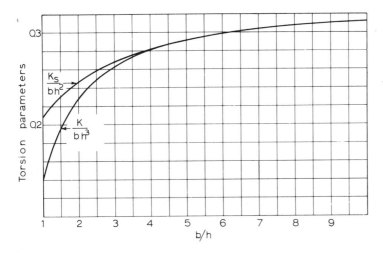

Fig. 30.1 Torsional parameters for calculating stress and angle of twist for rectangular cross sections.

When in a transversely loaded member the curvature and twist are expressed in terms of the deflection Y and the angle of twist η, we get

$$\frac{d^3Y}{d\theta^3} + \frac{dY}{d\theta} = -R^2 \left(\frac{1}{EI}\frac{dM}{d\theta} + \frac{T}{GK} \right) \tag{30.7}$$

and

$$\eta = \frac{MR}{EI} + \frac{1}{R}\left(\frac{d^2Y}{d\theta^2}\right) \tag{30.8}$$

TRANSVERSELY LOADED ARCHED CANTILEVERS

The two major types of problems which can be solved with the aid of the foregoing equations include arched cantilevers and circular rings loaded out of plane [11, 171–173]. For instance, consider an arched cantilever, shown in Fig. 30.2, which carries a transverse load P at its free end. Let $\bar{P}$ and $\bar{T}_0$ represent two fictitious forces which are assumed to act at the point of the desired displacements. The expressions for the bending and twisting moments at a point defined by θ are

$$M = -PR\sin\theta - \bar{P}R\sin(\theta - \alpha) + \bar{T}_0 \sin(\theta - \alpha) \tag{30.9}$$

and

$$T = PR(1 - \cos\theta) + \bar{P}R[1 - \cos(\theta - \alpha)] + \bar{T}_0 \cos(\theta - \alpha) \tag{30.10}$$

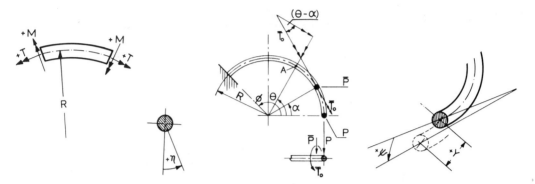

Fig. 30.2 Loading and notation for an arched cantilever in out-of-plane response.

Utilizing Eqs. (30.9) and (30.10) in the solution of Eq. (30.3) between the limits of 0 and ϕ and making the fictitious loads equal to zero, after obtaining the relevant partial derivatives, we get

$$Y = \frac{PR^3}{EI}[0.5(\phi - \alpha)\cos\alpha + 0.25\sin\alpha - 0.25\sin(2\phi - \alpha)]$$
$$+ \frac{PR^3}{GK}[0.75\sin\alpha + 0.25\sin(2\phi - \alpha) + 0.5(\phi - \alpha)\cos\alpha$$
$$+ \phi - \alpha - \sin\phi - \sin(\phi - \alpha)] \qquad (30.11)$$

Differentiating Eq. (30.11) with respect to α and dividing the result by R gives the slope equation

$$\psi = \frac{PR^2}{EI}[0.25\cos(2\phi - \alpha) - 0.5(\phi - \alpha)\sin\alpha - 0.25\cos\alpha]$$
$$+ \frac{PR^2}{GK}[\cos(\phi - \alpha) - 0.25\cos(2\phi - \alpha) - 0.5(\phi - \alpha)\sin\alpha$$
$$+ 0.25\cos\alpha - 1] \qquad (30.12)$$

Integrating Eq. (30.5) yields

$$\eta = \frac{PR^2}{EI}[0.25\sin(2\phi - \alpha) - 0.5(\phi - \alpha)\cos\alpha - 0.25\sin\alpha]$$
$$+ \frac{PR^2}{GK}[\cos(\phi - \alpha) - 0.25\cos(2\phi - \alpha) - 0.5(\phi - \alpha)\sin\alpha$$
$$+ 0.25\cos\alpha - 1] \qquad (30.13)$$

Proceeding in this manner, several typical cases have been worked out. The summary of pertinent results is given in Table 30.2.

Out-of-Plane Response

Table 30.2 Design Equations for Traversely Loaded Arched Cantilevers

---

Concentrated End Load

$Y = \dfrac{PR^3}{EI}(B_1 + \lambda B_2)$

$\psi = \dfrac{PR^2}{EI}(\lambda B_3 + B_4)$

$\eta = \dfrac{PR^2}{EI}(\lambda B_5 - B_1)$

$M = -PR \sin\theta$

$T = PR(1 - \cos\theta)$

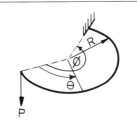

End Twisting Moment

$Y = \dfrac{T_o R^2}{EI}(\lambda B_5 - B_1)$

$\psi = \dfrac{T_o R}{EI}(\lambda - 1) B_4$

$\eta = \dfrac{T_o R}{EI}(\lambda B_6 + B_1)$

$M = T_o \sin\theta$

$T = T_o \cos\theta$

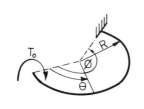

Uniform Loading

$Y = \dfrac{qR^4}{EI}(\lambda B_7 - B_3)$

$\psi = \dfrac{qR^3}{EI}(\lambda B_8 + B_5)$

$\eta = \dfrac{qR^3}{EI}(\lambda B_9 + B_3)$

$M = -qR^2(1 - \cos\theta)$

$T = qR^2(\theta - \sin\theta)$

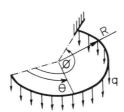

$B_1 = (2\phi - \sin 2\phi)/4$
$B_2 = (6\phi + \sin 2\phi - 8\sin\phi)/4$
$B_3 = (4\cos\phi - \cos 2\phi - 3)/4$
$B_4 = (\cos 2\phi - 1)/4$
$B_5 = (4\sin\phi - \sin 2\phi - 2\phi)/4$
$B_6 = (2\phi + \sin 2\phi)/4$
$B_7 = (2\phi^2 - \cos 2\phi - 4\phi\sin\phi + 1)/4$
$B_8 = (4\sin\phi + \sin 2\phi - 4\phi\cos\phi - 2\phi)/4$
$B_9 = (4\cos\phi + \cos 2\phi + 4\phi\sin\phi - 5)/4$

RING UNDER TWISTING MOMENTS

To illustrate the application of the differential equation approach, consider the case of a free circular ring subjected to two twisting moments T_0 acting at the end of a diameter as shown in Fig. 30.3. Because of symmetry, one-half of the ring can be analyzed separately. The bending and twisting effects at a section defined by θ can be stated here as follows:

$$M = \frac{T_0 \sin \theta}{2} \tag{30.14a}$$

and

$$T = \frac{T_0 \cos \theta}{2} \tag{30.14b}$$

Note that one-half of the twisting moment is attributed to one-half of the ring and that the necessary components represented by Eqs. (30.14a) and (30.14b) are obtained by resolution of the vectorial quantity $T_0/2$ along and perpendicular to the particular ring section defined by θ. It also follows from Eqs. (30.14a) and (30.14b) that $T_0/2 = (M^2 + T^2)^{1/2}$. Differentiation of the function given by Eq. (30.14a)

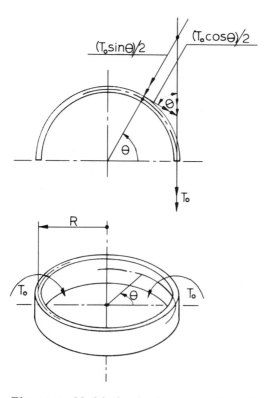

Fig. 30.3 Model of a circular ring under twist.

Out-of-Plane Response

with respect to θ yields

$$\frac{dM}{d\theta} = \frac{T_0 \cos\theta}{2} \tag{30.14c}$$

Substituting Eqs. (30.14a) to (30.14c) into Eq. (30.7) gives

$$\frac{d^3Y}{d\theta^3} + \frac{dY}{d\theta} = B\cos\theta \tag{30.14d}$$

where

$$B = -\frac{R^2 T_0}{2EI}(1+\lambda) \tag{30.14e}$$

The constant term B can be calculated if λ is known. The auxiliary relation for the solution of Eq. (30.14d) can be expressed as follows:

$$n_0(n_0^2 + 1) = 0 \tag{30.14f}$$

The roots of the foregoing equation are 0, $+i$, and $-i$, so that the complementary function becomes

$$F_c = A_1 + A_2 \cos\theta + A_3 \sin\theta \tag{30.14g}$$

By observation of Eq. (30.14d) one can guess the form of the particular solution after some practice

$$F_p = \frac{-B(\sin\theta + \theta\cos\theta)}{2} \tag{30.14h}$$

To check that the function above is indeed the particular solution of Eq. (30.14d), find $dY/d\theta$ and $d^3Y/d\theta^3$ and substitute these values in Eq. (30.14d). The result should be equal to the right-hand side of this equation, and the general solution becomes the sum of F_c and F_p, giving the following expression:

$$Y = A_1 + A_2\cos\theta + \left(A_3 - \frac{B}{2}\right)\sin\theta - \frac{B\theta\cos\theta}{2} \tag{30.14i}$$

In analyzing the symmetry of the ring under load shown in Fig. 30.3, we find the following boundary conditions for the problem:

$$Y = 0 \quad \text{when } \theta = 0$$

and

$$\frac{dY}{d\theta} = 0 \quad \text{for } \theta = 0 \text{ and } \theta = \frac{\pi}{2}$$

These conditions give a sufficient number of simultaneous equations for the evaluation of the constants A_1, A_2, and A_3. Utilizing Eq. (30.14i) and making the necessary substitutions, we get

$$Y = \frac{T_0 R^2}{8EI}(1 + \lambda)(2\theta \cos\theta - \pi \cos\theta + \pi - 2\sin\theta) \qquad (30.15)$$

Since the slope in the plane perpendicular to the plane of curvature is given by Eq. (30.4), differentiating Eq. (30.15) with respect to θ and substituting the result in Eq. (30.4) yields

$$\psi = \frac{T_0 R}{8EI}(1 + \lambda)(\pi - 2\theta)\sin\theta \qquad (30.16)$$

Next, differentiating Eq. (30.15) twice with respect to θ and substituting the result together with Eq. (30.14a) into Eq.(30.8) gives the following equation for the angle of twist at any angle θ:

$$\eta = \frac{T_0 R}{8EI}[2(1 - \lambda)\sin\theta + (1 + \lambda)(\pi - 2\theta)\cos\theta] \qquad (30.17)$$

Substituting in turn $\theta = 0$ and $\theta = \pi/2$ in Eq. (30.17) discloses an interesting phenomenon according to which, midway between the points of application of the moments T_0, the ring twist decreases to zero and then changes its sign. This finding can be verified easily by means of testing a simple paper model.

TRANSVERSELY LOADED RINGS

Two other important cases of out-of-plane loading on circular rings are shown in Table 30.3. The formulas given in the table indicate that there is a remarkable similarity between the deflections for the distributed and concentrated loading. Also, for the same total load, that is, $P = \pi q R$, the maximum deflection for the distributed load is equal to one-half of that due to the concentrated load. No such conclusions, however, can be drawn for the angle of twist.

The behavior of the elastic rings on multiple supports and on elastic foundation has also received attention [174, 175]. Some of the formulas and design factors for these problems have been correlated with other simplified solutions involved special out-of-plane response [11].

GIMBAL RING DESIGN

The analysis of gyroscope gimbals, handling equipment, structural support frames, flanged components, and similar configurations can be conducted with the aid of the principle of superposition and ring equations quoted so far for out-of-plane loading. Probably the two most frequently encountered cases involve uniform and concentrated loads disposed symmetrically with respect to the trunnion axis, as shown in Tables 30.4 and 30.5. The rings are assumed to have uniform cross sections and relatively rigid trunnions. All design factors are expressed in terms

Out-of-Plane Response

Table 30.3 Design Equations for Rings Loaded Out-of-Plane

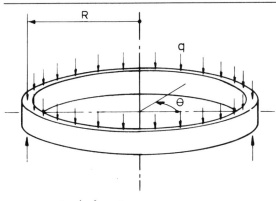

$$M = qR^2 \left(\frac{\pi \sin \theta}{2} - 1 \right)$$

$$T = qR^2 \left(\frac{\pi \cos \theta}{2} + \theta - \frac{\pi}{2} \right)$$

$$Y = \frac{qR^4}{8EI} \{2\pi(\theta \cos \theta - \sin \theta) + \pi^2(1 - \cos \theta) + \lambda[2\pi(\theta \cos \theta - 3 \sin \theta) + \pi^2(1 - \cos \theta) + 4\theta(\pi - \theta)]\}$$

$$\psi = \frac{qR^3}{8EI} \{\pi(\pi - 2\theta) \sin \theta + \lambda[\pi(\pi - 2\theta) \sin \theta + 4(\pi - 2\theta - \pi \cos \theta)]\}$$

$$\eta = \frac{\pi q R^3}{8EI}(1 + \lambda) \left[2 \sin \theta + (\pi - 2\theta) \cos \theta - \frac{8}{\pi} \right]$$

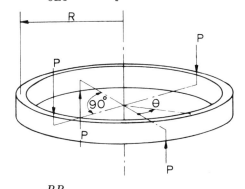

$$M = \frac{PR}{2}(\sin \theta - \cos \theta)$$

$$T = \frac{PR}{2}(\sin \theta + \cos \theta - 1)$$

$$Y = \frac{PR^3}{8EI} \{2(\theta - 1) \sin \theta + (2\theta + 2 - \pi) \cos \theta + \pi - 2 + \lambda[(6 + 2\theta - \pi) \cos \theta - 2(3 - \theta) \sin \theta + \pi + 4\theta - 6]\}$$

$$\psi = \frac{PR^2}{8EI} \{2\theta \cos \theta + (\pi - 2\theta) \sin \theta + \lambda[(\pi - 2\theta - 4) \sin \theta + 2(\theta - 2) \cos \theta + 4]\}$$

$$\eta = \frac{PR^2}{8EI}(1 + \lambda)[2(1 - \theta) \sin \theta + (\pi - 2 - 2\theta) \cos \theta]$$

Table 30.4 Trunnion-Supported Ring Under Uniform Loading

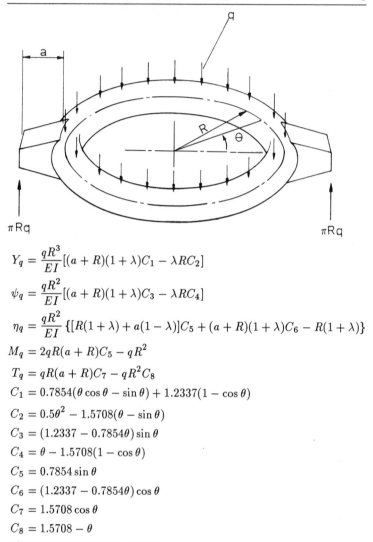

$$Y_q = \frac{qR^3}{EI}[(a+R)(1+\lambda)C_1 - \lambda R C_2]$$

$$\psi_q = \frac{qR^2}{EI}[(a+R)(1+\lambda)C_3 - \lambda R C_4]$$

$$\eta_q = \frac{qR^2}{EI}\{[R(1+\lambda) + a(1-\lambda)]C_5 + (a+R)(1+\lambda)C_6 - R(1+\lambda)\}$$

$$M_q = 2qR(a+R)C_5 - qR^2$$

$$T_q = qR(a+R)C_7 - qR^2 C_8$$

$$C_1 = 0.7854(\theta\cos\theta - \sin\theta) + 1.2337(1 - \cos\theta)$$

$$C_2 = 0.5\theta^2 - 1.5708(\theta - \sin\theta)$$

$$C_3 = (1.2337 - 0.7854\theta)\sin\theta$$

$$C_4 = \theta - 1.5708(1 - \cos\theta)$$

$$C_5 = 0.7854\sin\theta$$

$$C_6 = (1.2337 - 0.7854\theta)\cos\theta$$

$$C_7 = 1.5708\cos\theta$$

$$C_8 = 1.5708 - \theta$$

of the location angle θ, measured from the trunnions. It should be noted that out-of-plane displacements can be controlled by trunnion design. For instance, the supports can be placed inside the ring to reduce the magnitude of the deflection. In addition, the ratio of the flexural to torsional rigidity denoted by λ offers the designer an early opportunity to select the most appropriate cross-sectional geometry for a particular structural application. However, the design should not use thin-walled sections, which are liable to buckle or warp under stress. The formulas presented in Tables 30.4 and 30.5 do not account for this effect.

Table 30.5 Trunnion-Supported Ring Under Symmetrical Concentrated Loads

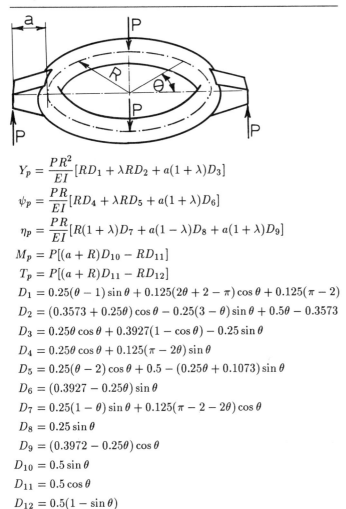

$$Y_p = \frac{PR^2}{EI}[RD_1 + \lambda RD_2 + a(1+\lambda)D_3]$$

$$\psi_p = \frac{PR}{EI}[RD_4 + \lambda RD_5 + a(1+\lambda)D_6]$$

$$\eta_p = \frac{PR}{EI}[R(1+\lambda)D_7 + a(1-\lambda)D_8 + a(1+\lambda)D_9]$$

$$M_p = P[(a+R)D_{10} - RD_{11}]$$

$$T_p = P[(a+R)D_{11} - RD_{12}]$$

$$D_1 = 0.25(\theta - 1)\sin\theta + 0.125(2\theta + 2 - \pi)\cos\theta + 0.125(\pi - 2)$$

$$D_2 = (0.3573 + 0.25\theta)\cos\theta - 0.25(3 - \theta)\sin\theta + 0.5\theta - 0.3573$$

$$D_3 = 0.25\theta\cos\theta + 0.3927(1 - \cos\theta) - 0.25\sin\theta$$

$$D_4 = 0.25\theta\cos\theta + 0.125(\pi - 2\theta)\sin\theta$$

$$D_5 = 0.25(\theta - 2)\cos\theta + 0.5 - (0.25\theta + 0.1073)\sin\theta$$

$$D_6 = (0.3927 - 0.25\theta)\sin\theta$$

$$D_7 = 0.25(1 - \theta)\sin\theta + 0.125(\pi - 2 - 2\theta)\cos\theta$$

$$D_8 = 0.25\sin\theta$$

$$D_9 = (0.3972 - 0.25\theta)\cos\theta$$

$$D_{10} = 0.5\sin\theta$$

$$D_{11} = 0.5\cos\theta$$

$$D_{12} = 0.5(1 - \sin\theta)$$

SYMBOLS

A_1, A_2, A_3	Integration constants
a	Length of trunnion, in. (mm)
B	Constant
B_1 through B_9	Design factors for arched cantilevers
C_1 through C_8	Gimbal factors for uniform loading
C	Torsional distance, in. (mm)
D_1 through D_{12}	Gimbal factors for concentrated loads
E	Modulus of elasticity, psi (N/mm^2)
F_c	Complementary solution
F_p	Particular integral

G	Modulus of rigidity, psi (N/mm^2)
I	Moment of inertia, in.^{4} (mm^4)
K	Torsional shape factor, in.^{4} (mm^4)
K_s	Section modulus for torsion, in.^{3} (mm^3)
M, M_p, M_q	Bending moments, lb-in. (N-mm)
n_0	Root of characteristic equation
P	Concentrated load, lb (N)
$\bar{P}$	Fictitious load, lb (N)
q	Uniform load, lb/in. (N/mm)
R	Radius of curvature, in. (mm)
T, T_p, T_q	Twisting moments, lb-in. (N-mm)
T_0	Twisting couple, lb-in. (N-mm)
$\bar{T}_0$	Fictitious twisting couple, lb-in. (N-mm)
U	Strain energy, lb-in. (N-mm)
X	Horizontal deflection, in. (mm)
Y, Y_p, Y_q	Vertical deflections, in. (mm)
Z	Section modulus, in.^{3} (mm^3)
α	Angle at which displacement is sought, rad
η, η_p, η_q	Angle of twist, rad
θ	Angle at which forces are considered, rad
λ	Ratio of flexural to torsional rigidity
τ	Torsional stress, psi (N/mm^2)
ϕ	Angle subtended by arched cantilever, rad
ψ, ψ_p, ψ_q	Slope, rad

V
PLATES AND FLANGES

31
Design Fundamentals

SIMPLIFYING ASSUMPTIONS

This section is concerned with the fundamental concepts of analysis and design related to the conventional plates and similar flat components subjected to transverse loading. This type of response implies out-of-plane behavior, and as such, it has been the subject of numerous mathematical studies during the past 100 years. From the practical point of view, then, such problems as the bending of flat panels, circular closures, brackets, or flanged configurations can be analyzed by means of stress and deformation equations for a plate element.

At times, some portions of the plate structures are thought of as two-dimensional beams, particularly when a plate bends in one direction. In a rigorous analysis, however, requiring the solution of the differential equations and complex boundary conditions, the general subject of thin plates represents a wide field of interest to the theoreticians. But even in this purely theoretical approach, the number of simplifying assumptions necessary for finding an elegant, closed-form solution is staggering. Some of these points may be summed up as follows [38]:

The plate is initially flat.
The material is elastic and homogeneous.
Thickness is small compared to other dimensions.
Deflection does not exceed the thickness dimension of the plate.
Strains due to bending are predominant.
Transverse shear strain is negligible.
Direct stress normal to the middle surface is neglected.

Stress resultants, not stress distributions, are considered.
Bending slopes are very small.
Boundaries are constant.

The foregoing restrictions have led to the classification of plate behavior in terms of the relative plate thickness. The plates may be put into the following four categories:

Thick plates and slabs (shear predominant).
Plates with average thickness (flexure important).
Thin plates (bending and tension).
Membranes (tension only).

Although the selection of pertinent methods depends on these categories, strict lines of demarcation between the various groups do not exist. The choice is made more difficult because of the relative scarcity of experimental data that can be applied directly to design. These considerations, coupled with continuing uncertainties of the material's behavior, often result in overdesign.

Since the appearance of the classical treatment of plates in this country [178], numerous handbooks have compiled tables and formulas for the various conditions of loading and support. In practice, however, only a few typical plate models are found to be used extensively in design.

BASIC PLATE EQUATIONS

To review briefly some of the fundamental concepts related to the general plate behavior, we can consider a square plate of constant thickness subjected to bending loading as shown in Fig. 31.1. If a long strip $ABCD$ is imagined to be cut out of this plate, then according to the elementary beam theory, the differential equation for the deflection curve is

$$EI\frac{d^2w}{dy^2} = M_2 \qquad (31.1)$$

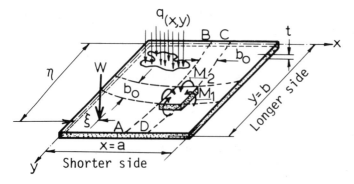

Fig. 31.1 Notation in bending analysis for a thin plate.

Design Fundamentals

Here w denotes the transverse or out-of-plane displacement of the middle surface of the plate and y is measured along the AB or CD line. The relevant moment of inertia of the cross section of this beam is $b_0 t^3/12$. The absolute value of the bending stress caused by the bending moment M_2 is obtained from the simple beam theory

$$S_2 = \frac{M_2}{Z} \tag{31.2}$$

Since in this analysis we consider strips of unit length, the bending stress can be stated as follows:

$$S_2 = \frac{6M_2}{t^2} \tag{31.3}$$

When a complete plate is bent like a beam in one plane only, lateral restriction due to the presence of adjacent strips provides additional stiffness. This consideration results in a differential equation

$$D \frac{d^2 w}{dy^2} = M_2 \tag{31.4}$$

where

$$D = \frac{Et^3}{12(1-\nu^2)} \tag{31.5}$$

The parameter defined by Eq. (31.5), known as the *plate flexural rigidity*, differs from the rigidity of the conventional beam by the factor $1/(1-\nu^2)$. If we now consider the complete theoretical range of Poisson's ratio for all the known materials, this factor can vary between 1 and 1.33. For the great majority of the metals ν is 0.3, so that the plate, bent like a beam in one plane only, should be stiffer than a simple equivalent beam by about 10%.

In a more general case, the plate can be bent in the two perpendicular directions simultaneously, so that the two different slopes are developed. If a small deflection of the curved surface of the plate is denoted here by w, and if w can be defined as a function of x and y coordinates, measured in the middle plane of the plate, then according to the theory of elasticity we get

$$D\left(\frac{\partial^2 w}{\partial x^2} + \nu \frac{\partial^2 w}{\partial y^2}\right) = -M_1 \tag{31.6}$$

and

$$\left(\frac{\partial^2 w}{\partial y^2} + \nu \frac{\partial^2 w}{\partial x^2}\right) = -M_2 \tag{31.7}$$

BENDING TO SPHERICAL SHAPE

When the two bending moments M_1 and M_2 are numerically equal, the curvatures of the plate surface in the two perpendicular directions must also be equal and the plate develops a spherical shape, for which the radius of curvature can be obtained from the expression

$$\zeta = \frac{D(1+\nu)}{M} \tag{31.8}$$

It can be argued that a plate of any shape should attain a spherical curvature as long as its edges are loaded by a uniformly distributed moment M. It is noted, however, that the spherical shape is undevelopable in contrast to a cylindrical surface which can be generated by a bending action in one plane only.

The fundamental feature expressed by Eq. (31.8) is useful in establishing the formula for the effect of a thermal stress in the plate with clamped edges. For instance, if the plate of any shape is heated in such a way that a linear thermal gradient is established across its thickness, the curvature resulting from such heating of an entirely free plate is

$$\frac{1}{\zeta} = \frac{\alpha \Delta T}{t} \tag{31.9}$$

where ζ is the radius of curvature of the plate caused by the temperature gradient ΔT.

If a clamped support is now introduced to prevent edge rotation due to the thermal gradient, the required bending moment per unit length of the edge becomes

$$M = \frac{\alpha \Delta T (1+\nu) D}{t} \tag{31.10}$$

For a rectangular element of the plate having unit width and thickness t, the maximum bending stress occurring at the top and bottom surfaces can be obtained using Eqs. (31.3), (31.5), and (31.10)

$$S = \frac{\alpha E \Delta T}{2(1-\nu)} \tag{31.11}$$

This result is indicated in Table 17.5, giving various approximate and exact formulas for thermal stresses due to internal constraint. The curvature of the free plate, given by Eq. (31.9), produces no stresses as long as the clamping effect is not present.

Expansion or contraction due to temperature is caused by thermal strains that alter the shape of the body. By forcing this body back to its initial shape, we must add mechanical forces that cause stresses. Therefore free shape does not experience such forces unless there are areas of the body where some incompatibility of natural expansion or contraction exists.

Design Fundamentals

THEORY OF RECTANGULAR PANELS

The response of rectangular panels under transverse loading is highly complex [179, 181]. One of the more general equations applicable to these panels can be stated as

$$\frac{\partial^4 w}{\partial x^4} + 2\frac{\partial^4 w}{\partial x^2 \partial y^2} + \frac{\partial^4 w}{\partial y^4} = \frac{q(x,y)}{D} \tag{31.12}$$

This classical expression, first obtained in 1815, is sometimes called the *Sophie Germain equation*. In this notation, x and y are the coordinates of the middle or neutral plane of the plate, as shown in Fig. 31.1. The surface load $q(x,y)$ can be either constant or some function of the same coordinate system. The solution of this equation must satisfy the prescribed boundary conditions for a given manner of loading and support. A summary of some typical boundary conditions is given in Table 31.1.

One of the special characteristics of bending of rectangular plates is the development of the *anticlastic surface*, which can be simulated by the forces concentrated at the corners. At one time this particular feature was used in experiments to verify the basic theory of bending of plates. The corners of a square plate, for instance, uniformly loaded and simply supported, have a tendency to rise.

The deflected surfaces of the elastic rectangular plates in general can be expressed by means of double periodic functions such as the trigonometric series, as long as they satisfy the prescribed boundary conditions. For a uniform constant pressure q, for instance, the following double series is applicable:

$$w = \frac{16q}{\pi^2 D} \sum_{m=1}^{\infty} \sum_{n=1}^{\infty} \frac{\sin\left(\frac{m\pi x}{a}\right) \sin\left(\frac{n\pi y}{b}\right)}{mn\left(\frac{m^2}{a^2} + \frac{n^2}{b^2}\right)^2} \tag{31.13}$$

The corresponding expression for a single concentrated load W is

$$w = \frac{4W}{\pi^4 abD} \sum_{m=1}^{\infty} \sum_{n=1}^{\infty} \frac{\sin\left(\frac{m\pi\xi}{a}\right)\sin\left(\frac{n\pi\eta}{b}\right)\sin\left(\frac{m\pi x}{a}\right)\sin\left(\frac{n\pi y}{b}\right)}{\left(\frac{m^2}{a^2} + \frac{n^2}{b^2}\right)^2} \tag{31.14}$$

In Eq. (31.13) m and n indicate a series of values 1, 3, 5, ... In Eq. (31.14) the m and n terms take on values of 1, 2, 3, ..., while ξ and η denote the coordinates of a point at which the concentrated load W is applied. In both equations, a is the

Table 31.1 Boundary Conditions for Plates

Clamped edge at $x = 0$	$w = 0$	$\dfrac{\partial w}{\partial x} = 0$
Simply supported edge at $y = 0$	$w = 0$	$\dfrac{\partial^2 w}{\partial y^2} + \nu \dfrac{\partial^2 w}{\partial x^2} = 0$
Free edge at $x = a$	$\dfrac{\partial^3 w}{\partial x^3} + (2-\nu)\dfrac{\partial^3 w}{\partial x \partial y^2} = 0$	$\dfrac{\partial^2 w}{\partial x^2} + \nu \dfrac{\partial^2 w}{\partial y^2} = 0$

CIRCULAR PLATE

By far the most important applications are found in the theory of circular plates of constant thickness. If the slope of the deformed plate is denoted by $\phi = -dw/dx$, the derivative of ϕ with respect to x, measured radially from the plate center, gives directly the relevant plate curvature. The complementary curvature is ϕ/x, so that the original expressions, Eqs. (31.6) and (31.7), can be restated as

$$M_1 = D\left(\frac{d\phi}{dx} + \nu\frac{\phi}{x}\right) \tag{31.15}$$

and

$$M_2 = D\left(\frac{\phi}{x} + \nu\frac{d\phi}{cx}\right) \tag{31.16}$$

Considering the equilibrium of forces on an element of the plate, cut out by two cylindrical and two diametral sections, neglecting the relevant terms of the higher order, and eliminating the bending moments with the aid of Eqs. (31.15) and (31.16) yields

$$\frac{d^2\phi}{dx^2} + \frac{d\phi}{x\,dx} - \frac{\phi}{x^2} = -\frac{Q}{D} \tag{31.17}$$

In Eq. (31.17), Q denotes the shearing force per unit length for a symmetrically loaded circular plate. The magnitude of this force can be determined from static equilibrium of the externally applied forces. The procedure gives

$$Q = \frac{qx}{2} + \frac{W}{2\pi x} \tag{31.18}$$

Note that in this definition x is the only independent variable, indicating polar symmetry of loading and geometry. The general plate expression then becomes

$$\frac{d^2\phi}{dx^2} + \frac{d\phi}{x\,dx} - \frac{\phi}{x^2} = -\frac{1}{D}\left(\frac{qx}{2} + \frac{W}{2\pi x}\right) \tag{31.19}$$

or

$$\frac{d}{dx}\left[\frac{1}{x}\frac{d}{dx}(x\phi)\right] = -\frac{1}{D}\left(\frac{qx}{2} + \frac{W}{2\pi x}\right) \tag{31.20}$$

Two consecutive integrations of Eq. (31.20) give the following two results:

$$\frac{1}{x}\frac{d}{dx}(x\phi) = -\frac{1}{D}\left(\frac{qx^2}{4} + \frac{W}{2\pi}\log_e x\right) + C_1 \tag{31.21}$$

Design Fundamentals

and

$$\phi = -\frac{qx^3}{16D} - \frac{Wx}{8\pi D}(2\log_e x - 1) + \frac{C_1 x}{2} + \frac{C_2}{x} \quad (31.22)$$

Since for small deflections $\phi \cong -dw/dx$, substituting in Eq. (31.22) and integrating once more gives

$$w = \frac{qx^4}{64D} + \frac{Wx^2}{8\pi D}(\log_e x - 1) - \frac{C_1 x^2}{4} - C_2 \log_e x + C_3 \quad (31.23)$$

The expression given by Eq. (31.23) is then the basic deflection equation for a symmetrically loaded, flat, circular plate of constant thickness.

Design Problem 31.1

Develop a design formula for the maximum stress at the center of the clamped plate shown in Fig. 31.2 in terms of the ratio R_o/t, assuming the central deflection δ to be equal to the plate thickness. The plate is subjected to a uniform, transverse pressure q. The edges of the plate are considered to be fixed as to slope, but are not otherwise constrained. The modulus of elasticity and Poisson's ratio are given as E and ν, respectively.

Solution

Since the outer edge of the plate is prevented from rotation and it does not permit any transverse displacement, the boundary conditions from Table 31.1 are $w = 0$ and $dw/dx = 0$. The applicable deflection expression follows from Eq. (31.23) for $w = 0$

$$w = \frac{qx^4}{64D} - \frac{C_1 x^2}{4} - C_2 \log_e x + C_3$$

The slope follows from Eq. (31.22) and is equal to

$$\phi = \frac{qx^3}{16D} + \frac{C_1 x}{2} + \frac{C_2}{x}$$

Since this slope is zero at $x = 0$, we get

$$-\frac{q(0)}{16D} + \frac{C_1(0)}{2} + \frac{C_2}{(0)} = 0$$

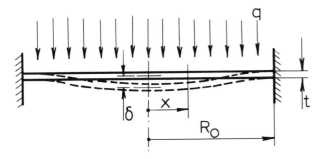

Fig. 31.2 Circular plate with clamped edges.

The limit of the last term in this expression is indeterminate. Using L'Hospital's rule [180], however, gives constant $C_2 = 0$. When $x = R_o$ and $\phi = 0$, the slope equation above yields

$$-\frac{qR_o^3}{16D} + \frac{C_1 R_o}{2} + \frac{C_2}{R_o} = 0$$

from which

$$C_1 = \frac{qR_o^2}{8D}$$

Since the deflection must also be equal to zero at $x = R_o$, we obtain the following additional relation for finding C_3

$$\frac{qR_o^4}{64D} - \frac{qR_o^4}{32D} + C_3 = 0$$

or

$$C_3 = \frac{qR_o^4}{64D}$$

Hence, the maximum deflection at $x = 0$ is $w = \delta$, which gives

$$\delta = \frac{qR_o^4}{64D} \tag{31.24a}$$

The formula for the bending slope in terms of the foregoing constants is then

$$\phi = \frac{qx}{16D}(R_o^2 - x^2)$$

To make use of Eq. (31.15) or (31.16), calculate first the derivative of the slope with respect to x

$$\frac{d\phi}{dx} = \frac{q}{16D}(R_o^2 - 3x^2)$$

so that

$$M_1 = D\left[\frac{q}{16D}(R_o^2 - 3x^2) + \nu \frac{q}{16D}(R_o^2 - x^2)\right]$$

When $x = 0$,

$$M_1 = M_2 = \frac{qR_o^2(1+\nu)}{16}$$

and the bending stress at the center becomes

$$S = \frac{3(1+\nu)qR_o^2}{8t^2} \tag{31.24b}$$

Combining Eqs. (31.24a), (31.24b), and (31.5) and making $\delta = t$ gives

$$S = \frac{2Et^2}{(1-\nu)R_o^2} \quad \blacklozenge \tag{31.24c}$$

Design Fundamentals

Design Problem 31.2

An elliptical plate of constant cross section carries a transverse uniform pressure q. Develop the design formula for the maximum deflection of the middle plane of the plate assuming built-in edge conditions where the bending slope and deflection are zero. The minor and major axes of the elliptical plate boundary are a and b, respectively. The plate has thickness t and the material properties E and ν are constant.

Solution

Let the deflection of the plate be described by the expression

$$w = w_0 \beta^2$$

where

$$\beta = \frac{4x^2}{a^2} + \frac{4y^2}{b^2} - 1$$

and x and y are measured from the center of the plate. When $x = a/2$ and $y = 0$, $\beta = 0$ and $w = 0$. On the other hand, when $x = y = 0$, $\beta = -1$ and $w = w_0$. The derivatives necessary for the application of Eq. (31.12) are

$$\frac{\partial w}{\partial x} = 2w_0 \beta \frac{\partial \beta}{\partial x}$$

$$\frac{\partial w}{\partial y} = 2w_0 \beta \frac{\partial \beta}{\partial y}$$

$$\frac{\partial \beta}{\partial x} = \frac{8x}{a^2}$$

$$\frac{\partial^2 \beta}{\partial x^2} = \frac{8}{a^2}$$

$$\frac{\partial \beta}{\partial y} = \frac{8y}{b^2}$$

$$\frac{\partial^2 \beta}{\partial y^2} = \frac{8}{b^2}$$

$$\frac{\partial^2 w}{\partial x^2} = \frac{128 w_0 x^2}{a^4} + \frac{16 w_0 \beta}{a^2}$$

$$\frac{\partial^2 w}{\partial y^2} = \frac{128 w_0 y^2}{b^4} + \frac{16 w_0 \beta}{b^2}$$

$$\frac{\partial^2 w}{\partial x \, \partial y} = \frac{128 w_0 xy}{a^2 b^2}$$

$$\frac{\partial^3 w}{\partial x^3} = \frac{384 w_0 x}{a^4}$$

$$\frac{\partial^3 w}{\partial y^3} = \frac{384 w_0 y}{b^4}$$

$$\frac{\partial^3 w}{\partial x^2 \, \partial y} = \frac{128 w_0 y}{a^2 b^2}$$

$$\frac{\partial^4 w}{\partial x^2 \, \partial y^2} = \frac{128 w_0}{a^2 b^2}$$

$$\frac{\partial^4 w}{\partial x^4} = \frac{384 w_0}{a^4} \quad \text{and} \quad \frac{\partial^4 w}{\partial y^4} = \frac{384 w_0}{b^4}$$

Substituting the appropriate terms into Eq. (31.12) gives the value of the constant w_0

$$w_0 = \frac{q}{16D \left(\frac{24}{a^4} + \frac{24}{b^4} + \frac{16}{a^2 b^2} \right)}$$

Hence, the general equation for the deflection is

$$w = \frac{q \left(1 - \frac{4x^2}{a^2} - \frac{4y^2}{b^2} \right)^2}{16D \left(\frac{24}{a^4} + \frac{24}{b^4} + \frac{16}{a^2 b^2} \right)} \tag{31.24d}$$

At the center where $x = y = 0$, the formula for the maximum deflection becomes

$$\delta = \frac{3qa^4 b^4 (1 - \nu^2)}{32 E t^3 (3a^4 + 3b^4 + 2a^2 b^2)} \quad \blacklozenge \tag{31.24e}$$

APPLICATIONS OF PLATE THEORY TO FLANGES

Although the theory of plates is useful not only in the study of circular closures and panels of various shapes but also in the extensive area of circular flanges, until quite recently engineers appeared to be reluctant to use it [182]. Several classical approaches to the solution of problems in pipe flanges assumed that a circular plate with a central hole was either loaded at the edges or loaded uniformly over the whole surface [183, 184]. When a circular plate is loaded and supported in the manner shown in Fig. 31.3, the maximum hoop stress is developed at the inside corners of the plate. According to the flat ring theory of Waters [183], the corner stress becomes

$$S_{\text{h}} = \frac{W(R - R')}{(R_{\text{o}} - R_{\text{i}}) t^2} \left[\frac{1.242 R_{\text{o}}^2 \log_e (R_{\text{o}}/R_{\text{i}})}{R_{\text{o}}^2 - R_{\text{i}}^2} + 0.335 \right] \tag{31.25}$$

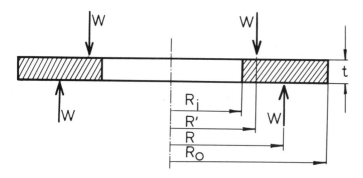

Fig. 31.3 Simply supported plate under concentric loading.

Design Fundamentals

The corresponding Holmberg-Axelson formula for this stress is

$$S_{\mathrm{h}} = \frac{W}{(R_{\mathrm{o}}^2 - R_{\mathrm{i}}^2)\,t^2}\left[0.35(R^2 - R'^2) + 1.195 R_{\mathrm{o}}^2 \log_e\left(\frac{R}{R'}\right)\right] \tag{31.26}$$

Examination of Eq. (31.25) shows certain natural limitations when the $R_{\mathrm{o}}/R_{\mathrm{i}}$ ratio is large, since such a case would correspond to the theory of plates rather than that for flat rings on which Eq. (31.25) is based. It appears that Eq. (31.26) is correct for all values of $R_{\mathrm{o}}/R_{\mathrm{i}}$. The results of the study also indicate that a plate pierced by a small hole in the center has the maximum circumferential stress twice as large as if the plate had been solid. This finding may be of special importance in those cases where the plate is not made out of fracture-tough material.

It appears that in treating certain plate configurations, which resemble machine and pipe flanges rather than circular closures, the analyst has the choice of following either the treatment of flat rings or flat plates. Probably the best method demonstrated in industry so far [184] involves breaking down the flange or plate structures into a series of concentric rings, each of which has a simple loading. The boundary conditions are solved by making the slope and radial moments continuous. Since, however, in this type of a treatment the number of constants is always equal to twice the number of rings, a large number of simultaneous equations may be involved. This should however present no special problems where large electronic computers are available.

APPROXIMATION OF LARGE DEFLECTION IN PLATES

The fundamentals discussed so far are generally applicable to plates with average and small thickness where deflections do not exceed the magnitude of plate thickness. When this is not the case, the analysis should include the effect of the strain of the middle plane of the plate [178, 179]. This procedure leads to differential equations of impractical nature where a considerable amount of numerical calculation is required to solve the problems of plate stresses and deflections. An approximate solution of this problem however, can be based on the following system of differential equations [178] due to Nadai:

$$\frac{d^2u}{dr^2} + \frac{1}{4}\frac{du}{dr} - \frac{u}{r^2} = -\frac{1-\nu}{2r}\left(\frac{dw}{dr}\right)^2 - \frac{dw}{dr}\frac{d^2w}{dr^2} \tag{31.27}$$

$$\frac{d^3w}{dr^3} + \frac{1}{r}\frac{d^2w}{dr^2} - \frac{1}{r^2}\frac{dw}{dr} = \frac{12}{t^2}\frac{dw}{dr}\left[\frac{du}{dr} + \nu\frac{u}{r} + \frac{1}{2}\left(\frac{dw}{dr}\right)^2\right] + \frac{1}{Dr}\int_0^r qr\,dr \tag{31.28}$$

Here r denotes the radius at any point and u is the component of the displacement in the radial direction. The terms D, w, and q have the same meaning as previously. The solution of the preceding system of equations can be obtained by a series of approximations. For example, we can start with a suitable expression for w as a function of the plate radius. Substituting its derivatives into the right-hand side of Eq. (31.27), we obtain a linear equation in u which can be integrated to give the first value for u. Substituting next the first approximations for u and w in the right-hand side of Eq. (31.28) gives a linear differential equation in w which after integration gives the second approximation for w. This second approximation can be used with Eq. (31.27) to get further refined values. This procedure gives the formula for the maximum deflection of a circular diaphragm with fixed edges, small flexural stiffness, and a Poisson's ratio of $\nu = 0.25$.

$$\frac{\delta}{t} + 0.583\left(\frac{\delta}{t}\right)^3 = 0.176\frac{q}{E}\left(\frac{R_o}{t}\right)^4 \tag{31.29}$$

When the deflection becomes rather large while bending rigidity of the plate is small, the following expression can be used:

$$\delta = 0.662 R_o \left(\frac{qR_o}{Et}\right)^{1/3} \tag{31.30}$$

The respective tensile membrane stresses at the center of the plate and at the built-in boundary are

$$S_t = 0.423\left(\frac{Eq^2 R_o^2}{t^2}\right)^{1/3} \tag{31.31}$$

and

$$S_t = 0.328\left(\frac{Eq^2 R_o^2}{t^2}\right)^{1/3} \tag{31.32}$$

It may be of interest to note that plate deflection from Eq. (31.30) is not directly proportional to the load intensity q but varies as a cube root of this quantity. To make the deflections proportional to q, it would be necessary to develop a corrugated membrane, as is frequently done in the field of instrumentation.

Thick plates are seldom used except in the reinforced concrete design, where they are generally known as slabs. The design of such structures is governed largely by practical experience. Also, for the first approximation their design can be related to the theory of short beams. In other applications, the concept of slabs of elastic and viscoelastic nature on a horizontal substratum can be found, for example, in geological studies dealing with the response of the earth's crust to ice loads. The mathematical problem in such a case can be based on the theory of beams on an elastic foundation and the viscoelastic behavior of materials [132, 179].

DESIGN CHARTS FOR LARGE DEFLECTION OF PLATES

The general approach to the study of the effect of large deflections on structural design criteria of plates has been described in the foregoing section. A number of practical formulas and rules for this purpose have been developed over the past 50 years. Many of these involve procedures where we begin with an assumed value of

Design Fundamentals

δ sufficiently close to that which can be calculated on the basis of small-deflection theory. Since a number of large-deflection formulas [186] are expressed in terms of δ/t ratios, taking δ/t somewhat smaller than that which is compatible with the elastic deflection, the correct value of δ/t can be found by a relatively few successive approximations.

It should be pointed out here that when the deflection of a plate becomes comparatively large, the middle surface is additionally strained due to the diaphragm effect. The degree of this additional strain will depend on the conditions of the plate support. The total stress will be equal to the sum of the maximum stress due to flexure and diaphragm tension. The load-deflection and load-stress relations in this particular case become nonlinear.

The majority of practical design applications involve circular plates and diaphragms under uniform pressure. The relevant deflection criterion can be expressed in dimensionless form as

$$\frac{qR_o^4}{Et^4} = H(\delta, t) \qquad (31.33)$$

The functional relationship $H(\delta, t)$ given by Eq. (31.33) is plotted in Fig. 31.4 for the three boundary conditions denoted by H_1 through H_3. The dimensionless parameter contains four quantities that have to be known a priori. The design

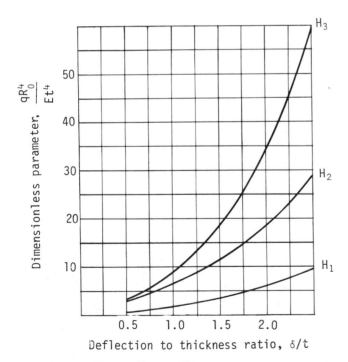

Fig. 31.4 Design chart for large deflections of circular plates. H_1, simple support; H_2, edge restrained in vertical plane and zero tension; H_3, fixed support with full diaphragm tension.

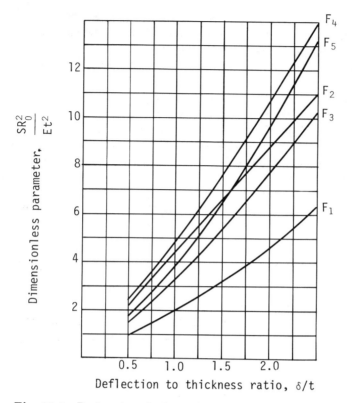

Fig. 31.5 Design chart for large deflection stresses in circular plates: F_1, center stress, zero constraint; F_2, edge restrained in vertical plane, edge stress; F_3, edge restrained in vertical plane, center stress; F_4, fixed support with full diaphragm tension, edge stress; F_5, fixed support with full diaphragm tension, center stress.

problem is then reduced to one unknown, which can be found with the aid of Eq. (31.33) and Fig. 31.4.

Similarly, the stress criterion for circular plates, subjected to large deflections and diaphragm behavior, can be defined as shown below:

$$\frac{SR_o^2}{Et^2} = F(\delta, t) \tag{31.34}$$

Again, depending on the manner of plate support and the location of stress, five parameters F_1 through F_5 can be used in the calculation. These parameters are plotted in Fig. 31.5. The curves denoted by F_1, F_3, and F_5 refer to the state of stress at the center of the plate. The remaining two factors F_2 and F_4 characterize the response of the edge of the plate.

Design Problem 31.3

Calculate the maximum deflection and stress in a circular aluminum plate 0.375 in. thick, having a 24-in. radius and carrying the transverse pressure of 6 psi. The edges of the

Design Fundamentals

plate are fixed as to slope and radial displacement. Assume the modulus of elasticity to be equal to 10×10^6 psi.

Solution

From Eq. (31.33),

$$\frac{qR_o^4}{Et^4} = \frac{6 \times 24^4}{10^7 \times 0.375^4} = 10$$

From Fig. 31.4, the above parameter and curve H_3 give

$$\frac{\delta}{t} = 1.075$$

so that

$$\delta = 1.075 \times 0.375 = 0.403 \text{ in.} \quad (10.2 \text{ mm})$$

The value of the required stress parameters follows from Fig. 31.5 for $\delta/t = 1.075$. This gives

$$F_4 = 5.3$$

Hence, using Eq. (31.34), the maximum stress is found at the fixed support

$$S = \frac{Et^2 F(\delta,t)}{R_o^2}$$

and since

$$F(\delta,t) = F_4 = 5.3$$

we get

$$S = \frac{10 \times 10^6 \times 0.375^2 \times 5.3}{24 \times 24} = 12{,}939 \text{ psi} \quad (89.2 \text{ N/mm}^2) \quad \blacklozenge$$

SPECIAL PLATE PROBLEMS

The fundamentals of plate theory and design discussed in this chapter so far cover the more likely areas of interest to mechanical engineers. This section will include two additional topics that can come up occasionally in design. They involve large deformation of rectangular plates and the ultimate strength of circular plates.

The theoretical problem of large deflection of a rectangular plate has been the subject of numerous investigations, particularly during the past 50 years. Various solutions obtained have been expressed in terms of dimensionless parameters similar to those used in conjunction with circular plates. Many solutions have also been compared with experimental data during the various phases of research

Table 31.2 Limit Loads for Circular Plates

$$q = \frac{3R_o t^2 S_y}{2x^2(3R_o - 2x)}$$

For $\dfrac{x}{R_o} \leq 0.61$: $\log_e\left(\dfrac{x}{R_o}\right) = \log_e\left(\dfrac{3}{2} - \dfrac{3S_y t^2}{4qx^2}\right) + \dfrac{t^2 S_y}{t^2 S_y - 2qx^2}$

For $\dfrac{x}{R_o} \geq 0.61$: $\log_e\left(\dfrac{x}{R_o}\right) = \dfrac{2qx^2 - 5S_y t^2 + S_y t^2 \log_e\left(\dfrac{3S_y t^2}{2qx^2}\right)}{4qx^2 - 2S_y t^2}$

$$q = \frac{3t^2 S_y}{2(R_o^2 + xR_o - 2x^2)}$$

---

sponsored by the National Advisory Committee for Aeronautics. A convenient summary of deflections and stresses for rectangular plates under uniform load is given by Roark [35]. The boundaries for these plates include simple supports, riveting constraints, and completely fixed conditions. The plate-loading parameter qa^4/Et^4 covers the range 0 to 250. The maximum ratio of deflection to thickness in Roark's summary is 2.2.

The customary approach to sizing of plates for a particular condition of loading and support is via the elastic theory. However, if the plate material is capable of sustaining plastic stresses, the loads can be increased to a higher value while the structure becomes free to assume a larger deformation at failure. The method of predicting such loads is based on the theory of limit design [187, 188]. A brief summary of formulas for limit loads on typical circular plates [21] is given in Table 31.2.

SYMBOLS

a	Shorter side of plate (also minor axis of ellipse), in. (mm)
b	Longer side of plate (also major axis of ellipse), in. (mm)
b_0	Width of element, in. (mm)
C_1, C_2, C_3	Integration constants
D	Plate flexural rigidity, (lb-in.^{2})/in. (N-mm^2/mm)
E	Modulus of elasticity, psi (N/mm^2)

Design Fundamentals

$F(\delta, t)$	Dimensionless parameter
F_1 through F_5	Plate stress factors
$H(\delta, t)$	Dimensionless parameter
H_1 through H_3	Plate deflection factors
I	Moment of inertia, in.^{4} (mm^4)
M, M_1, M_2	Plate bending moments, (lb-in)/in. (N-mm/mm)
m	Integer
n	Integer
Q	Shearing force per unit length, lb/in. (N/mm)
$q, q(x,y)$	Transverse pressures, psi (N/mm^2)
R	Radius to central load, in. (mm)
R'	Radius to edge of load, in. (mm)
R_o	Outer radius of plate, in. (mm)
R_i	Inner radius of plate, in. (mm)
r	Radius to any point on plate, in. (mm)
S, S_1, S_2	Plate bending stresses, psi (N/mm^2)
S_h	Hoop stress, psi (N/mm^2)
S_t	Membrane tensile stress, psi (N/mm^2)
S_y	Yield strength, psi (N/mm^2)
ΔT	Temperature gradient, °F (°C)
t	Plate thickness, in. (mm)
u	Displacement in the plane of plate, in. (mm)
W	Concentrated or ring load, lb (N)
w	Transverse displacement, in. (mm)
x	Coordinate or radius in plates, in. (mm)
y	Coordinate, in. (mm)
Z	Section modulus, in.^{3} (mm^3)
α	Coefficient of linear expansion, °F^{-1} (°C^{-1})
β	Displacement function
δ	Maximum deflection of plate, in. (mm)
ζ	Radius of curvature, in. (mm)
η	Coordinate, in. (mm)
ν	Poisson's ratio
ξ	Coordinate, in. (mm)
ϕ	Slope, rad

32
Panels and Closures

PROBLEM DEFINITION

Some of the fundamentals outlined in Chap. 31 point to the degree of complexity of the various plate solutions. This topic is seldom covered in practical engineering texts. When plate applications arise, a good deal of specialization is required, backed up by experimental work. This type of information is not easy to come by and the designer has to fall back on the classical solutions and the conservative assumptions of elasticity. This section will attempt to summarize some of the more basic practical data related to those plate configurations that occur most frequently and which can be used as approximate models for more complex solutions.

A typical structural panel may be defined as a flat piece of material, usually rectangular, elliptical, or similar in shape, which forms a part of the surface of a wall, door, cabinet, duct, machine component, fuselage window, floor, or similar component. The design boundaries illustrated in Fig. 32.1 may involve some degree of fixity or freedom when a given panel is subjected to uniform loading. One of more difficult considerations in estimating the panel strength and rigidity is the choice of the correct boundary condition. This process depends entirely on a knowledge of loading and support, which varies from problem to problem. According to the general design theory, the boundary conditions can vary from a completely built-in to a simple roller-type support allowing a full freedom of rotation. In the great majority of practical situations, some intermediate conditions exist, requiring a good deal of engineering judgment in selecting the most realistic model for panel support. The design criteria for uniform transverse loading can be governed by

Panels and Closures

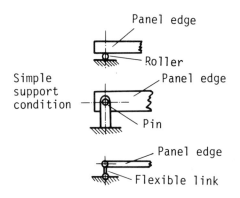

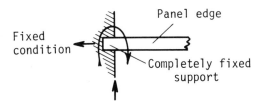

Fig. 32.1 Examples of edge conditions for panel design.

either the maximum bending strength or the allowable maximum deflection. The purpose of this brief survey is to provide a set of working equations and charts suitable for the preliminary design [185].

DESIGN CHARTS FOR PANELS

Simple rectangular panels are often supported by structural shapes whose bending stiffness is relatively high compared with that of the panels themselves. Under these conditions, fixed edges can be assumed in the calculations. However, when the supporting shapes are such that a finite slope can develop in the plane perpendicular to the panel, the design should be based on a simple support criterion. A summary of some of the more commonly used design equations for the rectangular and the elliptical panels is compiled in Table 32.1.

Design factors A_1 through A_8 from Table 32.1 are plotted in Figs. 32.2 and 32.3 against the panel length ratio a/b. For a rectangular panel, a and b denote the smaller and larger sides, respectively. For an elliptical geometry, a and b are the minor and major axes, respectively. While the maximum bending stress is found at the center for the simply supported rectangular and elliptical panels, built-in panels are stressed more at the supports. For a rectangular built-in panel, this point is at the midpoint of the longer edge. In the built-in elliptical panel, the maximum bending stress is at the ends of the minor axis a. As shown in Design Problem 32.1, the maximum deflection is a function of the a/b ratio. For instance, taking $b = 2a$

Table 32.1 Design Equations for Simple Panels Under Uniform Loading

Type of Panel	Maximum Stress	Maximum Deflection
Rectangular simply supported	$S = \dfrac{qa^2 A_1}{t^2}$	$\delta = \dfrac{qa^4 A_2}{Et^3}$
Rectangular built-in	$S = \dfrac{qa^2 A_3}{t^2}$	$\delta = \dfrac{qa^4 A_4}{Et^3}$
Elliptical simply supported	$S = \dfrac{qa^2 A_5}{t^2}$	$\delta = \dfrac{qa^4 A_6}{Et^3}$
Elliptical built-in	$S = \dfrac{qa^2 A_7}{t^2}$	$\delta = \dfrac{qa^4 A_8}{Et^3}$

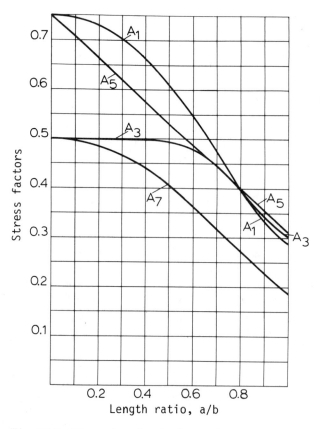

Fig. 32.2 Stress chart for simple panels.

Panels and Closures

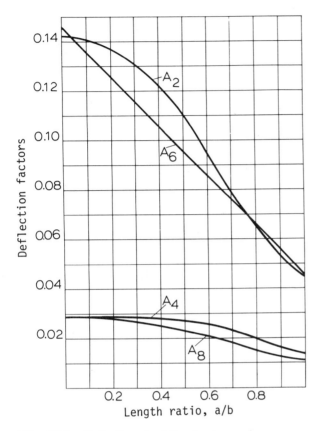

Fig. 32.3 Deflection chart for simple panels.

gives $A_8 = 3(1 - \nu^2)/118$, that is, reasonably close to the value which can be read from Fig. 32.3 for $a/b = 0.5$.

SIMILARITIES OF RECTANGULAR AND ELLIPTICAL PANELS

It is not by coincidence that the rectangular and elliptical panels can be grouped together. The charts given in Figs. 32.2 and 32.3 indicate a definite correlation of structural behavior. For this reason, a great number of panel shapes that fall between the rectangular and elliptical boundaries can be designed with the help of the charts given in Figs. 32.2 and 32.3. For instance, the arbitrary profile shown in Fig. 32.4 should display strength and rigidity characteristics which might be termed as intermediate between those of the elliptical and rectangular configurations, provided that the overall a and b dimensions remain the same.

The design engineer concerned with such a problem can develop an individual method of interpolation between the relevant results. For instance, the ratio of the unused corner area F to the total area difference between the rectangular and elliptical geometries can be used as a parameter. In terms of the dimensions indi-

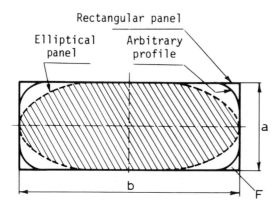

Fig. 32.4 Comparable plate configurations.

cated in Fig. 32.4, this parameter may be defined as $16F/ab(4 - \pi)$. It should be emphasized, however, that such a linear interpolation can be justified only because of the inherent similarities of structural behavior of the rectangular and elliptical configurations. The error introduced by this procedure is expected to be relatively small and certainly acceptable within the scope of the preliminary design, which under normal conditions involves ample margins of safety.

Design Problem 32.1

A pressure plate of rectangular geometry with rounded-off corners is shown in Fig. 32.5. It is simply supported and carries a uniform transverse loading of 200 psi. Assuming the dimensions shown in Fig. 32.5 and steel as the material, calculate the maximum stresses and deflections using the interpolation method described above.

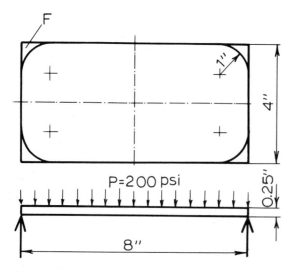

Fig. 32.5 Panel of arbitrary profile.

Panels and Closures

Solution

The unused corner area F from Fig. 32.5 is

$$F = 1 - \frac{\pi}{4} = 0.215 \text{ in.}^2$$

The total unused corner area between the rectangular and elliptical boundaries is

$$ab\frac{(4-\pi)}{16} = 8 \times \frac{4(4-\pi)}{16} = 1.717 \text{ in.}^2$$

The dimensionless ratio is then

$$\frac{0.215}{1.717} = 0.125$$

For $a/b = 4/8 = 0.5$, Fig. 32.2 gives approximately

$$A_1 = 0.61$$
$$A_5 = 0.53$$

The equation for interpolating the required stress factor can now be set up as follows

$$A_1 - \frac{(A_1 - A_5)16F}{ab(4-\pi)}$$

Hence,

$$0.61 - (0.61 - 0.53) \times 0.125 = 0.600$$

and using the formula for a rectangular plate from Table 32.1 gives

$$S = \frac{200 \times 16 \times 0.600}{09.25 \times 0.25} = 30,720 \text{ psi} \quad (212 \text{ N/mm}^2)$$

From Fig. 32.3,

$$A_2 = 0.11$$
$$A_6 = 0.096$$

Again, the interpolation formula for this case is

$$A_2 - \frac{(A_2 - A_6)16F}{ab(4-\pi)}$$

and since the parameter $16F/ab(4-\pi) = 0.125$, as before, we get

$$0.110 - (0.110 - 0.096)0.125 = 0.1083$$

Hence, using the plate deflection formula from Table 32.1 yields

$$\delta = \frac{200 \times 256 \times 0.1083}{30 \times 10^6 \times 0.25^3} = 0.012 \text{ in.} \quad (0.30 \text{ mm}) \quad \blacklozenge$$

CIRCULAR OPENINGS

A special class of a plate problem related to flanges, rings, and circular closures in structural applications is concerned with central openings. The fundamental expressions in this type of analysis are given by Eqs. (31.22) and (31.23). For instance, consider a plate with central opening and uniformly distributed edge moments as shown in Fig. 32.6. Since there is no transverse loading on the plate such as W or q, Eqs. (31.22) and (31.23) reduce to the form

$$\phi = \frac{C_1 x}{2} + \frac{C_2}{x} \tag{32.1}$$

and

$$w = -\frac{C_1 x^2}{4} - C_2 \log_e x + C_3 \tag{32.2}$$

Since from Eq. (32.1) $d\phi/dx = (C_1/2) - (C_2/x^2)$, substituting this expression together with Eq. (32.1) into Eq. (31.15) gives

$$M_1 = D\left[\frac{C_1}{2} - \frac{C_2}{x^2} + \nu\left(\frac{C_1}{2} + \frac{C_2}{x^2}\right)\right] \tag{32.3}$$

Hence, for $x = R_i$, $M_1 = M_i$ and Eq. (32.3) yield

$$M_i = \left[\frac{C_1}{2}(1+\nu) - \frac{C_2}{R_i^2}(1-\nu)\right]D \tag{32.4}$$

When $x = R_o$, $M_1 = M_o$ and Eq. (32.3) give

$$M_o = \left[\frac{C_1(1+\nu)}{2} - \frac{C_2}{R_o^2}(1-\nu)\right]D \tag{32.5}$$

Solving Eqs. (32.4) and (32.5) for C_1 and C_2 gives

$$C_1 = \frac{2(R_o^2 M_o - R_i^2 M_i)}{(1+\nu)D(R_o^2 - R_i^2)} \tag{32.6}$$

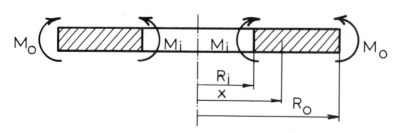

Fig. 32.6 Flat plate with central opening.

Panels and Closures

and

$$C_2 = \frac{R_o^2 R_i^2 (M_o - M_i)}{(1-\nu)D(R_o^2 - R_i^2)} \tag{32.7}$$

When the deflection is zero at the simple support located at $x = R_o$, Eq. (32.2) gives the remaining condition for calculating C_3. This gives

$$C_3 = \frac{R_o^2(R_o^2 M_o - R_i^2 M_i)}{2(1+\nu)D(R_o^2 - R_i^2)} + \frac{R_o^2 R_i^2 (M_o - M_i)\log_e R_o}{(1-\nu)D(R_o^2 - R_i^2)} \tag{32.8}$$

Substituting the values of the constants C_1 and C_2 into the bending moment expression, Eq. (32.3), yields the radial bending moment at any radius x shown in Fig. 32.6.

$$M_1 = \frac{1}{R_o^2 - R_i^2}\left[R_o^2 M_o - R_i^2 M_i - \frac{R_o^2 R_i^2 (M_o - M_i)}{x^2}\right] \tag{32.9}$$

It is clear from the above that for $x = R_i$, $M_1 = M_i$ and at $x = R_o$, $M_1 = M_o$. The slope and deflection formulas for the case shown in Fig. 32.6 can also be written with the aid of Eqs. (32.6), (32.7), and (32.8) as

$$\phi = \frac{1}{D(R_o^2 - R_i^2)}\left[\frac{(R_o^2 M_o - R_i^2 M_i)x}{1+\nu} + \frac{R_o^2 R_i^2 (M_o - M_i)}{x(1-\nu)}\right] \tag{32.10}$$

and

$$w = \frac{1}{D(R_o^2 - R_i^2)}\left[\frac{(R_o^2 M_o - R_i^2 M_i)(R_o^2 - x^2)}{2(1+\nu)} + \frac{R_o^2 R_i^2 (M_o - M_i)(\log_e R_o - \log_e x)}{1-\nu}\right] \tag{32.11}$$

If now the inner edge of the plate is restrained from rotation as shown in Fig. 32.7, the slope is zero at $x = R_i$. This gives, from Eq. (32.1),

$$C_1 = -\frac{2C_2}{R_i^2} \tag{32.12}$$

Also, at $x = R_o$, $M_1 = M_o$, which yields the second necessary condition for determining C_1 and C_2.

$$M_o = D\left[\frac{C_1(1+\nu)}{2} - \frac{C_2(1-\nu)}{R_o^2}\right] \tag{32.13}$$

Hence, solving Eqs. (32.12) and (32.13) yields

$$C_1 = \frac{2M_o R_o^2}{D[R_o^2(1+\nu) + R_i^2(1-\nu)]} \tag{32.14}$$

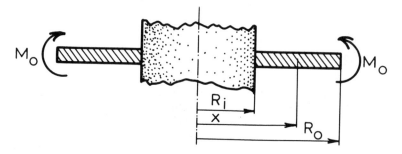

Fig. 32.7 Flat plate with central constraint.

Table 32.2 Formulas for Plates with Central Openings

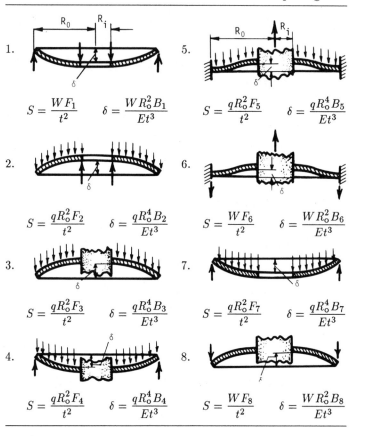

1. $S = \dfrac{W F_1}{t^2}$ $\delta = \dfrac{W R_o^2 B_1}{E t^3}$

2. $S = \dfrac{q R_o^2 F_2}{t^2}$ $\delta = \dfrac{q R_o^4 B_2}{E t^3}$

3. $S = \dfrac{q R_o^2 F_3}{t^2}$ $\delta = \dfrac{q R_o^4 B_3}{E t^3}$

4. $S = \dfrac{q R_o^2 F_4}{t^2}$ $\delta = \dfrac{q R_o^4 B_4}{E t^3}$

5. $S = \dfrac{q R_o^2 F_5}{t^2}$ $\delta = \dfrac{q R_o^4 B_5}{E t^3}$

6. $S = \dfrac{W F_6}{t^2}$ $\delta = \dfrac{W R_o^2 B_6}{E t^3}$

7. $S = \dfrac{q R_o^2 F_7}{t^2}$ $\delta = \dfrac{q R_o^4 B_7}{E t^3}$

8. $S = \dfrac{W F_8}{t^2}$ $\delta = \dfrac{W R_o^2 B_8}{E t^3}$

Panels and Closures

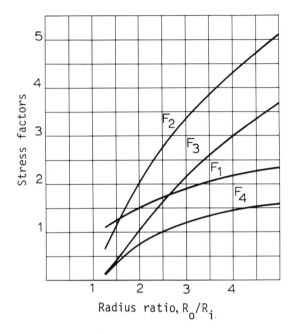

Fig. 32.8 Plate stress factors F_1 through F_4 for central openings and constraints.

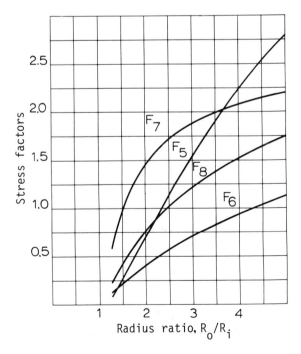

Fig. 32.9 Plate stress factors F_5 through F_8 for central openings and constraints.

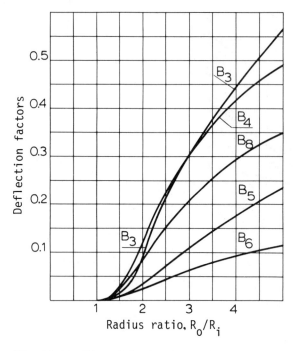

Fig. 32.10 Plate deflection factors B_3, B_4, B_5, B_6, and B_8 for central openings and constraints.

and

$$C_2 = -\frac{M_o R_o^2 R_i^2}{D[R_o^2(1+\nu) + R_i^2(1-\nu)]} \qquad (32.15)$$

Substituting Eqs. (32.14) and (32.15) into Eq. (32.3) gives

$$M_1 = \frac{M_o R_o^2}{R_o^2(1+\nu) + R_i^2(1-\nu)}\left[1+\nu+(1-\nu)\frac{R_i^2}{x^2}\right] \qquad (32.16)$$

The slope of any point is found by Eq. (32.1) by introducing constants C_1 and C_2

$$\phi = \frac{M_o R_o^2}{D[R_o^2(1+\nu) + R_i^2(1-\nu)]}\left(x - \frac{R_i^2}{x}\right) \qquad (32.17)$$

To find the general expression for the deflection, it is first necessary to determine C_3 from the condition that $w = 0$ at $x = R_i$. Hence, Eq. (32.2) gives

$$C_3 = \frac{M_o R_o^2 R_i^2(\frac{1}{2} - \log_e R_i)}{D[R_o^2(1+\nu) + R_i^2(1-\nu)]} \qquad (32.18)$$

Panels and Closures

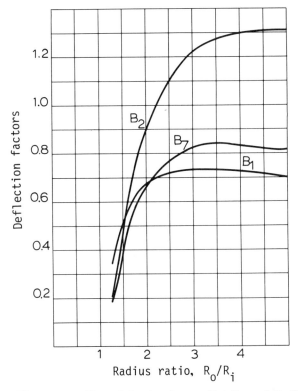

Fig. 32.11 Plate deflection factors B_1, B_2, and B_7 for central openings and constraints.

The formula for deflection is obtained by combining Eqs. (32.2), (32.14), (32.15), and (32.18).

$$w = \frac{M_o R_o^2}{2D\left[R_o^2(1+\nu) + R_i^2(1-\nu)\right]} \left[R_i^2 - x^2 + 2R_i^2(\log_e x - \log_e R_i)\right] \qquad (32.19)$$

FORMULAS AND CHARTS FOR CIRCULAR PLATES

The derivations given so far illustrate the basic theory needed in the solution of plate problems involving central openings and constraints. In general, there is no need for this type of a mathematical work in a design office, since engineering books provide a sufficient variety of formulas to suit a particular case of loading and support. Examples of the most commonly quoted cases in the literature for central openings and constraints are given in Table 32.2. The relevant design curves are plotted in Figs. 32.8 to 32.11. In addition, a summary of useful plate formulas is included in Table 32.3. The basic parameter x can be varied between zero and R_o to obtain a variety of stress and deflection relations on the premise that Poisson's

Table 32.3 Typical Formulas for Complete Plates

$$S = \frac{qx^2}{t^2}\left[1.500 + 1.950\log_e\left(\frac{R_o}{x}\right) - 0.263\left(\frac{x}{R_o}\right)^2\right] \quad \text{(at center)}$$

$$\delta = \frac{qx^2}{Et^3}\left[1.733R_o^2 - 0.683x^2\log_e\left(\frac{R_o}{x}\right) - 1.037x^2\right]$$

$$S = \frac{W}{t^2}\left[0.167 + 0.621\log_e\left(\frac{R_o}{x}\right) - 0.167\left(\frac{x}{R_o}\right)^2\right] \quad \text{(at center)}$$

$$\delta = \frac{W}{Et^3}\left[0.551(R_o^2 - x^2) - 0.434x^2\log_e\left(\frac{R_o}{x}\right)\right]$$

$$S = \frac{qx^2}{t^2}\left[1.5 - 0.75\left(\frac{x}{R_o}\right)^2\right] \quad \text{(at edge); max when } x > 0.58R_o$$

$$S = \frac{qx^2}{t^2}\left[1.950\log_e\left(\frac{R_o}{x}\right) + 0.488\left(\frac{x}{R_o}\right)^2\right] \quad \text{(at center);}$$

$$\text{max when } x < 0.58R_o$$

$$\delta = \frac{qx^2}{Et^3}\left[0.683R_o^2 - 0.683x^2\log_e\left(\frac{R_o}{x}\right) - 0.512x^2\right]$$

$$S = \frac{0.310W}{t^2}\left[2\log_e\left(\frac{R_o}{x}\right) + \left(\frac{x}{R_o}\right)^2 - 1\right] \quad \text{(at center);}$$

$$\text{max when } x < 0.31R_o$$

$$S = \frac{0.477W}{t^2}\left(1 - \frac{x^2}{R_o^2}\right) \quad \text{(at edge); max when } x > 0.31R_o$$

$$\delta = \frac{W}{Et^3}\left[0.217(R_o^2 - x^2) - 0.434x^2\log_e\left(\frac{R_o}{x}\right)\right]$$

ratio is 0.3. When x becomes very small compared to R_o, the formulas can be reduced to standard expressions involving concentrated loading.

SYMBOLS

A_1 through A_8	Factors for panels
a	Smaller side or minor axis, in. (mm)
B_1 through B_8	Plate deflection factors
b	Larger side or major axis, in. (mm)
C_1, C_2, C_3	Integration constants
D	Plate flexural rigidity, (lb-in.^{2})/in. (N-mm^2/mm)
E	Modulus of elasticity, psi (N/mm^2)
F	Corner area, in.^{2} (mm^2)
F_1 through F_8	Plate stress factors
M_1, M_i, M_o	Bending moments, lb-in./in. (N-mm/mm)

Panels and Closures

q	Uniform load, psi (N/mm^2)
R_i	Inner radius of plate, in. (mm)
R_o	Outer radius of plate, in. (mm)
S	Stress, psi (N/mm^2)
S_y	Yield strength, psi (N/mm^2)
t	Thickness of plate, in. (mm)
w	Transverse displacement, in. (mm)
W	Total load on plate, lb (N)
x	Arbitrary distance, in. (mm)
δ	Maximum deflection, in. (mm)
ν	Poisson's ratio
ϕ	Slope, rad

33
Flanges and Brackets

GENERAL BACKGROUND

An extensive list of analysis and design problems can be compiled when dealing with flanges, support rings, and structural brackets of various description and geometry. Loads on these members can come from any direction, but perhaps those acting out of plane are hardest to describe in mathematical terms. To make these matters worse, some standard textbooks attack these problems in bits and pieces, while others relegate the methods of solution to the industrial codes and manuals. The problems involving circular flanges with reinforcing gussets are never found in the open literature because such configurations include three-dimensional analysis, which, even with the help of modern computers, represents a tedious and costly procedure. Yet the design engineer has to make some basic decisions in many practical cases despite the lack of the well-established design methodology.

This undoubtedly can lead to gross overdesign and excessive cost, especially where larger-diameter pipes and greater pipe lengths are involved. In underground nuclear testing and similar applications, the trend appears to be toward the greater depths and larger overall size of the equipment, making a review of flange design theory and practice mandatory.

As evidenced by the theoretical and experimental work of the past 50 years, flange analysis can be a highly controversial and time-consuming task, even in the cases of simpler flange configuration and the manner of loading. It is helpful, therefore, not only to suggest a simplified approach to the problem at hand but also to review some of the more commonly accepted theoretical concepts and formulas.

Flanges and Brackets

Although the prime purpose of this study is a practical application, the background of some of the theoretical information concerned with generic flange problem may be particularly useful as it relates to a simplified approach to evaluating rib-stiffened configurations. The basis for such a review rests on the selected references and flange design standards developed in this country, England, and Germany.

The material presented in this chapter includes a number of practical derivations and formulas for design of a conventional or a rib-reinforced flange, where W is intended to denote the flange load referred to at times as the total bolt or tensile load. The individual bolt load should then be the flange load divided by the number of bolts required. It is further assumed that the number of bolts is equal to the number of stiffening ribs in a reinforced-flange configuration. However, the final selection of an individual bolt preload for a given flanged joint should be based on a design relationship governed by the bolt, flange, and gasket rigidities [277].

KEY STRESS CRITERIA

One of the key messages included in this section concerns the idea of elastic versus plastic stresses. Since the great majority of practicing engineers have been brought up in the tradition of the theory of elasticity and the concept of the elastic strength of materials, it is relatively easy to misinterpret the true meaning of the computed stresses. In fact, it is often presumed that the calculated values exceeding the elastic limit must necessarily be dangerous. This seems to be particularly misleading where the design formulas give the sum of the bending and membrane stresses without due allowance for material ductility and stress redistribution.

In a typical integral-type flange, that is, where the flange is butt-welded to the wall of the pipe, the adjacent portion of the wall is considered to act as a hub. The accepted design practice calls for calculation of the three major stresses: maximum axial stress in the hub, radial stress in the flange ring at its inside diameter, and the corresponding tangential stress at the same point. The theoretical and experimental evidence indicates that the axial stress in the hub is frequently by far the highest and it is often used as the basic design criterion for sizing the wall thickness. Some applications of this general rule are considered here in evaluating the maximum theoretical stress in the hub of a rib-stiffened flange.

The selection of a suitable design criterion and the corresponding calculation procedure depends in general upon the flange geometry and the materials involved. Various theories and design methods in the past utilized straight beam, cantilever, circular ring, and plate model approaches for the purpose of checking the flange stresses. The method of rib sizing proposed in this section is based on the theory of beams on elastic foundation.

EARLY DESIGN METHODS

The development of pressure vessels for higher pressures and temperatures has been largely responsible for increased interest in flange-stress formulas during the past 60 years in this country, England, and Germany [27, 182, 189–190]. Early flange designs involved hubs of approximately uniform thickness and the designs were checked by calculating the tangential stress at the inner diameter of the flange, ignoring entirely the possibility of the hub stresses. Further limitations of the early methods involved their narrow range off applicability, as they were developed for specific types and proportions of the flanges. This general situation persisted until publication of the Waters-Taylor formulas [183], which were based on the theoretical and experimental results. This classical paper marked the start of extensive deliberations of various approaches to flange design.

THIN-HUB THEORY

When the hub is relatively thin and a critical section is assumed to exist along one of the flange diameters, the maximum stress can be calculated from a simple beam formula. This approach, which is probably one of the earliest and best known, is illustrated in Fig. 33.1, where we assume that the flange is clamped along this radial cross section. The design is based on bending due to the external moment obtained by lumping together all bolt loads and utilizing the concept of a moment arm. In effect, a simple beam model is postulated where the net cross section is found by subtracting the projected areas of bolt holes. According to the notation indicated in Fig. 33.1, the available section modulus for the flange ring becomes

$$Z = \frac{(B_f - d_b)H^2}{3} \tag{33.1}$$

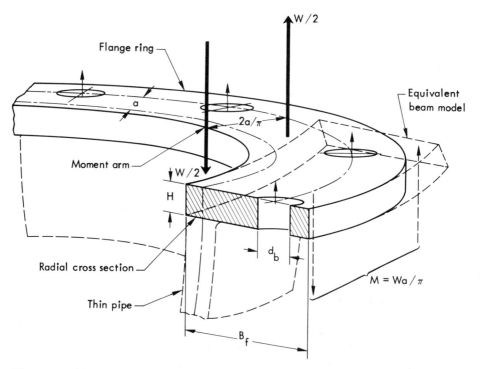

Fig. 33.1 Flange fixed along radial cross section.

Flanges and Brackets

Utilizing the moment arm shown in Fig. 33.1, we obtain the maximum bending stress from the elementary beam formula. Note that the term $2a/\pi$ follows from a consideration of the centers of gravity for the two concentric, semicircular arcs. Hence,

$$S_b = 0.95 \frac{Wa}{(B_f - d_b)H^2} \tag{33.2}$$

Obviously, Eq. (33.2) is only approximate, since the curvature of the flange ring and the effect of the pipe wall have been ignored. Nevertheless, the method is a rather ingenious use of the theory of straight beams and it gives surprisingly good results when applied to loose flanges or flanges welded to thin pipes. The effect of radial stresses in such flanges can of course be neglected.

THICK HUB AND RING MODEL

When the pipe is relatively thick and the circumferential stresses are ignored by assuming a number of radial slots, a cantilever beam method was sometimes employed. The corresponding notation and configurational details for this analytical model are given in Fig. 33.2. This method of calculation yields radial stresses only and it is expected to have a rather limited field of application. In reality, it can only be used in the case of a relatively thin flange made integral with the thick pipe of a large diameter. Under these conditions the maximum radial stress becomes

$$S_r = 0.95 \frac{Wa}{(R_i + T)H^2} \tag{33.3}$$

Where radial stresses are expected to be relatively low, a significant refinement is achieved by utilizing the theory of rings [5]. The corresponding mode of deformation and the basic notation are given in Fig. 33.3, where the cross section of the flange ring is assumed to rotate through angle θ, shown in an exaggerated manner. The cross-sectional dimensions of the flange ring are relatively small compared to the ring diameter, and it is assumed that the rectangular shape of the cross section does not change under stress. The latter assumption is consistent with the idea of neglecting radial stresses, which suggests that this theory applies to flanges attached to relatively thin pipes.

CRITERION OF FLANGE ROTATION

In establishing the equations for calculating the bending moment and the shearing force per unit length of the inner circumference of the pipe, where the flange ring and the pipe are joined, radial deflection is assumed to be zero and the angle of rotation of the edge of the pipe is made equal to the angle of rotation of the flange cross section. In Fig. 33.3, this angle is denoted by θ, and use is made of the theory of local bending and discontinuity stresses in thin shells [192], together with the theory of a circular ring subjected to toroidal deformation.

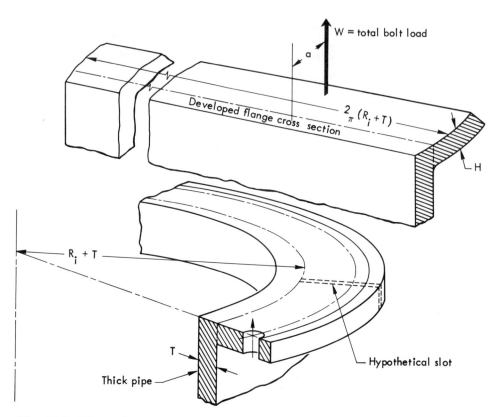

Fig. 33.2 Flange clamped circumferentially as a cantilever.

The maximum stress in the pipe, using this theory and adhering to the notation in Fig. 33.3, becomes

$$S_b = \frac{6M_0}{T^2} \tag{33.4}$$

where

$$M_0 = \frac{W_i(R_o - R_i)}{1 + \frac{\beta_s H}{2} + \frac{1-\mu^2}{2\beta_s R_i}\left(\frac{H}{T}\right)^3 \log_e\left(\frac{R_o}{R_i}\right)} \tag{33.5}$$

the corresponding shear force is

$$Q_0 = \beta_s M_0 \tag{33.6}$$

where

$$\beta_s = \frac{1.285}{(R_i T)^{1/2}} \tag{33.7}$$

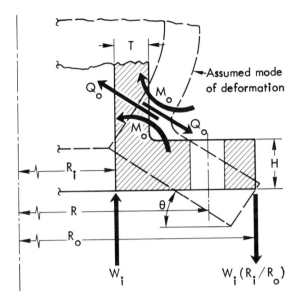

Fig. 33.3 Flange treated as a circular ring.

Recall that parameter β_s is useful in the analysis of beams on an elastic foundation and thin shells, indicating the extent of stress-affected zones in the vicinity of edge or local loading.

According to Timoshenko [5], W_i denotes the force per unit length of the inner circumference of the flange corresponding to radius R_i. The external bending moment applied to the flange involves the *moment arm*, which in Timoshenko's case is defined as $R_o - R_i$. A brief comparison with other methods indicates that the assumption of different moment arms is bound to significantly affect the calculated results. It is quite likely that Eq. (33.5) will always overestimate the bending moment M_0, because of the maximum moment arm used. Under the actual conditions, the loading may be found to be significantly removed from the inner and outer edges of the flange. Nevertheless, integral flanges with relatively thin pipes illustrated in Fig. 33.3 have been used with success and form the basis of some of the existing design standards in industry.

USE OF PLATE THEORY IN FLANGES

Further refinement of the flange analysis is achieved by applying the theory of plates where radial and circumferential stresses are taken into account. Radial stresses may be of importance in flanges integral with the thick pipes, which can resist the angle of tilt much better. This angle is shown in Fig. 33.4.

In applying the plate theory to the solution of a flange problem or to the stresses in a cylinder with rigidly attached flat heads [182–184], a strip may be cut out of the cylinder and treated as a beam on an elastic foundation while the flange is regarded as a flat plate with a central hole. The slopes and deflections at the end of the cylinder can be expressed in terms of the unknown moments and

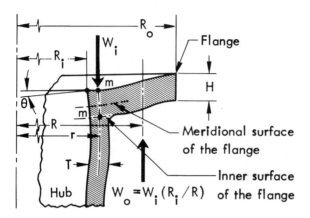

Fig. 33.4 Flange treated as a circular plate.

shear forces. These displacements are then equated to the slopes and deflections similarly determined for the plate. In this manner we establish a sufficient number of equations which are solved simultaneously to yield the unknown reactions and displacements.

The classical approach to the flange problem can be based on the premise that the bending moment existing at the root of the hub acts at the meridian plane of the flange instead of at its inner plane. When the flange portion is deformed, these planes become curved, as shown in Fig. 33.4. Furthermore, it can be assumed that the cylindrical surface containing points m–m does not alter its original curvature and that the expansion of the hub due to the internal pressure can be ignored.

FORMULA FOR HUB STRESS

One of the important phases of flange analysis concerns the maximum hub stress due to the total bolt load W. This section presents a simple formula used at one time in stress analysis of compressor casing flanges in jet engines.

The basic notation for this case is defined in Fig. 33.5. Denoting the maximum bending stress in the pipe wall by S_b as before, we obtain

$$S_b = \frac{0.48 W \beta_s T (R_o + R_i)}{\phi_0 \beta_s R_i T^3 (2 + \beta_s H) + H^3} \tag{33.8}$$

where β_s is determined from Eq. (33.7) and ϕ_0 is given in Fig. 33.6 as a function of flange ring ratio R_o/R_i. For a typical Poisson's ratio of 0.3, the flange factor ϕ_0 can be calculated as

$$\phi_0 = \frac{0.77 + 1.43 k^2}{k^2 - 1} \tag{33.9}$$

where $k = R_o/R_i$. By substituting Eqs. (33.7) and (33.9) into Eq. (33.8) and introducing convenient nondimensional numbers $m = R_i/T$ and $n = H/T$, we

Flanges and Brackets

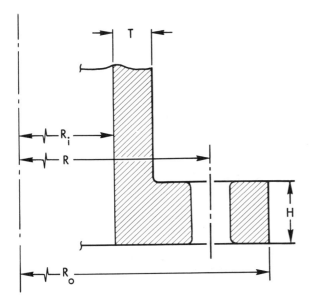

Fig. 33.5 Proposed simplified notation for analysis of straight flanges.

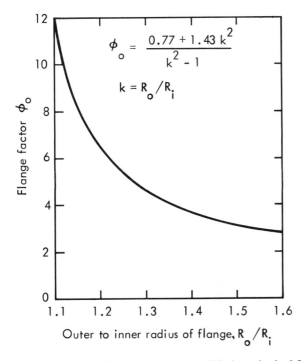

Fig. 33.6 Auxiliary chart for simplified method of flange design.

obtain

$$\frac{S_b T^2}{W} = \frac{0.614(k+1)(k^2-1)(m)^{1/2}}{(0.77+1.43k^2)[2.57(m)^{1/2}+1.65n]+n^3(k^2-1)} \quad (33.10)$$

The flange formulas, such as Eq. (33.10), give "apparent" rather than "actual" stresses and can therefore be considered conservative, particularly in designs involving ductile materials.

GERMAN PRACTICE OF FLANGE DESIGN

According to machine design practice in Germany [27], the calculation of maximum bending stresses in the pipe wall adjacent to the flange ring can be accomplished in a straightforward manner. This procedure recognizes the two basic modes of failure depending on the relative thickness of the flange ring and the pipe, and it can be applied to tapered as well as straight hubs. When the flange ring is thicker than the pipe, failure is expected in the hub. The reverse is true for heavy pipe walls. In both instances, the flange ring surface under strain is assumed to conform to a spherical shape.

Notation and overall proportions of the flanged section given by the German flange standards are shown in Fig. 33.7. The pipe section is assumed to fail at an angle α generally considered to be between 20 and 30°. The depth of the fractured surface can be related to the nominal thickness of the pipe

$$(33.11a)$$

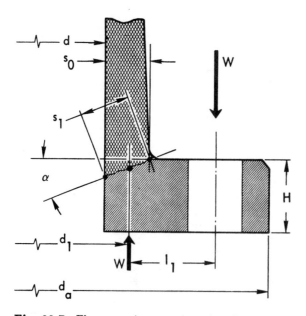

Fig. 33.7 Flange section notation after German standards.

Flanges and Brackets

In terms of the notation given in Fig. 33.7, we postulate the following approximate relationships

$$d_1 = 2R_i + T \tag{33.11b}$$

$$s_1 = \frac{T}{\cos \alpha} \tag{33.11c}$$

$$l_1 = \frac{R_o - R_i}{2} \tag{33.11d}$$

Utilizing then Eqs. (33.11a) to (33.11d) and taking the average value of $\alpha = 25°$, the maximum bending stress in the pipe adjacent to the flange is

$$S_b = \frac{W(k-1)[0.228 + 0.035n \sin^2(\pi/n)]}{nT^2} \tag{33.12}$$

The derivation of this equation involves a number of simplified steps and symbols consistent with other formulas given in this chapter [267].

WATERS-TAYLOR FORMULA

If flange design according to American practice is presented in similar terms [183], the so-called Waters-Taylor formula can be stated as

$$S_b = \frac{W}{T^2} \left\{ \frac{0.25(m)^{1/2}[k^2(1 + 8.55 \log k) - 1]}{(1.05 + 1.94k^2)[(m)^{1/2} + 0.64n] + 0.53n^3(k^2 - 1)} \right\} \tag{33.13}$$

where the logarithmic term is calculated to base 10. The formula consistent with the German code, Eq. (33.12), should be applicable to all cases for which $n \neq 0$ and $k < 1.8$. Parametric studies also indicate that Eqs. (33.10) and (33.12) yield lower numerical values of hub stresses than those which can be predicted on the basis of Eq. (33.13).

CIRCUMFERENTIAL STRESS

It may be of interest to note a simplified method of predicting the circumferential stresses in a standard flange ring in terms of the maximum bending hub stress [190]. In this type of analysis we assume that the radial expansion of the flange ring due to the discontinuity shear force and the bulging of the pipe due to any internal pressure can be ignored. Because of the flange rotation, one can expect to find the maximum tensile stress on the outside of the pipe wall, and pipe bulging due to internal pressure. Hence, the estimate should be conservative, and assuming that the pipe in the vicinity of the flange ring can be stressed to the yield point of the material, we obtain

$$\frac{S_F}{S_b} = \frac{n(2m+1)}{3.64mn + 4m(2m+1)^{1/2}} \tag{33.14}$$

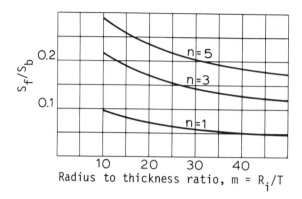

Fig. 33.8 Ratio of flange to hub stress.

where S_F denotes the flange-ring circumferential stress produced by the dishing under the axial bolt load and S_b is made equal to the yield stress. For $2m \gg 1$, Eq. (33.14) is plotted in Fig. 33.8. The theoretical limits for the stress ratio are 0.55 and 0 for zero and infinite m values, respectively. The intermediate range indicates clearly that the maximum circumferential stresses in the flange ring are always considerably smaller than the hub stresses. Conversely, the theory indicates that if yielding of the pipe in the vicinity of the flange is to be avoided, the flange ring would have to be extremely thick and therefore unacceptable for all practical purposes. The major conclusion drawn from this finding is that, for truly economic design in ductile materials, plastic deformation of the pipe in the vicinity of the flange ring can be permitted. The reserve of strength beyond the onset of yield can be quite significant. For instance, the theoretical collapse load of a beam of rectangular cross section is 1.5 times the load causing yield in the outer fibers, assuming a rigid-plastic stress-strain characteristic for the material.

APPARENT STRESS CRITERIA

Because there has been much discussion over the past 60 years with reference to the allowable stress levels, the preceding section was devoted largely to the problem of selecting the most rational method of analyzing hub stresses in the light of modern flange practice. In general, hoop stress in the flange ring or radial stress across the junction of the hub and the ring may also be significant, but in the majority of cases involving maximum axial loading on the flange, the magnitude of the longitudinal bending stress in the hub is the highest. In a standard flange application, the distributed loading consists of bolt load, gasket pressure, hydrostatic pressure of the flange leakage area, and the hydrostatic end force. All the loading is represented by an equivalent bending couple consisting of two equal and opposite loads. When the internal pressure is absent and a flange gasket is relatively close to the bolt circle, these loads can be appreciably removed from the inner and outer edges of the flange. Equation (33.10) corresponds to the loads which are placed sufficiently far from the flange edges.

Flanges and Brackets

The theoretical hub stresses calculated from Eq. (33.10), (33.12), or (33.13) are based on the elastic behavior and should be considered as apparent rather than actual stresses. Generally, this situation persists in many areas of engineering analysis and should be reviewed continually with reference to practical design requirements. The numerical values of apparent stresses, interpreted as strains multiplied by the relevant moduli of elasticity, bear little relation to the actual material stresses and can be evaluated only with special regard to the stress-strain curve. The concept of apparent stresses is a very real one and is fully supported by practical experience. It is generally recognized that in many actual flange designs the calculated hub stresses are extremely high and yet the flanges are satisfactory. In such circumstances, of course, the classical elastic formulas show limited validity and flange design by test is recommended [193]. Actually, such formulas are basically correct, but the relevant interpretation of the numerical results is misleading if the yield strength and the reserve of plastic strength in a flange, or other machine part, are ignored. This leads directly to the analysis in terms of the maximum allowable stresses—an engineering concept that is so often abused and misunderstood.

That calculated apparent stress exceeds the proportional limit of a material does not necessarily mean that failure is imminent. Seemingly, engineers and designers have been inclined to think in terms of the vertical axis of a stress-strain diagram and have paid less attention to the problem of strain. It may be fitting then in many areas of design analysis to think in terms of the maximum allowable strain rather than in terms of stress.

Along the same lines of reasoning, one should consider the problem of safety factors. A more conventional approach to flange design usually consists of three phases: (1) determination of gasket size and material (when applicable), (2) calculation of bolting required, and (3) design of flange proportions. Sizing is often based on the apparent stresses, so that heavy bolting and unnecessary thick flanges result. Specifically, regarding the level of hub stresses, it is advisable to recognize the existence of local yielding as long as there is a sufficient reserve of strength in the adjacent sections of the flange to take care of the increased loading placed upon these sections by the region of local yielding.

PLASTIC CORRECTION

To illustrate the problem of apparent stresses and to suggest an approximate plastic correction for interpreting calculated stresses higher than the yield strength of the material, consider Fig. 33.9. The stress-strain curve shown is typical of a low-strength ductile material, although the strain at yield, ϵ_y, has been exaggerated for the purpose of clarity. Hence, the slope of the curve below the proportional limit in Fig. 33.9 should not be used for direct numerical evaluation of the modulus of elasticity. The values of stresses and strains in Fig. 33.9 are merely to indicate that the material characteristics discussed are close to those found in low-carbon steel, aluminum alloy, or other materials that may behave similarly.

Let us first assume that the stress-strain diagram has been established experimentally and that the curve obtained can be approximated by the two straight lines from zero to the yield point and from the yield region to the highest point on the curve, as shown by the line x–x. Let the corresponding elastic moduli be E and

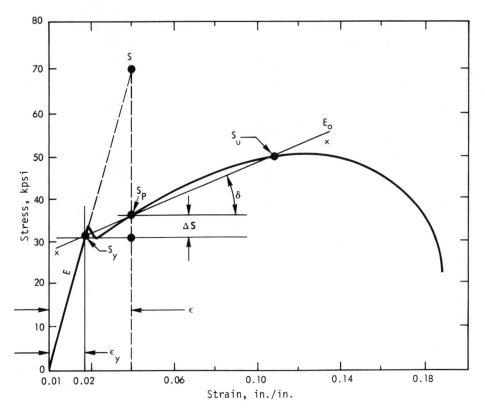

Fig. 33.9 Typical stress-strain curve for a low-strength ductile material.

E_0, respectively. In the study of plasticity of materials, a bilinear approximation of this kind is well known and indicates that the material in question conforms to the law of linear strain hardening [194].

In terms of the calculated stress S, shown as 70,000 psi in Fig. 33.9, Hooke's law gives

$$\epsilon = \frac{S}{E} \tag{33.15a}$$

The corresponding strain at yield stress S_y is

$$\epsilon_y = \frac{S_y}{E} \tag{33.15b}$$

By the definition of the elastic modulus

$$\tan \delta = E_0 = \frac{\Delta S}{\epsilon - \epsilon_y} \tag{33.15c}$$

Flanges and Brackets

Denoting the actual stress by S_p, we get

$$\Delta S = S_p - S_y \tag{33.15d}$$

Hence, substituting Eq. (33.15d) into (33.15c), the elastic modulus for the x–x portion of the curve becomes

$$E_0 = \frac{S_p - S_y}{\epsilon - \epsilon_y} \tag{33.15e}$$

Solving Eq. (33.15e) for S_p and eliminating strains with the aid of Eqs. (33.15a) and (33.15b) gives

$$S_p = S\frac{E_0}{E} + S_y\left(1 - \frac{E_0}{E}\right) \tag{33.16}$$

It is evident that when E_0 tends to zero, a stress-strain curve representing ideal plastic material is obtained, and $S_p = S_y$ for all values of strain. On the other hand, when $E_0 = E$, there is only elastic behavior, and the stress-strain curve represents purely elastic action up to the point of failure. The concept of ideal plastic material is often employed in the theory of plasticity because a typical mild steel stress-strain diagram is close to that of a perfect plastic material. With various alloying elements, the material still exhibits desirable ductility, but, in addition, there may be some strain hardening.

Considering the strain-hardening characteristic to be that illustrated in Fig. 33.9, the following appraisal of the apparent versus actual stress can be made. Assume that the calculated hub stress in a flange is 70,000 psi, as shown in Fig. 33.9. If material knowledge gives the yield strength $S_y = 31,000$ psi, and the ultimate strength $S_u = 50,000$ psi, but the shape of the stress-strain curve is unknown, then on the basis of our calculation we may conclude that the pipe section will fail. Specifically, the safety factors 0.44 and 0.71 are based on the given values of yield and ultimate strength, respectively. To increase these factors it would appear that either a better material is needed or that the nominal wall thickness of the pipe should be increased. Either approach would not be in the interest of economy and could lead to gross overdesign.

Assume now some knowledge of the actual stress-strain curve, and take $E_0/E = 0.25$—not an unusual ratio for a material with linear strain hardening—so that using Eq. (33.16) yields

$$S_p = 0.25 \times 70{,}000 + 31{,}000(1 - 0.25) = 40{,}700 \text{ psi}$$

The corresponding safety factors are then changed to 0.76 and 1.23, respectively. If the flange in question is not subject to fatigue, as is normally the case with many flanged configurations, the pipe failure becomes less likely.

If the material assumed in the above numerical illustration can be considered perfectly plastic, then $E_0/E = 0$ and the maximum factor of safety becomes unity, because the actual hub stress cannot exceed the value of yield. The results of this discussion point clearly to the need for a realistic approach to the interpretation of

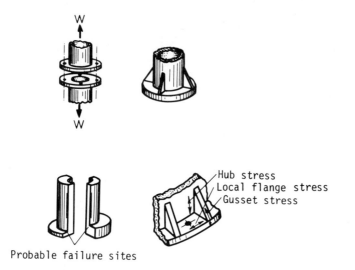

Probable failure sites

Fig. 33.10 Typical flanges and reinforcements.

calculated elastic stresses in conjunction with the stress-strain characteristics of the materials involved. Hence if the maximum bending stress in the hub is restricted to the pipe surface any higher stress than the yield strength of the material at that point should cause stress redistribution.

Several examples of flanges used to connect piping and tubular structural members [195, 196] are shown in Fig. 33.10 with the features of progressive complexity and the regions of more significant stress components. The correction for plastic action described by Eq. (33.16) can be applied to any of the stress components provided their elastic values can be estimated. The most difficult problem, however, is the first estimate of the elastic response. This situation has not changed markedly despite the significant progress in numerical techniques.

HEAVY-DUTY FLANGES

In the development of heavy-duty pipe flanges and similar hardware components, it has been customary to utilize the concept of a compound flanging where the two concentric flanged rings on a pipe are joined by external ribs parallel to the pipe axis. Although this is not an unusual concept where the overall flange rigidity against the toroidal deformation is required, a rigorous analytical approach to the design of such a three-dimensional structure must of necessity be highly involved. A typical approach to the theoretical problem would be through finite-element techniques and rather lengthy experimental verification of the stress picture using strain gages, photostress, or three-dimensional photoelasticity. If, for economic or scheduling reasons, a more fundamental approach to this problem is not feasible, a relatively simple approximate solution can be developed utilizing the existing knowledge of structural mechanics. Such solutions appear to be on the conservative side and are not far removed from reality as shown by theoretical and experimental investigations conducted by Werne [268].

Flanges and Brackets

To reach a rational compromise as to the selection of flange thickness in the case of double-ring ribbed design, consider Fig. 33.11. The action of a double-flanged ring may be simulated by a twist of a circular ring loaded by couples M_0 which are uniformly distributed along the center line of the ring. To simplify the calculation it is assumed that the cross-sectional dimensions of the ring are relatively small when compared with the mean radius R. Timoshenko [5] shows that the angle of rotation of flange cross section θ and the normal stress are inversely proportional to the moment of inertia about the ring axis x–x.

EQUIVALENT DEPTH FORMULA

The assumption of the cumulative value H, sometimes used in the calculations, cannot be accurate because the material, indicated as the dotted areas in Fig. 33.11, is really nonexistent except in the regions of the ribs. However, to deal with this characteristic dimension, the equivalent depth H_e may have to be calculated with reference to Fig. 33.12 to provide a solid rectangular cross section that would yield the same angle of rotation as the original channel section of depth H. The possibility of some local distortion of the channel section in between the ribs exists; however, in the simplified approach, this feature will be excluded.

Although the composite channel section indicated in Fig. 33.12 is not symmetrical about the axis x–x, the moment of inertia can be found directly without calculating the position of the center of gravity of the section [11]. For this purpose, a convenient baseline, m–m, is selected as shown in Fig. 33.12, so that the following equation can be written

$$I_x = I_b + I_g - \frac{J^2}{A_t} \tag{33.17}$$

Here J and I_b denote the first and second moments of area with respect to the baseline m–m, I_g stands for the sum of the moments of inertia of all component areas about their own centers of gravity, and A_t defines the total cross-sectional area. The problem is handled in the usual way by breaking the whole area into geometrically convenient component areas and applying Eq. (33.17). Invoking this rule and adhering to the notation shown in Fig. 33.12 gives

$$I_x = \frac{H^4T^2 + 36B_r HTT_0^3 - 30B_r H^2 TT_0^2 + 12B_r H^3 TT_0 + 81B_r^2 T_0^4}{12(HT + 3B_r T_0)} \\ + \frac{2B_r^2 H^2 T_0^2 - 6B_r^2 HT_0^3}{HT + 3B_r T_0} \tag{33.18}$$

Hence, the equivalent depth of the ribbed flange consisting of the flange and backup rings may be stated as

$$H_e = 2.29 \left(\frac{I_x}{B_r + T}\right)^{1/3} \tag{33.19}$$

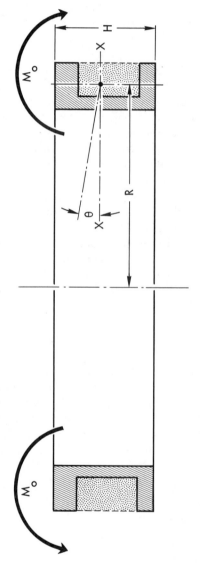

Fig. 33.11 Model of a twisted flange ring.

Flanges and Brackets

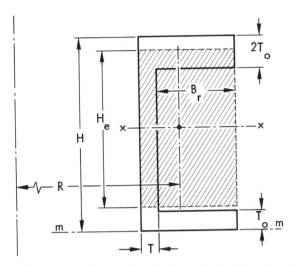

Fig. 33.12 Section details for equivalent depth calculation.

A view of the compound flange with the main flange ring, backup ring, rectangular ribs, and the triangular gussets is given in Fig. 33.13. Appropriate size and location of the flange reinforcement should be determined on the basis of hub, gusset, and flange stresses. The equivalent depth of the ribbed flange H_e can be used in conjunction with such formulas as those given by Eqs. (33.10), (33.12), or (33.13). When the backup ring is not used and $2T_0 = h$, the formula for calculating the

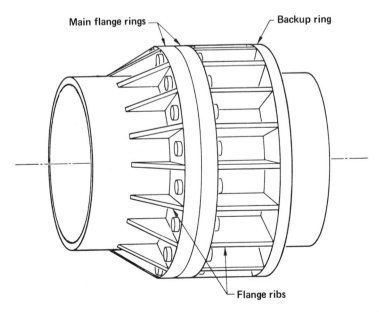

Fig. 33.13 Pipe flange geometry with rectangular and triangular ribs.

equivalent depth of the ribbed flange becomes

$$H_e = \left[\frac{HT^3(HT + 4hB_r) - 3HTh^2B_r(2H - h)}{(B_r + T)(HT + hB_r)} \right]^{1/3} \qquad (33.20)$$

LOAD SHARING IN RIBBED FLANGES

A partial view of a rib-stiffened main flange ring is illustrated in Fig. 33.14. In evaluating the moment-carrying capacity of a stiffening rib in relation to that of the main flange ring, the theory of beams on elastic foundation and toroidal deformation

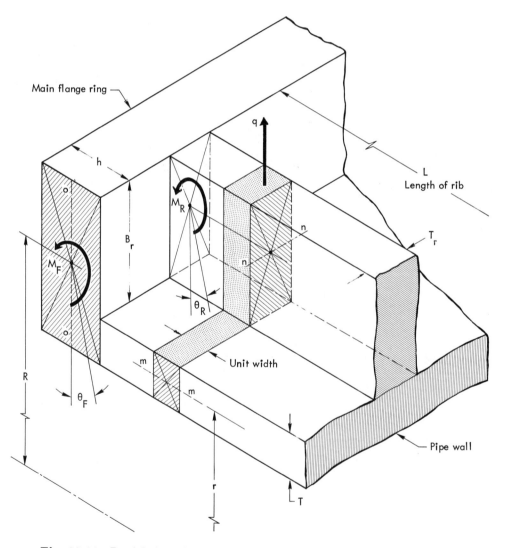

Fig. 33.14 Partial view of a rib-stiffened flange ring.

Flanges and Brackets

of a circular ring may be employed. If the moment arm is taken as $R - r$, the total external moment due to the bolt load, referred to unit length along the bolt circle, may be defined as

$$M = \frac{W(R-r)}{2\pi R} \tag{33.21a}$$

If N denotes the total number of ribs supporting the flange ring, the external bending moment per length of the circumference corresponding to one rib spacing is

$$M_c = \frac{2\pi R M}{N} \tag{33.21b}$$

Substituting Eq. (33.21a) in Eq. (33.21b) gives

$$M_c = \frac{W(R-r)}{N} \tag{33.21c}$$

Denoting the bending moments M_F and M_R carried by the main flange ring and rib, respectively, gives

$$M_c = \frac{2\pi R M_F}{N} + M_R \tag{33.21d}$$

Hence Eqs. (33.21c) and (33.21d) yield

$$W(R-r) = 2\pi R M_F + N M_R \tag{33.21e}$$

Although the foregoing algebraic operations are rather elementary, it is advisable to follow the basic derivations to assure correct dimensional identities before developing subsequent working formulas. We note, therefore, that M_F is expressed in lb-in./in. and M_R is given in lb-in.

The angle of twist of an elastic ring undergoing a toroidal deformation under the action of a twisting moment M_F may be expressed by the classical formula as

$$\theta_F = \frac{M_F R^2}{E I_0} \tag{33.21f}$$

The bending slope of the stiffening rib at the rib-flange junction, treated as a beam of finite length resting on the elastic foundation and acted upon by a concentrated end moment M_R, can be calculated from the following relations. Let Y denote the deflection of a beam on an elastic foundation with hinged ends that is bent by a couple M_R applied at the end as shown in Fig. 33.15. Then the general expression for the deflection line is

$$Y = \frac{2M_R \beta^2 [\cosh \beta L \sin \beta x \sinh \beta(L-x) - \cos \beta L \sinh \beta x \sin \beta(L-x)]}{K(\cosh^2 \beta L - \cos^2 \beta L)} \tag{33.21g}$$

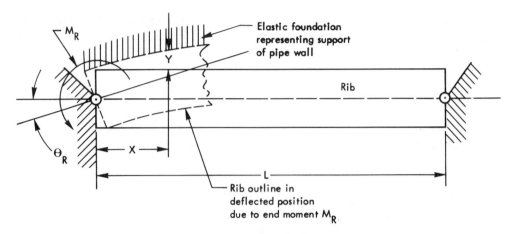

Fig. 33.15 Mode of rib deformation as a beam on an elastic foundation.

The required slope is found by calculating dy/dx from Eq. (33.21g) and making $x = 0$. This gives

$$\theta_R = \frac{2M_R \beta^3 (\cosh \beta L \sinh \beta L - \cos \beta L \sin \beta L)}{K(\cosh^2 \beta L - \cos^2 \beta L)} \qquad (33.21\text{h})$$

where

$$\beta = \left(\frac{K}{4EI_n}\right)^{1/4} \qquad (33.21\text{i})$$

Eliminating K with the aid of Eq. (33.21i) and introducing an auxiliary dimensional function characterizing the above beam on elastic foundation yields

$$\theta_R = \frac{M_R \phi(\beta, L)}{EI_n} \qquad (33.21\text{j})$$

Here

$$\phi(\beta, L) = \frac{\cosh \beta L \sinh \beta L - \cos \beta L \sin \beta L}{2\beta(\cosh^2 \beta L - \cos^2 \beta L)} \qquad (33.22)$$

To compute this value it is first necessary to calculate the parameter β from Eq. (33.21i), which contains term K, defined as the modulus of the elastic foundation. As the ribs are supported by the pipe wall, we can consider a slice of the pipe wall together with the rib as shown in Fig. 33.14. A partial view of the slice is shown in the figure as a dotted area for the sake of clarity. The complete circumferential slice can then be represented by a mathematical model of a radially loaded circular ring as illustrated in Fig. 33.16. Radial load intensity q is shown in Figs. 33.14 and 33.16. For a slice of unit width load intensity, q is numerically equal to the load per linear inch of the rib acting upon the ring at N equidistant

Flanges and Brackets

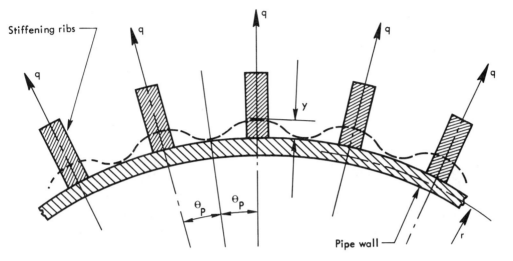

Fig. 33.16 Circular ring model for modulus of foundation.

points. The modulus of the foundation then can be interpreted directly as a spring constant K, since the load-deflection relation for a particular ring loading [35] may be given as

$$y = \frac{qr^3(\theta_p^2 + \theta_p \sin\theta_p \cos\theta_p - 2\sin^2\theta_p)}{4EI_m \theta_p \sin^2\theta_p} \qquad (33.21\text{k})$$

Since by the definition $K = q/y$ and $\theta_p = \pi/N$, substituting for the modulus of foundation in Eq. (33.21k) yields

$$\beta = \left\{ \frac{2\pi N I_m \sin^2(\pi/N)}{r^3 I_n [2\pi^2 + \pi N \sin(2\pi/N) - 4N^2 \sin^2(\pi/N)]} \right\}^{1/4} \qquad (33.21\ell)$$

A parametric investigation of Eq. (33.21ℓ) within the range $8 < N < 100$ leads to the simplification

$$\beta = 0.69 \left(\frac{N}{r}\right) \left(r\frac{I_m}{I_n}\right)^{1/4} \qquad (33.23)$$

Assuming next that the angle of twist of the main flange ring is equal to the bending slope of the rib at $x = 0$, Eqs. (33.21f) and (33.21j) give

$$\omega = \frac{M_R}{M_F} = \frac{I_n R^2}{I_0 \phi(\beta, L)} \qquad (33.21\text{m})$$

Hence, solving Eqs. (33.21e) and (33.21m) simultaneously yields

$$M_R = \frac{W(R-r)\omega}{2\pi R + \omega N} \qquad (33.21\text{n})$$

and

$$M_F = \frac{W(R-r)}{2\pi R + \omega N} \tag{33.21o}$$

The main flange ring can now be viewed as consisting of N equal sectors in which the load distribution between the ribs and the corresponding sectors of the flange can be calculated from a load ratio that follows directly from Eq. (33.21e) and Fig. 33.14

$$f = \frac{0.16 N R T_r B_r^3}{(B_r + T)h^3 \phi(\beta, L)} \tag{33.24}$$

Since the auxiliary function given by Eq. (33.22) is a dimensional quantity expressed in inches, the load-sharing equation above is nondimensional, in contrast with the ratio ω defined by Eq. (33.21m).

STRENGTH OF FLANGE RIBS

The analysis of load sharing between the stiffening ribs and the main flange ring must necessarily be considered as conservative because of the assumed pin-jointed supports (Fig. 33.15) and because the shearing stresses between the pipe wall and the rib have been neglected. In fact, the analysis appears to indicate that the weld between the rib and the wall might possibly be omitted in the double flange ring design for several manufacturing reasons. Some of these reasons include cost reduction, metallurgical control of welding procedure, reliability of weld inspection, and residual stress effects due to welding.

While sizing a new compound flange, Eq. (33.24) may be useful in establishing the first criterion for sharing of the bending moment between the rib and the corresponding sector of the main flange. Combining Eqs. (33.24) and (33.21e) yields

$$M_R = \frac{fW(R-r)}{N(1+f)} \tag{33.25}$$

Hence, the corresponding bending stress in the rib becomes

$$S_{bR} = \frac{6fW(R-r)}{NT_r(1+f)(B_r+T)^2} \tag{33.26}$$

Equation (33.26) is applicable to the stiffening ribs sharing the external load with the main flange ring on the premise that the weld between the rib and the pipe wall is sufficiently strong. Here again, our criterion should be conservative because the additional cross-sectional area $T_r \times T$ included in Eq. (33.26) does not truly represent the effect of the pipe wall at the welded junction. However, for the purpose of the preliminary design, Eq. (33.26) is satisfactory, provided the maximum computed stress is elastic. Further interpretation of the maximum stress value may be made utilizing the concept of plastic correction.

Flanges and Brackets

Although the load-sharing capacity for a typical stiffening rib has been established only on the basis of flange ring rotation and beam bending due to a couple applied at its end, the effect of direct tension on the maximum stress can be included. The maximum bending stress and the tensile stress in the rib may be added directly. Hence, the total stress is as follows: In direct tension, load sharing between the rib and the pipe wall will be established in direct proportion to the working areas. Denoting by W_R the portion of the tensile load carried by the rib as illustrated in Fig. 33.17 gives

$$W_R = \frac{W B_r T_r}{N B_r T_r + 2\pi r T} \qquad (33.27)$$

Since the simple tensile stress is $W_R/B_r T_r$, combining Eqs. (33.26) and (33.27) yields

$$S_{TR} = \frac{6fW(R-r)}{NT_r(1+f)(B_r+T)^2} + \frac{W}{NB_r T_r + 2\pi r T} \qquad (33.28)$$

The estimate of load sharing capability based on Eq. (33.24) indicates that in the majority of design situations involving rib-stiffened flanges of the usual proportions, only a limited toroidal moment will be expected to be carried by the main flange ring, and in many such cases the corresponding stresses can be ignored. However, in those instances where the calculated number f from Eq. (33.24) is found to be rather small (of the order of 5 or less), the maximum flange stress due to the

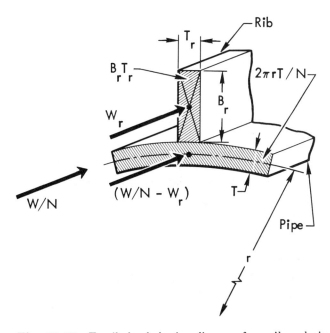

Fig. 33.17 Tensile load-sharing diagram for a rib and pipe sector.

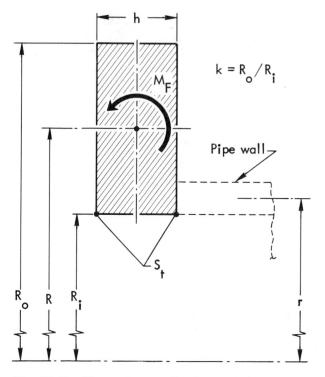

Fig. 33.18 Flange ring notation for stress check under toroidal moment.

twist may be calculated from the simplified expression

$$S_t = \frac{0.96W(R-r)R}{h^2 r^2 (1+f) \log_e k} \tag{33.29}$$

The flange ring notation applicable to Eq. (33.29) is shown in Fig. 33.18.

LOCAL BENDING OF FLANGE RING

When the rib system is relatively rigid and the resulting toroidal deformation of the main flange ring is limited, it is still recommended that the order of the local stresses likely to exist in the flange ring under the individual bolt loads be checked. Because of the symmetry, only one portion of the flange, held by the two consecutive ribs and the corresponding portion of the pipe wall, needs to be examined. Basically, this is a three-dimensional, complex problem, and only an experimental program can provide a reliable answer. As such test data appear to be very scarce, and since a formal three-dimensional solution to this problem would be very lengthy indeed, a compromise is suggested here based on the elastic theory of plates. This method may be used in preliminary design and data reduction in support of experimental work.

Utilizing as far as practicable the notation already employed, Fig. 33.19 illustrates the proposed mathematical model. Some theoretical solutions are avail-

Flanges and Brackets

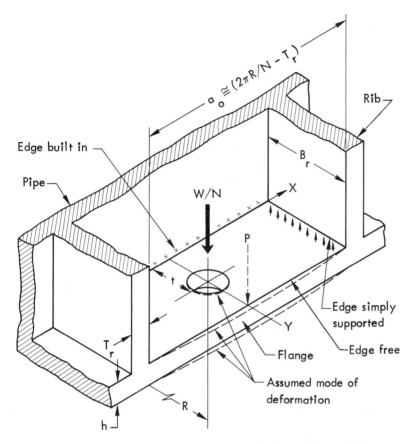

Fig. 33.19 Approximate plate model for main flange ring analysis.

able [178] which involve a rectangular plate having one long edge free, and both short edges simply supported. The boundary conditions along the shorter edges are likely to be closer to those of a built-in character, and therefore the stresses calculated according to the model illustrated in Fig. 33.19 should be considered as approximate at best.

This analysis is based on the deflection of the free edge of a relatively long plate under the action of a concentrated force P. From Fig. 33.19 the edge force may be taken as proportional to Wt/NB_r. Hence, the design equation becomes

$$Y_p = 1.83 \frac{WtB_r}{ENh^3} \tag{33.30}$$

The edge deflection for a uniformly loaded plate for the same boundary conditions is

$$Y_q = 1.37 \frac{WB_r^3}{a_0 ENh^3} \tag{33.31}$$

The maximum bending stress in the flange ring (Fig. 33.19) should be at the middle of the built-in edge (i.e., where $x = y = 0$), as shown in the sketch. If the bolt load W/N is first assumed to be uniformly distributed, the maximum bending moment at the midpoint of the built-in edge in a classical solution varies as a function of B_r/a_0, as shown in Fig. 33.20. Here M_y denotes the moment acting about the built-in edge and it is expressed in lb-in. per inch of circumference. Hence, the parameter $V = a_0 N M_y / W B_r$ must be nondimensional. The dimensional quantity a_0 is defined in Fig. 33.19.

The bending stress due to the equivalent uniform load W/Na_0B_r may be given as

$$S_b = \frac{6VWB_r}{Na_0h^2} \tag{33.32}$$

In order to make a conservative correction for the effect of a concentrated loading W/N, the bending stress can be assumed to be roughly proportional to Y_p/Y_q as determined by Eqs. (33.30) and (33.31). The criterion for the flange stress, located at the midpoint between the two adjoining ribs and very close to the outer surface of the pipe, can now be defined as

$$S_b = \frac{8VWt}{NB_rh^2} \tag{33.33}$$

When the outer radius R_o of the flange is relatively large compared with the mean radius of the pipe, the approximate stress in the flange ring becomes

$$S_b = \frac{0.95Wt}{NB_rh^2} \tag{33.34}$$

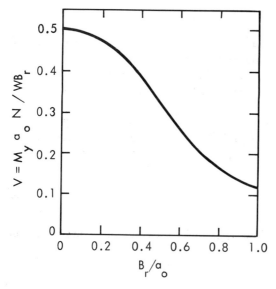

Fig. 33.20 Maximum moment for a rectangular plate simulating ribbed flange.

Flanges and Brackets

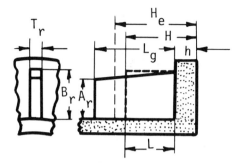

Fig. 33.21 Tapered gusset.

The above stresses are shown to decrease with an increase in the number of gussets.

CORRECTION FOR TAPERED GUSSETS

When gussets have a linear change in depth from B_r to A_r as shown in Fig. 33.21, the equations derived so far for the rectangular gussets can also be applied to the tapered geometry by determining an equivalent length L for a gusset of total length L_g. This equivalent length L can be found from Fig. 33.22 for a given value of the ratio A_r/B_r. Once L is known, the overall depth H can be taken as $L + h$. Then H_e and the stresses can be estimated as for rectangular gussets. The calculation can now proceed with Eqs. (33.19) and (33.20) using the parameter $n = H_e/T$.

The development of the equivalent gusset length, Fig. 33.22, was based on the resolution of the external forces acting on the gusset. The assumption was also made

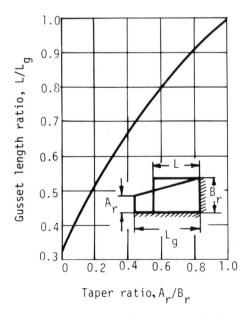

Fig. 33.22 Design chart for length of tapered gussets.

that the average shearing stresses, applied along the gusset and perpendicular to it, were numerically equal. The derivation involved the summation of the external moments for the tapered and "equivalent" rectangular ribs.

The foregoing analysis of a compound circular flange suggests bending rather than shear as the basic mechanism of potential structural failure. The approach then seems to be particularly sensitive with regard to the application of the principles of fracture-safe design.

Design Problem 33.1

Estimate load sharing between the ribs and the main flange ring assuming the following dimensions for a compound flange design.

Length of stiffening rib	$L = 4$ in.
Number of ribs	$N = 16$
Thickness of pipe wall	$T = 0.375$ in.
Mean radius of pipe	$r = 11.8$ in.
Depth of rib	$B_r = 2$ in.
Thickness of rib	$T_r = 0.25$ in.
Mean radius of flange ring	$R = 13.5$ in.
Thickness of flange ring	$h = 1.75$ in.

Solution

$$I_m = \frac{T^3}{12} = \frac{0.375^3}{12} = 0.0044 \text{ in.}^4/\text{in.}$$

$$I_n = \frac{T_r B_r^3}{12} = 0.25 \times \frac{2^3}{12} = 0.167 \text{ in.}^4$$

From Eq. (33.23),

$$\beta = 0.69 \left(\frac{16}{11.8}\right) \left(11.8 \times \frac{0.0044}{0.167}\right)^{1/4} = 0.698$$

$$\beta L = 0.698 \times 4 = 2.79$$

For a relatively high value of βL, Eq. (33.22) simplifies considerably to give

$$\phi(\beta, L) \cong \frac{1}{2\beta} = \frac{1}{2 \times 0.698} = 0.716$$

Hence, from Eq. (33.24) the load-sharing factor becomes

$$f = \frac{0.16 \times 16 \times 13.5 \times 0.25 \times 8}{(2 + 0.375) \times 1.75^3 \times 0.716} = 7.6$$

In other words, the load carried by the rib is $(7.6/8.6) \times$ (total load), or in our case, $M_R = 0.88 M_c$, where M_c denotes the external moment corresponding to one sector of the flange ring. In terms of the derived equations,

$$M_R = M_c \left(\frac{f}{1+f}\right) \tag{33.35}$$

Here M_c follows from Eq. (33.21c). Additional parametric studies also indicate that in most practical cases $f \gg 1$, so that with only a small error, the ribs can be designed on the basis of the total external moment. For the same reason, the flange ring would be subject to only a limited toroidal moment. ◆

ELEMENTS OF BRACKET DESIGN

Structural configurations such as pipe supports, motor mountings, machinery brackets, and landing seats of various types are frequently encountered in hardware design. They involve rolled shapes, plate components, and prefabricated structural elements to suit the requirements of strength, rigidity, appearance, and low cost of manufacture.

Despite their apparent simplicity, certain design criteria must be imposed which have their basis in the theory of strength of materials and elastic stability. The choice of the critical dimensions may also be governed by the elastoplastic response and local buckling resistance, requiring the help of experimental stress analysis. All these configurations may be of importance in sizing support systems and estimating the degree of structural safety and quality of performance. Yet it is surprising how little information is readily available on this subject in the open literature. One of the reasons for the lack of standard data in this area may be the inherent diversity of configurations, loading conditions, and safety requirements for individual support systems.

This account is primarily concerned with the basic elements of bracket design which can be applicable to a variety of mechanical systems. In many cases, such a generalization is admissible because different structural brackets can have similar geometrical features. They often involve some sort of bearing plates to distribute the load and edge-loaded plate elements acting as stiffeners and gussets. Such components can be designated for weld fabrication from standard structural shapes of simple geometry, making the application of the general design rules more acceptable.

Several representative types of brackets are shown in Figs. 33.23–33.29 [197]. These by no means exhaust all the possibilities, but they certainly illustrate some of the more important structural features that affect the design choice and methods of stress analysis. The examples selected indicate welded configurations which, with modern fabricating techniques, are likely to be reliable and economic. However, this statement is not intended to imply that welding processes never cause problems. Despite significant progress during the past 50 years, strict quality control of welding should be maintained at all times. Fracture-safe design, for instance, can easily be compromised by a change in material properties in the heat-affected zone due to welding, flame cutting, or other operation.

The mechanical characteristics of the various support brackets can be summarized as follows. The short bracket shown in Fig. 33.23 is made of a standard angle with equal legs. This component can be designed on the basis of bending and transverse shear. When loading arm d is relatively short, the structural element is rigid and the effect of bending may be neglected. A box-type support bracket (Fig. 33.24) can be made out of two channels using butt-welding techniques. The

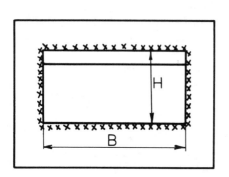 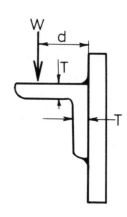

Fig. 33.23 Shear-type bracket.

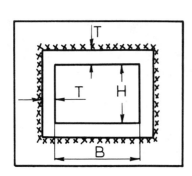 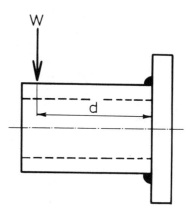

Fig. 33.24 Box-type support bracket.

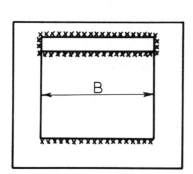 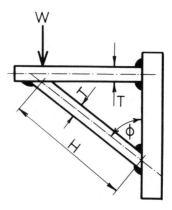

Fig. 33.25 Heavy-duty plate bracket.

Flanges and Brackets

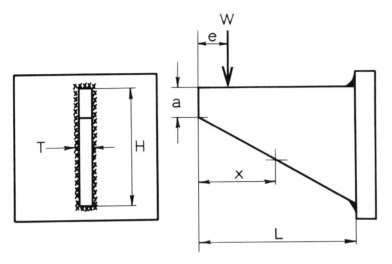

Fig. 33.26 Tapered plate bracket.

strength check here is performed using a simple beam model under bending and shear. Rugged bracket construction is illustrated in Fig. 33.25, where heavy loads have to be supported. Because of the frame-type appearance and mechanics of this type of a support, external loading can be resolved into tensile and compressive forces for design purposes. In this design case the cross sections of the tensile and compression members are large enough to carry substantial loads. A simple and light construction is illustrated in Fig. 33.26. When the plate is relatively long, the bracket must be designed to resist bending, shear, and local buckling loads. A more conventional type of a bracket design is shown in Fig. 33.27. This bracket can be made either from separate plate members by flame cutting and welding or by cutting standard rolled shapes such as I or T beams. For larger loads, a double-T

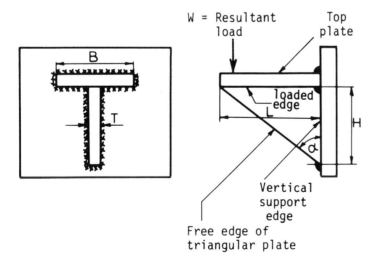

Fig. 33.27 T-section bracket.

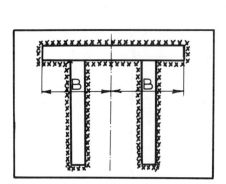 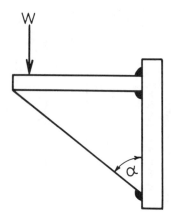

Fig. 33.28 Double-T section bracket.

configuration bracket design, shown in Fig. 33.28, may be recommended. The design should be checked, however, for bending effects, shear strength, and stability of the free edges due to the compressive stresses. Yet another version, shown in Fig. 33.29, can be flame-cut from a standard channel and welded to the base plate to form a solid unit. The design analysis in this case is similar to that employed for the configuration given in Fig. 33.28.

THEORY OF WELD STRESSES

In considering the manner in which working stresses will be applied to the brackets in service, some typical weld calculations should be made. This section includes a discussion of a few important rules to use with the calculations, although for more details the reader is referred to welding handbooks and up-to-date publications of the Welding Research Council. Major findings in this area are reported by the American Welding Society.

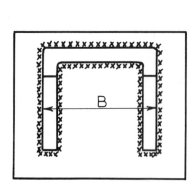 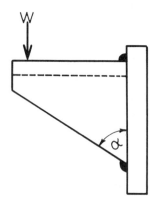

Fig. 33.29 Channel-type heavy-duty bracket.

Flanges and Brackets

In looking over the designs illustrated by Figs. 33.23 to 33.29, we find that we have to deal with the transverse and parallel welds subjected to bending moments. To examine the general principle involved here, consider the case shown in Fig. 33.30. For the fillet weld shown, the size of the weld leg is h. The overall linear dimensions of the weld are B and H for the transverse and longitudinal welds, respectively. The bending moment M_1 on the transverse welds can be imagined to be a couple consisting of two equal forces F acting at the center of the weld legs, as shown in Fig. 33.30. Since it is a standard practice to calculate the stresses on the basis of a weld-throat section, the area on which the component force F is acting must be approximately equal to $Bh/\sqrt{2}$. This is somewhat conservative because of the additional weld material found at the corner which is not accounted for in calculating the weld area. We now have

$$M_1 = F(H + h) \tag{33.36a}$$

and the tensile stress across the throat section is

$$S_1 = \frac{F\sqrt{2}}{Bh} \tag{33.36b}$$

Combining Eqs. (33.36a) and (33.36b) gives

$$S_1 = \frac{\sqrt{2}M_1}{Bh(H+h)} \tag{33.36c}$$

The effect of the external load W on the parallel welds can be treated with the help of a simple beam theory. The section modulus of the parallel weld throat is approximately equal to

$$Z = \frac{bH^2}{6\sqrt{2}} \tag{33.36d}$$

Since the two longitudinal sections are involved in resisting M_2, we get

$$S_2 = \frac{3\sqrt{2}M_2}{hH^2} \tag{33.36e}$$

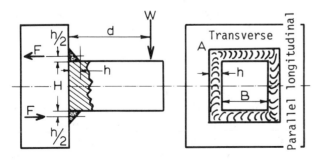

Fig. 33.30 Example of fillet weld in bending.

As stated above, the maximum stress at a common point A must be the same for the transverse and longitudinal welds, that is, $S_1 = S_2$. Hence, combining Eqs. (33.36c) and (33.36e) gives the relationship between the bending moments M_1 and M_2

$$M_1 = \frac{3B(H+h)M_2}{H^2} \tag{33.36f}$$

Since $M = M_1 + M_2$, Eq. (33.36f) yields

$$M = M_2 \left[1 + \frac{3B(H+h)}{H^2}\right] \tag{33.36g}$$

Finally, utilizing Eqs. (33.36e), (33.36g), and $M = Wd$ from Fig. 33.30, the following bending stress formula is obtained

$$S = \frac{3\sqrt{2}Wd}{h[H^2 + 3B(H+h)]} \tag{33.37}$$

By analogy to a hollow box configuration, the average shear stress under load W becomes

$$\tau = \frac{\sqrt{2}W}{2h(H+h)} \tag{33.38}$$

The maximum principal stresses can be found from the strength-of-materials formulas for the combined stresses

$$\tau_{max} = \left[\left(\frac{S}{2}\right)^2 + \tau^2\right]^{1/2} \tag{33.39}$$

and

$$S_{max} = \frac{S}{2} + \tau_{max} \tag{33.40}$$

Experiments indicate that the stress at the throat section is essentially tensile. On this basis, it is customary to use the following design equation consistent with the model given in Fig. 33.31.

$$S = \frac{\sqrt{2}P}{2Bh} \tag{33.41}$$

where B denotes the length of one of the welds supporting the horizontal plate shown in Fig. 33.25. It is also instructive to look at the equilibrium of a double fillet weld illustrated in Fig. 33.32. We assume here a weld seam of unit length at an arbitrary section defined by θ. The normal and shear components are denoted by F_n and F_s, respectively. The throat dimension measured at angle θ is taken to

Flanges and Brackets

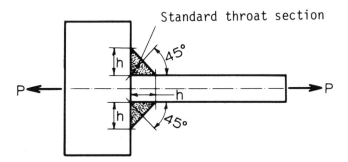

Fig. 33.31 Symmetrical fillet weld in tension.

be equal to t. In terms of the specified quantities, this dimension can be expressed as

$$t = \frac{h}{\sin\theta + \cos\theta} \tag{33.42a}$$

In the conventional calculations, $\theta = 45°$, so that Eq. (33.42a) gives $t = h/\sqrt{2}$. The parameter θ is considered to be a variable quantity for the purpose of the mathematical model. Summation of the forces along the line of P gives

$$2F_s \sin\theta + 2F_n \cos\theta = P \tag{33.42b}$$

Since the components F_s and F_n must be balanced in order not to produce a horizontal reaction on the system, we get

$$F_s \cos\theta - F_n \sin\theta = 0 \tag{33.42c}$$

Solving Eqs. (33.42b) and (33.42c) for the shear component F_s yields

$$F_s = \frac{P \sin\theta}{2} \tag{33.42d}$$

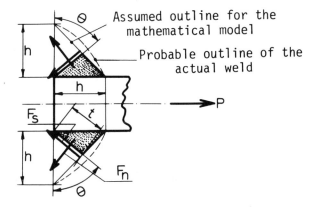

Fig. 33.32 Free-body diagram for symmetrical fillet weld.

For a length of weld seam equal to B, the shear stress is calculated using Eqs. (33.42a) and (33.42d). This gives

$$S_s = \frac{P \sin\theta(\sin\theta + \cos\theta)}{2Bh} \qquad (33.42e)$$

The corresponding normal stress becomes

$$S_n = \frac{P \cos\theta(\sin\theta + \cos\theta)}{2Bh} \qquad (33.43)$$

Examination of Eq. (33.43) indicates that at no point of the weld does the theoretical stress exceed the value computed from Eq. (33.41). The experimental evidence does not shown any typical shear failure in tensioning a symmetrical fillet weld. On the theoretical basis then Eq. (33.42e) assures a more conservative weld design.

SELECTION OF FORMULAS FOR BRACKETS

The bracket illustrated in Fig. 33.23 is very simple in design and manufacture. If the line load W is located d inches from the main plate, the bending and shear stresses across the plate section can be computed as follows.

$$S_b = \frac{6W(d-T)}{BT^2} \qquad (33.44)$$

$$S_s = \frac{W}{BT} \qquad (33.45)$$

The relevant weld stresses for the case in Fig. 33.23 are

$$S_b = \frac{4.24Wd}{h(H^2 + 3BH + 3Bh)} \qquad (33.46)$$

and

$$S_s = \frac{0.7071W}{h(H+h)} \qquad (33.47)$$

Design of the box structure shown in Fig. 33.24 can be accomplished with the aid of the elementary theory of bending and shear stresses

$$S_b = \frac{3Wd}{HT(H + 2B + 4T)} \qquad (33.48)$$

$$S_s = \frac{W}{2(H + 2T)} \qquad (33.49)$$

The corresponding equations for the weld stresses are

$$S_b = \frac{4.24Wd}{h[H(H+4T) + 3(B+2T)(H+h)]} \qquad (33.50)$$

Flanges and Brackets

and

$$S_s = \frac{0.7071W}{h(H + 2T + h)} \tag{33.51}$$

The analysis of a heavy-load plate bracket such as that shown in Fig. 33.25 is based on the assumption that the external load W can be resolved into two components acting along the central plane of the two plates. The primary stresses in this case can be defined as tensile for the horizontal member and compressive for the inclined plate, respectively. The relevant formulas for the plate members are

$$S_t = \frac{W \sin \phi}{BT \cos \phi} \tag{33.52}$$

$$S_c = \frac{W}{BT \cos \phi} \tag{33.53}$$

The calculation of weld stresses involves the following expressions.

$$S_b = \frac{0.7071W \tan \phi}{Bh} \tag{33.54}$$

and

$$S_s = \frac{0.5W \tan \phi}{Bh} \tag{33.55}$$

It is not practical to use large angles ϕ, because the corresponding plate forces become relatively high, as can be seen from the foregoing theoretical expressions. In addition, the bracket having high ϕ loses its frame character of structural behavior and tends to becomes a cantilevered member for which even small transverse loads can cause substantial bending stresses.

A bracket angle of $\phi = 45°$ is often selected in practical design. With the typical proportions of plate members in use, Eqs. (33.52) and (33.53) suffice for sizing calculations. However, it should be appreciated that the compressive member of the bracket can become elastically unstable if its thickness is drastically reduced. Since in the angle brace shown in Fig. 33.25 the two edges of the plate are free to deform, we have the case of buckling of a relatively wide beam subjected to axial compression. Denoting the width and length of this beam by B and H, respectively (Fig. 33.25), and assuming end fixity due to welding, the following expression for the critical buckling stress can be used.

$$S_{Cr} = \frac{3.62ET^2}{H^2} \tag{33.56}$$

This formula is limited to elastic behavior, and therefore the yield strength of the material S_y can be used to determine the maximum allowable value of H/T to avoid failure by buckling. The corresponding critical ratio is

$$\frac{H}{T} = 1.9 \left(\frac{E}{S_y}\right)^{1/2} \tag{33.57}$$

The term E/S_y may be called the *inverse strain parameter* because it follows directly from Hooke's law. For the conventional metallic materials the ratio E/S_y varies between 100 and 1000 for high-strength and low-strength materials, respectively.

STRENGTH AND STABILITY CONSIDERATIONS

A somewhat different set of theoretical problems is encountered with the plate-type bracket given in Fig. 33.26. Essentially, we have here a short, tapered cantilever plate loaded on edge. The normal stresses on a section, such as that defined by distance x, must vary from tension to compression. The maximum bending stress depends on the taper and it can be calculated as follows.

$$S_b = \frac{6WL^2(x-e)}{T[aL + x(H-a)]^2} \tag{33.58}$$

The distance x at which the highest bending stresses develop can be found from the condition $dS_b/dx = 0$, calculated from Eq. (33.58). This yields

$$x = e + (e^2 + c)^{1/2} \tag{33.59}$$

where

$$c = \frac{aL[2e(H-a) + aL]}{(H-a)^2} \tag{33.60}$$

The procedure is therefore to compute x from Eqs. (33.59) and (33.60), and to substitute this value in Eq. (33.58) to obtain the maximum stress value. With the usual proportions of brackets found in practice, the aspect ratio H/L can be used to make a rough estimate of the relevant buckling coefficient K_b from Fig. 33.33. This coefficient is then used in calculating the critical elastic stress of the free edge of the bracket using the expression

$$S_{Cr} = K_b E \left(\frac{T}{H}\right)^2 \tag{33.61}$$

The plate buckling coefficient K_b given in Fig. 33.33 can be determined experimentally for each case of plate proportions, boundary conditions, and type of stress distribution. It represents the tendency of a free edge of the plate element to local instability when the compressive stresses reach a certain critical value. The consequence of local buckling then may be interpreted in the following two ways:

Flanges and Brackets

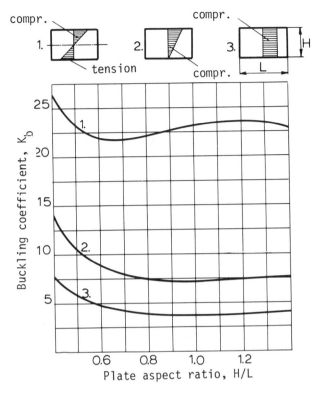

Fig. 33.33 Buckling parameters for simply supported plates under nonuniform longitudinal stresses.

Overall collapse by rendering the plate element less effective in the postbuckling region of structural response.

Detrimental stress redistribution influencing the load-carrying capacity of the system.

The design values given in Fig. 33.33 depend largely on the type of stress distribution in compression. Although K_b values are sensitive to the type of stress distribution and vary in a nonlinear fashion, their dependence on the aspect ratio H/L is only moderate.

When the actual compressive stress given by Eq. (33.58) exceeds that given by Eq. (33.61), it is customary to assume that the free edge of the bracket is susceptible to local elastic buckling. To make a conservative allowance for the critical buckling stress in the plastic range, the following set of the design formulas may be used:

$$S_{Cr} = K_b E \eta \left(\frac{T}{H} \right)^2 \tag{33.62}$$

where

$$\eta = \left(\frac{E_t}{E} \right)^{1/2} \tag{33.63}$$

and

$$E_t = \frac{dS}{d\epsilon} \tag{33.64}$$

In Eq. (33.64) the terms S and ϵ denote the normal stress and uniaxial strain, respectively. Therefore, Eq. (33.64) defines the tangent modulus of the stress-strain characteristics of the material at a specified level of stress.

The strength of the weld in bending is estimated as follows.

$$S_b = \frac{4.24W(L-e)}{h(H^2 + 3HT + 3hT)} \tag{33.65}$$

The numerical value of shear stress for this case can be obtained from Eq. (33.47).

The conventional plate bracket and its design derivatives are shown in Figs. 33.27, 33.28, and 33.29. Although triangular-plate support brackets are frequently used in practice, comparatively few theoretical formulas, applicable to this type of a support, are available.

The design shown in Fig. 33.27 contains the two basic elements of structural support. One is the top support plate, which helps to distribute the load. The other is a triangular plate loaded on edge and designed to carry the major portion of the load. The two plates acting together form a relatively rigid tee configuration.

Experience indicates that the free edge carries the maximum compressive stress $X_{\max}$, which depends on the aspect ratio L/H. Practical design situations give aspect ratios somewhere between 0.5 and 2.0 [40]. For this particular case the maximum allowable total load W near the center of the upper plate can be estimated using the following equation.

$$W = S_{\max}(0.60H - 0.21L)\frac{TL}{H} \tag{33.66}$$

When the working load W is specified the maximum corresponding stress $S_{\max}$ can be calculated from Eq. (33.66). It is then customary to make $S_{\max} < S_y$, where S_y denotes the yield compressive strength of the material. The design condition for the critical values of L/T can be represented by the following set of criteria. For $0.5 \leq L/H \leq 1.0$,

$$\frac{L}{T} \leq \frac{180}{S_y^{1/2}} \tag{33.67}$$

For $1.0 \leq L/H \leq 2.0$,

$$\frac{L}{T} \leq \frac{60 + 120(\frac{L}{H})}{S_y^{1/2}} \tag{33.68}$$

Equations (33.66) to (33.68) are valid when the resultant load W is located reasonably close to the center of the top plate and when the yield strength S_y is expressed in ksi. However, when this load moves out toward the edge of the plate, the analytical method described above loses its degree of conservatism and an alternate

Flanges and Brackets

approach should be based on the concept of increased eccentricity. The strength of a welded connection in this design may be checked from Eqs. (33.54) and (33.55).

Some of the specific features of the triangular-plate bracket can be analyzed with reference to Fig. 33.34. The maximum stress at the free edge of the triangular part may be calculated on the basis of elementary beam theory by combining the stresses due to the bending moment $W \times e$ and the compressive load equal to $W/\cos\alpha$. This gives

$$S_{\max} = \frac{W(L+6e)}{TL^2 \cos^2 \alpha} \quad (33.69)$$

A conservative check on free-edge stability can be made by assuming that the shaded portion of the plate acts as a column with a cross section equal to $(TL\cos\alpha)/4$ and length equal to $H/\cos\alpha$.

When a relatively small value of L/T must be used, there is little danger of elastic instability and the bracket can be designed to undergo a certain amount of local yielding. For $0.5 \leq L/H \leq 2.0$, the recommended L/T ratio is

$$\frac{L}{T} \leq \frac{48 + 24\left(\frac{L}{H}\right)}{S_y^{1/2}} \quad (33.70)$$

Here the yield strength S_y is expressed in ksi [40].

The maximum permissible load on the bracket under fully plastic conditions can be calculated from the expression

$$W_{\text{pl}} = TS_y \cos^2\alpha \left[(L^2 + 4e^2)^{1/2} - 2e\right] \quad (33.71)$$

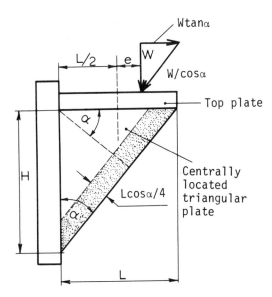

Fig. 33.34 Approximate model for a triangular plate bracket.

The cross-sectional area of the top plate should be designed for the horizontal component of the external load

$$A = W_{pl} \frac{\tan \alpha}{S_y} \tag{33.72}$$

The design formulas given by Eqs. (33.44) to (33.72) are applicable to various practical situations wherever a particular structure can be modeled as a support bracket similar to one of the configurations illustrated in Figs. 33.23 to 33.29. By checking the weld strength, beam strength, and stability, the structural integrity of a bracket can be assured, provided that material and fabrication controls are not compromised.

SYMBOLS

A	Area of cross section, in.² (mm²)
A_r	Depth of tapered rib, in. (mm)
A_t	Total cross-sectional area, in.² (mm²)
a	Moment arm; depth of tapered plate, in. (mm)
a_0	Mean length of flange sector, in. (mm)
B	Width of bracket, in. (mm)
F_f	Width of flange cross section, in. (mm)
B_r	Depth of rib, in. (mm)
c	Tapered plate parameter, in.² (mm²)
d	Distance to loaded point, in. (mm)
d_1	Mean pipe diameter, in. (mm)
d_b	Bolt hole diameter, in. (mm)
E	Elastic modulus, psi (N/mm²)
E_0	Reduced modulus of elasticity, psi (N/mm²)
E_t	Tangent modulus, psi (N/mm²)
e	Eccentricity of load application, in. (mm)
F	Load on weld seam, lb (N)
F_n	Normal force component, lb (N)
F_s	Shear force component, lb (N)
f	Load-sharing ratio
H	Depth of standard flange; maximum depth of bracket, in. (mm)
H_e	Equivalent depth of flange, in. (mm)
h	Thickness of flange ring; size of weld leg, in. (mm)
I_b	Second moment of area, in.⁴ (mm⁴)
I_g	Moment of inertia of a component section, in.⁴ (mm⁴)
I_m	Moment of inertia of wall element of unit width, in.⁴/in. (mm⁴/mm)
I_n	Moment of inertia of a rib cross section, in.⁴ (mm⁴)
I_0	Moment of inertia of main flange section, in.⁴ (mm⁴)
I_x	Moment of inertia about central axis, in.⁴ (mm⁴)
J	First moment of area, in.³ (mm³)
K	Modulus of elastic foundation, psi (N/mm²)

Flanges and Brackets

K_b	Buckling coefficient
$k = R_o/R_i$	Flange ring ratio
L	Length of rib of constant depth; length of bracket, in. (mm)
L_g	Length of tapered rib, in. (mm)
ℓ_1	Moment arm, in. (mm)
M	General symbol for bending moment, lb-in. (N-mm)
M_1, M_2	Bending moment components, lb-in. (N-mm)
M_c	Bending moment per one rib spacing, lb-in. (N-mm)
M_0	Discontinuity bending moment, lb-in./in. (N-mm/mm)
M_y	Bending moment about longer edge of plate, lb-in./in. (N-mm/mm)
M_F	Toroidal moment on flange ring, lb-in./in. (N-mm/mm)
M_R	Bending moment on rib, lb-in. (N-mm)
$m = R_i/T$	Ratio of inner radius to wall thickness
N	Number of ribs
$n = H/T$	Ratio of flange to pipe thickness
P	Tensile load on bracket; edge force on plate, lb (N)
Q_0	Discontinuity shearing force, lb/in. (N/mm)
q	Radial load intensity, lb/in. (N/mm)
R	Radius to bolt circle; mean flange radius, in. (mm)
R_i	Inner radius of pipe, in. (mm)
R_o	Outer radius of flange, in. (mm)
r	Mean radius of pipe, in. (mm)
S	General symbol for stress, psi (N/mm^2)
S_1, S_2	Weld stress components, psi (N/mm^2)
S_b	Bending stress, psi (N/mm^2)
S_{bR}	Rib bending stress, psi (N/mm^2)
S_c	Compressive stress, psi (N/mm^2)
S_{Cr}	Critical compressive stress, psi (N/mm^2)
S_F	Flange dishing stress, psi (N/mm^2)
S_N	Normal stress, psi (N/mm^2)
S_p	Plastic stress, psi (N/mm^2)
S_R	Radial stress in flange ring, psi (N/mm^2)
S_s	Shear stress, psi (N/mm^2)
S_t	Tensile stress; toroidal stress in flange, psi (N/mm^2)
S_{TR}	Total stress in rib, psi (N/mm^2)
S_u	Ultimate strength, psi (N/mm^2)
S_y	Yield strength, psi (N/mm^2)
S_{max}	Maximum principal stress, psi (N/mm^2)
s_0	Wall thickness, in. (mm)
s_1	Depth of section at failure, in. (mm)
T	Thickness of pipe; thickness of plate, in. (mm)
t	Distance from bolt circle to outer pipe surface; thickness of fillet weld, in. (mm)
T_r	Thickness of rib, in. (mm)
T_0	Thickness of backup ring, in. (mm)
V	Moment factor in plate analysis
W	Total bolt load; external load on bracket, lb (N)

W_i	Load per inch of pipe circumference, lb/in. (N/mm)
W_0	Load per inch of bolt circle, lb/in. (N/mm)
W_R	Tensile load on rib, lb (N)
W_{pl}	Plastic load on bracket, lb (N)
x	Arbitrary distance, in. (mm)
Y	Deflection of beam on elastic foundation, in. (mm)
Y_p	Plate edge deflection under concentrated load, in. (mm)
Y_q	Plate edge deflection under uniform load, in. (mm)
y	Coordinate; ring deflection, in. (mm)
Z	Section modulus, in.^{3} (mm^3)
α	Angle of fractured part; bracket angle, rad
β	Elastic foundation parameter, in.$^{-1}$ (mm^{-1})
β_s	Shell parameter, in^{-1} (mm^{-1})
δ	Slope of stress-strain curve, rad
ϵ	Strain, in./in. (mm/mm)
ϵ_y	Uniaxial strain at yield, in./in (mm/mm)
η	Modulus ratio
θ	Angle of twist; angle in weld analysis, rad
θ_F	Angle of twist of main flange ring, rad
θ_p	Rib half-angle, rad
θ_R	Bending slope at end of rib, rad
μ	Poisson's ratio
τ	Shear stress component, psi (N/mm^2)
τ_{max}	Principal shear stress, psi (N/mm^2)
ϕ	Plate angle, rad
ϕ_0	Flange factor
$\phi(\beta, L)$	Auxiliary function for a beam on elastic foundation
ω	Ratio of rib to flange moment

34
Special Configurations

PERFORATED PLATES

Conventional design of flat plates can be extended to plates with perforations and reinforcements if the equivalent material and geometrical parameters can be defined. In the case of thin and thick plates with regular hole patterns, the ASME code presents a detailed method of analysis [198], provided that the following conditions are met:

There is an equilateral-triangular array of holes.
The holes are circular (19 or more).
The ligament efficiency is higher than 5%.
The plate thickness is more than twice the hole pitch.
Local reinforcement effects are included in the ligament efficiency.

In principle, the perforated plate can be replaced by a solid plate with modified elastic constants E^* and ν^*. Modified constants are considered to be functions of the ligament efficiency within the range 0.05 to 1.0, and specific detailed design charts are available that feature stress multipliers as a function of hole orientation angle in relation to the direction of loading. The actual plate stresses are obtained by multiplying the design factors by the nominal stresses calculated for the equivalent solid plate. All conditions treated by the codes are axisymmetric. The effects of temperature are included in consideration of structural interaction with the adjacent members. Where thin or irregular ligament patterns are involved, the code recommends using the average stress intensities.

The concept of a ligament efficiency and its effect on the modified elastic constants has been the subject of various theoretical and experimental investigations. Ample literature on this specialized topic is available [199]. Engineering applications can involve both square and triangular patterns of penetration, as shown in Fig. 34.1. Out of these, the triangular pattern appears to be more widely used, particularly in the construction of boiler feedwater heaters, steam generators, heat exchangers, and similar systems. The condition of such mechanical parts can often be simulated by perforated circular plates carrying uniform external pressure. The design parameter called *ligament efficiency* can be defined by the ratio h_0/P_0, as shown in Fig. 34.1. Since boiler tubes attached to the plate may have some influence on plate rigidity, the concept of virtual ligament efficiency can be defined as the actual ligament plus an effective portion of the tube wall divided by the tube hole pitch. When the total thickness of the tubes is assumed effective in stiffening the perforated plates, the calculated strains and deflections appear to agree well with the measurements [200]. Ligament stresses are found to be higher than those for an unperforated plate. Peak stresses can exist immediately adjacent to the tube wall. Such stresses, however, are damped out rather rapidly as a function of the distance from the tube wall. An example of modified elastic constants for a triangular pattern according to the ASME code [198] is shown in Fig. 34.2. This chart has been developed for a Poisson's ratio of 0.3 and t/P_0 values greater than 2, where t is the thickness of a perforated plate. Design curves for other values of Poisson's ratio are also available [199]. Various stress formulas for use with the modified elastic constants are given in the code [198].

In general, the code suggests performing the analysis on the basis of the modified elastic constants E^* and ν^*. With the usual proportions of perforated plates, such calculations can lead to rather low values of ligament efficiency. It should

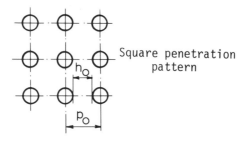

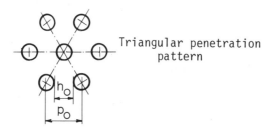

Fig. 34.1 Typical plate perforations.

Special Configurations

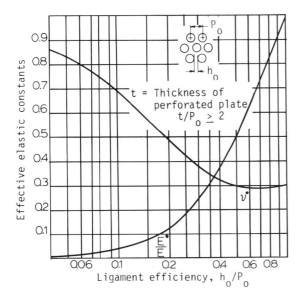

Fig. 34.2 Modified elastic constants for triangular patterns ($\nu = 0.3$).

be pointed out here that in addition to the ligament efficiency, one could define the deflection efficiency, which, however, would have to be represented by a much more complicated expression than h_0/P_0. There is a general opinion among workers in the field that for all practical purposes, the ligament and deflection efficiencies are approximately equal. Also, some investigators suggest a different formula for estimating the relevant ligament efficiency, which can be stated as follows:

$$\text{Ligament efficiency} = \frac{P_0^2 - A}{P_0^2} \tag{34.1}$$

where A denotes the area of the hole and P_0 is the pitch, as shown in Fig. 34.1. Furthermore, this and other formulas for ligament efficiency appear to be equally applicable to square and triangular hole patterns. The advantage of Eq. (34.1) is that the relevant magnitudes of efficiency fall closer, between 0.5 and 0.8, which seems to be more realistic than the smaller values resulting from h_0/P_0. A preliminary estimate of the maximum stress in the perforated plate can be obtained by dividing the maximum stress in the homogeneous plate by the ligament efficiency. Similarly, the deflections of the perforated plate can be calculated if the flexural rigidity of the homogeneous plate D is multiplied by the ligament efficiency. Since the deflection is inversely proportional to D, ligament correction tends to increase the magnitude of the deflection for a perforated plate compared to that of a solid plate.

REINFORCED PLATES

Reinforcement of circular plates may be accomplished by an equidistant orthogonal system of ribs, either symmetrical or nonsymmetrical with respect to the middle plane of the plate. For a symmetrical system, the plate rigidity can be expressed as

$$D = \frac{Et^3}{12(1-\nu^2)} + \frac{EI}{d} \qquad (34.2)$$

where d denotes the spacing between the rib centers and I is the moment of inertia of a rib with respect to the middle axis of the plate. For a nonsymmetrical case involving a T shape of the wall and rib combination (rib on one side of the wall only), the flexural plate rigidity can be taken as follows:

$$D = \frac{EI}{d} \qquad (34.3)$$

where I is the average moment of inertia of the T section of width d about its centroid. Grillage-type plates are used in the nuclear core reactor vessels and other applications where support and cover plate size requirements are such that these members cannot be procured as a solid plate. The fabrication is accomplished by welding together a complex web system.

In some cases the plates can be reinforced by the use of a concentric stiffening ring which reduced the stresses and deflections due to its toroidal stiffness. Such a concentric ring has the tendency to turn inside out as the plate deflects under the load. Working design charts have been developed for the effect of such a reinforcing ring on the maximum deflection and radial stress for a circular plate with a built-in edge and the transverse uniform load [201]. The most effective location of a ring of a specified size is not the same for the stress and deflection criteria, and a design compromise may be needed. Generally, placing the ring at about 0.6 value of the radius, measured from the plate center, represents a satisfactory compromise.

Where the major design criterion is deflection rather than stress, an alternative to a thick cover plate would be a relatively thin plate heavily reinforced. One such design for instance involves a system of straight, radial ribs radiating to the outer edge of the plate as shown in Fig. 34.3. While a relatively exact mathematical model can be applied to the bending of plate with orthogonal ribs, no flexural theory has yet been established for accurate calculation of the deflection of a plate with radial reinforcement [202]. Radial ribs may be of constant depth or tapered geometry with gradually diminishing depth toward the outer edge of the plate. Because of this, such a reinforcement can only be analyzed with the aid of the three-dimensional theory of elasticity, presenting almost unmanageable boundary conditions. Experimental evaluation of the stresses and deflections can be made, but many models have to be tested prior to determining the optimum criteria of strength and rigidity. In such cases, however, where the rib system is relatively stiff compared with the plate to which these ribs are attached, the approximate solutions may be possible on the basis of flexible sectorial plates.

In the case of a cover plate shown in Fig. 34.3, an individual radial rib can be regarded as a simple beam subjected to a bending moment at the junction with the circular stiffener, a supporting reaction at the other end, and a distributed load along the rib length according to a linear function. The basic difficulty in the analysis, however, is the fact that ribs behave as T beams of variable stiffness and

Special Configurations

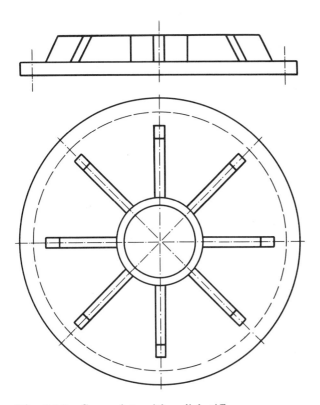

Fig. 34.3 Cover plate with radial stiffeners.

that the portions of the adjacent plate act as flanges. Because of these constraints the only tractable approach has been, so far, through experimental techniques [202]. Ribbed configurations in this type of an experiment are machined from a solid flat plate and, in the sequence of tests, the depth of the webs is progressively reduced, providing the test samples with different combinations of web shape and size. All plates in a quoted experiment had eight ribs and the deflections under uniform load were measured by optical means involving an interferometer. The plate models were about 8 in. in diameter, and measured deflections were assessed in relation to the theoretical deflections of solid plates of equal weight calculated for a simply supported boundary. The results of this research indicated that the most effective use of material could be achieved with deep and slender ribs in such a way as to make the cumulative mass of the rib system equal to about 40% of the total mass of the plate. An empirical formula for the central deflection of the eight-ribbed design shown in Fig. 34.3 was established as follows:

$$Y = \frac{21.6qa^{10}\lambda^3}{EW^3} \tag{34.4}$$

In Eq. (34.4), the Poisson's ratio was assumed to be equal to 0.3, which is good for the majority of metallic materials. For the more heavily reinforced ribs, 40% may not be possible to achieve. This type of reinforcement, as well as the elimination of any potential waviness of the plate boundary, could be achieved by increasing the number of ribs. Although further experimental analysis is required for finding the precise effect of the number of ribs on the deflection and local stresses, the eight-rib system analyzed so far can be used as a rough guide for establishing the deflection criteria for other designs. For instance, a plate of 12 ribs of the same size relative to the basic plate and one of the 8-rib type are expected to have similar deflection ratios.

PIN-LOADED PLATES

In a riveted and bolted connection a plate can be loaded through a pin in the hole which causes a complex stress distribution [203]. Such a stress pattern may be of special interest in the determination of the fatigue strength of a joint made of high-strength alloy steels or lightweight alloys. For these materials the endurance limits are well within the elastic range, in contrast to mild steel, which has the endurance limit close to the yield strength of the material. In the study quoted here [203], strain gage and photoelastic measurements were made to determine the stress concentration factor for the various ratios of hole diameter to the width of plate, ranging from 0.086 to 0.76. The notation for this problem is given in Fig. 34.4. If the ratio of the maximum tensile stress S at the edge of the hole to the average bearing stress on the pin is denoted by K_b, then a useful parameter for presentation

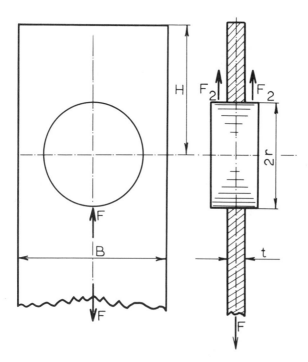

Fig. 34.4 Notation for a pin-loaded plate.

Special Configurations

of the results may be postulated:

$$K_b = \frac{2rtS}{F} \tag{34.5}$$

This simple parameter plotted versus $2r/B$ is shown in Fig. 34.5. It is a convenient method of relating the stress factor to the average bearing stress. For example, taking $2r/B = 0.5$, Figure 34.5 gives $K_b = 2.4$, and the relevant maximum tensile stress becomes $S = 1.2F/rt$. The problem with the empirical data such as compiled in Fig. 34.5 is, however, that a direct extrapolation of the curves beyond, say, $2r/B = 0.8$ is likely to be in error. Also, the study indicates that the stress factor K_b should increase between the pin and the hole. This effect was already noted in Chap. 29 for the case of eye-bar design. The effect of clearance also depends on the head ratio H/B shown in Fig. 34.4. For example, the smaller the ratio H/B, the greater the effect of clearance. The influence of clearance on the maximum stress is, of course, caused by the changes in curvature. Experiments also indicate that in every case involving neat-fitting pins the maximum stresses occurred at the ends of the horizontal diameter. These stresses moved away from this location whenever a final clearance was present.

It should also be added that stress factors due to a single pin in the plate are the highest. For instance, when the load is equally divided between the two

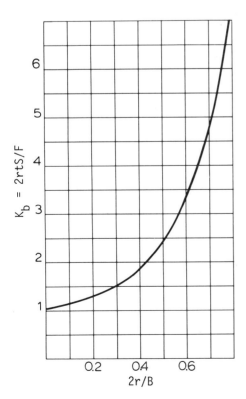

Fig. 34.5 Experimental stress factor for a pin-loaded plate. (From Ref. 203.)

symmetrically located pins, having individual diameters equal to one-half the single-pin diameter, the stress concentration can be reduced by about 20%. Pin lubrication can also produce a small decrease in the K_b values.

BELLEVILLE WASHER

Although the conical disk spring known as a *Belleville washer* is not a flat plate, it is included in this chapter on flanges and brackets because its response under load resembles the behavior of a plate-ring or a pipe flange. This resemblance to a flat plate, such as that illustrated by case 1 of Table 32.2, may lead to errors when the Belleville spring is analyzed with the aid of formulas developed for flat plates.

The Belleville spring is an important machine element where, among other features, space limitation, high load, and relatively small deflections are required. Although calculations of load-deflection characteristics by the majority of methods available shows generally satisfactory correlation with practice, the problems of stress still remain speculative in the various design applications.

This state of spring knowledge is not in the least surprising, however, if we consider the fundamental mechanics of a Belleville washer involving the mathematical theory of large deflections of a shallow conical shell in the elastic-plastic range. The reason for bringing the elastic-plastic concept into this discussion is dictated by past spring art and experience, which appear to show that under certain circumstances the Belleville spring can operate successfully at a calculated elastic stress exceeding 600,000 psi, which obviously no presently available engineering material can withstand.

Some 60 years ago, an account was published of a photoelastic investigation of the tangential and radial stresses along a radial section of a celluloid model of a Belleville washer spring [204]. The original results appeared to be of a doubtful nature, and no stress formula was recommended. Later, however, Almen and Laszlo [205] derived very useful design formulas for the maximum stresses at the inner edges of the washer, and their method has been widely accepted throughout the industry. Their derivation was based on elastic behavior on the assumption that radial stresses were negligible. The cross section of the disk was allowed to rotate about a neutral point such as O in Fig. 34.6. Later work on Belleville springs [206]

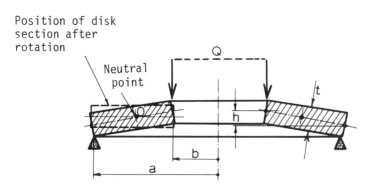

Fig. 34.6 Notation for a Belleville spring.

Special Configurations

culminated in equations for the tensile and compressive circumferential stresses which agreed well with those of Almen and Laszlo. Ashworth [206] also reported on tests conducted in Germany, according to which there was a good correlation between theory and experiment with regard to the maximum compressive stresses. The same experiments confirmed that the radial strains, measured at various points of the disk spring, were rather small in comparison with the circumferential strains. A more recent investigation [207] resulted in a proposed refinement of the Almen and Laszlo solution. In this work the effect of radial strains was included. The relevant stresses, however, appeared to be only some 10% higher. Hence, for all practical purposes, and despite some of the limitations, the Almen and Laszlo theory remained unchanged. Their key design situations for the stress and deflection have been stated as follows:

$$S_c = \frac{YE}{2(1-\nu^2)a^2 C_1}[C_2(2h - Y) + 2tC_3] \qquad (34.6)$$

where Y can be determined from the load formula

$$Q = \frac{YE}{(1-\nu^2)a^2 C_1}\left[(h - Y)\left(h - \frac{Y}{2}\right)t + t^3\right] \qquad (34.7)$$

The dimensionless constants involved in Eqs. (34.6) and (34.7) were

$$C_1 = \frac{6\left(\frac{a}{b} - 1\right)^2}{\pi \log_e \left(\frac{a}{b}\right)} \qquad (34.8)$$

$$C_2 = \frac{6}{\pi \log_e \left(\frac{a}{b}\right)} \left[\frac{\frac{a}{b} - 1}{\log_e \left(\frac{a}{b}\right)} - 1\right] \qquad (34.9)$$

and

$$C_3 = \frac{3\left[\left(\frac{a}{b}\right) - 1\right]}{\pi \log_e \left(\frac{a}{b}\right)} \qquad (34.10)$$

Evaluation of the relative effect of C_1, C_2, and C_3 on the magnitude of the stresses calculated from Eq. (34.6) indicates that with a relatively small numerical error, Eq. (34.6) can be reduced to the following expression, on the premise that $\nu = 0.3$.

$$S_c = \frac{CYE(2h + 2t - Y)}{a^2} \qquad (34.11)$$

The load-deflection formula for $\nu = 0.3$ can also be written as

$$Q = \frac{GYE\left[(h - Y)\left(h - \frac{Y}{2}\right)t + t^3\right]}{a^2} \qquad (34.12)$$

Nondimensional constants C and G are plotted in Figs. 34.7 and 34.8, respectively. The corresponding stress and load formulas for the limiting case of a Belleville

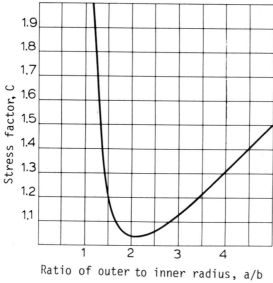

Fig. 34.7 Design chart for a modified stress factor.

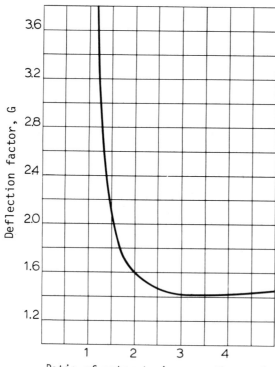

Fig. 34.8 Design chart for a modified deflection factor.

Special Configurations

spring flattened out completely are then

$$S_{hc} = \frac{CEh(h+2t)}{a^2} \tag{34.13}$$

and

$$Q_h = \frac{GEht^3}{a^2} \tag{34.14}$$

An important design relation may be obtained by dividing Eq. (34.12) by Eq. (34.14) and by introducing $k = Y/h$ and $m = h/t$.

$$\frac{Q}{Q_h} = \frac{km^2}{2}(1-k)(2-k) + k \tag{34.15}$$

Equation (34.15) represents a family of curves known as the load-deflection characteristics of conical-disk springs of various proportions. Such curves are given in engineering handbooks and spring design manuals for direct design applications. When $0 < m < 1.4$, the spring rate is always positive. For the interval $0.6 < k < 1.2$ with $1.4 < m < 2.8$, the spring rate is negative, and the actual load decreases as the deflection increases. The Belleville washer becomes unstable for $m > 2.8$; that is, at a particular compressive load Q, the conical disk shape snaps into a new position. Also, for $m \cong 1.6$, a zero spring rate can be achieved for an appreciable range of k.

The ratio of the stress at solid to that at any deflection smaller than the free height h follows from Eqs. (34.11) and (34.13). Utilizing the same parameters k and m as before gives

$$\frac{S_{hc}}{S_c} = \frac{m+2}{k(2+2m-km)} \tag{34.16}$$

This stress ratio is shown in Fig. 34.9 for a few typical values of m. The chart also indicates that S_{hc}/S_c is relatively sensitive to changes in parameter k, as might have been anticipated. It also appears that the correct expression for stresses could only be obtained by either setting up the problem as a nonlinear one in a shallow shell theory, followed by a numerical solution, or conducting an experimental study for the purpose of developing an empirical formula. None of the Belleville spring formulas used by engineers to date have been of that origin.

In practice, many Belleville springs operate at deflections smaller than those corresponding to the load at solid. The amount of permanent set that can be tolerated is about 2% of the maximum working deflection. If it is expected that permanent set can exceed 2%, manufacturers specify a setting-out operation. This consists of loading the washer to solid and noting the amount of decrease in the free height after unloading. The experience shows that the first loading cycle causes the maximum amount of permanent deformation and it considerably stabilizes the washer geometry [208].

As more stringent conditions are imposed on a particular design, the more difficult it is to find a solution. One approach may be to assume the material strength and the working stress and to start the calculation with Eq. (34.13). Simplified

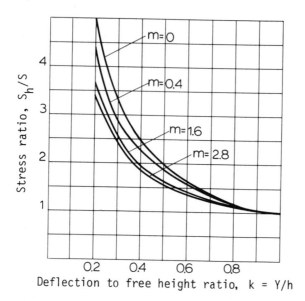

Fig. 34.9 Parametric study of Eq. (34.16).

tables for a single washer and for nests of washers have also been worked out [209]. These tables help reduce the trial-and-error procedures involved in the application of standard Belleville spring formulas.

SYMBOLS

a	Radius of plate, in. (mm)
B	Width of rectangular plate, in. (mm)
b	Inner radius, in. (mm)
C	Approximate stress factor
C_1, C_2, C_3	Almen and Laszlo factors
D	Flexural rigidity of plate, (lb-in.^{2})/in. (N-mm^2/mm)
d	Spacing of parallel ribs, in. (mm)
E	Modulus of elasticity, psi (N/mm^2)
E^*	Equivalent modulus, psi (N/mm^2)
F	Force on pin, lb (N)
G	Deflection factor
H	Head distance from edge of hole, in. (mm)
h	Free height of Belleville spring, in. (mm)
h_0	Distance between holes, in. (mm)
I	Moment of inertia, in.^{4} (mm^4)
K_b	Stress factor in pin-loaded plate
$k = Y/h$	Deflection-to-free height ratio
$m = h/t$	Free height-to-thickness ratio
P_0	Pitch of circular holes, in. (mm)
Q	Load on Belleville spring, lb (N)

Special Configurations

Q_h	Load at solid, lb (N)
q	Uniform pressure, psi (N/mm^2)
r	Radius of pin, in. (mm)
S	Tensile stress in plate, psi (N/mm^2)
S_c	Compressive stress at inner edge of Belleville spring, psi (N/mm^2)
S_{hc}	Compressive stress at solid, psi (N/mm^2)
t	Thickness of plate or washer, in. (mm)
W	Weight of ribbed plate, lb (N)
Y	Deflection, in. (mm)
λ	Weight density, lb/in.^{3} (N/mm^3)
ν	Poisson's ratio
ν^*	Equivalent Poisson's ratio

VI
PIPING AND VESSELS

35
Internal Pressure

INTRODUCTION

Chapter 34 was concerned with plane-walled members. However, when such walls are shaped to curved surfaces, we deal with a very extensive field of shells. The theoretical principles involved in shell analysis are applicable to pressure vessel and piping design. These structural configurations of relative geometrical simplicity have been under investigation for a great many years and the literature on the subject is truly enormous. No single chapter or even an entire textbook can do justice to all advancements in the field, particularly where the internal pressure, external pressure, and other modes of loading are concerned [178, 198, 200, 210–215].

Although the majority of piping and vessel components are designed for internal pressure and receive a major attention, numerous cases may also involve external pressure loading where stress, elastic stability, and structural collapse must be evaluated. Since this area of design is seldom treated, several sections of this book contain the relevant formulas for vessels under external pressure.

MEMBRANE THEORY

The more common forms of pressure vessels involve spheres, cylinders, and ellipsoids, although conical and toroidal configurations are also found. When such components have small thickness compared with the other dimensions and offer a limited resistance to bending, the stress can be calculated with the aid of the membrane theory. Such stresses, taken as average tension or compression over the thickness of the vessel wall, act in the direction tangential to the surface. Since the middle surface of the wall extends in two dimensions, the analysis can become complicated where more than one expression for the curvature is required to describe the displacement of a particular point. In a more rigorous sense, then, it would be necessary to define a normal force, two transverse shearing forces, two bending moments and a torque in order to describe the entire state of stress. Fortunately, the membrane theory allows us to neglect the bending, shearing, and twisting effects. In a number of elementary but practical cases, the simple equations of equilibrium of forces are sufficient for deriving the necessary design formulas.

THIN CYLINDERS

When a thin circular cylinder is subjected to internal pressure, stresses must be induced on the longitudinal and circumferential sections as shown in Fig. 35.1. By analogy to a circular ring under internal pressure, the total force normal to the y axis is equal to the summation of all the force components resolved on the x axis. This gives

$$2F_t = 2 \int_0^{\pi/2} PR_i \sin\theta \, d\theta$$

from which

$$F_t = PR_i \tag{35.1}$$

It should be noted that in deriving Eq. (35.1), the length term on the longitudinal section drops out, and the circumferential force can be assumed to act on the unit length of the section. Hence, dividing Eq. (35.1) by $T \times$ the unit length gives the

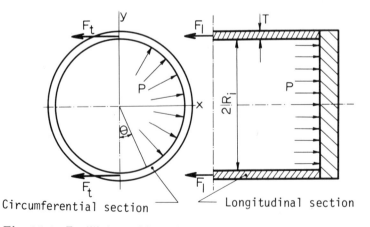

Fig. 35.1 Equilibrium of forces in a thin cylinder.

Internal Pressure

familiar pressure vessel formula

$$S_t = \frac{PR_i}{T} \tag{35.2}$$

Despite its simplicity, Eq. (35.2) has extremely wide applications in many practical situations involving boiler drums, accumulators, piping, casing, chemical processing vessels, and nuclear pressure vessels, to mention a few. Equation (35.2) is often referred to as the *hoop stress formula* and represents the maximum hoop (tangential) stress in the vessel wall on the assumption that end closures provide no support, such as is the case with a long cylinder. It is also evident that the cylinder radius R_i can be replaced by a mean radius R without any significant error provided that the cylinder is relatively thin. In further calculations the R/T ratio, denoted throughout this chapter by m, will be a useful parameter in many cases involving structural response of piping and pressure vessel components. In line with this notation, the hoop stress formula is

$$S_t = mP \tag{35.3}$$

The equilibrium of forces in the axial direction gives

$$F_1 = P\pi R^2 \tag{35.4}$$

and the corresponding stress in the axial sense is then

$$S_1 = \frac{mP}{2} \tag{35.5}$$

In the analysis of a general shell of revolution, the term "meridional" is sometimes used instead of "axial" or "longitudinal." It is now evident from Eqs. (35.3) and (35.5) that the efficiency of the circumferential joints need only be half that of the longitudinal joints. It is also clear from the equilibrium of forces in a spherical shell that the relevant maximum stress there is represented by Eq. (35.5). This simple deduction is of great importance in the design of pressure vessels because the thickness requirement for a spherical vessel, for the same material strength and parameter m, is only one half that necessary for a cylinder. Hence, the sphere is the most efficient configuration for a pressure vessel.

RADIAL GROWTH

An important criterion in pipe and vessel analysis is radial growth, or dilation under internal pressure. For a long cylindrical vessel this dimensional change is given by

$$\delta = \frac{(2-\nu)PR^2}{2ET} \tag{35.6}$$

The corresponding dilation of a spherical vessel is

$$\delta = \frac{(1-\nu)PR^2}{2ET_s} \tag{35.7}$$

When a vessel is made out of a cylindrical portion and a hemispherical head, equating radial growth for the two cases shown by Eqs. (35.6) and (35.7) gives the following relation between the two thicknesses T_s and T.

$$T_s = \frac{T(1-\nu)}{2-\nu} \tag{35.8}$$

For instance, when Poisson's ratio is 0.3, Eq. (35.8) gives $T_s = 0.41T$. For the complete theoretical range of Poisson's ratio between $\nu = 0$ and $\nu = 0.5$, the corresponding thickness ratios T_s/T vary between 0.50 and 0.33.

The analysis of a general shell of revolution leads to other formulas, such as those required in the calculation of the conical and ellipsoidal vessels. For the cone configurations subtending angle 2α, the hoop and meridional stresses are

$$S_t = \frac{mP}{\cos \alpha} \tag{35.9}$$

and

$$S_l = \frac{mP}{2\cos \alpha} \tag{35.10}$$

As α approaches zero, the hoop stress for a conical shell approaches that for a cylinder. However, when the cone begins to flatten and α approaches 90°, the stress becomes unreasonably large, indicating that a flat membrane cannot support loads perpendicular to its plane. Similar reasoning can be applied to Eq. (35.10).

ELLIPSOIDAL SHELLS

The problem of an ellipsoidal shell, used frequently as a closure for pressure vessels and storage tanks, is complicated because of the expressions for curvature, which varies from point to point as shown in Fig. 35.2. The ABC portion is frequently used as a vessel head with an axis of revolution coinciding with the cylinder axis. The meridional stress S in this case is the same as the longitudinal stress in the adjoining cylinder

$$S_l = \frac{Pa}{2T} \tag{35.11}$$

Since $a = R$, Eqs. (35.5) and (35.11) must be identical. The hoop stress (not indicated in Fig. 35.2), since it acts in the plane perpendicular to the paper, is different

$$S_t = \frac{Pa(2b^2 - a^2)}{2Tb^2} \tag{35.12}$$

Internal Pressure

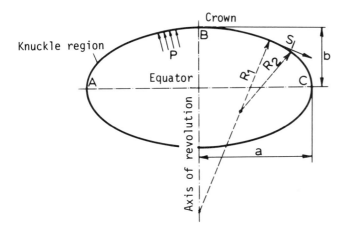

Fig. 35.2 Notation for an ellipsoid.

It is of some interest to note that the hoop stress can change its sign and become compressive when $a/b > \sqrt{2}$, despite the fact that the vessel is loaded internally. For this condition also, the maximum shear stress found at the equator (plane AC) becomes

$$\tau = \frac{Pa(a^2 - b^2)}{4Tb^2} \tag{35.13}$$

The criterion defined by Eq. (35.13) is of importance for vessels made from ductile materials. Experiments also indicate that a buckle can develop in the knuckle region under internal pressure because of the change from the tensile to the compressive hoop stress, Eq. (35.12), for certain values of a/b [216].

By analogy to a cylindrical vessel, the radial growth of a conical vessel is

$$\delta = \frac{(2-\nu)PR^2}{2ET\cos\alpha} \tag{35.14}$$

The equatorial dilation of an ellipsoidal head is not always positive and may be expressed by the formula

$$\delta = \frac{PR^2}{ET}\left(1 - \frac{a^2}{2b^2} - \frac{\nu}{2}\right) \tag{35.15}$$

When $\nu = 0.3$ and $a/b > 1.3$, the equatorial dilation can become negative, and it can cause an increase in the discontinuity stresses when a purely ellipsoidal head is used with a cylindrical shell of the same thickness.

TOROIDAL VESSEL

In many cases a torus or a doughnut-shaped pressure vessel is used in construction. This could be a steam generator, a bent tube, or a containment vessel in a nuclear reactor system. The stress analysis formulas can be developed from the equilibrium of forces with reference to the sketch shown in Fig. 35.3. For instance, vertical load on a shaded portion of the torus is

$$V = \pi(x^2 - \zeta^2)P \tag{35.16a}$$

Also, from the component of hoop stress this load is

$$V = 2\pi x T \cos\theta S_t \tag{35.16b}$$

Equating Eqs. (35.16a) and (35.16b) yields

$$S_t = \frac{P(x^2 - \zeta^2)}{2xT\cos\theta} \tag{35.16c}$$

Since

$$x = \zeta + R\cos\theta \tag{35.16d}$$

substituting this term in Eq. (35.16c) gives the general design formula for hoop stress

$$S_t = \frac{PR}{2T}\left(\frac{2\zeta + R\cos\theta}{\zeta + R\cos\theta}\right) \tag{35.17}$$

We can now have three conditions for the various values of S_t. For instance, when $\theta = 0$, Eq. (35.17) gives the minimum possible value as

$$S_t = \frac{PR}{2T}\left(\frac{2\zeta + R}{\zeta + R}\right) \tag{35.18}$$

For the case of the torus centerline, that is, at $\theta = \pi/2$, one obtains the standard hoop formula given by Eq. (35.3). However, when the θ line rotates in such a

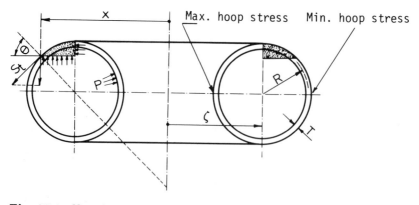

Fig. 35.3 Notation for a circular torus.

Internal Pressure

manner that $\theta = \pi$ and $x = \zeta - R$, the hoop stress in the torus becomes

$$S_t = \frac{PR}{2T}\left(\frac{2\zeta - R}{\zeta - R}\right) \tag{35.19}$$

The meridional stress is given by Eq. (35.5), and its value is independent of location on the torus. The same formulas can also be obtained from the Laplace equation for a thin axisymmetrical shell [178]. The form of this equation is

$$\frac{S_l}{R_1} + \frac{S_t}{R_2} = \frac{P}{T} \tag{35.20}$$

where R_1 and R_2 are the relevant radii of curvature given by $x/\cos\theta$ and R, respectively. The variation of the maximum hoop stress with the ratio of bend radius ζ to cross-sectional radius R is shown in Fig. 35.4. The vertical axis indicates a nondimensional parameter obtained by dividing the stress in torus by the hoop stress in the corresponding straight pipe. Theoretically, hoop stress becomes infinitely large for very small bend radii, that is, when the doughnut hole is essentially closed. Experience indicates that caution must be exercised when applying torus theory to the process of tube bending to a small radius. Strain hardening of the material and wall thickening on the inside of the sharp bend may cause the pipe to fail on the centerline of the bend instead where the stress is essentially the same as that in a straight cylinder. In such a case, Eq. (35.19) would not be recommended.

The majority of pipe and vessel configurations currently used can be classified as relatively thin, and the dividing line between the thin and thick vessels may be set at an m value of about 10. The lower practical limit of m is about 2.5 for cylinders and 3.5 for spheres. With these types of m ranges, it is customary to use the membrane stress criteria for thin shells and Lamé theory for thick shells. Use

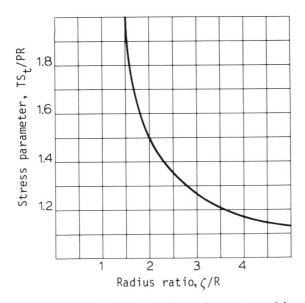

Fig. 35.4 Variation of maximum hoop stress with torus geometry.

THICK-CYLINDER THEORY

Although it is beyond the scope of this book to discuss various details of the mathematical theory of elasticity, it is perhaps of some value to briefly review some of the fundamental concepts of strains in terms of displacement in a two-dimensional system. These concepts are necessary for derivation of the expressions for stresses and deflections in thick-walled vessels.

The deformation characteristics of a two-dimensional element shown in Fig. 35.5 indicate radial and tangential movements u and v, respectively. The change of element thickness in the radial direction divided by the element thickness gives radial strain

$$\epsilon_r = \frac{\partial u}{\partial r} \tag{35.21a}$$

In the tangential direction the strain is affected by the change of v with respect to θ and by the displacement to the new radius $r + u$. This then gives

$$\epsilon_t = \frac{\partial v}{r\,\partial \theta} + \frac{(r+u)\delta\theta - r\,\delta\theta}{r\,\delta\theta} \tag{35.21b}$$

or

$$\epsilon_t = \frac{\partial v}{r\,\partial \theta} - \frac{u}{r} \tag{35.21c}$$

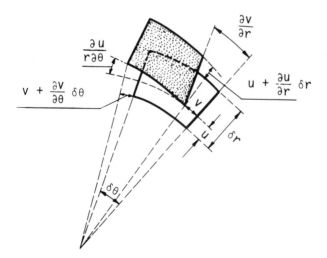

Fig. 35.5 Deformation of a two-dimensional element.

Internal Pressure

The corresponding shear strain is

$$\lambda_{rt} = \frac{\partial v}{\partial r} + \frac{\partial u}{r\, \partial \theta} - \frac{v}{r} \tag{35.21d}$$

For all the axisymmetric problems, the tangential displacement v and the shear strain λ_{rt} are zero, so that the above equations reduce to

$$\epsilon_r = \frac{du}{dr} \tag{35.21e}$$

and

$$\epsilon_t = \frac{u}{r} \tag{35.21f}$$

Because of the special importance of thick-cylinder theory in gun barrels, hydraulic ram cylinders, heavy piping, and similar applications, some of the details of the axisymmetric stress system for a thick-walled vessel are shown in Fig. 35.6. In a general case, the wall can be subjected to an internal pressure P_i and external pressure P_o, corresponding to the radii, R_i and R_o, respectively. Furthermore, because of the symmetry of loading, it is sufficient to analyze one element of the cylinder subtended by a small angle $d\theta$ and a small radial thickness dr. Summing up all the forces in the direction of the bisector of the angle $d\theta$ and making $\sin d\theta = d\theta$ gives

$$S_r r\, d\theta + S_t\, dr\, d\theta - \left(S_r + \frac{dS_r}{dr} dr\right)(r + dr)\, d\theta = 0 \tag{35.21g}$$

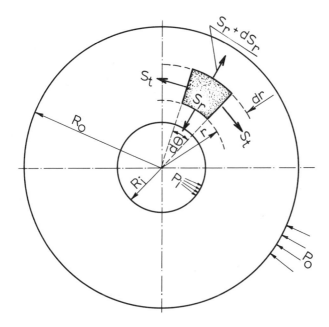

Fig. 35.6 Notation for a stress system in a thick cylinder.

Neglecting small quantities of the higher order in Eq. (35.21g) yields the differential equation

$$S_t - S_r - r\frac{dS_r}{dr} = 0 \tag{35.21h}$$

According to the theory of elasticity [2], the stresses can be expressed in terms of strains given by Eqs. (35.21e) and (35.21f). This yields

$$S_r = \frac{E}{1-\nu^2}\left(\frac{du}{dr} + \nu\frac{u}{r}\right) \tag{35.21i}$$

and

$$S_t = \frac{E}{1-\nu^2}\left(\frac{u}{r} + \nu\frac{du}{dr}\right) \tag{35.21j}$$

Substituting the foregoing stress values into Eq. (35.21h) gives

$$\frac{d^2u}{dr^2} + \frac{1}{r}\frac{du}{dr} - \frac{u}{r^2} = 0 \tag{35.21k}$$

The general solution of this differential equation is

$$u = C_1 r + \frac{C_2}{r} \tag{35.21l}$$

The first derivative of u with respect to r follows from Eq. (35.21l)

$$\frac{du}{dr} = C_1 - \frac{C_2}{r^2} \tag{35.21m}$$

Substituting Eqs. (35.21l) and (35.21m) into Eqs. (35.21i) and (35.21j) provides

$$S_r = \frac{E}{1-\nu^2}\left[C_1(1+\nu) - C_2\frac{1-\nu}{r^2}\right] \tag{35.21n}$$

and

$$S_t = \frac{E}{1-\nu^2}\left[C_1(1+\nu) + C_2\frac{1-\nu}{r^2}\right] \tag{35.21o}$$

The constants C_1 and C_2 are now determined from the known boundary conditions defining normal stresses P_i and P_o at R_i and R_o, respectively. Substituting these conditions into Eq. (35.21n) gives the two simultaneous equations for finding C_1 and C_2

$$C_1 = \frac{1-\nu}{E}\frac{R_i^2 P_i - R_o^2 P_o}{R_o^2 - R_i^2} \tag{35.21p}$$

Internal Pressure

and

$$C_2 = \frac{1+\nu}{E} \frac{R_i^2 R_o^2 (P_i - P_o)}{R_o^2 - R_i^2} \tag{35.21q}$$

Introducing the values of these constants into Eq. (35.21n) and (35.21o) yields

$$S_r = \frac{R_i^2 P_i - R_o^2 P_o}{R_o^2 - R_i^2} - \frac{R_i^2 R_o^2 (P_i - P_o)}{r^2 (R_o^2 - R_i^2)} \tag{35.22}$$

and

$$S_t = \frac{R_i^2 P_i - R_o^2 P_o}{R_o^2 - R_i^2} + \frac{R_i^2 R_o^2 (P_i - P_o)}{r^2 (R_o^2 - R_i^2)} \tag{35.23}$$

Equations (35.22) and (35.23) are known as *Lamé formulas*, named after a French engineer who obtained this solution in 1833. It is quite remarkable how this theory has withstood the test of time, advancing technology, and modern trends of numerical analysis.

The maximum shearing stress at any point of a thick cylinder follows from Eqs. (35.22) and (35.23) and is equal to

$$\tau = \frac{(P_i - P_o) R_o^2 R_i^2}{(R_o^2 - R_i^2) r^2} \tag{35.24}$$

Finally, the general expression for the radial displacement at any radius r follows from Eqs. (35.21l), (35.21p), and (35.21q)

$$u = \frac{1-\nu}{E} \frac{R_i^2 P_i - R_o^2 P_o}{R_o^2 - R_i^2} r + \frac{1+\nu}{E} \frac{R_i^2 R_o^2 (P_i - P_o)}{(R_o^2 - R_i^2) r} \tag{35.25}$$

When the cylinder is subjected to internal pressure only, we have

$$S_r = \frac{R_i^2 P_i}{(R_o^2 - R_i^2)} \left(\frac{r^2 - R_o^2}{r^2} \right) \tag{35.26}$$

$$S_t = \frac{R_i^2 P_i}{(R_o^2 - R_i^2)} \left(\frac{r^2 + R_o^2}{r^2} \right) \tag{35.27}$$

$$\tau = \frac{P_i R_o^2 R_i^2}{(R_o^2 - R_i^2) r^2} \tag{35.28}$$

and

$$u = \frac{1-\nu}{E} \frac{R_i^2 P_i r}{R_o^2 - R_i^2} + \left(\frac{1+\nu}{E} \right) \frac{R_i^2 R_o^2 P_i}{(R_o^2 - R_i^2) r} \tag{35.29}$$

It follows from the Lamé theory that the sum of the two stresses given by Eqs. (35.22) and (35.23) remains constant, suggesting that the deformation of all the elements in the axial direction is the same and the cylinder cross sections remain plane after the deformation. Equations (35.26) and (35.27) also indicate that both stresses reach maximum values at $r = R_i$ and that S_r is always compressive and smaller than S_t. The minimum value of the tangential stress is found at $r = R_o$, which is smaller than that at the inner surface.

So far the equations considered have been applicable to an infinitely long cylinder, and no axial stress was present. However, when the cylinder contains rigid closures and no change in length is possible, the axial stress under internal pressure is

$$S_l = \frac{2\nu P_i R_i^2}{R_o^2 - R_i^2} \tag{35.30}$$

When the vessel contains closures but is free to change its length under strain, the axial stress is

$$S_l = \frac{P_i R_i^2}{R_o^2 - R_i^2} \tag{35.31}$$

The corresponding axial strain is

$$\epsilon_l = \frac{(1 - 2\nu) P_i R_i^2}{E(R_o^2 - R_i^2)} \tag{35.32}$$

Finally, when the pressure is maintained by a piston-type closure at each end of the cylinder and there is no connection between the cylinder and the piston, the axial stress is zero and the strain is

$$\epsilon_l = -\frac{2\nu P_i R_i^2}{E(R_o^2 - R_i^2)} \tag{35.33}$$

THICK-WALLED SPHERE

In considering a thick-walled spherical shell, cylinder equations are not applicable, and it is therefore necessary to start the problem from first principles. For instance, the element of the vessel wall shown in Fig. 35.7 is subjected to a system of six normal forces, giving one equation of equilibrium. Because of the symmetry involved, there will be no shearing stresses on the faces of the element. This assumption is the same as that for thick cylinders. Hence, summing up the forces on a radial plane through the center of the element [217] gives

$$\left(S_r + \frac{dS_r}{dr} dr\right)(r + dr)^2 (d\theta)^2 - S_r r^2 (d\theta)^2 - 4 S_t r \, dr \, d\theta \sin\left(\frac{d\theta}{2}\right) = 0 \tag{35.34a}$$

Simplifying this expression and neglecting higher-order terms yields

$$\frac{dS_r}{dr} + \frac{2}{r}(S_r - S_t) = 0 \tag{35.34b}$$

Internal Pressure

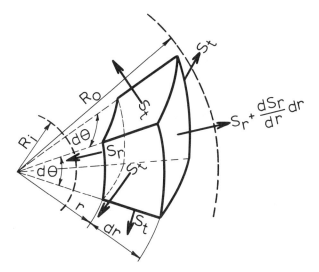

Fig. 35.7 Equilibrium of an element for a spherical shell.

Since S_t is the same in all directions, as indicated in Fig. 35.7, the radial and tangential strains can be expressed as

$$\epsilon_r = \frac{S_r}{E} - \frac{2\nu S_t}{E} \tag{35.34c}$$

and

$$\epsilon_t = \frac{(1-\nu)S_t}{E} - \frac{\nu S_r}{E} \tag{35.34d}$$

Since $\epsilon_r = du/dr$ and $\epsilon_t = u/r$, the stresses can now be written in terms of u and substituted into the equilibrium relation, Eq. (35.34b). On performing this operation and simplifying the result, the following equation in u is obtained

$$\frac{d^2u}{dr^2} + \frac{2}{r}\frac{du}{dr} - \frac{2u}{r^2} = 0 \tag{35.34e}$$

The general solution of this equation is

$$u = \frac{(1-2\nu)Ar}{E} + \frac{(1+\nu)B}{Er^2} \tag{35.34f}$$

If the stresses are now obtained from Eqs. (35.34c) and (35.34d) in terms of strains, we get

$$S_r = E\frac{2\nu\frac{u}{r} + (1-\nu)\frac{du}{dr}}{1-\nu-2\nu^2} \tag{35.34g}$$

and

$$S_t = E\frac{\frac{u}{r} + \nu\frac{du}{dr}}{1 - \nu - 2\nu^2} \tag{35.34h}$$

Now taking u and du/dr from Eq. (35.34f), substituting these values into Eqs. (35.34g) and (35.34h), and solving for the boundary conditions P_i and P_o yields

$$A = \frac{P_i R_i^3 - P_o R_o^3}{R_o^3 - R_i^3} \tag{35.34i}$$

and

$$B = \frac{R_i^3 R_o^3 (P_i - P_o)}{2(R_o^3 - R_i^3)} \tag{35.34j}$$

The final equations can be now obtained with the aid of u, du/dr, A, and B to give

$$S_r = \frac{P_i R_i^3 - P_o R_o^3}{R_o^3 - R_i^3} - \frac{R_i^3 R_o^3 (P_i - P_o)}{r^3 (R_o^3 - R_i^3)} \tag{35.35}$$

$$S_t = \frac{P_i R_i^3 - P_o R_o^3}{R_o^3 - R_i^3} + \frac{R_i^3 R_o^3 (P_i - P_o)}{2r^3 (R_o^3 - R_i^3)} \tag{35.36}$$

The corresponding expression for the radial displacement is

$$u = \frac{(1 - 2\nu) r (P_i R_i^3 - P_o R_o^3)}{E(R_o^3 - R_i^3)} + \frac{(1 + \nu) R_i^3 R_o^3 (P_i - P_o)}{2E r^2 (R_o^3 - R_i^3)} \tag{35.37}$$

For the case of internal pressure alone, the design formulas become

$$S_r = \frac{P_i R_i^3}{R_o^3 - R_i^3}\left(1 - \frac{R_o^3}{r^3}\right) \tag{35.38}$$

$$S_t = \frac{P_i R_i^3}{R_o^3 - R_i^3}\left(1 + \frac{R_o^3}{2r^3}\right) \tag{35.39}$$

and

$$u = \frac{P_i R_i^3}{E(R_o^3 - R_i^3)}\left[(1 - 2\nu)r + \frac{(1 + \nu)R_o^3}{2r^2}\right] \tag{35.40}$$

DESIGN CHARTS FOR THICK CYLINDERS

One of the more common design situations is to require a quick estimate of the maximum stresses and displacements at the surfaces of the thick cylinders working

Internal Pressure

under internal pressure. For these purposes, the maximum tangential stress at the inner surface can be stated as

$$S_t = P_i K_1 \tag{35.41}$$

Here K_1 is a dimensionless parameter given in Fig. 35.8 for a number of thick cylinders, and it is expressed as a function of the dimensionless ratio obtained by dividing the inner radius by the thickness of the wall. It is noted that K_1 is always larger than 1, indicating that the tangential stress should always exceed the applied internal pressure. The corresponding displacement of the inner surface is

$$u_i = \frac{P_i(R_o - R_i)}{E} K_2 \tag{35.42}$$

Similarly, the displacement of the outer surface of the cylinder under internal pressure is

$$u_o = \frac{P_i(R_o - R_i)}{E} K_3 \tag{35.43}$$

The factors given in Fig. 35.8 have been calculated for a Poisson's ratio of 0.3, which is common to various metals. It appears to have a relatively small effect on the magnitude of the displacements, so that the value of 0.3 also holds reasonable true for the majority of nonmetallic structural materials exhibiting elastic properties.

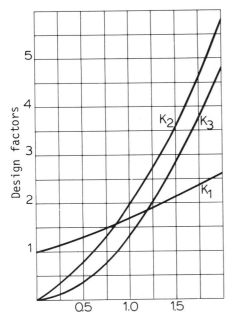

Fig. 35.8 Stress and deflection factors for thick cylinders under internal pressure.

ULTIMATE STRENGTH CRITERIA

So far, various formulas have been discussed that allow reasonably accurate design estimates of the elastic stresses and deflections in thin and thick pressure vessels or piping. The complete problem, however, may also require the prediction of the ultimate pressures in cylinders and spheres at which these components can fail by bursting.

The general criteria of shell failure under internal pressure have been developed from the classical theory of plasticity, which has been shown to be suitable for explaining burst characteristics of thick-walled vessels made of ductile materials. The essential difference between the elastic and plastic response of a thick shell can be illustrated as follows. In a purely elastic response of a thick shell, the maximum stress under internal pressure develops at the bore. On further load increase at this point, the material reaches the yield but, contrary to what one might expect, failure of ductile fibers does not start at the bore. The reason for this is that the strain at the inner zone of the wall is held at a constant level by the restraint of the outer fibers until the region of plastic flow moves radially outward. The motion of the elastic-plastic interface causes the tangential stresses in the outer fibers to increase until a complete state of plastic strain is established throughout the wall and the fibers along the outer surface of the shell begin to fail. Researchers in the field of plasticity have defined the *bursting pressure* as the internal pressure required to move the elastic-plastic interface into the outer radius of the vessel.

The state of plastic stress defined above as a criterion of bursting pressure has several important implications. First of all a vessel made of a ductile material has a considerable amount of strength beyond the onset of yielding at the inner fiber and should retain its usefulness up to the very point of fracture. This reserve of strength, however, must be dependent on the entire history of stress and temperature at a particular critical location. It is therefore necessary to assign a correct level of design stress in predicting the relevant bursting pressures. The formulas for this purpose contain either yield or ultimate strength terms of the material involved. In the case of the majority of vessel applications, work-hardening materials are used which require knowledge of the stress-strain curves. This condition introduces additional functional relationships in the development of pressure-deformation equations for cylinders and spheres. It has been customary to employ a Ludwik type of a stress-strain curve in which the stress is represented by an exponential function of strain. The work-hardening capacity of the material is said to increase when the exponent of the stress-strain curve increases. The bursting pressure, in turn, decreases with the increase in working hardening.

The technical literature contains a multitude of design formulas for the calculation of bursting strength of thick- and thin-walled vessels, derived from the assumed theories of failure and containing various limitations. One development in this area comes from England [218]. This approach is based on the pressure-deformation response of a vessel and on an idealized stress-strain curve of the Ludwik type. The formulas resulting from this work are intended for work-hardening materials, provided that the strain-hardening exponent for the particular material can be established. In the case of commonly employed low- and intermediate-strength low-alloy and carbon steels, a simple relation can be obtained between the strain-hardening exponents and the materials strength ratio β. This ratio is calculated

Internal Pressure

by dividing the yield strength of the material by its ultimate strength. It is well to recall here that the definition of the yield point on a particular stress-strain plot can lead to some difficulties. To circumvent this problem, it has been a custom in industry to accept a 0.20% offset yield strength for most engineering calculations. This offset strength value is usually higher than the elastic limit of the material by about 10%, but the exact spread between the two values will depend on the type of material involved.

BURST PRESSURE OF CYLINDERS AND SPHERES

Introducing the relation between the strain-hardening exponent and the strength ratio β into Svensson's equations for thick cylindrical and spherical vessels gives the design formulas

$$P_c = S_y \psi B_1 \quad \text{(cylinder)} \tag{35.44}$$

and

$$P_s = S_y \psi B_2 \quad \text{(sphere)} \tag{35.45}$$

The use of Eqs. (35.44) and (35.45) involves calculating the geometry factor ψ and burst factors B_1 and B_2. To help with the design process, the charts in Figs. 35.9 and 35.10 can be used after the dimensionless parameters m and β have been established. It should be noted that the well-known formulas of Faupel [21] can be reduced to essentially the same form as that illustrated by the foregoing equations. The equivalent B functions of Faupel, however, are linear in parameter β making

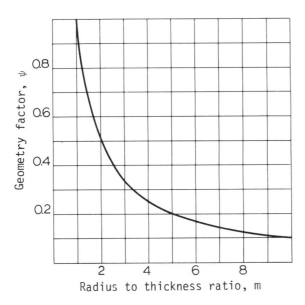

Fig. 35.9 Geometry factor for thick cylinders and spheres.

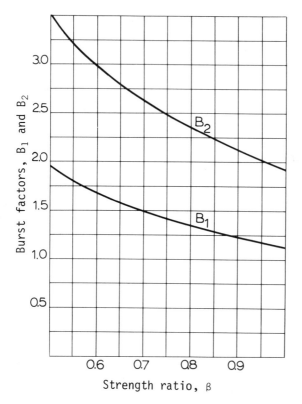

Fig. 35.10 Burst factors for thick cylinders and spheres.

the results at lower β values more conservative. Both formulas, Eqs. (35.44) and (35.45), as well as the Faupel equations, are especially convenient in design because of the usual availability of the material properties necessary for calculating the dimensionless ratio β. The formula selection in this chapter is essentially based on the assumption that the burst pressure/strength ratio relation is more likely to be nonlinear over the major part of the β range. As reported by various investigators in the past, both Svensson's and Faupel's formulas have been found to be useful in correlating experimental data.

When the cylindrical and spherical vessels can be classed as relatively thin, Svensson's theory leads to the working equations

$$P_c = \frac{S_y}{m} B_3 \quad \text{(cylinder)} \tag{35.46}$$

and

$$P_s = \frac{S_y}{m} B_4 \quad \text{(sphere)} \tag{35.47}$$

The design chart given in Fig. 35.11 can be used for finding burst factors B_3 and B_4 when the strength ratio β is known. The formulas given by Eqs. (35.46) and (35.47) are intended for all values of m greater than 10.

Internal Pressure

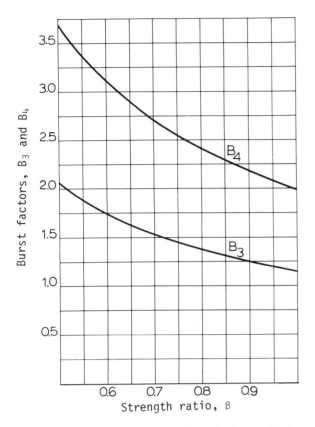

Fig. 35.11 Burst factors for thin cylinders and spheres.

SHRINK-FIT DESIGN

The fundamental objective of a shrink-fit construction is to introduce residual or initial stresses into the material in order to control the critical features of the stress field. This process can, for instance, increase the elastic resistance of a multiwall pressure vessel, strengthen the extrusion die, or enhance the fatigue life of a wheel mounted on the shaft. Such a shaft can be either hollow or solid. Also a shrunk-on shell, applied to the liner of a pressure vessel, should help to retard crack propagation. The shrink-fit design then can mitigate the peak stresses and enhance utilization of the materials.

The basic design formulas for a shrink-fit assembly are derived from the theory of heavy-walled cylinders. This brief section is limited to the two-shell construction, shown in Fig. 35.12, which is widely used in industry. The inner cylinder defined by radii R_s and R_i may represent a pressure vessel or a hollow shaft. The outer cylinder bounded by R_i and R_o may act as a pressure vessel liner or a component in a double-wall cylinder construction. In either of these cases the system is characterized by the amount of radial interference. When the inner radius of the outer cylinder is made smaller than the outer radius of the inner part, the system can be assembled either by heating the outer cylinder or cooling the inner component. This process results

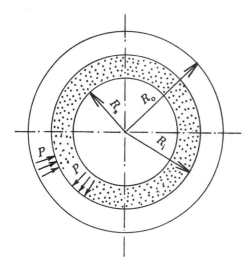

Fig. 35.12 Shrink-fit assembly.

in a contact pressure P often described as the interference or shrink-fit pressure. The amount of radial interference (or lack of it) δ at the common boundary defined by R_i is equal to the sum of the decrease of the outer radius of the inner part and the increase of the inner radius of the outer cylinder.

Although this sounds complicated, the two surfaces in contact seek only a common boundary in relation to their individual stiffnesses. The resulting design formula [21] becomes

$$P = \frac{E\delta}{2R_i^3} \frac{(R_i^2 - R_s^2)(R_o^2 - R_i^2)}{(R_o^2 - R_s^2)} \tag{35.48}$$

The foregoing expression applies to same materials. When the inner radius R_s is reduced to zero we obtain the case of a solid shaft held by a hub. The relevant formula is

$$P = \frac{E\delta(R_o^2 - R_i^2)}{2R_i R_o^2} \tag{35.49}$$

Note that this formula was also used in Design Problem 14.3. When the two shrink-fitted cylinders have different mechanical properties such as E_o and ν_o for the outer cylinder and E_i and ν_i for the inner member, the interference pressure can be calculated as

$$P = \frac{\delta}{\frac{R_i}{E_o}\left(\frac{R_i^2 + R_o^2}{R_o^2 - R_i^2} + \nu_o\right) + \frac{R_i}{E_i}\left(\frac{R_s^2 + R_i^2}{R_i^2 - R_s^2} - \nu_i\right)} \tag{35.50}$$

Internal Pressure

When R_s tends to zero we have the case of a solid shaft defined by a simpler formula

$$P = \frac{\delta}{R_i \left[\frac{1}{E_o} \left(\frac{R_o^2 + R_i^2}{R_o^2 - R_i^2} + \nu_o \right) + \frac{1-\nu_i}{E_i} \right]} \tag{35.51}$$

Finally making the $E_o = E_i = E$ and $\nu_o = \nu_i = \nu$, Eqs. (35.50) and (35.51) reduce to Eqs. (35.48) and (35.49). The foregoing equations define the four basic design situations of shrink-fit pressure where other effects such as internal pressure or centrifugal forces are not involved. These lead to much more complicated expressions. One of the cases illustrated in some detail in Chapter 38 is that of a pressure attenuation in nested cylinders.

SYMBOLS

A	Constant of integration
a	Major half-axis, in. (mm)
B	Constant of integration
B_1 through B_4	Burst factors
b	Minor half-axis, in. (mm)
C_1, C_2	Integration constants
E	Modulus of elasticity, psi (N/mm^2)
E_o	Modulus of elasticity of sleeve, psi (N/mm^2)
E_i	Modulus of elasticity of shaft, psi (N/mm^2)
F_l	Longitudinal force, lb (N)
F_t	Tangential force, lb (N)
K_1, K_2, K_3	Thick-cylinder factors
$m = R/T$	Radius-to-thickness ratio
P	General symbol for pressure, psi (N/mm^2)
P_o	External pressure, psi (N/mm^2)
P_i	Internal pressure, psi (N/mm^2)
P_c	Burst pressure of cylinders, psi (N/mm^2)
P_s	Burst pressure of spheres, psi (N/mm^2)
R	Mean radius, in. (mm)
R_i	Inner radius, in. (mm)
R_o	Outer radius, in (mm)
R_1, R_2	Radii of curvature, in. (mm)
R_s	Inner radius of shaft, in. (mm)
r	Arbitrary radius, in. (mm)
S_l	Longitudinal or axial stress, psi (N/mm^2)
S_r	Radial stress, psi (N/mm^2)
S_t	Hoop or tangential stress, psi (N/mm^2)
S_u	Ultimate strength, psi (N/mm^2)
S_y	Yield strength, psi (N/mm^2)
T	Thickness of cylinder wall, in. (mm)
T_s	Thickness of spherical wall, in. (mm)
u	Radial displacement, in. (mm)

u_i	Radial displacement of inner surface, in. (mm)
u_o	Radial displacement of outer surface, in. (mm)
V	Vertical load, lb (N)
v	Tangential displacement, in. (mm)
x	Arbitrary distance, in. (mm)
α	Cone half-angle, rad
$\beta = S_y/S_u$	Strength ratio
δ	Dilation, also radial interference, in. (mm)
ϵ_l	Longitudinal strain, in./in. (mm/mm)
ϵ_r	Radial strain, in./in. (mm/mm)
ϵ_t	Tangential strain, in./in. (mm/mm)
ζ	Major radius of torus, in. (mm)
θ	Arbitrary angle, rad
λ_{rt}	Shear strain, in./in. (mm/mm)
ν	Poisson's ratio
ν_o	Poisson's ratio of sleeve
ν_i	Poisson's ratio of shaft
τ	Shear stress, psi (N/mm^2)
ψ	Geometry factor

36
External Pressure

INTRODUCTION

Although the design of vessels subjected to external pressure is found in many industrial applications, the majority of publications devote rather limited attention to this important topic. Lack of emphasis in this area of mechanics may lead to design errors when stress criteria intended for the treatment of internally pressurized vessels are extended to cover the externally loaded containers. Such errors are due to the fact that the externally pressurized vessels can fail by elastic instability long before the relevant compressive stresses can reach a critical magnitude. In addition, the effect of shape imperfections on the externally loaded vessels results in a number of analytical and computational difficulties. Often the problem is accentuated by the presence of manufacturing imperfections and the variations in material properties. These constraints make it mandatory to pursue the development of idealized models and the approximate design formulas based on the empirical data. This chapter is devoted to such simplified design solutions and practical considerations of vessel sizing criteria.

THINNESS FACTOR

Intuition tells us that there must be a definite difference between the responses of a short, thick-walled vessel and a long, thin-walled vessel subjected to external pressure. One hundred years of investigation of the theoretical and experimental aspects of the response of cylindrical shells to external pressure has produced many technical papers. The specific question of relating vessel geometry to the mode of failure under external pressure was reviewed [219] with special regard to the concept of the "thinness ratio" [220]. This study was supported by careful experiments on steel cylinders.

The collapse mechanism under external pressure appears to fall into two distinct patterns. In one type of response we can observe circumferential lobes and localized buckles as soon as the material begins to yield. In other cases the mechanical response may be characterized by the development of an hourglass shape sometimes described as a waisted configuration.

In mathematical terms the *thinness ratio* can be defined as

$$\lambda = 1.2(m)^{1/4}(k\phi)^{-1/2} \tag{36.1}$$

where k, m, and ϕ denote dimensionless parameters which can be obtained from the geometrical and physical constraints of the problem. To date, experimental data on cylindrical vessels [211, 219] have led to the establishment of a number of bracketing design criteria which can be obtained from Eq. (36.1).

STRESS RESPONSE

The value of the distinguishing boundary parameter λ between the lobing (local buckling) and nonlobing response was found to be equal to about 0.35. In practice, such low values of λ can be expected only when m is relatively small and the $k\phi$ product is relatively high. The value of 0.35 is compatible with the characteristics of a short and thick cylinder made of a low-strength material. Hence, for all λ values smaller than 0.35, simple hoop stress should govern the design exclusively, and there is no need for calculating the collapse pressures by the stability formulas. The critical compressive stress reaches the yield point midway between such discontinuities as vessel heads or stiffening rings. Local yielding is also possible in the vicinity of the stiffeners and similar transitions. However, high longitudinal discontinuity effects in these areas are not expected to precipitate the overall collapse. A remarkably simple design formula based on the concept of midbay collapse may be stated as

$$P_y = \frac{0.9S_y}{m} \tag{36.2}$$

Formula (36.2), developed more than 60 years ago [221], resulted from a thorough stress analysis of a thin-walled cylindrical vessel reinforced by stiffening rings and subjected to the combined effect of uniform radial and axial pressures. Detailed expressions leading to the establishment of the formulas for this type of loading involved the dimensions of the stiffeners and the vessel itself. Equation (36.2) indicates that the collapse pressure corresponding to the yield strength is only about 10% lower than that which can be obtained from the elementary membrane stress formula. This formula is very useful in preliminary design calculations involving ring-stiffened cylindrical vessels having m values higher than about 10. It should

External Pressure

also be noted that the parameter m is determined here on the basis of a mean cylinder radius, although the actual pressure is applied at the outer surface. In most practical cases, however, the effect of this difference will be found to be negligible.

STABILITY RESPONSE

When the response of a cylindrical vessel under external pressure is purely in a stability mode, preferential yielding at some region of the wall away from stiffeners is initiated, which in turn leads to unstable geometry and sudden formation of a lobe. Such a pattern of deformation is nonaxisymmetric and the parameter λ corresponding to this condition should exceed 2.5. It may be said, then, that above the limit $\lambda = 2.5$, elastic stability should govern the design. In this category of calculations, the classic *long cylinder formula* is often used. For a Poisson's ratio of 0.3, the long cylinder formula becomes

$$P_c = \frac{0.275E}{m^3} \tag{36.3}$$

The formula given by Eq. (36.3) is intended for perfectly round cylinders where collapse takes place at stresses significantly lower than the elastic limit of the material. Such a mode of failure is essentially due to the insufficient flexural rigidity of the cylinder, and the pressure at which the initially circular form becomes unstable may be defined as the critical buckling pressure P_c. The cylinder is assumed to buckle into an elliptical shape. Some of the elementary design formulas for buckling loads in arches and rings were given in Table 10.7. When the theory of a buckled ring is extended to a long vessel by considering an elementary ring cut out of the cylinder by two cross sections a unit distance apart, Eq. (36.3) is obtained. This formula can also be modified for the materials with a pronounced yield point [31] to give

$$P_c = \frac{\phi S_y}{m(\phi + 3.64m^2)} \tag{36.4}$$

As the parameter m in Eq. (36.4) increases, the critical pressure approaches the limiting theoretical value given by Eq. (36.3). For other values of m, the critical pressure calculated by Eq. (36.4) is less than that obtained from the classical formula, Eq. (36.3). In design situations where the collapse pressure is required beyond the elastic limit of the material, the tangent modulus of elasticity E_t may be used in Eq. (36.3) instead of the customary Young's modulus denoted by E.

Design Problem 36.1

Evaluate the extent of the error involved in estimating the external collapse pressure by means of a membrane stress criterion instead of the stability formulas on the assumption that the material's ratio ϕ corresponds to that of a mild steel known as A-36, having a yield strength of $S_y = 36,000$ psi. Assume the minimum and maximum m values to be 10 and 50, respectively.

Solution

By definition of the inverse strain parameter,

$$\phi = \frac{30 \times 10^6}{36 \times 10^3} = 833$$

The membrane stress criterion for a cylindrical vessel is

$$P_m = \frac{S_y}{m}$$

Hence, using Eq. (36.3) gives

$$\frac{P_m}{P_c} = \frac{3.64 m^2}{\phi}$$

and

$$\frac{P_m}{P_c} = 0.00436 m^2$$

This ratio is shown as a function of m in the following table.

m	10	20	30	40	50
P_m/P_c	0.436	1.744	3.924	6.976	10.900

◆

It is evident from Design Problem 36.1 that the use of the elementary membrane stress criterion for predicting collapse pressures can sometimes be rather misleading, particularly in the case of a relatively thin vessel or piping. On the other hand when the specific value of λ is 0.35 or less, the use of the membrane stress criterion appears to be quite appropriate. This statement requires, however, that the relevant length of the cylindrical vessel be well defined, together with the other parameters necessary for the solution of Eq. (36.1).

MIXED MODE RESPONSE

The bracketing values of λ equal to 0.35 and 2.5 cannot be exact, and a large "gray area" must exist between these limits. Experience also indicates that for values of λ up to 1.0, the stress criterion is still reasonably accurate. The hoop stress criteria in general should give conservative results even without due corrections for small initial imperfections and manufacturing tolerances. The reader should be cautioned, however, that gross initial imperfections can be very detrimental and may precipitate the overall cylinder collapse. This aspect of the analysis therefore requires a good deal of engineering judgment, since the precise definitions of what may be considered as "small" or "gross" imperfections have not been fully established.

The intermediate region bracketed by the values of λ equal to 1.0 and 2.5 represents a complicated picture of cylinder behavior under external pressure where some combination of stress and stability effects is involved. Relevant tests show

External Pressure

that under these conditions, some local yielding may take place before the critical elastic instability load is reached, precipitating the onset of ultimate collapse. Furthermore, with the simultaneous involvement of stress and stability response, the stresses arising from the imperfections may prove to be rather significant.

The material parameter ϕ numerically representing the inverse of the elastic strain, varies normally between about 100 and 1000 for the great majority of metallic engineering materials. The medium range of yield strength is about 60,000 to 150,000, with the corresponding ϕ values of 500 and 200, respectively. In this range we may find such steels as A537, HY80, or 4330 series, which are recognized for their superior fracture toughness characteristics. The practical ranges of the dimensionless parameters k and m can be assumed to be as follows:

$$10 \leq m \leq 100$$

and

$$0.001 \leq k \leq 0.200$$

CLASSICAL FORMULA FOR SHORT CYLINDERS

Predictions based on the elastic behavior are quite common and can be considered as upper-bound estimates [31]. A well-known example of this type of solution is based on the short-cylinder theory describing the buckling response of the cylinder wall midway between the stiffeners. A simplified version of the relevant formula is

$$P_c = \frac{0.87 E k}{m^{3/2}} \quad (36.5)$$

Equation (36.5) applies to vessel materials that have a Poisson's ratio equal to about 0.3. This theory does not include the effect of the circumferential stiffener on the strength of the shell, and it is assumed that the initial out-of-roundness of the shell is zero. In general, the design of hardware for the various applications of cylindrical vessels and piping, involving external pressure loading, is complicated despite the availability of a variety of design formulas and their respective ranges of practical use. This situation is not surprising when one considers the modes of failure in the elastic or elastoplastic range, in addition to the geometrical and manufacturing features of pressure vessels.

For all values of $\lambda > 1.0$, stability-oriented design formulas should be used. Here the length of the cylinder in relation to the cylinder radius determines what type of a design model is likely to be applicable. Because of the stability criteria affecting the design, the analysis should consider possible effects of manufacturing out-of-roundness on the collapse resistance of a particular cylindrical vessel.

MODIFIED FORMULA FOR SHORT CYLINDERS

The classical short-cylinder formula can be modified to give the following simplified expression for calculating the ultimate collapse pressure

$$P_u = S_y F(k) F(m) \tag{36.6}$$

This equation is essentially based on Timoshenko's theory of ellipticity [31], and it will be assumed that Eq. (36.6) conforms to the intermediate region of cylinder response, where λ values fall somewhere between 1.0 and 2.5. It may again be recalled that this region of λ represents the behavior in a mixed mode involving stress and stability.

In Eq. (36.6), $F(m)$ depends on the R/T ratio and $F(k)$ is a geometrical parameter containing the effect of out-of-roundness. The design formula expressed by Eq. (36.6) reduces essentially to the short-cylinder formula (36.5) when the effects of out-of-roundness are neglected. The relevant detailed expressions for the various design parameters [222] can now be stated as

$$F(k) = K_1' - (K_1'^2 - K_2')^{1/2} \tag{36.7}$$

$$K_1' = m^{1/2} + 0.87k\phi(1 + 6n) \tag{36.8}$$

$$K_2' = 3.48k\phi(m)^{1/2} \tag{36.9}$$

$$F(m) = \frac{1}{2m^{3/2}} \tag{36.10}$$

Here n is the out-of-roundness parameter, defined as the ratio of radial deviation from the perfect circular shape to the thickness of the vessel. It is emphasized that this deviation is measured on the radius and has a linear dimension similar to a tolerance on the drawing.

The range of applicability of Eq. (36.6) may be approximated using the following general criteria, in which we seek out the appropriate ratios of cylinder length to radius. The length can be measured between the stiffeners or cylinder heads

$$\left(\frac{L}{R}\right)_{min} = 0.63(m)^{-1/2}\left(\frac{1.44\phi}{m} + 1\right) \tag{36.11}$$

and

$$\left(\frac{L}{R}\right)_{max} = 3.1(m)^{1/2} \tag{36.12}$$

When the actual L/R ratio, given by the particular design constraints, is found to fall between the two values computed from Eqs. (36.11) and (36.12), Eq. (36.6) is deemed to apply. The designer should be cautioned, however, that under certain circumstances, Eqs. (36.11) and (36.12) may indicate that the range of applicability of Eq. (36.6) is nonexistent. This should not be surprising when one considers the complex nature of the functions and the number of parameters involved. The range of applicability of Eq. (36.6) appears to decrease substantially with the simultaneous

External Pressure

decrease in the R/T ratio and the yield strength of the material. Fortunately, in the majority of real design situations, R/T ratios are found to be higher than about 20 and ϕ values lower than 500. Within these combinations of boundary values, there is sufficient scope for a relatively wide usage of Eq. (36.6).

SIMPLIFIED CRITERION FOR OUT-OF-ROUNDNESS

The out-of-roundness parameter n can be defined in terms of the extreme diametral measurements and wall thickness of a cylindrical pressure vessel as

$$n = \frac{D_{max} - D_{min}}{4T} = \frac{e}{T} \qquad (36.13)$$

It should be mentioned here that Eq. (36.13) only applies to cases where radial deviation can be related to the even number of circumferential lobes of the out-of-roundness pattern. For a conservative estimate based on Eqs. (36.6) and (36.13), a purely elliptical mode of cylinder collapse may be recommended for design calculations. Unfortunately, the actual out-of-roundness parameter n will seldom be known before manufacture. Nevertheless, as the first rational step in design, the extent of the anticipated maximum out-of-roundness can be deduced from knowledge of the customary manufacturing tolerances. For instance, when the tolerance on the diameter of a particular cylindrical canister is given, say, as ± 0.05 in., the corresponding value of n can be taken as $n = 0.025/T$.

For a long time now the effect of out-of-roundness on the collapse strength of vessels and piping under external pressure has been duly recognized by the ASME code. For the purpose of eliminating the potential of any gross out-of-roundness, the ASME code recommends that the ratio $e/2R$ not exceed 1%, where R is the nominal radius. This conservative rule is at times used as the upper limit in establishing the design criteria. The actual manufacturing and field experience tends to indicate, however, that 1% grossly overestimates the extent or radial deviation from the perfect circularity found by the measurements. Advances in mechanical technology result in relatively small increases in values of n, even for large-diameter vessels.

LONG CYLINDER WITH OUT-OF-ROUNDNESS

When the pressure vessel is relatively long and its characteristics fall outside the range of short-cylinder geometry dictated by Eq. (36.6), a conservative formula for the collapse strength under external pressure is

$$P_{ul} = S_y A_m A_n \qquad (36.14)$$

where

$$A_m = \frac{1}{2m^3} \qquad (36.15)$$

$$A_n = A_1 - (A_1^2 - A_2)^{1/2} \qquad (36.16)$$

$$A_1 = m^2 + 0.275\phi(1 + 6n) \tag{36.17}$$
$$A_2 = 1.1 m^2 \phi \tag{36.18}$$

When the effect of out-of-roundness is neglected by making $n = 0$, Eq. (36.14) reduces to the classical formula for the elastic buckling of a long cylinder given by Eq. (36.3).

EFFECTIVE OUT-OF-ROUNDNESS

Although there is always a good deal of uncertainty as to the extent of manufacturing imperfections which can be defined ahead of time, it is possible to make a reasonable assumption of the out-of-roundness parameter for design purposes. It is important to consider this feature because it can become significant in relation to other effects, such as variation in wall thickness or the residual stress patterns. It should be added here that in the case of well-known structural members such as columns, the collapse is essentially of the bending type, leading to the formation of a plastic hinge. Unfortunately, as far as the vessels and piping subjected to external pressures are concerned, the collapse mechanism is much more sensitive because even small local deformations can give rise to significant bending moments, because of the presence of large compressive forces. For example, when conducting a typical structural test of a pressure vessel, it is easy to note that the relevant load deflection curve develops smoothly almost up to the level of external collapse pressure, at which point a rather violent failure suddenly occurs. This type of structural behavior is known as an *implosion*, in contrast to the "explosion" caused by internal pressure.

It is also fair to state that none of the rigorous theoretical treatments available at present can allow for all the major factors influencing the collapse, and none of them are capable of predicting the collapse pressures with too great a degree of reliance. It is also unlikely that this situation will drastically change in the immediate future, and therefore the best course of action for the present appears to be to include the out-of-roundness corrections in prototype design and to verify the integrity of a pressure vessel by experiment. To date, correlations of the measured and calculated collapse pressures appear to indicate a trend of improved agreement with the decreasing m ratio. The scatter in the area of thick-walled piping and vessels may be caused mainly by the variation in material properties and the initial imperfections. It is less severe for thick-walled tubes and many test results for thick-walled piping should indicate collapse pressures that exceed the calculated values. While thick-walled vessels and tubes need perhaps less restrictive criteria, the results on thin-walled tubes may still require somewhat larger factors of safety.

The foregoing discussion indicates that the collapse of thick-walled and thin-walled vessels is based on different criteria, and it is often difficult to determine the exact boundary between the response of thick and thin vessels. Short and thick-walled vessels fail at the yield point of the material, while long and thin-walled vessels tend to become elastically unstable at wall stresses far below the yield strength.

The concept of thinness factor λ can now be utilized in developing the method of assessing the effective out-of-roundness. Such an approach may be justified on

External Pressure

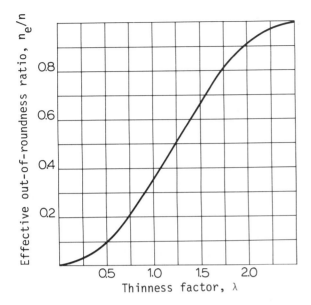

Fig. 36.1 Chart for effective out-of-roundness.

the grounds that the effect of out-of-roundness on the collapse of a long thin-walled vessel, for example, is expected to be more significant than the same effect on a short, thick-walled vessel. A brief review of the elementary mathematical models indicates that radial deviation-to-thickness ratio, entering the collapse-strength formulas Eqs. (36.6) and (36.14) can be represented by the following expression:

$$n_e = n \sin^2(36\lambda) \tag{36.19}$$

where λ, termed the thinness factor, is given by Eq. (36.1) and n_e is the ratio of the effective radial deviation to thickness. Assuming an elliptical shape of the vessel as the most likely mode of failure under external pressure, the first step is to estimate the extent of out-of-roundness parameter from the rule given by Eq. (36.13)

For a specified vessel geometry and selected material properties, the thinness factor λ can be obtained from Eq. (36.1). Hence, the ratio n_e/n follows directly from Eq. (36.19) or Fig. 36.1. Any functional relationship other than the one shown may not be sufficiently realistic since only the S-curve of Fig. 36.1 appears to fit the boundary conditions for the short, intermediate, and long cylinder criteria.

Design Problem 36.2

Determine the maximum allowable out-of-roundness for a large cylindrical canister at a design pressure of 750 psi, assuming a clear distance between the stiffeners equal to 60 in., a mean radius of 35 in., and a wall thickness of 0.875 in. The material is HY80 with a minimum expected yield strength of 80,000 psi and the modulus of elasticity of 30×10^6 psi. Plot the variation of the collapse pressure with radial deviation from a perfect circular shape, and calculate the corrected design pressure based on the theory of effective out-of-roundness.

Solution

The relevant dimensionless parameters are

$$k = \frac{0.875}{60} = 0.0146$$

$$m = \frac{35}{0.875} = 40$$

$$\phi = \frac{30 \times 10^6}{80 \times 10^3} = 375$$

Performing a "λ check" based on Eq. (36.1) gives

$$\lambda = 1.2(40)^{1/4}(0.0146 \times 375)^{-1/2} = 1.29$$

Hence, the canister is likely to behave as a short cylinder where a mixed mode of failure involving the stress and stability may be expected. The minimum critical length criterion follows from Eq. (36.11)

$$\left(\frac{L}{R}\right)_{\min} = 0.63(40)^{-1/2}\left(\frac{1.44 \times 375}{40} + 1\right) = 1.44$$

so that

$$L_{\min} = 1.44 \times 35 = 50.4 \text{ in.}$$

Since $50.4 < 60$, the theoretical short-cylinder formula, Eq. (36.6), should apply. It is obvious from a brief look at Eq. (36.12) that the critical maximum value of applicable length will be considerably higher than 60 in. With this decision in mind, we can compute the next needed parameter from Eq. (36.8)

$$K_1' = (40)^{1/2} + 0.87 \times 0.0146 \times 375\left(1 + \frac{6e}{0.875}\right)$$

$$= 11.09 + 32.66e$$

Since we have not as yet assigned the level of radial deviation, the term e can be carried further with K_1'. From Eq. (36.9),

$$K_2' = 3.48 \times 0.0146 \times 375(40)^{1/2} = 120.5$$

Hence utilizing the above results according to Eq. (36.7) the $F(k)$ formula becomes

$$F(k) = 11.09 + 32.66e - 1.58(1 + 291e + 429.8e^2)^{1/2}$$

From Eq. (36.10),

$$F(m) = \frac{1}{2 \times 40^{1.5}} = 0.00198$$

Hence, using Eq. (36.6) gives the final parametric equation for the collapse pressure

$$P_u = 1756.7 + 5173.3e - 250.3(1 + 291e + 429.8e^2)^{1/2}$$

External Pressure

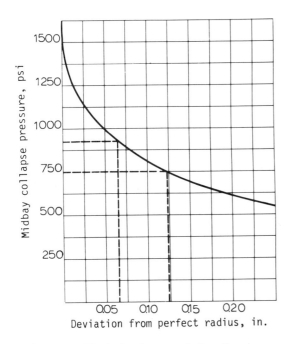

Fig. 36.2 Deviation from perfect radius, in.

This functional relationship is illustrated in Fig. 36.2, from which the maximum allowable deviation from a perfect radius for the design pressure of 750 psi is 0.123 in. However, at $\lambda = 1.29$ the effective out-of-roundness for this case can be obtained from Eq. (36.19). This calculation yields

$$n_e = 0.52n = \frac{0.52 \times 0.123}{T} = \frac{0.064}{T}$$

The corrected design pressure in Fig. 36.2 depends on the assumed value of cylinder thickness $T = 0.875$ in. ♦

This simple method of correction relaxes the original degree of conservatism of Eq. (36.6) because of the assumption of the elliptical mode of failure. In the actual numerical illustration of Design Problem 36.2, this amounts to increasing the estimated design pressure above 750 psi. The proposed critical pressure formulas given so far in this chapter, together with the calculational results and discussion, are offered as an illustration of the process of application of relatively simple solid mechanics principles to a complex design problem. The formulas given by Eqs. (36.6), (36.14), and (36.19) may be used in the preliminary design and development testing, but only within the limits defined by Eqs. (36.1), (36.11), and (36.12). The proposed method is offered here as an alternative model rather than a substitute for the established and proven code practices such as, for instance, those recommended by the ASME Boiler and Pressure Vessel Code.

EMPIRICAL DEVELOPMENTS

Because of rather substantial theoretical difficulties in the treatment of ring-stiffened cylinders subjected to external pressure, considerable effort has been expended by pressure vessel specialists in this country and abroad in gathering well-documented experimental data [211]. A nondimensional plot of such results, representing a lower-bound curve of midbay collapse pressures in cylindrical, ring-stiffened vessels, is given in Fig. 36.3. The plot includes both internally and externally stiffened cylinders, where the parameter $(1.05mP_e/S_y)$ is shown as a function of $0.92k\phi/m^{1/2}$, in line with the definitions and notation used so far in this chapter. P_e denotes the experimentally determined interstiffener collapse pressures for reliable published data having m values between 6 and 250 and the wide range of L/R ratios between 0.04 and 50. In the majority of test cylinders reported in the literature, the out-of-roundness was much less than 1% of the radius. In some cases, however, the values of out-of-roundness were found to be on the order of 1%, representing what can be judged to be relatively poor manufacturing practices. Because of this series of relatively low experimental collapse pressures, however, a lower-bound curve was selected here for illustration.

It should be recognized that the lower-bound curve represents a wide range of geometrical proportions and that the correlation must break down for cylinders with oversized and very closed spaced stiffeners. Naturally, the critical buckling pressures for such structures would become much higher than those which the simple hoop-stress criterion at midbay would tend to indicate.

Figure 36.3 contains the elements of maximum hoop stress at the midpoint between the stiffeners indicated on the vertical axis. It represents the ratio of the

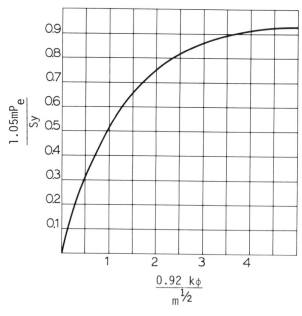

Fig. 36.3 Lower-bound empirical curve for ring-stiffened cylinders under external pressure.

External Pressure

critical pressure sought and that expressed by Eq. (36.2). The parameter on the abscissa is obtained by dividing Eq. (36.5) by Eq. (36.2) and by simplifying the expression in the usual range of the geometries where the nondimensional ratio $m \gg 1$. For example, in many practical design applications, the numerical value of ratio m is seldom found to be lower than 20. Under such circumstances, the error introduced by the foregoing simplification should not exceed 3%.

The advantage of referring to an empirical curve such as that shown in Fig. 36.3 is that the results cover a relatively wide range of L/R ratios without drawing any particular distinction among short, intermediate, or long cylinders. The disadvantages of the empirical curve, however, are as follows: (1) lack of exact details of the out-of-roundness corresponding to the chart, and (2) the practical necessity of relying on the lower-bound curve, resulting in a conservative prediction of the ultimate collapse pressure.

For the empirical data to be of direct computational use in designing cylindrical vessels for external pressure, the following approximate formulas may be postulated

$$P_{\text{CR}} = S_y Z_1 (Z_2 - n Z_3) \tag{36.20}$$

$$Z_1 = e^{-\frac{0.815 m^{1/2}}{k\phi}} \tag{36.21}$$

$$Z_2 = \frac{1}{m^{0.95}(k\phi)^{0.10}} \tag{36.22}$$

$$Z_3 = \frac{50}{m^{1.95}(k\phi)^{0.10}} - \frac{33}{m^2} \tag{36.23}$$

The essential variables such as S_y, k, m, n, and ϕ quoted here make the comparison with other existing formulas relatively straightforward. However, the main reason for developing a general formula of the type given by Eq. (36.20) is to have a continuous mathematical model which could be used with short, intermediate, and long cylinders alike, thereby eliminating the problems associated with the transitions and gray areas. The formula represented by Eq. (36.20) is intended for the following ranges of the parameters

$$0 \leq n \leq \frac{m}{100}$$
$$10 \leq m \leq 100$$
$$0.001 \leq k \leq 0.200$$
$$100 \leq \phi \leq 1000$$

In order to make a brief comparison between some of the theoretical and published experimental results, a number of geometries and mechanical properties have been randomly selected from various design records. Intentionally, however, the examples also involve some typical as well as some extreme configurations, for the sake of a more general assessment of the potential differences. The detailed data and the results of the correlation for these examples are given in Table 36.1. The last three columns of the table indicate reasonable agreement among the various estimates, despite the marked variations in cylinder proportions. The actual discrepancies

Table 36.1 Examples of Collapse Pressures Calculated for Cylindrical Vessels

Type of Vessel	R (in.)	T (in.)	L (in.)	e (in.)	S_y (ksi)	E (psi)	P_u (psi)[a]	P_{CR} (psi)[b]	P_e (psi)[c]
Experimental canister	35.88	1.125	82.5	0.0625	80	30×10^6	1260	1000	1000
Experimental canister	35.75	0.500	72	0.0625	36	30×10^6	180	152	160
Line-of-sight pipe	8.81	0.375	96	0.0450	35	30×10^6	500	430	470
Well casing	24.38	0.750	96	0.1220	36	30×10^6	410	490	530
Well casing	27.00	0.625	28	0.0400	100	30×10^6	1190	1010	1060
Experimental canister	40.50	1.500	272	0.1000	36	30×10^6	340	510	540
Experimental canister	27.75	0.750	48	0.1425	36	30×10^6	600	570	690
Experimental canister	3.1	0.4	42	0.0050	152	26×10^6	8550	4980	5230
Concrete room	91	27	240	0.4550	12	4×10^6	2920	2640[d]	3120

[a] See Eq. (36.6).
[b] See Eq. (36.20).
[c] See Fig. 36.3.
[d] See Eq. (36.2).

often appear to be on the conservative side, and in general the differences are not too serious when one considers the inherent complexity of the collapse problem.

The approximate formula (36.20) is easy to use but requires further refinements. It is hoped that its practical use will provide, in due course, the necessary background for improving the proposed model. It is offered here as an example of a plausible starting point in the search for a general formula applicable to all proportions of cylindrical vessels.

In the region of relatively low values of k and m, the minimum collapse pressure of casing, tubing, or drill pipe can be estimated on the basis of specifications recommended by the American Petroleum Institute [223]. The formulas suggested by the institute include experimental corrections and apply to yield strengths between 40,000 and 150,000 psi. They are basically of the Lamé [5] or Stewart [220] type.

The classical buckling response of a thin and long cylinder depends primarily on the modulus of elasticity of the material and the cube of the ratio of mean radius to thickness, as shown by Eq. (36.3). In other cases, the pressure vessels, well casing, or piping may fall into the category of intermediate or short cylinders. In this range, governed by Eq. (36.3), the collapse strength may increase or decrease as a function of raising or lowering the yield strength of the material. Also, when an externally pressurized and relatively thick pipe is subjected to axial tension, such as may be the case with a long string of casing emplaced vertically in the ground, the question may be raised as to the effect of tension on collapse. This effect is briefly analyzed in the next section.

EFFECT OF AXIAL STRESSES ON COLLAPSE

It appears that by way of a simplified model based on purely geometrical considerations, it is possible to arrive at a clue as to the effect of axial loading on collapse. For example, whereas the axial compression of a cylinder tends to increase wall thickness through the Poisson's ratio effect, axial tension does the opposite. Hence, "wall thinning" should contribute to lowering of the classical buckling pressure, which is directly proportional to the cube of thickness. Naturally, such geometrical effects must be relatively small, and one has to examine additional aspects of the problem such as the role of wall thinning in accentuating manufacturing imperfections as well as the influence of biaxial loading on the yield strength of the material.

Despite considerable progress over the past half a century, theoretical work on externally pressurized cylinders has not been fully successful in providing rigorous solutions to stability problems in the presence of manufacturing imperfections, superimposed axial stresses, or local plastic deformations. However, the need for practical solutions to some of these problems has prompted extensive experimental studies of the effect of the combined longitudinal loading and the external pressure [224]. This particular work was conducted in support of the requirements of the oil drilling industry. It involved more than 200 tests on seamless tubing loaded simultaneously by external pressure and longitudinal tension. The ratios of tube radius to wall thickness for this experiment ranged from 5 to 11. The yield strength of the tube material varied from about 30,000 to 80,000 psi. Unlike some earlier speculations, the new study has clearly established that the effect of combined loading can substantially reduce both the collapse strength and the tensile strength of the tubing. The results of this study have since been utilized in the development of practical handbook data for the oil drilling industry [225].

Conventional theories of the strength of materials indicate that in a biaxial state of stress, where tension and compression act at right angles to each other, the effective yield strength of the material appears to be lowered. The particular case of the maximum strain-energy theory gives a good approximation to the experimental data where ductile materials are involved. This strength theory is based on the assumption that the quantity of strain energy stored in a unit volume of an elastic material attains a maximum at the instant of the material's failure. Therefore, knowing the energy required to cause the failure in a simple tensile test specimen, the approximate limiting stresses for the combined loading can probably be estimated.

The general theoretical problem of stress interaction can best be described by the ellipse of biaxial yield stress, as shown in Fig. 36.4. The effect of biaxial loading in the case under consideration corresponds, therefore, to the functional relationship depicted by the right lower quadrant of the ellipse. In this region the longitudinal tensile stress is represented by the horizontal axis set out to the right of the zero point, and the compressive hoop stress, resulting from the external pressure, conforms to the lower portion of the vertical line. The ellipse diagram shown in Fig. 36.4 is expressed in terms of the nondimensional ratios to ensure its generality. The experimental points derived from tests on steel tubing were found to correlate quite well with the elliptical curve drawn in the compression-tension quadrant [224, 225]. This rather remarkable agreement between the maximum

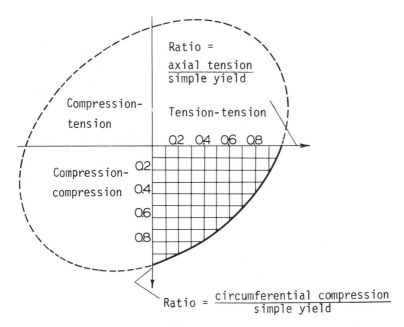

Fig. 36.4 Diagram for theory of maximum strain energy.

strain energy formulation and experimental evidence provides a firm basis for design of well casing employed in the oil industry and in similar engineering ventures.

Whereas investigations involving piping are likely to be successful, the problem of sizing a large-diameter vessel may present a somewhat different task because the intermediate range of mean radius-to-thickness ratio is higher and varies between about 10 and 40. There seem to be no available experimental data in this range of the geometry which could be directly used in evaluating the effect of the tensile stresses on collapse, although it is known that the yield strength in circumferential compression is definitely one of the principal factors in controlling the collapse resistance of oil well casing characterized by relatively low-radius-to-thickness ratios. Our intuition and experience may suggest that thin-walled, high-strength casing should be less affected by changes in the yield strength due to the biaxiality of loading. The opposite argument can be advanced for the case of a thick-walled, low-strength casing. However, the response of a large-diameter casing, characterized by an intermediate ratio of radius to thickness and the average yield strength, can only be inferred from the oil industry experimental data discussed above and a suitable theoretical model, such as that defined by the theory of the maximum strain energy.

It may be instructive at this point to briefly analyze the combined effect of thickness and material strength parameters which should have a finite influence on the susceptibility of the externally loaded pressure vessels to biaxial effects. This can be approximated with the aid of the thinness factor λ, expressed in terms of the geometry and materials parameters k, m, and ϕ, as before.

Equation (36.1) may be used to determine the basic range of the failure characteristics for a cylindrical vessel or piping subjected to external pressure. It is

External Pressure

recalled that when λ exceeds 2.5, stability criteria are predominant and the effect of biaxiality may, therefore, be less critical. On the other hand, when λ is less, say, than 0.35, the circumferential stress is likely to govern the collapse, making the vessel more susceptible to biaxial effects.

The criterion for a relatively long cylinder is such that its length must be at least equal to or higher than the length obtained from Eq. (36.12). This gives

$$L = 3.1 R(m)^{1/2} \qquad (36.24)$$

Combining Eqs. (36.1) and (36.24) yields the thinness factor formula for a relatively long pipe or casing

$$\lambda = \frac{2.11 m}{\phi^{1/2}} \qquad (36.25)$$

The expressions for the bracketing values of m follow directly from Eq. (36.25) when $\lambda = 0.35$ and $\lambda = 2.5$ are substituted. The chart defining relative sensitivity of the collapse pressure to superimpose axial tensile stresses is shown in Fig. 36.5. For instance, consider two pressure vessel designs corresponding to a high-strength steel, thin-walled cylinder and a low-yield-strength vessel with a thicker wall:

$\phi = 100$ and $m = 35$ (point A, Fig. 36.5)

$\phi = 833$ and $m = 4$ (point B, Fig. 36.5)

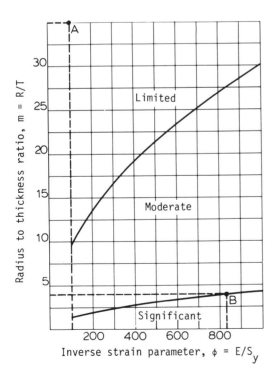

Fig. 36.5 Bracketing values for effect of tension on collapse pressure.

The points describing these two conditions are indicated in Fig. 36.5. In the first case, the effect of tensile stresses is likely to be rather limited, while in the second instance, involving a low-strength material such as A 36, the effect of tension may be important. This example and the chart given in Fig. 36.5 are intended only as a general guide for designing relatively long cylindrical components. Similar interpretation of the sensitivity of the collapse resistance to tension can be developed for intermediate and short cylinders using the relevant limits of applicability of specific formulas.

The theoretical correction factor for the effect of tension can be derived from analytical expressions for the maximum strain energy. The work done on the elastic material in the state of biaxial tension may be described as

$$U = \frac{S_1^2 + S_2^2}{2E} - \frac{\nu S_1 S_2}{E} \tag{36.26}$$

The maximum amount of the elastic work that can be done in direct tension is obtained when the corresponding stress approaches the yield strength for a given material. The formula representing this amount of work is

$$U = \frac{S_y^2}{2E} \tag{36.27}$$

The state of midbay hoop stress and axial wall stress in a casing subjected to external pressure and longitudinal tension simultaneously can be defined by $S_2 = -S_c$ and $S_1 = S_t$, respectively. Here S_c denotes hoop compressive stress due to external pressure and S_t is the average axial tensile stress. Substituting these terms and combining Eqs. (36.26) and (36.27) gives

$$S_y^2 = S_c^2 + S_t^2 + 2\nu S_c S_t \tag{36.28}$$

For the majority of metallic materials we have $\nu = 0.3$. Using this value and putting $a = S_t/S_y$, Eq. (36.28) yields

$$\frac{S_c}{S_y} = F_a \tag{36.29}$$

where

$$F_a = (1 - 0.91a^2)^{1/2} - 0.3a \tag{36.30}$$

Since the absolute value of the compressive wall stress is proportional to S_y and F_a, as shown by Eq. (36.29), the correction for tension can be obtained by multiplying the conventionally calculated collapse pressure by F_a. The design chart based on Eq. (36.30) is given in Fig. 36.6.

In field applications involving piping or casing emplaced vertically underground and backfilled, the ratio a is expected to be relatively small, so that the correction factor F_a does not differ significantly from unity. However, the chart given in Fig. 36.6 indicates that the influence of tension can be significant. This effect

External Pressure

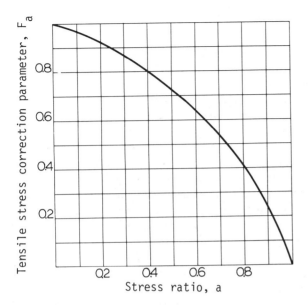

Fig. 36.6 Design chart for effect of tension on collapse.

manifests itself as that which would pertain if the pipe or tubing were made of a lower-strength material.

A brief correlation can now be made between the theoretical values obtained on the basis of Eq. (36.30) and some of the existing experimental values in the lower range of m found in the literature [224]. The relevant comparison is given in Table 36.2.

Table 36.2 Effect of Axial Stress on Collapse by Test and by Theory

m	a	Yield strength	Experimental Collapse Pressure	Calculated Collapse Pressure
9.1	0.138	500	44.4	45.2
9.1	0.276	500	43.1	41.9
9.1	0.404	500	35.8	38.1
6.6	0.107	290	43.4	43.8
6.6	0.317	290	35.8	37.0
10.9	0.135	560	28.9	29.1
10.9	0.385	560	29.3	24.9
5.6	0.289	565	96.5	96.4
5.6	0.474	565	75.8	82.7

Note: All values in megapascals: 1 megapascal = 1 N/mm^2 = 145 psi = 1 MPa.

ELASTIC BUCKLING OF SPHERICAL SHELLS

The definition of "spherical shell" used in the context of this chapter is assumed to include complete spherical configurations, hemispherical heads such as pressure vessel domes, and shallow spherical caps. The analysis of a spherical cap may also be used in modeling the behavior of a complete spherical vessel containing thickness discontinuities, reinforcements, and penetrations.

Although the response of a spherical shell to external pressure has received a good deal of attention in this country and elsewhere, particularly during the past 50 years, the calculation of the collapse pressure still represents substantial difficulties in the presence of geometrical discontinuities and manufacturing imperfections. The bulk of the theoretical work carried out thus far has had a rather limited effect on the methods of engineering design, and therefore much experimental support is still needed. At the same time, the application of spherical geometry to the optimum vessel design has continued to be attractive in many branches of industry dealing with submersibles, satellite probes, storage tanks, pressure domes, diaphragms, and similar systems. This chapter deals with the mechanical response and working formulas for spherical shell design in the elastic and plastic ranges of collapse which could be used for underground and aboveground applications. The material presented is based on state-of-the-art knowledge in pressure vessel design and analysis.

By way of introduction, it may be helpful to recall the original theoretical work of Zoelly and Van der Neut [31], who utilized the classical theory of small deflections and the solution of linear differential equations. Based on this theory, the elastic buckling pressure for a complete, thin spherical shell was known to be

$$P_{\text{CR}} = \frac{1.21 E}{m^2} \tag{36.31}$$

In Eq. (36.31), m denotes the ratio of mean radius to thickness of a sphere and E is the elastic modulus. The numerical constant given in Eq. (36.31) is based on the average value of Poisson's ratio ν equal to 0.3. For the more common structural materials, Poisson's ratio varies between 0.25 and 0.35, and hence the error induced by the use of a constant factor of 1.21 instead of $2/\sqrt{3(1-\nu^2)}$ is only on the order of 1 to 2%.

CORRECTED FORMULA FOR SPHERICAL SHELLS

At the time of the development of the classical theory which led to Eq. (36.31), no systematic experimental work was done. Several years later, however, some tests reported at the California Institute of Technology [226] showed that the experimental buckling pressure could be as low as 25% of the theoretical value given by Eq. (36.31). The value derived by means of Eq. (36.31) was then considered as the upper limit of the classical elastic buckling, while several investigators embarked on special studies with the aim of explaining these rather drastic differences between the theory and experiment. There was no reason to doubt the classical theory of elasticity, which worked well for flat plates, and it was soon suspected that the

External Pressure

effect of curvature and spherical shape imperfections could have been responsible for the discrepancies.

This thesis led to the realization that the classical theory must have failed to reveal the fact that for a vessel configuration, not far away but somewhat different from the perfect geometry, lower total potential energy was involved, and therefore a lower value of buckling load could be expected, such as that indicated by tests. The theoretical challenge then became to formulate a solution compatible with such a lower boundary of collapse pressure at which the spherical shell could undergo the "oil canning" or "Durchschlag" process.

After making a number of necessary simplifying assumptions, von Kármán and Tsien [226] developed a formula for the lower elastic buckling limit for collapse pressure, which for $\nu = 0.3$ was found to be

$$P_{\text{CR}} = \frac{0.37E}{m^2} \tag{36.32}$$

This level of collapse pressure may be said to correspond to the minimum theoretical load necessary to keep the buckled shape of the shell with finite deformations in equilibrium. The lower limit defined by Eq. (36.32) appeared to compare favorably with experimental results, also given in the literature [226]. On the other hand, the upper buckling pressure given by Eq. (36.31) could be approached only if extreme manufacturing and experimental precautions were taken. In practice, then, the buckling pressure is found to be closer to the value obtained from Eq. (36.32), and therefore this formula is often recommended for design.

The exact calculation of the load-deflection curve for a spherical segment subjected to uniform external pressure is known to involve nonlinear terms in the equations of equilibrium, which cause substantial mathematical difficulties [227].

PLASTIC STRENGTH OF SPHERICAL SHELLS

Design formulas (36.31) and (36.32) represent the results of elastic theory. Subsequent studies [228–230] were aimed at the development of inelastic solutions by applying plasticity reduction factors to classical linear theory. To extend the knowledge of both the elastic and inelastic behavior of full or partial spherical shells made of strain-hardening materials, Krenzke [230] conducted a series of experiments on 26 hemispheres bounded by stiffened cylinders. The materials were 6061-T6 and 7075-T6 aluminum alloys, and all the test pieces were machined with great care at the inside and outside contours. The junctions between the hemispherical shells and the cylindrical portions of the model provided good natural boundaries for the problem. The relevant physical properties for the study were obtained experimentally. The best correlation was arrived at with the aid of the following expression:

$$P_{\text{CR}} = \frac{0.84(E_s E_t)^{1/2}}{m^2} \tag{36.33}$$

In Eq. (36.33), E_s and E_t denote the secant and tangent moduli, respectively, at the specific stress levels. These values can be determined from the experimental stress-strain curves in standard tension tests. The relevant test ratios of radius to thickness in Krenzke's work varied between 10 and 100 with a Poisson's ratio of 0.3. The correlation based on Eq. (36.33) gave the agreement between experimental data and the predictions within +2 and −12%.

The extension of the Krenzke results to other hemispherical vessels should be qualified. Although his test models were prepared under controlled laboratory conditions, the following detrimental effects should be considered in a real environment:

Local and/or overall out-of-roundness
Thickness variation
Residual stresses
Penetration and edge boundaries

These effects are likely to be more significant when spherical shells are formed by spinning or pressing rather than by careful machining.

EFFECT OF INITIAL IMPERFECTIONS

In a subsequent series of collapse tests, Krenzke and Charles [231] aimed at evaluating the potential applications of manufactured spherical glass shells for deep submersibles. Because of the anticipated elastic behavior of glass vessels, the emphasis was placed on verifying the linear theory that resulted in Eq. (36.31). Prior to this series of tests, very limited experimental data existed which could be used to support a rational, elastic design with special regard to the influence of initial imperfections.

The formula for the collapse pressure of an imperfect spherical shell can be expressed in terms of a buckling coefficient K and a modified ratio m_i

$$P_{CR} = \frac{KE}{m_i^2} \tag{36.34}$$

In this formulation, we can assume, based on the work of Krenzke and Charles [231], that because of local thickness variation, the relevant radius-to-thickness ratio can be approximated as

$$m_i = \frac{R_i}{h} \tag{36.35}$$

The notation needed in the calculation of the parameter m_i for the locally imperfect shell geometry is shown in Fig. 36.7.

According to the results obtained by Krenzke and Charles on glass spheres, the buckling coefficient K in Eq. (36.34) was equal to about 0.84. Their study showed that the elastic buckling strength of initially imperfect spherical shells must depend on the local curvature and the thickness of a segment of a critical arc length. For a Poisson's ratio of 0.3, this critical length can be estimated as

$$L_c = 2.42 h (m_i)^{1/2} \tag{36.36}$$

In a related study conducted at the David Taylor Model Basin Laboratory for the Department of the Navy, the effect of clamped edges on the response of a

External Pressure

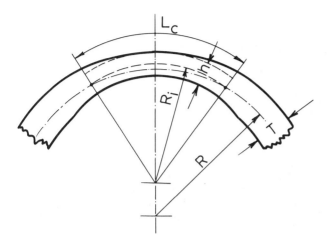

Fig. 36.7 Notation for defining a local change in wall thickness.

hemispherical shell was evaluated. The relevant collapse pressure was found to be about 20% lower than that for a complete spherical shell having the same value of the parameter m and the elastic modulus E. Although these tests on accurately made glass spheres tended to support the validity of the small-deflection theory of buckling, there appeared to be little hope that metallic shells would yield a similar degree of correlation, even under controlled conditions.

The investigations reviewed above may be of particular interest to designers dealing with complete spherical vessels as well as domed-end configurations. From a practical point of view, the most satisfactory method of predicting the collapse pressure would be to use a plot of experimental data as a function of the following well-defined dimensional quantities:

Experimental collapse pressure, P_e
Pressure to cause membrane yield stress, P_m
Classical linear buckling pressure, P_{CR}

EXPERIMENTS WITH HEMISPHERICAL VESSELS

Based on available experimental data involving the collapse of hemispherical vessels under external pressure, a nondimensional plot was developed [211] suitable for the preliminary design purposes. This plot contains a lower-bound curve, shown in Fig. 36.8, in which the following dimensionless ratios of the above-listed dimensional quantities are employed:

$$\frac{P_e}{P_{CR}} = \frac{0.83 P_e m^2}{E} \qquad (36.37)$$

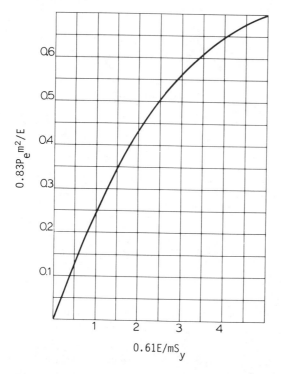

Fig. 36.8 Lower-bound design curve for hemispherical vessels under external pressure: P_e, collapse pressure; m, mean radius divided by thickness; E, modulus of elasticity; S_y, yield strength.

and

$$\frac{P_{CR}}{P_m} = \frac{0.61E}{mS_y} \tag{36.38}$$

The accuracy with which the collapse pressure can be predicted on the basis of experimental data must be influenced by the maximum scatter band involved. Since this scatter is sensitive to material and geometry imperfections, their probable extent should be known before a more reliable, lower-bound curve can be developed. The results given in Fig. 36.8 include hemispherical vessels in the stress-relieved and as-welded condition without, however, specifying the extent of geometrical imperfections, which, in this particular case, were known to be less pronounced. It follows, then, that Fig. 36.8 is applicable only to the design of hemispherical vessels, where good manufacturing practice can be assured. Further research work is recommended to narrow the scatter band to assure better correlation for the lower bound.

The dimensionless plot given in Fig. 36.8 is sufficiently general for practical design purposes. For instance, consider a titanium alloy hemisphere with $m = 60$, $E = 117,200$ N/mm^2, and the compressive yield strength $S_y = 760$ N/mm^2. From Eq. (36.38) we get $0.61E/mS_y = 1.57$. Hence, Fig. 36.8 yields $0.83P_e m^2/E = 0.36$, from which $P_e = 14.1$ N/mm^2.

External Pressure

It may now be instructive to look briefly at this empirical result in relation to the theoretical limits defined by Eqs. (36.31) and (36.32) for the complete spherical vessels.

Making $P_e = P_{CR} = 14.1$ N/mm^2 and solving Eq.(36.34) for the magnitude of the buckling coefficient gives $K = 0.43$. This value is close to the theoretical lower limit of 0.37 given by Eq. (36.32) for a complete spherical vessel, and it appears to suggest that certain portions of such a vessel under uniform external pressure may behave in a manner similar to that of a complete vessel. This observation may be of special importance in dealing with the spherical shells containing local reinforcements and penetrations. It is also generally consistent with the elastic theory of shells, according to which the influence of geometrical discontinuities is local and does not extend significantly beyond the range determined by the value of the parameter $T(m)^{1/2}$.

RESPONSE OF SHALLOW SPHERICAL CAPS

The problem of a relatively thin and shallow spherical cap fully clamped at the edge and subjected to uniform external pressure [38] has several additional considerations. The basic nature of structural response in this case can be defined in three ways, depending on the cap geometry. For a typical Poisson's ratio of 0.3, these conditions can be described as follows:

$\lambda_0 < 2.08$ continuous deformation with buckling

$\lambda_0 > 2.08$ axisymmetric snap-through

$\lambda_0 > 6$ local buckling

Provided that the spherical cap can be classed as a shallow and relatively thin shell, the numerical value of the parameter λ_0 can be obtained from either of the following two formulas:

$$\lambda_0 = \frac{1.82 a_0}{T(m)^{1/2}} \tag{36.39}$$

and

$$\lambda_0 = 2.57 \left(\frac{H}{T}\right)^{1/2} \tag{36.40}$$

According to the notation depicted in Fig. 36.9, the dimension H can be expressed as a function of a_0 and R in the following way:

$$a_0 = R \sin \theta \tag{36.41a}$$
$$H = R(1 - \cos \theta) \tag{36.41b}$$

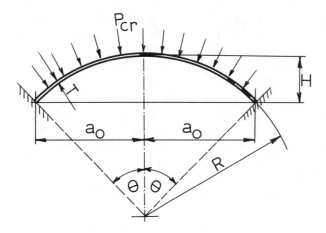

Fig. 36.9 Notation for shallow spherical caps.

Squaring and adding both sides of Eqs. (36.41a) and (36.41b) eliminates angle θ and yields

$$H^2 - 2HR + a_0^2 = 0 \tag{36.41c}$$

Assuming that for a small value of H, the square of H is markedly lower than $2HR$, Eq. (36.41c) gives

$$H = \frac{a_0^2}{2R} \tag{36.41d}$$

Substituting Eq. (36.41d) into Eq. (36.40) now yields Eq. (36.39). The two forms of the λ_0 equation, however, give identical results only when the spherical cap is shallow.

As a rough guide, a spherical cap can be regarded as thin when $m > 10$. Shallow geometry in this calculation can be approximately defined by the relation $a_0/H \geq 8$, with due regard to the notation given in Fig. 36.9. Once the geometrical parameter λ_0 is calculated from either of the two formulas given above, the estimate of the critical buckling pressure can be made with the help of Fig. 36.10. The design curve given in Fig. 36.10 is based on the numerical data quoted by Flügge [38]. The curve has been smoothed out somewhat in the mid-region of the parameter λ_0, which involves a transition between the theoretical and experimental data used to simplify the mathematical process of fitting an appropriate algebraic equation. For instance, using the design curve from Fig. 36.10, the following expression for the critical buckling pressure can be developed:

$$P_{\text{CR}} = 0.075 E n_0^{-4} \lambda_0^{4.15} e^{-0.095 \lambda_0} \tag{36.42}$$

In this expression, n_0 denotes the dimensionless ratio a_0/T. As an example of application of this formula, consider $R = 127$ mm, $a_0 = 31.8$ mm, $T = 2.1$ mm, and $E = 117{,}200$ N/mm². From this we get $m = 60.5$ and $n_0 = 31.8/2.1 = 15.1$.

External Pressure

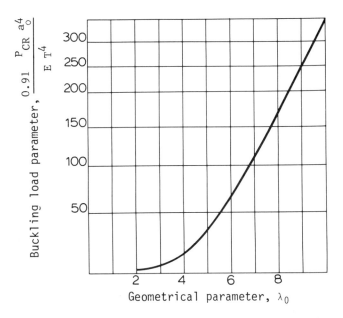

Fig. 36.10 Design chart for a shallow spherical cap under external pressure.

Using Eq. (36.39) yields $\lambda_0 = 3.53$. Hence, substituting into Eq. (36.42) gives

$$P_{CR} = 0.075 \times 117,200 \times 15.1^{-4} \times 3.53^{4.15} \times e^{-0.095 \times 3.53}$$
$$= 22.7 \text{ N/mm}^2$$

In a special situation where a spherical cap is very thin, with a range of m values between 400 and 2000, the following empirical formula has been suggested for the relevant buckling pressure [232]:

$$P_{CR} = \frac{(0.25 - 0.0026\theta)(1 - 0.000175m)E}{m^2} \qquad (36.43)$$

In this expression, θ denotes the central half-angle of a spherical cap in degrees. This angle is also shown in Fig. 36.9. The empirical equation given by Eq. (36.43) is intended for θ values between 20 and 60 degrees. Although this formula is useful within the indicated brackets of m, it may not be quite suitable for bridging the boundaries between the shallow caps and hemispherical shells without a careful study. Ideally, the formula for the collapse pressure of a spherical shell should be reduced to the form of Eq. (36.34) with the K value representing a continuous function of the shell geometry and manufacturing imperfections. For inelastic behavior, the parameter $(E_s E_t)^{1/2}$ appears to have the best chance of success for a meaningful correlation of theory and experiment. In the interim, however, the formulas given in this chapter are recommended for the preliminary design and experimentation.

STRENGTH OF THICK CYLINDERS

When the stability criteria do not apply to the vessels or piping subjected to external pressure, Lamé's theory gives the most convenient approach for design of thick and moderately thick cylinders. For instance, the maximum tangential stress at the inner surface of the cylinder can be expressed in the following simplified way:

$$S_t = P_0 K_4 \tag{36.44}$$

This stress is compressive and is given as a function of the ratio of the inner radius to thickness, as shown in Fig. 36.11. It is of interest to note that when the cylinder is very thick to the point of becoming a solid shaft, the maximum tangential stress tends to a value equal to twice the externally applied pressure.

The displacement of the inner surface of the cylinder toward the central axis can be calculated as

$$u_i = \frac{P_0 T}{E} K_3 \tag{36.45}$$

Similarly, the outer surface of the cylinder displaces radially under pressure according to the following rule:

$$u_o = \frac{P_0 T}{E} K_5 \tag{36.46}$$

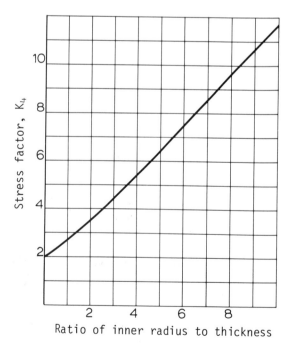

Fig. 36.11 Thick cylinder under external pressure.

External Pressure

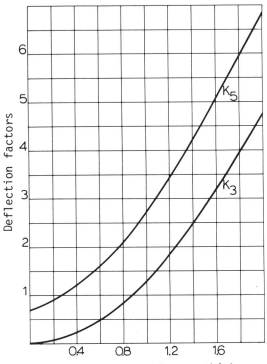

Fig. 36.12 Deflection factors for thick cylinders under external pressure.

The displacement factors K_3 and K_5 are illustrated in Fig. 36.12 for the lower ratios of inner radius to wall thickness. Unlike the factor K_2 in Eq. (35.42) and K_3 repeated in Figs. 35.8 and 36.12, the displacement factor K_5 does not vanish when the inner radius tends to zero. For a solid shaft, then, using this theory, the displacement of the outer surface is equal to $0.7 P_0 T/E$ and represents the maximum amount of radial compression of the shaft under elastic conditions.

Design Problem 36.3

A straight cylindrical hub and a solid shaft assembly are shown in Fig. 36.13. Assuming that both components are made of steel, calculate the shrinkage allowance Δ to correspond with the design shrink-fit pressure $P = 10,000$ psi.

Solution

The amount of radial deformation of the hub follows from Eq. (35.42) for $T_h = R_o - R_i$.

$$u_i = \frac{P T_h K_2}{E}$$

The amount of radial compression of the shaft is given by Eq. (36.46):

$$u_o = \frac{P T_s K_5}{E}$$

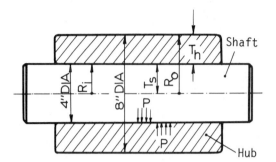

Fig. 36.13 Shrink-fit assembly for Design Problem 36.3.

The absolute sum of the two individual displacements of the inner and outer surfaces in contact must be equal to Δ. Hence,

$$\Delta = \frac{P}{E}(T_h K_2 + T_s K_5)$$

For the hub, $R_i/T_h = 1$. For the shaft, $R_i/T_s = 0$. Therefore, Figs. 35.8 and 36.12 give

$$K_2 = 1.97$$
$$K_5 = 0.70$$

so that

$$\Delta = \frac{10,000(2 \times 1.97 + 2 \times 0.70)}{30 \times 10^6} = 0.00178 \text{ in.} \quad (0.045 \text{ mm}) \quad \blacklozenge$$

STRENGTH OF THICK SPHERES

When a thick-walled spherical vessel is subjected to external pressure, the maximum stress develops at the inner surface equal to

$$S = \frac{3P_0 R_o^3}{2(R_o^3 - R_i^3)} \tag{36.47}$$

The displacement of the inner surface toward the center of the vessel is

$$u_i = \frac{3P_0 R_i R_o^3 (1-\nu)}{2E(R_o^3 - R_i^3)} \tag{36.48}$$

The corresponding displacement of the outer surface under external pressure is

$$u_o = \frac{P_0 R_o}{2E(R_o^3 - R_i^3)}\left[(1-\nu)\left(2R_o^3 + R_i^3\right) - 2\nu\left(R_o^3 - R_i^3\right)\right] \tag{36.49}$$

External Pressure

For a solid sphere subjected to external pressure, the amount of radial compression in the elastic range becomes

$$u_o = \frac{P_0 R_o (1 - 2\nu)}{E} \tag{36.50}$$

APPROXIMATE STRESS CRITERION FOR CYLINDERS

Manufacturing imperfections such as local out-of-roundness or a slightly oval shape of a cylindrical vessel can hardly be avoided. Furthermore, this problem is especially important in the design and construction of large containers, for which the industry has developed certain practical rules for the allowed deviations. So far this section referred to the various formulas for correcting the calculated collapse pressures for cylindrical containers subjected to uniform external loading. The corrections outlined were intended for the criteria of elastic stability and an elliptical mode of cylinder failure. This also implies that the majority of containers found in industry can be characterized by relatively thin walls and predominantly elastic response. Although this statement is largely true, there are exceptions where the vessels may have thicker walls made of a lower-strength material and where stress rather than buckling may govern the design. Hence, substituting $\lambda \cong 0.35$ into Eq. (36.1) results in the following approximate criterion:

$$\frac{100m}{k^2 \phi^2} \leq 0.75 \tag{36.51}$$

Here k and m denote the dimensionless ratios as before and ϕ is the material's characteristic defined as the inverse strain parameter. For instance, taking $m = 10$, $k = 0.1$, and $\phi = 1,000$, Eq. (36.51) gives 0.1, which is less than 0.75. According to this rule, then, the selected cylindrical geometry and the material are such that stress criteria should govern the design. Hence, any combination of k, m and ϕ which gives a numerical quantity smaller than 0.75 determines the condition of the governing stress for the cylinders loaded externally. However, when the vessels are loaded internally, the design is always governed by stress. The only differentiation needed in the latter type of loading concerns the relative thickness of the vessel wall, which determines the choice of the analytical approach.

OUT-OF-ROUNDNESS CORRECTION FOR STRESS

The maximum wall stress in an out-of-round cylinder can be obtained through superposition of the tangential stress calculated for a perfectly circular cylinder and the offset bending stress caused by the out-of-roundness. The offset stress follows from the bending moment, which may be determined by multiplying the average hoop load times the amount of radial deviation. For a relatively thin cylinder for which $m > 10$, this procedure gives

$$S = Pm(1 + 6n) \tag{36.52}$$

Usually, the magnitude of n defining the particular out-of-roundness is markedly less than unity. Nevertheless, Eq. (36.52) shows how quickly the total stress can increase as a function of n. It is also well to note that in the case of internal pressure, the hoop load tends to make the cylinder more circular. The actual amount of this straightening effect, however, is not easy to estimate because of the nonlinear relation between the hoop load and radial deformation. Hence, Eq. (36.52) in its present form is likely to be conservative. Also, the model implied by Eq. (36.52) may be related to the stress theory of oval tubes subjected to internal pressure [233], with the relevant agreement appearing to be good for n values smaller than 1. This bracketing condition should certainly cover the range of radial deviations most likely to be encountered in pressure vessel design.

Since Eq. (36.52) is based on the membrane stress theory, there is no need to differentiate between the internal and external loading on a cylindrical shell. When, however, the cylinder ratio m is less than 10, thick-shell theory dictates the choice of the design formulas. In such a case, the total tangential stress should be calculated as the sum of the maximum Lamé stress and the offset bending stress. The latter may be obtained from the hoop load, which can be taken as the product of the average stress and the longitudinal cross-sectional area of the cylinder. In this manner, the corrected stress for the case of internal pressure can be described as

$$S = P\left[\frac{4m^2 + 1}{4m} + \frac{3n(4m^2 - 2m + 1)}{2m}\right] \tag{36.53}$$

Similarly, for the condition of the external pressure, we obtain

$$S = P\left[\frac{(2m+1)^2}{4m} + \frac{6n(m^2 + 0.5m + 0.25)}{m}\right] \tag{36.54}$$

It is evident from Eqs. (36.53) and (36.54) that when $m \gg 1$, both expressions can be reduced to the thin-cylinder formula given by Eq. (36.52). In this notation m defines the ratio of mean radius of the cylinder to wall thickness.

SYMBOLS

A_1, A_2	Long cylinder factors
A_m	Long cylinder parameter
A_n	Long cylinder parameter
$a = \frac{S_t}{S_y}$	Stress ratio
a_0	Projected radius of spherical cap, in. (mm)
E	Elastic modulus, psi (N/mm^2)
E_s	Secant modulus, psi (N/mm^2)
E_t	Tangent modulus, psi (N/mm^2)

External Pressure

e	Deviation from perfect radius, in. (mm)
F_a	Axial load correction factor
$F(k)$	Short cylinder parameter
$F(m)$	Short cylinder parameter
H	Depth of shallow cap, in. (mm)
h	Reduced thickness of shell, in. (mm)
K	Buckling coefficient
K'_1, K'_2, K'_3	Short cylinder factors
K_1 through K_5	Thick cylinder design factors
$k = T/L$	Dimensionless ratio
L	Length of cylinder, in. (mm)
L_c	Critical arc length, in. (mm)
$m = R/T$	Ratio of mean radius to thickness
$m_i = R_i/h$	Ratio of mean radius to local thickness
$n = e/T$	Ratio of radial deviation to thickness (out-of-roundness)
n_e	Effective out-of-roundness
$n_0 = a_0/T$	Shallow cap ratio
P_c	Classical buckling pressure, psi (N/mm^2)
P_{CR}	General symbol for buckling pressure, psi (N/mm^2)
P_m	Pressure to cause membrane yield stress, psi (N/mm^2)
P_0	External pressure, psi (N/mm^2)
P_e	Experimental collapse pressure, psi (N/mm^2)
P_u	Short cylinder collapse pressure, psi (N/mm^2)
P_{ul}	Long cylinder collapse pressure, psi (N/mm^2)
P_y	External pressure at yield, psi (N/mm^2)
R	Mean radius, in. (mm)
R_o	Outer radius, in. (mm)
R_i	Inner radius, in. (mm)
S	General symbol for stress, psi (N/mm^2)
S_1, S_2	Principal stresses, psi (N/mm^2)
S_c	Compressive stress, psi (N/mm^2)
S_t	Tensile stress, psi (N/mm^2)
S_y	Yield strength, psi (N/mm^2)
T	Thickness of wall, in. (mm)
T_h	Thickness of hub, in. (mm)
T_s	Thickness of hollow shaft, in. (mm)
U	Elastic strain energy, (lb-in.)/in.^{3} (N-mm/mm^3)
u_i	Displacement of inner surface, in. (mm)
u_o	Displacement of outer surface, in. (mm)
Z_1, Z_2, Z_3	Collapse pressure formula parameters
Δ	Shrinkage allowance, in. (mm)
θ	Central half-angle of spherical cap, deg
λ	Thinness factor
λ_0	Shallow cap parameter
ν	Poisson's ratio
$\phi = E/S_y$	Inverse strain parameter

37
Axial and Bending Response

INTRODUCTION

The general problem of stability and buckling resistance of various structural components under axial load was outlined in Chap. 10. This particular part of the book, dealing with piping and pressure vessels, examines a number of theories and formulas related to the axial response of cylindrical components. The material also includes a number of special topics, such as the axial response of a pipe constrained in the transverse direction and a rolling diaphragm theory. These topics, although seldom treated in standard engineering texts, involve interesting mathematical models and can be useful in many practical applications.

COMMENT ON SECTIONAL PROPERTIES

If a given length of a pipe can be regarded as a slender column under a concentric loading, the critical buckling load can be expressed by Eq. (10.1). Assuming that R and T represent the mean radius and wall thickness, respectively, the moment of inertia of pipe cross section can be taken as $\pi R^3 T$. It may be instructive to recall here how this simplified form is obtained. Consider, for example, a standard formula for the moment of inertia, where R_o and R_i denote outer and inner radii of the cylinder.

$$I = \frac{\pi}{4}\left(R_o^4 - R_i^4\right) \qquad (37.1a)$$

Axial and Bending Response

Rewriting Eq. (37.1a) gives

$$I = \frac{\pi}{4}\left(R_o^2 + R_i^2\right)(R_o + R_i)(R_o - R_i) \tag{37.1b}$$

Putting $R = (R_o + R_i)/2$ and $T = R_o - R_i$ into Eq. (37.1b) gives

$$I = \frac{\pi RT}{2}\left(2R_i^2 + 2R_i T + T^2\right) \tag{37.1c}$$

If the pipe, tubing, or a cylindrical vessel is relatively thin, then ignoring term T^2 and making $R \cong R_i$ yields

$$I = \pi R^3 T \tag{37.2}$$

Similarly, the section modulus is

$$Z = \pi R^2 T \tag{37.3}$$

The effect of these approximations is illustrated in Fig. 37.1. It is clear that in the case of the moment of inertia, surprisingly accurate results can be obtained for a wide range of R_o/R_i. The two curves represent the ratios of the exact and the approximate values with the exact magnitudes placed in the numerators.

COLUMN BEHAVIOR OF PIPE

When a relatively long and flexible pipe of uniform cross section is loaded as a column between the pin-jointed ends, the buckling load and the corresponding critical buckling stress can be defined as follows

$$P_{\text{CR}} = \frac{\pi^3 E R^3 T}{L^2} \tag{37.4}$$

and

$$S_{\text{CR}} = \frac{\pi^2 E R^2}{2L^2} \tag{37.5}$$

The formulas given by Eqs. (37.4) and (37.5) correspond to the first, fundamental mode of buckling and require the smallest value of axial load to produce elastic instability. When some horizontal constraint at this inflection points is present, preventing development of the fundamental mode, the question of a higher buckling mode may arise [5]. Here the general expression for a pin-jointed response of a Euler-type column becomes

$$P = \frac{\pi^3 E R^3 T (1 + 2\alpha)^2}{L^2} \tag{37.6}$$

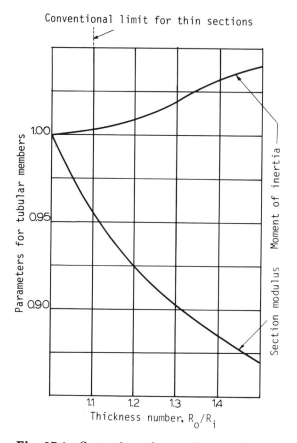

Fig. 37.1 Comparison of approximate section properties for cylinders.

Examples of higher-mode shapes for a pin-jointed column are given in Fig. 37.2, which indicates how rapidly the load can increase with an increase in α. Here α denotes the number of wave shapes. The modes shown in Fig. 37.2 are assumed to be symmetrical about the midpoint of the pipe. The response of the column corresponding to $\alpha = 0$ leads to Eq. (37.4), which also represents the condition of the minimum elastic energy. The formula given by Eq. (37.6) leads to an interesting statement concerning a higher mode of equilibrium consistent with the yield strength of the pipe in axial compression. For example, take

$$P_y = 2\pi R T S_y \tag{37.7}$$

Solving Eqs. (37.6) and (37.7) gives

$$\alpha = 0.225 \left(\frac{L}{R}\right)\left(\frac{S_y}{E}\right)^{1/2} - 0.5 \tag{37.8}$$

Axial and Bending Response

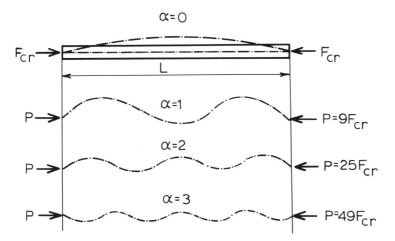

Fig. 37.2 Higher-mode equilibrium for a simple column.

For example, taking $L/R = 100$ and $E/S_y = 300$, Eq. (37.8) yields $\alpha = 0.8$. The nearest integer is 1, which, according to Fig. 37.2, would correspond to $P = 9P_{CR}$, where P_{CR} is given by Eq. (37.4).

PIPE ON ELASTIC FOUNDATION

When support of the pipe is virtually continuous, the problem may be solved by modeling the pipe as a bar of finite length resting on an elastic medium [31]. In this solution the energy method is used and the deflection curve of the bar is represented by a trigonometric series. The work done in compressing the bar axially is made equal to the elastic strain energy of bending of the bar and that of the deformation of the elastic medium supporting the bar. The critical value of the compressive axial force on the bar is then found by solving the energy balance equation for the axial force and making the resultant expression a minimum.

The results of this analysis can be expressed in terms of the usual parameters using a standard column as criterion

$$P_{CR} = \frac{\pi^3 E R^3 T}{L_e^2} \qquad (37.9)$$

It is noted that Eqs. (37.4) and (37.9) are almost identical. In the case of the elastic foundation, however, a reduced value of the column length L_e should be used instead of the total length L. The selection of the reduced length depends on parameter Z_0 which involves the flexural rigidity of the column and the modulus of foundation.

$$Z_0 = 0.02 \frac{\beta L^4}{E R^3 T} \qquad (37.10)$$

where L denotes the actual length of the column resting against the elastic foundation and the parameter β defines the modulus of the foundation in psi. The dimensional check of Eq. (37.10) shows that the result is compatible with the definition of β. The ratio of L_e/L depends on the parameter Z_0, and this relation can be conveniently represented in a graphical form or algebraic terms. For example, when the parameter Z_0 varies between 0 and 30, the relevant magnitudes of the ratio L_e/L follow directly from Fig. 37.3.

For the higher values of Z_0, the required length ratios can be calculated from the following approximate relations.
When $30 \leq Z_0 \leq 1000$,

$$\frac{L_e}{L} = 0.6675 - 0.1575 \log Z_0 \tag{37.11}$$

When $1000 \leq Z_0 \leq 10{,}000$,

$$\frac{L_e}{L} = 0.450 - 0.085 \log Z_0 \tag{37.12}$$

The theory of column buckling at higher modes may be related to the theory of a column of finite length supported by an elastic medium. The maximum axial stress at the point of instability must increase with the increase in the number of half waves into which the originally straight column transforms. Similarly, the axial

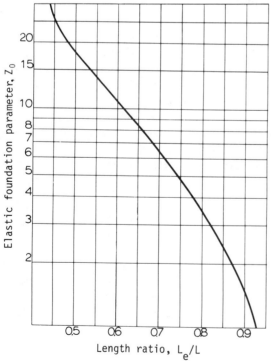

Fig. 37.3 Foundation parameter for a finite-length column.

stress in a finite-length column, supported by a continuous elastic medium, must increase with the increase in the spring constant of the elastic foundation. Hence, a definite correlation exists between the buckling mode and the foundation modulus. Assuming the end support conditions to be the same for both cases and combining Eqs. (37.6) and (37.9) leads to a very simple formula

$$\alpha = \frac{L - L_e}{2L_e} \quad (37.13)$$

When $L = L_e$, elastic parameter Z_0 is zero. For a finite pipe geometry this can be true only if $\beta = 0$, or in other words when the surrounding medium offers no resistance to the transverse pipe bending. Equation (37.13) shows that the mode factor α must also be zero, which corresponds to the fundamental buckling mode of a column with pin-jointed ends as shown in Fig. 37.2. In the case of the other bracketing assumption, where L_e tends to a very small value, the column offers progressively more resistance to the buckling deformation and, at least theoretically, the number of half-waves should become very high. This process has, of course, a natural boundary condition depending on the yield strength of the material or some other mode of failure, such as that of a cylinder in axial compression. Before the latter boundary is reached, however, Eq. (37.13) may be used to approximate the buckling characteristics of the pipe by calculating the mode factor α for a given ratio of L_e/L.

THEORY OF ONE-WAY BUCKLING

In the majority of problems involving concepts of elastic stability and buckling of axially loaded members, the assumption is made that the bar is perfectly straight before the end load is applied. The equilibrium based on a direct axial compression is considered stable until the critical load is reached, at which time even the slightest lateral force can immediately produce a lateral deflection which will not vanish on the removal of the lateral force. In other words, a new state of equilibrium is attained in a slightly bent configuration, and the critical load is just sufficient to maintain such equilibrium.

Since in various practical situations we may have to deal with pipe sections that are not perfectly straight to start with, and since such components as the utility pipe or a well casing can be supported by the surrounding medium in the transverse direction, the mathematical model of the pipe equilibrium should include the effects of eccentricity and the transverse resistance to buckling.

Standard solutions to column problems seldom include all such effects simultaneously. For instance, the analysis of buckling of a bar on an elastic foundation usually does not account for any initial curvature of the bar before the axial load is applied. In another instance the effect of the initial curvature on the transverse deflection of the column may be calculated without any allowance of the transverse loading distributed along the axis of the column. Such effects from the two separate solutions are not directly additive, and therefore the problem of a buckling column, with some residual curvature and transverse loading occurring simultaneously, should be formulated and developed from first principles. This form of the pipe behavior will be referred to in this discussion as *one-way buckling*.

The original problem of buckling of this type has been treated [234]. This analysis is made on the assumption that the transverse resistance of medium is constant along the pipe length and that the pipe behaves as a beam with built-in ends at which the deflection and the slope are zero. Furthermore, in addition to the shear reaction and axial forces, the fixing moments are considered at the supports.

In formulating the mathematical problem of one way buckling, the following basic assumptions are made:

1. The bending moment equation includes the change of the lever arm of the axial compressive load.
2. The tangential slope is relatively small, so that the pipe curvature may be represented by the strength-of-materials formula involving second derivative of the displacement.

To relate this analysis to a standard treatment of a beam-column problem, it will be noted that the equilibrium of forces in a one-way buckling type of solution includes the transverse loading and the original shape of the bar simultaneously. The procedure of setting up the differential equation of equilibrium for the portion of the pipe behaving as a prebent beam with the transverse restraint follows the general rule of the second-order theory of structural analysis. This rule simple states that the equations of equilibrium are written for the geometry of the deformed structure [38].

A simplified sketch of the half length of the pipe section is shown in Fig. 37.4. Note that L_0 defines the half-length of a symmetrically deformed portion of the pipe which contains three locations of zero slope, two of which are indicated in Fig. 37.4. Let Q represent the uniform resistance of the surrounding medium. End-fixing moment and shear reactions are denoted by M_0 and QL_0, respectively. The compressive forces P are shown to be offset by the amount δ. The deformed pipe is considered to remain in equilibrium under the assumed system of forces as long as the value of P is unchanged. Hence, the basic analytical problem in this instance is to develop a working formula for the limiting compressive force in terms of the

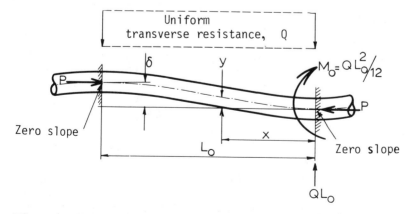

Fig. 37.4 Half length of pipe for analysis of one-way buckling.

Axial and Bending Response

transverse resistance of the medium, flexural rigidity of the pipe, and the maximum eccentricity δ. No stipulation is made here as to how this value of δ was originally established, and the basic reason for including this parameter is to indicate the extent of the existing eccentricity for the end load. Hence, from the practical point of view, δ may be looked upon as the manufacturing tolerance, the original bending deflection due to the lack of straightness, or a combination of these effects.

Bearing in mind the deformed shape of the pipe at any point x measured from the right-hand-side support in Fig. 37.4, the bending moment equilibrium may be given by the following expression

$$EI\frac{d^2y}{dx^2} = QL_0 x - \frac{Qx^2}{2} + \frac{QL_0^2}{12} - Py \tag{37.14a}$$

Putting $K^2 = P/EI$ and rearranging Eq. (37.14a) gives the differential equation

$$\frac{d^2y}{dx^2} + K^2 y + F(x) = 0 \tag{37.14b}$$

where

$$F(x) = \frac{QK^2}{12P}\left(6x^2 - 12L_0 x - L_0^2\right) \tag{37.14c}$$

When $F(x)$ is a polynomial of not more than fifth degree, the general solution of Eq. (37.14b) may be obtained directly with the aid of the relation

$$y = A\sin Kx + B\cos Kx - \frac{F(x)}{K^2} + \frac{d^2 F(x)}{K^4 dx^2} - \frac{d^4 F(x)}{K^6 dx^4} \tag{37.14d}$$

From Eq. (37.14c)

$$\frac{d^2 F(x)}{K^4 dx^2} = \frac{Q}{PK^2} \tag{37.14e}$$

$$\frac{d^4 F(x)}{K^6 dx^4} = 0 \tag{37.14f}$$

Substituting Eqs. (37.14c), (37.14e), and (37.14f) into (37.14d) yields

$$y = A\sin Kx + B\cos Kx - \frac{Q(6x^2 - 12L_0 x - L_0^2)}{12P} + \frac{Q}{PK^2} \tag{37.14g}$$

The constants of integration can be found with reference to the geometry depicted in Fig. 37.4. For example, $y = dy/dx = 0$ at $x = 0$ are the geometric boundary conditions, selected for the problem at hand, which have sufficient practical justification. Consider, for example, a long string of pipe emplaced in a borehole with the usual manufacturing and assembly tolerances. Since the string may contain several prebent portions of the pipe, the model illustrated in Fig. 37.4 corresponds to a real situation. The selection of the pipe length $2L_0$ will, of course, vary with different designs and the three-dimensional survey of the borehole. Hence, the length selected should be considered as a variable in parametric studies.

Evaluation of the boundary conditions in line with the geometry and loading details shown in Fig. 37.4 yields the following integration constants:

$$A = -\frac{QL_0}{PK} \tag{37.14h}$$

and

$$B = Q\left(\frac{L_0^2}{12} - \frac{1}{K^2}\right) \tag{37.14i}$$

Hence, substituting Eqs. (37.14h) and (37.14i) into Eq. (37.14g) gives

$$y = \frac{QEI}{P^2}\left[1 - KL_0 \sin Kx + \frac{K^2 L_0^2 - 12}{12}\cos Kx \right.$$
$$\left. - \frac{K^2}{12}(6x^2 - 12L_0 x - L_0^2)\right] \tag{37.14j}$$

The general expression for the slope follows directly from Eq. (37.14j).

$$\frac{dy}{dx} = \frac{QEI}{12P^2}[K(12 - K^2 L_0^2)\sin Kx - 12K^2 L_0 \cos Kx$$
$$- K^2(12x - 12L_0)] \tag{37.14k}$$

According to the model selected, $dy/dx = 0$ at the midpoint of the pipe, where $x = L_0$. Substituting this value in Eq. (37.14k) and solving for KL_0 gives

$$\tan KL_0 = \frac{12KL_0}{12 - K^2 L_0^2} \tag{37.14l}$$

The first solution of Eqs. (37.14l) is obtained when $KL_0 = 0$. Since this is a trivial solution, it will be ignored. A brief review of Eq. (37.14l) indicates that the first nonzero value can be found in the fourth quadrant, where tangent is negative. Putting $\tan KL_0 = F(K)$, where $F(K)$ denotes the right-hand side of Eq. (37.14l), the solution of Eq. (37.14l) can be found, for example, by the simple graphical procedure referred to in Chap. 22. The graphical solution gives $KL_0 = 4.917$. When $x = L_0$, the maximum load eccentricity is equal to δ in accordance with the model shown in Fig. 37.4. Hence, utilizing Eq. (37.14j), yields

$$\delta = \frac{QEI}{P^2}\left[1 - KL_0 \sin KL_0 + \frac{(K^2 L_0^2 - 12)\cos KL_0}{12} + \frac{7K^2 L_0^2}{12}\right] \tag{37.14m}$$

Substituting the relevant numerical values of the trigonometric functions involved into Eq. (37.14m) and solving for $P = P_{\text{CR}}$ gives

$$P_{\text{CR}} = 4.49\left(\frac{QEI}{\delta}\right)^{1/2} \tag{37.15}$$

Axial and Bending Response

When $I = \pi R^3 T$ and the cross section is $2\pi RT$, the corresponding axial stress in the pipe becomes

$$S_{\text{CR}} = 1.27 \left(\frac{QER}{\delta T} \right)^{1/2} \qquad (37.16)$$

This rather simple set of equations determines the load and stress conditions for a flexible pipe under a system of axial and transverse forces such as those that could exist under seismic or blast conditions. For instance, a section of utility duct or piping in earthquake environment can be subject to axial loading with surrounding soil acting as resistance until load Q tending to restrict the development of Euler buckling mode is overcome.

AXIAL RESPONSE OF CYLINDERS

The instability of thin pipes and containers subjected to end compression remains one of the most challenging problems, since there can still be a wide disparity between the theory and practice. One of the reasons for this state of knowledge is the complexity of mathematical formalism involved in the development of the theories of elastic stability. Also, experimental techniques required to prove the theories are full of difficulties and practical limitations. Last, but not least, the influence of manufacturing imperfections on the results is hard to estimate without experiencing a significant scatter band. This section is confined therefore, to some of the more proven rules and formulas suitable for direct application to design.

Essentially all typical responses of cylindrical components under axial compression due to static and pseudostatic loading can be put into three categories:

Diamond-shaped buckles of local character
Bellows-type wall deformation
Direct yield stress criterion

The experiments indicate that thin pipes, vessels, and cans develop isolated buckles of the diamond pattern. This is largely due to the fact that the local bending rigidity of a shell is proportional to T^3, while the resistance to membrane tension depends on the first power of thickness. The thinner the cylindrical surface therefore, the higher the tendency to develop a diamond-shaped pattern. Also a random appearance of diamond buckles suggests that initial imperfections must be responsible for this type of local buckling. Such defects may not be visible, yet they can cause sufficient variation in the compressive stress distribution around the cylinder to trigger the onset of instability. Furthermore, experiments show that diamond-shaped buckles can proceed with great rapidity, resulting often in a "snap-through" type of response.

The classical theory of elastic stability for a cylindrical component predicts [31] a buckling stress equal to

$$S_{\text{CR}} = \frac{0.605 E}{m} \qquad (37.17)$$

The corresponding critical load is

$$P_{\text{CR}} = 3.8ET^2 \tag{37.18}$$

Wide discrepancies between the theory and experiments have finally led to an empirical formula due to Donnell [31]:

$$S_{\text{CR}} = E\frac{0.605 - 10^{-7}m^2}{m(1 + 0.004\phi)} \tag{37.19}$$

Here, as before, ϕ denotes the inverse strain parameter defined by E/S_y. Formula (37.19) appears to correlate well with the experimental data, particularly in the range of higher values of m.

The classical theory of symmetrical buckling of a cylindrical shell under the action of uniform axial compression applies essentially to perfect thin shells of revolution for which the critical buckling stress does not exceed the proportional limit of the material. In this regard a cylindrical component may be regarded to be thin when its ratio of mean radius to thickness exceeds 20. The theoretical response in axial compression also indicates that the critical buckling load is independent of the pipe length. However, the number of half-ways into which the cylinder may buckle can be defined as

$$n_e = \frac{0.58L}{(RT)^{1/2}} \tag{37.20}$$

The corresponding elastic buckling stress is still given by Eq. (37.17).

Equations (37.17) and (37.20) are both intended for isotropic materials for which Poisson's ratio is 0.3, and only for the case of a purely elastic response.

PLASTIC BUCKLING IN AXIAL MODE

When the pipe ratio R/T is decreased, axial buckling stress can approach and exceed the proportional limit of the material. Under these conditions, the length of the half waves into which the pipe buckles becomes shorter. When short pieces of pipe are joined by means of couplings thicker than the pipe itself, lateral expansion of the pipe at the joints becomes restricted. This restriction causes local bending of the pipe wall which, when combined with the direct axial compression, gives rise to the formation of the first axisymmetric half-way buckle. An example of such a buckle is shown in Fig. 37.5 [235]. With further increase in axial compression of the pipe, the buckle splits open. The tests also confirm that the formation of the buckle and the onset of split is likely to be accentuated by the degree of load eccentricity.

For the type of deformation experienced with relatively thicker piping such as that found in the field and the quoted experiment, the theory indicates that buckling should occur beyond the proportional limit of the material. The relevant calculation of the number of the half waves and the corresponding plastic buckling

Axial and Bending Response

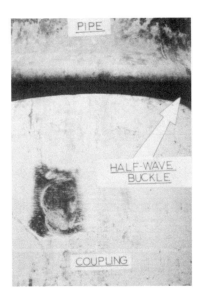

Fig. 37.5 Formation of compression buckle on pipe (9 5/8 in. o.d., 0.435 in. thick). (Courtesy of Lawrence Livermore National Laboratory)

load require the introduction of a reduced modulus concept [32]. This gives

$$n_p = \frac{0.58L(E)^{1/4}}{(RT)^{1/2}(E_r)^{1/4}} \tag{37.21}$$

and

$$S_p = \frac{0.605 E_r T}{R} \tag{37.22}$$

It should be noted that, depending on the shape of the actual stress-strain curve of the material, the reduced modulus E_r can be either constant or variable, as shown by the expression

$$E_r = \frac{4EE_t}{\left(E^{1/2} + E_t^{1/2}\right)^2} \tag{37.23}$$

where E_t denotes the tangent modulus of elasticity at a particular working stress. It also follows that for a constant value of E_t and fixed pipe dimensions R and T, the length of a half-wave remains constant for all the strains beyond the proportional limit of the material. The corresponding critical buckling stress also remains constant. These features are evident from a review of Eqs. (37.21) to (37.23).

The experimental work [235] consisted of the compression tests on three 7-in.-o.d. and three 9 5/8-in.-o.d. pipe specimens with the couplings manufactured from the API C75 grade of steel having a minimum value of ultimate strength equal to about 95,000 psi. The maximum average values of the plastic compressive loads obtained in the tests were 780,000 lb and 1,292,000 lb for the 7-in.-o.d. and 9 5/8-in.-o.d. pipe sections, respectively. Assuming a concentric type of loading, the direct compressive stress becomes

$$S_c = \frac{W}{2\pi RT} \tag{37.24}$$

Making $S_c = S_p$ by combining Eq. (37.22) with Eq. (37.24) gives

$$E_r = \frac{0.263W}{T^2} \tag{37.25}$$

The thicknesses of the pipe wall were 0.435 and 0.317 in. for the larger and smaller pipe diameters, respectively.

Utilizing Eq. (37.25), the effective values of the reduced modulus of elasticity E_r are found to be 1.80×10^6 and 2.04×10^6 psi for the larger and smaller pipes, respectively. The difference between the two values may be largely due to the experimental and material variations, and therefore it is reasonable to assume an average value of $E_r = 1.92 \times 10^6$ psi for the purpose of calculating the corresponding value of the tangent modulus E_t. Also, based on the load test data obtained on the six test pieces, the average ratio of the yield to the ultimate strength can be taken here as 0.84. This gives an estimated yield point for the material of 80,000 psi. The average tangent modulus of elasticity calculated from Eq. (37.23) is 0.45×10^6 psi. This information, together with the rated ultimate strength of the material at 95,000 psi, allows us to construct a bilinear type of stress-strain curve compatible with the plastic response of the tested pipe.

It may also be of interest to add that rather good agreement was found between the calculated half-wavelengths of the buckles and the actual values observed. As far as it can be established from the original records of the experiment, the discrepancy between the theoretical and test data was less than 25%. This is acceptable considering the complex nature of the mechanics of formation of a plastic convolution. Furthermore, a brief comparison of the elastic and plastic response models strongly suggests that the concept of the reduced modulus of elasticity E_r is an indispensable element in a realistic appraisal of the maximum buckling load. Since the majority of common structural steels are relatively ductile, the problem of prediction of the ultimate axial load on the pipe must involve prior knowledge of E_r.

ANALYSIS OF BELLOWS-TYPE BUCKLE

Based on the analysis of quoted experimental data and the concept of the reduced modulus of elasticity, the design estimate of the ultimate pipe capacity for a bellows type of buckling may be accomplished according to the following four-point procedure:

1. Fit a bilinear approximation to the actual stress-strain curve of the pipe material.
2. Obtain the tangent modulus of elasticity E_t from the bilinear characteristics.
3. Calculate the reduced modulus E_r.

Axial and Bending Response

4. Calculate the ultimate axial load W and the corresponding length of a half-wave buckle L_w using the following expressions:

$$W = 3.80 E_r T^2 \tag{37.26}$$

and

$$L_w = 1.72(RT)^{1/2} \left(\frac{E_r}{E}\right)^{1/4} \tag{37.27}$$

The minimum value of E_r for which Eqs. (37.26) and (37.27) are still applicable may be defined with the aid of Eqs. (37.24) and (37.26)

$$E_{r\min} = \frac{1.65 R S_y}{T} \tag{37.28}$$

The practical upper value of E_r can be taken as $E_{r\max} = 2E_{r\min}$, on the assumption that $S_u = 2S_y$. However, the four-step procedure is not recommended outside the range limited by $E_{r\min}$ and $E_{r\max}$.

It should be noted that the absolute size and strength of the standard pipe coupling or a similar local reinforcement does not appear to enter the calculations. Its presence in the pipe string, subjected to axial compression, only serves as the local constraint and the origin of the first perturbation in the continuous process of bellows-type buckling.

The simplest mode of response is defined by the yield stress in direct compression. This gives

$$P_y = 2\pi R T S_y \tag{37.29}$$

EXAMPLE OF LOAD ECCENTRICITY

Under special conditions of underground explorations and tests, a long string of piping can be subject to the effect of load eccentricity. This elementary concept is shown in Fig. 37.6, where we assume a superposition of the tensile and bending stresses in the wall. Hence, defining the amount of eccentricity by e with $n = e/T$ and $m = R/T$, the combined stress criterion becomes

$$S = \frac{W}{T^2}\left(\frac{0.16}{m} + 0.32\frac{n}{m^2}\right) \tag{37.30}$$

The variation of the stress given by Eq. (37.30) indicates that the effect of load eccentricity on the combined stress in the pipe wall decreases rather rapidly with an increase in the radius-to-thickness ratio. This is not surprising when we consider the effect of the increase of pipe radius on the numerical values of the wall bending stress. It is noted that the maximum combined stress is assumed to be tensile. For a relatively high eccentricity, however, it is possible to visualize another condition under which the resultant stresses are changed to compressive. When this happens,

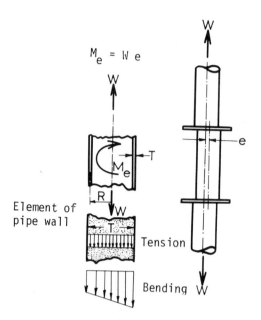

Fig. 37.6 Combined stress in offset pipe.

an additional bending criterion and its effect on the possibility of a local pipe buckling should be examined.

A brief description of this type of failure is included in this section as a matter of general illustration. It is very unlikely that this mode of failure will be induced. However, there may be special applications where this consideration could prove to be of value.

The eccentricity of axial loading then at which local buckling of pipe is likely to develop may be estimated from the expression

$$e = 1.1 \frac{E}{W} mT^3 + \frac{mT}{2} - 1.57 \frac{P_e}{W} m^3 T^3 \qquad (37.31)$$

In deriving Eq. (37.31), the assumptions were made that the theoretical local bending stress in the pipe wall producing elastic instability is equal to the algebraic sum of the following stresses:

Membrane compressive stress due to external pressure
Compressive bending stress due to eccentric loading
Tensile stress due to axial loading

Since the external pressure on the pipe is, at times, relatively low, the last term of Eq. (37.31) can be ignored, to give

$$e = 1.1 \frac{E}{W} mT^3 + \frac{mT}{2} \qquad (37.32)$$

It is quite obvious from Eqs. (37.31) and (37.32) that the extent of eccentricity and the proportions of the canisters or emplacement piping may be such that the

Axial and Bending Response

elastic instability due to bending would not take place. This may be due to the effect of the direct tension on the stress distribution across the pipe wall, as well as the influence of a relatively high ratio of E/W.

BUCKLING DUE TO BENDING

For the case of a negligible axial tension and the external pressure acting on the pipe or a cylindrical vessel, the maximum bending moment corresponding to the onset of instability may be calculated from the expression

$$M_b = 1.1 E m T^3 \tag{37.33}$$

The corresponding critical buckling stress in the pipe or vessel in bending is

$$S_{CR} = 0.35 \frac{E}{m} \tag{37.33a}$$

Equations (37.31) to (37.33a) are based on a Poisson's ratio of 0.3. Experience also indicates that typical cylindrical components in service are seldom exposed to bending moments of the magnitude suggested by Eq. (37.33). For buckling due to bending to develop, the relevant wall thickness and modulus E would have to be rather low. However, the ultimate bending moment is also sensitive to variations in wall thickness.

It should be added that the buckling behavior of a pipe or a cylindrical vessel subjected to bending is consistent with the mechanism of axial compression. The buckling stresses in both cases appear to be of similar magnitude. Consequently, a number of investigations were in the past directed toward the correlation of test data on cylinders in bending and axial compression.

Design Problem 37.1

Compare the critical buckling stresses for a pin-jointed steel column having a tubular cross section of mean radius 2 in. and wall thickness of 0.05 in. assuming bending and axial compression loading independently. The tubular column is 42 in. long and the modulus of elasticity is 28×10^6 psi.

Solution

From Eq. (37.5)

$$S_{CR} = \frac{\pi^2 E R^2}{2L^2} = \frac{\pi^2 \times 28 \times 10^6 \times 4}{2 \times 42 \times 42} = 313{,}322 \text{ psi}$$

From Eq. (37.33a)

$$S_{CR} = 0.35 \frac{E}{m} = \frac{0.35 \times 28 \times 10^6}{40} = 245{,}000 \text{ psi} \quad \blacklozenge$$

It is unlikely that the stresses of this magnitude will be reached in practice, because of the usually lower yield strength of the structural material at our disposal. In either case, then, the tubular member under consideration will be governed by stress rather than stability. Equations (37.5) and (37.33a) also suggest that pertinent parameter in the process of comparing the designs based on the overall and local buckling characteristics is

$$m = 0.07 \left(\frac{L}{R}\right)^2 \tag{37.33b}$$

This relation may be of interest in compression and bending tests on tubular columns. The foregoing criterion defines the tube thickness parameter for which the critical stresses due to overall and local buckling are equal.

THEORY OF ROLLING DIAPHRAGM

One of the more interesting phenomena of axial response of a cylindrical component is concerned with the formation of a convolution. Such a mechanism is found in rolling cylindrical diaphragms, inverted tube shock absorbers, and positive expulsion devices, to mention a few [236]. A simplified model of this mechanism can be established on the assumption that the neutral axis of the sheet material coincides with its centerline and that the strain energy due to axial and shear forces can be neglected as being small in relation to bending and hoop extension energy. Furthermore, it is assumed that the strain energy of bending and unbending during the process of diaphragm inversion is the same and that the material is bent continuously through a full angle of 180°. The strain energy of bending with a constant bending moment is obtained as the product of the plastic moment and the full angle of bend expressed in radians. The unit energy due to circumferential strain is taken as the product of the plastic hoop stress and the corresponding total hoop strain.

The basic geometry of the cylindrical diaphragm analyzed is illustrated in Fig. 37.7. The diaphragm is considered to be held rigidly in the plane of circumference A-A. The mean radii of diaphragm and bend are R and r, respectively. Assuming a continuous plastic deformation under pressure differential ΔP, the diaphragm end moves from plane B-B to C-C, as shown. During this process each element of the diaphragm (Fig. 37.8) undergoes longitudinal strains due to pure bending and hoop strains as the result of the overall increase in radius. At the same time, the plane $O'''O''G''G'''$ rotates through angle $\theta = \pi$ into a new position $O'OGG'$. For a small bend element of unit width cut out of the diaphragm, the cross-sectional geometry is approximately rectangular and the corresponding section modulus for a fully developed plastic condition is given by the elementary theory of plasticity

$$Z_p = \frac{t^2}{4} \tag{37.34}$$

Axial and Bending Response

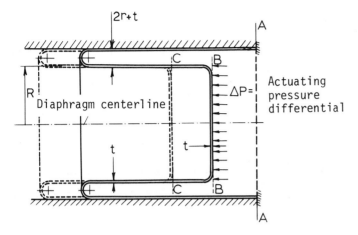

Fig. 37.7 Notation for a cylindrical diaphragm.

By the rules of strength-of-materials theory, the ultimate bending moment for a relatively wide beam is

$$M_p = \frac{S_y Z_p}{1 - \nu^2} \tag{37.35a}$$

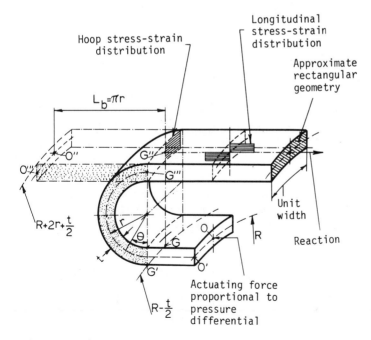

Fig. 37.8 Geometry of bent element.

Substituting Eq. (37.24) into (37.35a) gives

$$M_p = 0.275 S_y t^2 \qquad (37.35b)$$

The longitudinal strain due to pure bending, in terms of the mean radius of curvature of the bend and wall thickness, follows from the relation

$$\epsilon_b = \frac{t}{2r} \qquad (37.35c)$$

The length of the bend which undergoes a complete plastic deformation is obtained directly from the geometry depicted in Fig. 37.8

$$L_b = \pi r \qquad (37.35d)$$

The energy of deformation due to bending corresponding to this length is πM_p. Hence, utilizing Eq. (37.35b) gives

$$U_b = 0.87 S_y t^2 \qquad (37.35e)$$

The tensile hoop strain developed during the straightening out of the bend element may be defined as

$$\epsilon_h = \frac{\text{final circumference less original circumference}}{\text{original circumference}}$$
$$= \frac{2\pi \left(R - \frac{t}{2} + 2r + t\right) - 2\pi \left(R - \frac{t}{2}\right)}{2\pi \left(R - \frac{t}{2}\right)}$$

which yields

$$\epsilon_h = \frac{4r + 2t}{2R - t} \qquad (37.35f)$$

Since the unit energy due to hoop strain is $S_y \epsilon_h$, the corresponding strain energy per inch of width of the diaphragm becomes

$$U_h = S_y \epsilon_h L_b t \qquad (37.35g)$$

Combining Eqs. (37.35c), (37.35d), and (37.35g), the strain energy in the hoop extension can be expressed in terms of the hoop and longitudinal strains as follows:

$$U_h = 1.57 S_y t^2 \left(\frac{\epsilon_h}{\epsilon_b}\right) \qquad (37.35h)$$

Introducing nondimensional parameters $k = R/t$ and $m = r/t$ into Eqs. (37.35c) and (37.35f), the hoop energy per unit width of the diaphragm is

$$U_h = \frac{\pi m(4m + 2)}{2k - 1} S_y t^2 \qquad (37.35i)$$

Axial and Bending Response

Hence, the total energy of deformation for a bend element of unit width is obtained as the algebraic sum of the energies expressed by Eqs. (37.35e) and (37.35i). This yields

$$U = \left[0.87 + \frac{\pi m(4m + 2)}{2k - 1}\right] S_y t^2 \tag{37.35j}$$

The external work done on the diaphragm by the actuating pressure ΔP for one length of bend is simply the total actuating force multiplied by the distance traveled. This gives

$$W_e = \pi R^2 \Delta P L_b \tag{37.35k}$$

The pressure acting on the side of the diaphragm is excluded as being oriented normal to the path of the actuating pressure. Substituting Eq. (37.35d) in the above expression yields

$$W_e = \pi^2 \Delta P R^2 r \tag{37.35l}$$

The grand total of the deformation energy is obtained by multiplying Eq. (37.35j) by the average length of the circumference involved. This gives

$$U_t = 2\pi(R + r)U \tag{37.35m}$$

Making the external work done equal to the internal energy of deformation, Eqs. (37.35l) and (37.35m) yield

$$\Delta P = \frac{2(R + r)U}{\pi R^2 r} \tag{37.35n}$$

Finally, substituting Eq. (37.35j) into Eq. (37.35n) gives an approximate design formula for actuating pressure on the cylindrical diaphragm

$$\Delta P = \frac{(k + m)(0.55k + 4m^2 + 2m)S_y}{mk^3} \tag{37.36}$$

In many practical cases where m is small compared to k, Eq. (37.36) can be reduced to

$$\Delta P = \frac{(0.55k + 4m^2 + 2m)S_y}{mk^2} \tag{37.37}$$

To increase the capacity of the diaphragm, it is possible to choose an annular geometry such as that shown in Fig. 37.9. Utilizing the same type of analysis, the expression for the annular version becomes

$$\Delta P_a = \frac{2S_y}{\pi m(k_0^2 - n_0^2)} \left\{ \frac{k_0 + m}{2k_0 - 1}[0.87(2k_0 - 1) + \pi m(4m + 2)] \right. \\ \left. + \frac{n_0 + m}{(2n_0 - 1)}[0.87(2n_0 - 1) + \pi m(4m + 2)] \right\} \tag{37.38}$$

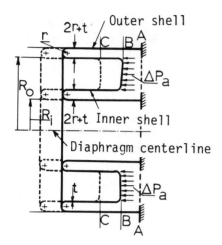

Fig. 37.9 Annular rolling diaphragm.

When parameter m is relatively small and when $(2k_0 - 1)$ and $(2n_0 - 1)$ are approximated by $2k_0$ and $2n_0$, respectively, Eq. (37.38) simplifies to

$$\Delta P_a = \frac{[0.55(k_0 + n_0) + 8m^2 + 4m] S_y}{m(k_0^2 - n_0^2)} \qquad (37.39)$$

The design calculations can be speeded up by providing dimensionless plots of various parameters [236]. The change from cylindrical to annular geometry can be governed by the criteria of equal volume or equal pressure for the two configurations. The analysis shows that for the condition of equal volume, the thickness of the annular diaphragm is

$$t_a = 0.87 \left(\frac{R_i r \Delta P_a}{S_y} \right)^{1/2} \qquad (37.40)$$

The approximate thickness for the equal pressure is

$$t_a = t \left(\frac{R_o - R_i}{R} \right)^{1/2} \qquad (37.41)$$

Although the inherent complexity of the mechanics of a convolution presents a limited opportunity for a more detailed and fully representative theoretical analysis, the study indicates at least that the materials selected for the rolling diaphragms should exhibit good elongation and ductility. However, since the characteristics of uniform elongation in a metallic material are closely associated with the work-hardening properties and are subject to local variations resulting from heat treatment and alloy content, it is desirable, whenever possible, to select non-heat-treatable alloys. On this basis, aluminum alloys with the limited alloy content appear to be most promising, provided that they are compatible with other metals in the system.

Axial and Bending Response

The equations derived indicate that the actuating pressures depend on geometric parameters and are directly proportional to the yield strength of the material. Hence, for the same thickness of a diaphragm wall, the higher the yield strength, the higher the pressure required to produce the rolling action. This simple relation was deduced on the assumption of the idealized stress-strain characteristics of material undergoing continuous plastic deformation. Under such conditions, the working plastic stress remains sensibly constant while the material is subjected to gradually increasing strains. To fulfill such stress-strain requirements, the relevant material must have good ductility and must be able to sustain appreciable elongation without rupture. From the point of view of practical design, it is well to keep in mind the fact that because of large strains associated with the process of a toroidal inversion of the cylinder wall, thin-gage materials and large diameters may be required for minimizing the working stresses. However, large ratios of diaphragm radius to wall thickness imply a decrease in the resistance of the wall to local buckling. Hence, a suitable design compromise may well be required.

SYMBOLS

A		Constant of integration
B		Constant of integration
E		Modulus of elasticity, psi (N/mm^2)
E_r		Reduced modulus, psi (N/mm^2)
$E_{r\min}$		Minimum reduced modulus, psi (N/mm^2)
$E_{r\max}$		Maximum reduced modulus, psi (N/mm^2)
E_t		Tangent modulus, psi (N/mm^2)
e		Load eccentricity, in. (mm)
$F(k)$		Auxiliary function
I		Moment of inertia, in.^{4} (mm^4)
K		Curvature parameter, in. (mm)
$k = R/t$		Cylindrical diaphragm ratio
$k_0 = R_o/t$		Annular diaphragm ratio
L		Length, in. (mm)
L_e		Reduced column length, in. (mm)
L_0		Half-length of pipe, in. (mm)
L_b		Developed length of corner, in. (mm)
L_w		Length of half-wave buckle, in. (mm)
M_e		Moment due to load offset, lb-in. (N-mm)
M_b		Bending moment to cause buckling, lb-in. (N-mm)
M_0		End moment, lb-in. (N-mm)
M_p		Plastic moment, lb-in. (N-mm)
m		Mean radius-to-thickness ratio
$n = e/T$		Eccentricity ratio
$n_0 = R_i/t$		Annular diaphragm ratio
n_p		Number of plastic half-waves
P		Axial load on pipe, lb (N)
P_{CR}		Critical axial load, lb (N)
P_e		External pressure, psi (N/mm^2)

P_y	Axial load at yield, lb (N)
ΔP	Pressure difference across cylindrical diaphragm, psi (N/mm^2)
ΔP_a	Pressure difference across annular diaphragm, psi (N/mm^2)
Q	Transverse resistance, lb/in. (N/mm)
R	Mean radius of cylindrical component, in. (mm)
R_o	Outer radius, in. (mm)
R_i	Inner radius, in. (mm)
r	Mean radius of convolution, in. (mm)
S	General symbol for stress, psi (N/mm^2)
S_c	Stress in direct compression, psi (N/mm^2)
S_{CR}	Critical column stress, psi (N/mm^2)
S_p	Plastic stress, psi (N/mm^2)
S_y	Yield stress, psi (N/mm^2)
S_u	Ultimate stress, psi (N/mm^2)
T	Thickness of pipe or vessel, in. (mm)
t	Thickness of diaphragm wall, in. (mm)
U	Total elastic energy per inch of convolution, lb-in./in. $\left(\frac{\text{N-mm}}{\text{mm}}\right)$
U_b	Unit energy due to bending, lb-in./in. $\left(\frac{\text{N-mm}}{\text{mm}}\right)$
U_h	Unit energy due to hoop strain, lb-in./in. $\left(\frac{\text{N-mm}}{\text{mm}}\right)$
U_t	Total elastic energy, lb-in. (N-mm)
W	Ultimate axial load, lb. (N)
W_e	External work done, lb-in. (N-mm)
x	Arbitrary distance, in. (mm)
y	Initial offset at any point, in. (mm)
Z	Elastic section modulus, in.^{3} (mm^3)
Z_0	Elastic foundation parameter
Z_p	Plastic section modulus, in.^{3} (mm^3)
α	Buckling mode number
β	Modulus of foundation, psi (N/mm^2)
δ	Maximum initial offset, in. (mm)
ϵ_b	Bending strain, in./in. (mm/mm)
ϵ_h	Hoop strain, in./in. (mm/mm)
ν	Poisson's ratio
$\phi = E/S_y$	Inverse strain parameter

38
Special Problems in Cylinders

DILATION OF CLOSED CYLINDERS

Design office experiences indicates that in addition to the questions involving stresses in pressurized components, there are a number of specialized topics related to this field. These may include such matters as dilation of cylinders, nested-cylinder effects, or the theory of circumferential stiffeners. In some of these cases even well-established formulas and practices contain speculative features, despite the general progress in design technology.

The dilation effect can be defined as radial growth of a vessel subjected to the internal pressure. The magnitude of the relevant dimensional change may be of interest to designers concerned with the limited assembly tolerances in a given mechanical system. Additionally, one of the more important applications in this area involves the development of pressure transducers in the field of instrumentation. Transducer manufacturers are concerned with the optimum use of the strain gages to assure an electrical signal proportional to the internal pressure in a closed-ended tube. It is clear, therefore, that some detailed knowledge of the dilation characteristics as a function of tube geometry is essential. Unfortunately, only a limited amount of information on this topic is available in the open literature [237].

If the outer and inner tube diameters are denoted by D and d, respectively, the amount of radial growth for a thin closed-end tube is

$$\delta = \frac{PD(2-\nu)(1+k)^2}{16E(1-k)} \tag{38.1}$$

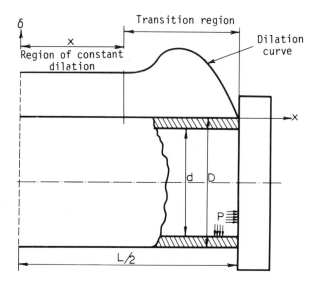

Fig. 38.1 Partial view of a closed-end tube under dilation.

This formula is applicable to the region of the tube not affected by the end closures. The transition region indicated by Fig. 38.1 is marked by a perturbation which has been observed during tests [237]. The region of uniform radial growth, defined by x in Fig. 38.1, depends on two dimensionless parameters, L/D and d/D. The design chart in Fig. 38.2 gives the variation of the span coordinate x/L with the above two parameters. The experimental evidence also suggests that the ratio of the maximum

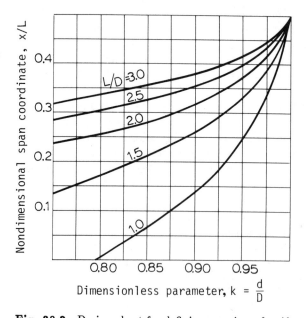

Fig. 38.2 Design chart for defining a region of uniform dilation.

Special Problems in Cylinders

dilation to the theoretical uniform dilation given by Eq. (38.1) should not exceed the value 1.10. However, this ratio must depend on the basic shell parameters and the accuracy of tube manufacturer. The overall effect of geometric irregularities such as out-of-roundness, bore eccentricity, and thickness variation has not as yet been accounted for in experimental and theoretical studies.

NESTED CYLINDERS

In most design situations the calculation of wall thickness for a cylindrical canister subjected to external pressure is performed well ahead of hardware development and manufacture. However, there may be special cases where an existing canister has to be modified to meet particular load specifications or increased safety considerations. Under these conditions, the following options are available:

1. Providing an entirely new canister design
2. Providing circumferential stiffeners
3. Providing a continuous structural sleeve

If the time is limited, solution (1) may not be acceptable. If manufacturing and metallurgical problems arise, solution (2) may have to be excluded. This leaves alternative (3). We have to assume here that sleeve reinforcement can be provided without undue fabrication difficulties. For the purpose of this analysis, we assume that the composite canister, consisting of the main shell and the reinforcing sleeve, can be assembled with zero or minimal initial interference. Thus, the practical question to be answered is: What degree of pressure attenuation can one reasonably expect in a reinforced system?

Although the elastic theory suitable for defining the attenuation factor is rather elementary, few practical design formulas are readily available in engineering handbooks.

Let us consider, for example, a double-wall system such as that sketched in Fig. 38.3. The displacement of the inner surface of the reinforcing cylindrical sleeve, in terms of a given external pressure P_o and the unknown contact pressure P, follows directly from Eq. (35.25)

$$U_{io} = \frac{(1-\nu)(P_o R_o^2 - PR^2)R + (1+\nu)(P_o - P)R_o^2 R}{E(R_o^2 - R^2)} \tag{38.2}$$

Here E and ν denote the modulus of elasticity and Poisson's ratio, as before. Other symbols are defined in Fig. 38.3.

The displacement of the outer surface of the canister is given by

$$U_{oi} = \frac{(1-\nu)(PR^2 - P_i R_i^2)R + (1+\nu)(P - P_i)RR_i^2}{E(R^2 - R_i^2)} \tag{38.3}$$

For the assumed condition of zero radial interference between the sleeve and the canister, $U_{io} = U_{oi}$. Hence, utilizing Eqs. (38.2) and (38.3) and making $\nu = 0.3$ and $P_i = 0$, we obtain the following expression for predicting the attenuation ratio

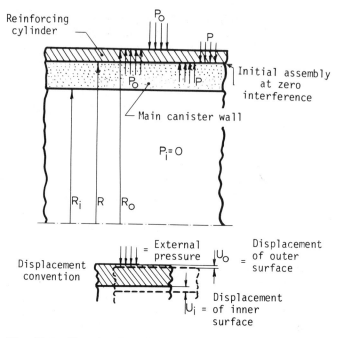

Fig. 38.3 Notation for reinforcement analysis.

ψ:

$$\psi = \frac{2(1 - k_1^2)}{(1.3 + 0.7k_2^2)(1 - k_1^2) + (0.7 + 1.3k_1^2)(1 - k_2^2)} \qquad (38.4)$$

The following nondimensional parameters are used in Eq. (38.4):

$$\psi = \frac{P}{P_o}$$
$$k_1 = \frac{R_i}{R}$$
$$k_2 = \frac{R}{R_o}$$

When $k_1 = 1$, the thickness of the canister wall vanishes and the interference pressure becomes equal to zero. When $k_2 = 1$, that is, the sleeve is assumed to be infinitely thin, Eq. (38.4) yields $\psi = 1$, indicating no pressure attenuation. When $k_1 = 0$ or $k_2 = 0$, real solutions can still be obtained from Eq. (38.4), although practical applications in cases reflecting such bracketing values are not very likely to exist.

For a suitable range of parameters k_1 and k_2, the values of attenuation ratio ψ can be presented in the form of a family of curves such as those shown in Fig. 38.4.

Special Problems in Cylinders

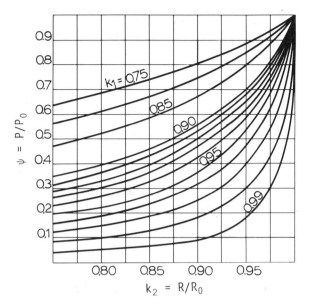

Fig. 38.4 Pressure attenuation chart for nested cylinders.

DESIGN OF RING STIFFENERS

As mentioned previously, reinforcement of the cylindrical vessels and large piping can be accomplished with the aid of circumferential stiffeners. This procedure requires the determination of spacing and cross-sectional dimensions of the stiffeners in such a manner as to prevent overall collapse of the vessel under external pressure. The relevant notation is shown in Fig. 38.5. The formula for unit external load, sufficient to buckle the ring elastically, is found from Table 10.7 to be

$$q_{CR} = \frac{3EI}{R^3} \tag{38.5}$$

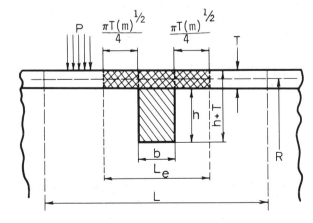

Fig. 38.5 Notation for a ring stiffener.

For the length of cylinder L, corresponding to one ring stiffener, the external load per inch of circumference is

$$q_{CR} = PL \tag{38.6a}$$

Combining Eqs. (38.5) and (38.6a) and solving for the moment of inertia gives

$$I = \frac{PLR^3}{3E} \tag{38.6b}$$

The effective second moment of the stiffener and a length of the cylindrical shell working with the stiffener is made equal to the moment of inertia defined by Eq. (38.6b). The effective length L_e, shown in Fig. 38.5, can be approximated by the rule [238]

$$L_e = b + 1.57(RT)^{1/2} \tag{38.6c}$$

Consider now the following expression for finding the moment of inertia:

$$I = I_b + I_g - \frac{J^2}{A} \tag{38.6d}$$

Here

$$A = (h+T)b + 1.57T^2(m)^{1/2} \tag{38.6e}$$

$$J = \frac{b(h+T)^2}{2} + \frac{1.57T^2(m)^{1/2}(2h+T)}{2} \tag{38.6f}$$

$$I_b = \frac{b(h+T)^3}{4} + \frac{1.57T^2(m)^{1/2}(2h+T)^2}{4} \tag{38.6g}$$

$$I_g \cong \frac{b(h+T)^3}{12} \tag{38.6h}$$

Introducing $\lambda = b/T$, $\epsilon = h/T$, and combining Eqs. (38.6d) to (38.6h), the resultant moment of inertia is obtained. Making this moment of inertia equal to that expressed by Eq. (38.6b) yields the final formula for sizing ring stiffeners:

$$\eta = \frac{0.24\lambda^2(\epsilon+1)^4 + 0.39\lambda(m)^{1/2}(\epsilon+1)(4\epsilon^2 + 2\epsilon + 1)}{m^3[\lambda(\epsilon+1) + 1.57(m)^{1/2}]} \tag{38.7}$$

where

$$\eta = \frac{PL}{ET} \tag{38.8}$$

Ring-stiffened vessels are generally most efficient. Although the behavior of stiffened cylinders has been studied by numerous investigators, the question of L_e given by Eq. (38.6c) has not yet been fully resolved, and it remains an interesting area for further research.

Special Problems in Cylinders

As indicated previously in connection with the definition of the effective out-of-roundness, Eq. (36.19), this influence becomes less important as the spacing of ring stiffeners decreases. Hence, the stiffening rings not only make a lighter design but also allow some relaxation of the fabrication tolerances. The experiments on ring-stiffened vessels show that the entire shell and stiffeners work together until local buckling of the panel takes place. Such buckling is often followed by a tear at the junction of the shell and the stiffener. The stiffener, however, seldom fails unless the panels on both sides of the ring become unstable and buckle prior to shell tear.

SYMBOLS

A	Cross-sectional area, in.^{2} (mm^2)
b	Width of stiffener cross section, in. (mm)
D	Outer diameter of cylinder, in. (mm)
d	Inner cylinder diameter; bolt diameter, in. (mm)
E	Modulus of elasticity, psi (N/mm^2)
h	Depth of ring stiffener, in. (mm)
I	Moment of inertia, in.^{4} (mm^4)
I_b, I_g	Second moments of area, in.^{4} (mm^4)
J	Static moment, in.^{3} (mm^3)
$k = d/D$	Diameter ratio
k_1, k_2	Radius ratios
L	Length of cylinder, in. (mm)
L_e	Effective length, in. (mm)
$m = R/T$	Radius-to-thickness ratio
P	Symbol for uniform pressure, psi (N/mm^2)
P_o	External pressure, psi (N/mm^2)
P_i	Internal pressure, psi (N/mm^2)
q_{CR}	Critical buckling load, lb/in. (N/mm)
R	Interface or average radius, in. (mm)
R_o	Outer radius, in. (mm)
R_i	Inner radius, in. (mm)
T	Thickness of wall, in. (mm)
U_{io} (or U_i)	Displacement of inner surface, in. (mm)
U_{oi} (or U_o)	Displacement of outer surface, in. (mm)
x	Arbitrary distance
δ	Dilation, in. (mm)
$\epsilon = h/T$	Depth ratio
η	Dimensionless parameter
$\lambda = b/T$	Width ratio
ν	Poisson's ratio
$\psi = P/P_o$	Attenuation ratio

VII
MISCELLANEOUS STRESS TOPICS IN DESIGN

39
Design of Bolted Joints

TORQUE FORMULA

One of the special topics selected for the last part of this book is related to flanges and vessels, and it is concerned with the well-known area of joint design. While the safety of a mechanical system often depends on the structural integrity of a simple part such as a common bolt, there is still no complete agreement on some of the elementary aspects of bolt design, and a good deal of research conducted in this field remains relatively unknown [239, 240, 269, 270, 271].

Established rules of individual industries and regulatory agencies have been, for many years, recommending specific torque values and gasket materials for bolted joints, based on the rule-of-thumb value $\mu = 0.15$ for the coefficient of friction. According to the design theory, the initial torque exerted on a bolt or nut can be calculated from the expression

$$M_t = W_i r \left(\frac{\cos\theta \tan\alpha + \mu}{\cos\theta - \mu \tan\alpha} + \frac{\mu R}{r} \right) \tag{39.1}$$

In Eq. (39.1), the symbol M_t represents the sum of the thread and collar torques developed simultaneously to overcome the frictional resistance at the respective contact surfaces. Since the helix angle α is relatively small, introduction of a small-angle approximation and substitution of some of the typical values of R/r, θ, and μ in Eq. (39.1) leads to the well-known approximate formula for the tightening torque given in standard engineering handbooks

$$M_t = \frac{W_i d}{5} \tag{39.2}$$

For ease of calculations, d represents the nominal bolt diameter and W_i is the initial bolt load in pounds.

The theoretical basis of Eq. (39.2) may be justified in calculating the approximate number of bolts required for a particular joint, average-bearing stress on the gasket, and sufficient preload for leakage, thermal, and shock conditions. It is evident, however, that for the usual proportions of a bolted joint, W_i can vary significantly as a function of μ [241]. For instance, the coefficient of friction for a steel bolt and nut under dry conditions can be between 0.15 and 0.25. Application of industrial-type lubricants of graphite or mineral-oil type can lower the coefficient to about 0.03. The design rationale based on the minimum value of μ would then be appropriate for calculating the highest stress intensity in the threaded portions and the most conservative factor of safety on the bearing stresses on seating or gasket surfaces. On the other hand, the application of the highest coefficient of friction should ensure that somewhat lower but still sufficient preload could be developed for a specific set of service conditions.

It should be added here for the sake of completeness of theory that Eq. (39.1) is essentially the same as that defining the magnitude of the turning moment required to advance a standard power screw against the working load. Therefore, the basic mechanics of common bolt tightening and power screw advance is the same. It is also well to note that Eq. (39.1) is applicable to bolts that have flat and rigid bearing surfaces, but it cannot be applied to the case of zero friction. Finally, some earlier experience also indicates [242] that faulty lubrication and manufacture can cause a variation in bolt tension perhaps as high as an order of magnitude. Although this extreme range between the maximum and minimum values may be considered too high for modern practice in machine design and manufacture, certainly the error factor in estimating the axial preload can easily be on the order of 2.

The variation, together with the unknowns in material strength and skills of mechanical technicians in making up the bolted joints, has led to the development of early practices in which the design stress for bolts was made a function of bolt size [243]. This approach resulted in a well-known rule of thumb designating the minimum practical bolt size of 1/2 in., as a safeguard against the case of unintentional overstressing.

EXTERNAL LOAD

General opinions still vary among practical engineers concerning the recommended design magnitude of the effective load W_e on the clamped components in relation to the initial tightening bolt load W_i. These basic loads in a bolted assembly such as that shown in Fig. 39.1 constitute some of the more important factors influencing the resultant bolt load W. For instance, in the case illustrated, the flexibility of four separate components may be involved in determining the resultant load. The elastic response of each component can be expressed in terms of the individual

Design of Bolted Joints

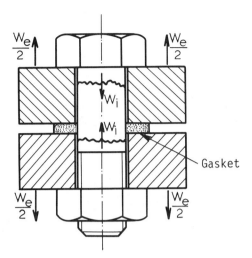

Fig. 39.1 Basic loads on a bolted joint.

spring constants, while the stiffness of the compressed assembly can be represented by means of the resultant spring constant K_c. Equating the change in length of the bolt to the change in length of the bolted components yields the following general equation for estimating the resultant bolt load.

$$W = W_i + \frac{W_e}{1 + \left(\frac{K_c}{K_B}\right)} \tag{39.3}$$

Here K_B denotes the spring constant of the bolt. For a bolt shank of constant cross section, this parameter becomes

$$K_B = \frac{A_B E_B}{L_B} \tag{39.4}$$

The resultant spring constant of the compressed assembly is

$$K_c = \frac{1}{\frac{1}{K_1} + \frac{1}{K_2} + \cdots \frac{1}{K_n}} \tag{39.5}$$

The individual terms $K_1, K_2, \ldots, K_n$ have essentially the same meaning as that indicated by Eq. (39.4) although there is an inherent difficulty in determining the effective cross-sectional areas of the flange portions undergoing compressive deformation [244]. Where relatively large washers and gaskets are used, the joints develop greater flange areas under compression, mitigating local yielding of the assembled parts. Experience shows that by changing the stiffness of the individual washers and gaskets, the overall spring constant of the assembled joint and the resultant bolt load can be changed.

EFFECTIVE AREA CONCEPT

The problem of determining the size of the effective cross-sectional area of the flange has, so far, been solved photoelastically [244]. The method for finding the effective area is illustrated in Fig. 39.2. Point O is located at the maximum radius of the bolt contact area. It is found from dimensions of a standard nut and, in the particular sketch, this radius corresponds to half the width across the flats. A line drawn through point O at the half-cone angle ϕ establishes a reference point at a distance $h/2$, as shown. Experimental work suggests that ϕ may vary between 25 and 33° for the most common joint materials. The required annulus dimension a follows from a condition of geometry.

$$a = \rho + 0.5(h \tan \phi - d) \tag{39.6}$$

and

$$\rho = R + g \tan \phi \tag{39.7}$$

The maximum radius of contact area between the bolt head and the washer is R. This dimension is usually known ahead of time for a given bolt size, so that the number of variables in the above equations can be reduced to three. The calculation starts with an assumed value of ρ or g, defining the washer dimensions as shown in Fig. 39.2. Hence, the solution of Eqs. (39.6) and (39.7) establishes the washer size and the annulus width a. This, in turn, determines the effective area in compression and the relevant spring constants.

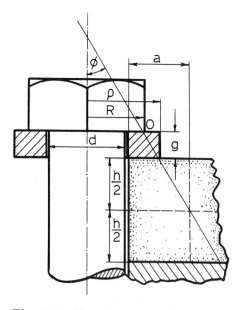

Fig. 39.2 Notation for the effective area method.

Design of Bolted Joints

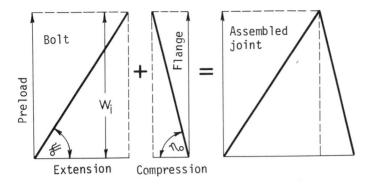

Fig. 39.3 Simplified diagram for bolt preload.

ELASTIC RESPONSE

Reference to Eq. (39.3) can clarify some of the theoretical aspects of bolt-joint elastic response. For instance, when the clamped components are very rigid compared to the bolt, K_c/K_B becomes rather high and the effect of the term W_e on the resultant bolt load virtually disappears. Conversely, when the components of the joint behave as a soft spring while bolt rigidity is high, the ratio K_c/K_B tends to zero and the total external load on the joint may contribute to the resultant bolt load. These bracketing conditions can be summed up as follows:

$$K_c \gg K_B \quad \text{(soft bolt clamps, rigid components)} \quad W = W_i \quad (39.8)$$
$$K_c = K_B \quad \text{(all components have equal rigidity)} \quad W = W_i + 0.5W_e \quad (39.9)$$
$$K_c \ll K_B \quad \text{(rigid bolt clamps, soft components)} \quad W = W_i + W_e \quad (39.10)$$

The initial elastic response of a complete bolted joint can be simulated with the aid of a simplified diagram such as that given in Fig. 39.3. As the tightening torque on the bolt is increased, the shank elongates according to the linear force-deformation curve, consistent with Hooke's law. The flange, representing a typical connected member, undergoes elastic compression. Since the same amount of preload acts on the bolt and the flange simultaneously, the two force-deformation characteristics can be assembled into one diagram, as shown in Fig. 39.3. This type of a complete diagram is found in books on machine design as a basis for discussing the elastic response of a bolted connection. The relevant spring constants are $K_B = \tan \psi_0$ and $K_c = \tan \eta_0$ for the bolt and the clamped component, respectively.

STRESSES IN BOLTS

Stress intensity across the bolt cross section under working conditions results from a combination of axial and torsional stresses. If in addition to axial loading, the bolts carry the transverse shear load, the analysis should be based on the *interaction method* [29]. This method is concerned with the structural failure under combined stresses. The combined loading may be represented by the load or stress ratios, and the relation between such ratios follows a continuous curve, the shape of which is determined by theory and tests.

In the case of an essentially axial loading on a bolted joint found in lifting and emplacement operations, the stress intensity depends primarily on the bolt diameter and the coefficient of friction on contact surfaces. Such a stress, determined for the combined tension and torsion, may be found from the expression [245]

$$S = \frac{0.024W \left[(1 + 36\mu^2)^{3/2} - 1\right]}{\mu^2 d^2} \tag{39.11}$$

Starting with Eq. (39.11), the dimensionless parameter Sd^2/W is plotted in Fig. 39.4 for the common range of the coefficient of friction corresponding to some typical industrial lubricants. In practical design situations involving stress and load calculations, the working strength of bolts may also be obtained from an empirical formula [246]

$$S = \frac{4W}{d(2.2d - 1)} \tag{39.12}$$

In this formula, bolt load W is expressed in pounds, nominal bolt diameter d in inches, and stress S in psi. It is also clear from Eq. (39.12) that this expression is not intended for bolt diameters smaller than about 5/8 in. This is in agreement with the well-known rule of thumb stating that normal torquing procedures can lead to overstressing and breakage of bolts having diameters equal or smaller than 1/2 in.

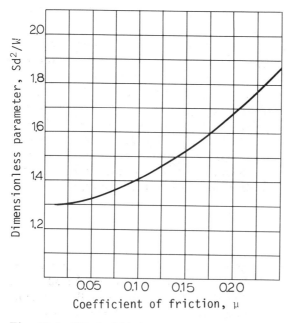

Fig. 39.4 Effect of friction on stress intensity.

Design of Bolted Joints

For a quick preliminary estimate, Eq. (39.2) and (39.12) can be combined to give the stress in terms of the tightening torque and the nominal bolt diameter. This procedure yields

$$S = \frac{20M_t}{d^2(2.2d - 1)} \tag{39.13}$$

The stress intensity per unit torque based on Eq. (39.13) is shown plotted in Fig. 39.5 as a function of nominal bolt diameter. For example, taking $d = 1.125$ in. and a torque of 3600 lb-in. gives the total axial stress of 36,000 psi. A change from 1.125 to 1.250 nominal dimension would cut the stress by about 30%, assuming proper thread engagement geometry and manufacturing tolerances. The chart in Fig. 39.5 indicates, therefore, that having any choice in the matter at all, it would always be advisable to select bolt diameters that are slightly oversized.

Loaded bolt threads are subject to a complex combination of nonuniform bending, shear, and compressive stresses. Because of the three-dimensional nature of loading, the effective stress concentration factors are sometimes used in the analysis [247]. These factors are known to vary somewhere between 3 and 7, depending on the type of thread design and material. However, they apply largely to joints

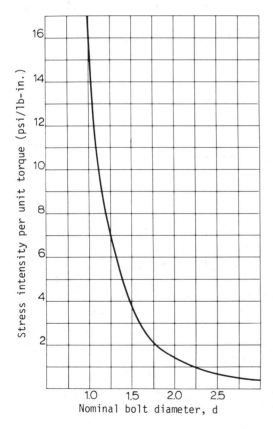

Fig. 39.5 Stress intensity based on experiments.

subjected to fatigue loading. Similar factors may be observed at the junction between the shank and the head, as well as at other locations involving abrupt changes of the diameter. Under essentially elastic and static loading of bolted joints made of ductile materials, there is little justification of high stress concentration factors. Experience also indicates that for the majority of bolts larger than about 3/4 in., the initial stresses can safely be on the order of 75% of the yield strength of the material.

IMPROVEMENTS IN BOLT PERFORMANCE

Ideally, the best bolt reliability can be achieved when the safety factors in the threaded portion and in other critical sections of the shank are equal. However, as the threaded portion of the bolt elongates during the nut compression, the pitch of the bolt threads increases while that of the nut decreases. Since the two parts still remain in contact, the net result is that load distribution among the individual threads becomes uneven. Past investigations show that for a conventional bolted-joint design, about 35% of the total load is taken by the thread closest to the compressed surface of the nut, provided that the total number of engaged threads is at least six. Progressive plastic deformation of individual threads tends to make load distribution somewhat more uniform. For example, a soft nut and a hard bolt result in an improved load distribution because soft threads can readily deform plastically and transfer some of the load to the less stressed threads.

Generally speaking, bolt performance can be in some cases improved without increasing the nominal bolt diameter and the overall size of a bolted assembly. The specific methods of improvement can be summarized as follows:

Decrease stress gradient in threads.
Decrease axial stiffness of the bolt.
Make load distribution on the threads more uniform.
Minimize bolt bending.
Control bolt preload.

Bending stresses in the bolt are caused by nonaxial application of the external loading on the joint, while the stress concentration can be decreased by using, for example, the rolled threads. Longitudinal bolt stiffness can be lowered by machining down the shank cross-sectional area, as long as the critical design stress at the minimum section is acceptable. Control of preload is also significant because a bolted joint tends to relax its initial preload in service. This relaxation can be caused by the metal flow under pressure at any uneven point of contact.

When an increase in longitudinal strain on the bolt causes a small decrease in bolt diameter due to the Poisson's ratio effect, some radial slippage of the engaged thread must take place. This action is reinforced by the nut flattening in direct compression. Under vibration, then, a certain amount of nut loosening can occur particularly where soft gasket materials are used. To reduce the tendency to vibration loosening of the joint, the amount of the initial strain due to preload should be as generous as possible. Failing this, however, friction and positive engagement devices may be employed to prevent nut loosening [248].

Design of Bolted Joints

ANALYSIS OF THREAD LOADING

The thread portion of the bolt or nut can be modeled as a short, cantilever beam. It was also found in the past that thread deflection based on this model was of the same order of magnitude as that due to the axial tension of the bolt or compression of the nut [249]. The compressive deformation of the nut can be affected by the following component deformations:

1. Axial extension of the bolt core
2. Axial contraction of the nut
3. Bending of bolt thread
4. Bending of nut thread
5. Circumferential strain in the nut wall

Components 1 and 2 are primarily responsible for the type of uneven load distribution between the consecutive threads as mentioned previously. Components 3 to 5 tend to counteract this effect, although they do not eliminate it completely. The analysis of load and stress distribution in the threads would then require a detailed examination of all types of flexibility as specified by items 1 through 5. The analysis also indicates that even under conditions of proper thread engagement, the nut thread can move radially by a distance V as shown in Fig. 39.6. This radial displacement, referred to as *thread recession* [249], is partly due to the lateral expansion under axial compression of the nut and partly influenced by the radial component of the thread load. Thread recession, then, combined with an increased bending flexibility of individual threads in contact, is likely to be of special importance in the case of unconventional thread engagement. For instance, 1 1/8-in.-diameter bolt in a 1 1/4-in.-diameter nut would have what might be termed excessive thread recession and increased thread bending flexibility, as shown in Fig. 39.7. Since the total axial load remains the same, contact load intensity and the resultant bending moment on the thread must increase with an increase in V_e. In the context of this discussion, $V_e = V + C$, where C may be defined as radial clearance due to the lack of proper thread engagement. Since the total bolt load is independent of V_e, the frictional force between the threads and that developed under the bolthead during

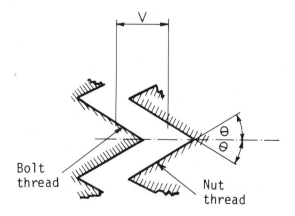

Fig. 39.6 Definition of thread recession.

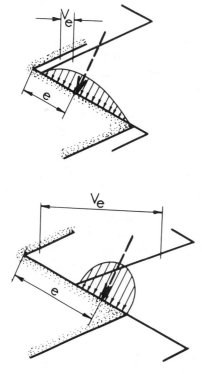

Fig. 39.7 Variation of contact load with thread recession.

the torquing procedure are essentially the same, giving no clue as to what might be the new state of stress and flexibility of the threads under load.

The resultant stress picture in an unconventional thread engagement, however, is not clear and requires careful study before meaningful observations can be made. In principle, it should be possible to set up a mathematical model of a short, tapered cantilevered beam under variable loading involving the quantities e and V_e. The model could then be used to calculate the effect of V_e on the stress distribution and flexibility of individual threads. This is indeed a promising area for further research.

JOINT RIGIDITY AND LEAKAGE

For more than 100 years now, countless tests have been conducted to better understand the behavior of structural joints [269]. Literally thousands of articles have been written in parallel with research and standard practice development aimed at specifications on an international scale. This brief section touches on some of the basic issues of joint rigidity, leakage control, and bolt characteristics still under debate [270, 271] in this country and abroad.

In conventional structural joints the forces can be transferred through shear, tension, or the combined effect of tension and shear of the bolts. In this category

Design of Bolted Joints

we have flat-plate splices, end plate, brace, butt joint, and bracket connections where in addition to the bolt clamping effect the frictional resistance of the contact surfaces must play a special role. A state-of-the-art summary of experimental and theoretical studies in this area of joint behavior is available for further reading [269].

The classical case of gasketed as well as metal-to-metal bolted joints applied to piping, pressure vessels, and similar components involves the rigidity and leakage considerations which continue to be under scrutiny.

The performance of a bolted joint with respect to the degree of tightness depends to a large extent on bolt spacing, flange thickness, and the properties of gasket materials. However, these cannot be the only criteria, because the rate of leakage must be a function of the pressure and characteristics of the fluid contained. It is not necessary, for instance, to have significant cracks or deformations to sustain a gas flow because the leakage can even exist through the process of diffusion. The diffusion of gases is possible in essentially all materials, including steel. The task, then, of containing the liquids or gases by means of a tight bolted joint, with or without the gaskets, is of unusual complexity. It can be stated without much debate that a truly leak-free joint does not exist. The best one can hope is to establish the maximum permissible leakage rate by controlling the gasket materials and joint design. This process should involve the selection of flange and bolt rigidity with due regard to the properties of gaskets. The main uncertainty, however, lies in the certification of gasket materials in order to comply with the regulatory practices. Bickford [270] has reported, for instance, that in certain applications the differences between the ASME recommended gasket factors and "values that work" can be very drastic.

Once the internal pressure and the gasket response are established, the problem at hand is the assignment of the flange contact area to a particular set of bolts. The theoretical choice is essentially between a few larger bolts or a greater number of bolts of smaller diameter. The analysis of the optimum number of bolts can be made with the help of the theory of beams on elastic foundation. However, the preliminary estimate is likely to be equally good when made on the basis of a practical formula derived from past experience. One of such formulas is credited to Taylor Forge Company [272], which recommended the maximum bolt spacing B to assure the tightness of a bolted joint for industrial purposes

$$B = 2d + \frac{6h}{m + 0.5} \tag{39.14}$$

where B is bolt spacing in inches, h denotes the thickness of the flange in inches, and m is the dimensionless quantity defined as the gasket factor. The original factors ranged from 1, for a rubber-type material, to 7 for stainless steel. There was also some evidence that the actual m values varied over a much wider range where the joint sealing was required against nitrogen or a head of water. The practice of using a distinctly lower gasket factor approaching $m = 1$ resulted in a wider-gasket concept which was expected to offer a decreased probability for developing complete leakage channels in service.

In addition to the correct choice of the gasket factor, it is necessary to determine the load per bolt consistent with the resistance of the flanges to toroidal deformation. Boardman [193] proposed the following ingenious formula for establishing a suitable number of bolts for joint design:

$$W = \frac{\pi P R_i^2 (R_o - R_i)}{N(R_o - R)} \qquad (39.15)$$

where W defines the load per bolt, P is the applied internal pressure to the flanged pipe, and N is the number of bolts. The outer and inner radii of the flanged assembly, together with the radius of bolt circle R, are the same as those shown in Fig. 33.5. The end force in a bolted joint sets up a bending moment about the bolt circle, causing the toroidal deformation of the flange which depends on the flange thickness. Boardman again [193] suggested a solution to this problem in terms of the radial stress S_r and other parameters already defined

$$h = 1.75 R_i \left\{ \frac{P(R - R_i)}{RS_r} \right\}^{1/2} \qquad (39.16)$$

In order to reduce the flange rotation to an absolute minimum [271], the flanges should have full-face metal-to-metal contact. This condition combined with the highly preloaded bolts can reduce the risk of bolt failure in service. However, the preload must not be too high. Highly torqued bolts tend to warp the flanges, which results in leakage. Hence the tightening torque should be just enough to control the leakage and it should include a reasonable margin for counteracting the relaxation of preload with time. Many conventional flanges with low resistance to warping, high preload, as well as short and stubby bolts are not the best for controlling the relaxation and leakage.

Since the use of even modern gaskets in a number of cases is not always the best tool for leakage control, a series of compact flanges was investigated by Webjörn [271] in order to obtain the following major characteristics:

Bolt length equal to at least six times the nominal bolt diameter
Preload above the 80% of the certified minimum yield of the material
Pipe joint with full-face flanges in metal-to-metal contact
Only hard washers under the bolt head and under the nut
Thickness of the flange equal to about three times the nominal bolt diameter

This particular type of a compact flange design should be able to counteract the toroidal deformation of the flange rings and it should also be characterized by a smaller size. For design details the reader is referred to the original paper. In addition, another type of a rigid flange design, utilizing a system of ribs, is described in some detail in Chapter 33.

SELECTION AND TESTING OF BOLT MATERIALS

Because of limitless combinations of service conditions, bolts can be made of ferrous, nonferrous, and nonmetallic materials [273]. When the bolt's main function is to deliver strength, steel comes to mind. For a corrosive environment one can have steel with a protective coating, stainless steel, or a nonferrous alloy. If magnetic permeability is involved, consider austenitic stainless steel, aluminum, copper, or plastic. Where high electrical conductivity is important we fall back on aluminum or copper. For a weight-saving task select aluminum, but where this feature is combined with high strength requirement the clear choice is titanium. Finally, for high- and low-temperature environments look for superalloys and stainless.

Since the cost of the raw material represents a high percentage of the total cost of manufacture, certain generic rules should be kept in mind. For instance, if the cost of medium-carbon steel corresponds to unity, 300 series of stainless costs $2\frac{1}{2}$ times as much, copper alloy 5 times, aluminum alloy 6 times, Monel 20 times, titanium 75 times, and super heat-resistant and low-temperature alloys about 100 times.

The practical strength limits vary over a wide range, as expected. Low-carbon steel bolts can be designed for about 60,000 psi in tension, while the super steel alloys reach 260,000 psi, with more exotic and technologically possible strengths exceeding 400,000 psi. In most industrial applications, however, there is seldom any need to consider tensile strengths higher than 180,000 psi.

The strength of aluminum and aluminum alloys ranges from 13,000 psi for pure aluminum to about 60,000 psi for the more popular alloys such as 2024 and 7075 grades. The tensile strength of copper-base alloys varies between 50,000 psi for No. 462 naval brass to 105,000 psi for No. 630 aluminum bronze. The nickel-base alloys should have a strength range between 80,000 and 130,000 psi. Titanium is universally accepted by the aerospace, missile, and chemical processing industries, despite the high cost of this material. Bolts of the two workhorse alloys, Ti-4Mn-7Al and Ti-6Al-4V, have tensile strengths of 150,000 psi. On the other hand, plastics, including the widely used nylon, have tensile strengths that seldom exceed 10,000 psi. For exact specifications the designer is advised to consult the particular manufacturer.

While the physical properties of the raw materials essentially remain unchanged during the manufacture, the mechanical properties, which identify the reaction of a bolt or a fastener to the applied loads, are subject to dramatic variations depending on the choice of fabrication techniques and metallurgical treatments [273].

The tensile strength that a bolt can exhibit without fracture should be verified using full-size bolts having the length equal to at least 2.25 times the bolt nominal diameter. Hex and hex flange screws and bolts, studs, and hex socket cap screws are wedge tensile tested. Other externally threaded fasteners are tested in an axial mode. The wedging action induces a severe bending stress concentrated at the junction of the head and the shank. This test also requires that the fracture occur at a point other than at the junction. The wedge angle normally varies between 4 and 10°.

When the bolt material is stressed beyond the elastic limit the yield strength can be established by a specified amount of permanent deformation. However, it is not easy to correlate the yield point of the material with that of the bolt because of the different strain rates possible in the threaded portions.

To circumvent this difficulty, the proof load concept has been introduced. It corresponds to the applied tension load which the bolt can support without evidence of any permanent deformation. The proof load represents an absolute value equal to

Table 39.1 Proof Load and Tensile Strength Capacity for Coarse Threads

Nominal diameter (inch)	Threads per inch	Grade 5		Grade 7		Grade 8	
		Proof load (lb)	Tensile capacity (lb)	Proof load (lb)	Tensile capacity (lb)	Proof load (lb)	Tensile capacity (lb)
0.250	20	2,700	3,800	3,350	4,250	3,800	4,750
0.375	16	6,600	9,300	8,150	10,300	9,300	11,600
0.500	13	12,100	17,000	14,900	18,900	17,000	21,300
0.625	11	19,200	27,100	23,700	30,100	27,100	33,900
0.750	10	28,400	40,100	35,100	44,400	40,100	50,100
0.875	9	39,300	55,400	48,500	61,400	55,400	69,300
1.000	8	51,500	72,700	63,600	80,600	72,700	90,900
1.125	7	56,500	80,100	80,100	101,500	91,600	114,400
1.250	7	71,700	101,700	101,700	127,700	116,320	145,400
1.375	6	85,500	121,300	121,300	153,600	138,600	173,200
1.500	6	104,000	147,500	147,500	186,900	168,600	210,800

about 90 to 93% of the expected minimum yield strength of the bolt material. The actual measurements allow only a small tolerance for error. The tensile strength and the proof load values are given in Tables 39.1 and 39.2 for the coarse and fine threads, respectively.

Hardness is a measure of a material's ability to resist mechanical abrasion and indentation. The quick and easy hardness test has tremendous importance in procurement and quality control of bolts. In the case of steel bolts there is also a good correlation between the Brinell hardness and tensile strength parameters. The

Table 39.2 Proof Load and Tensile Strength for Fine Threads

Nominal diameter (inch)	Threads per inch	Grade 5		Grade 7		Grade 8	
		Proof load (lb)	Tensile capacity (lb)	Proof load (lb)	Tensile capacity (lb)	Proof load (lb)	Tensile capacity (lb)
0.250	28	3,100	4,350	3,800	4,850	4,350	5,450
0.375	24	7,450	10,500	9,200	11,700	10,500	13,200
0.500	20	13,600	19,200	16,800	21,300	19,200	24,000
0.625	18	21,800	30,700	26,900	34,000	30,700	38,400
0.750	16	31,700	44,800	39,200	49,600	44,800	56,000
0.875	14	43,300	61,100	53,400	67,700	61,100	76,400
1.000	12	56,400	79,600	69,600	88,200	79,600	99,400
1.125	12	63,300	89,900	89,900	113,800	102,700	128,400
1.250	12	79,400	112,700	112,700	142,700	128,800	161,000
1.375	12	97,300	138,100	138,100	174,900	157,800	197,200
1.500	12	117,000	166,000	166,000	210,300	189,700	237,200

Design of Bolted Joints

same correlation in the case of nonferrous and stainless steel bolts is less precise. Rockwell, Brinell, or even Vickers (diamond pyramid) numbers can be used in hardness tests. The minimum hardness number in general corresponds to the known minimum tensile strength of the material. The routine hardness measurements are taken on the top or side face of the bolt head, on the shank, or, as required, on the end of the product. In the case of quenched and tempered bolts, the hardness at the surface in the threaded section is compared with that of the base metal. A significantly lower or higher surface hardness than the specified value represents a potential risk of inferior performance.

Ductility is the ability of the material to absorb a significant deformation before fracture. However, the process of measuring the magnitude of elongation and the reduction of area in a thread is totally impractical. The only meaningful indicators of good ductility have been observed when the specified maximum hardness has not been exceeded and when the wedge tensile test has shown the required tensile strength. Another indicator of ductility is the ratio of yield to the ultimate strength. The lower the ratio, the more ductile the bolt is likely to be.

The material's ability to sustain impact and shock loading represents toughness. The measure of this parameter is the amount of energy absorbed before fracture. Chapters 14 and 15 are devoted to the relevant problems of fracture mechanics and fracture control.

The shear strength of a bolt is equal to about 60% of the specified minimum tensile strength. In the case of industrial quality bolts, with the exception of transmission tower joints, the shear capacity is seldom a requirement. However, in aerospace practice the design strength in shear and the shear testing are frequently mandatory. Standard test methods for shear strength of threaded components are currently under development. The related problem of torsional strength is defined as the torque at which a fastener fails by being twisted off about its axis. Tapping screws and stainless steel metric machine screws are probably the only industrial fasteners for which torsional strength is specified as a requirement. The screw is assumed to pass the test under a given torque if its head does not twist off. For further details of the materials technology pertinent to threaded fasteners, the reader may wish to consult the Industrial Fasteners Institute and similar organizations [273].

It is fair to note that this chapter on bolted joint design represents only a sprinkling of general rules and ideas spread over many years of industrial experience.

Whether we turn to older formulas, newer design methodology, or better materials data, the overall number of variables whose effects can be measured only in terms of statistical quantities is still staggering.

In an effort to control and inspect some of the more critical aspects of bolting technology, the industry looks toward alternative designs and direct tension indicators [274]. The more active industries in this country include nuclear power, automobile and aerospace corporations. The remaining serious problem in the pressure vessel and piping arena is leakage, because it is still more difficult to select the correct preload for a gasketed rather than a metal-to-metal joint. Progress is being made in determining the optimum amount of assembly stress for a particular gasketed assembly, although the regulatory bodies are not quite ready yet to recommend the appropriate joint specifications. The need for this action is still with us because it is driven by the fast-changing and litigious world.

SYMBOLS

A_B	Bolt shank cross section, in.^{2} (mm^2)
a	Effective width of annulus, in. (mm)
B	Bolt spacing, in. (mm)
c	Radial clearance in threads, in. (mm)
d	Nominal diameter of bolt, in. (mm)
E_B	Modulus of bolt material, psi, (N/mm^2)
e	Distance to center of thread pressure, in. (mm)
g	Thickness of washer, in. (mm)
h	Thickness of flange, in. (mm)
K_B	Spring constant of bolt, lb/in. (N/mm)
K_c	Spring constant of joint, lb/in. (N/mm)
$K_1, \ldots, K_n$	Individual spring constants, lb/in. (N/mm)
L_B	Length of bolt shank in tension, in. (mm)
M_t	Tightening torque for preload, lb-in. (N-mm)
m	Gasket factor
N	Number of bolts
P	Internal pressure, psi (N/mm^2)
R	General symbol for radius, in. (mm)
R_i	Inner radius of flange, in. (mm)
R_o	Outer radius of flange, in. (mm)
r	Mean radius of thread, in. (mm)
S	Combined stress, psi (N/mm^2)
S_r	Radial stress, psi (N/mm^2)
V	Thread recession, in. (mm)
V_e	Effective recession of thread, in. (mm)
W	Total bolt load, lb (N)
W_e	External load on joint, lb (N)
W_i	Initial bolt load, lb (N)
α	Helix angle of thread at mean radius, deg
η_0	Slope, rad
θ	Half-angle of thread, deg
μ	Coefficient of friction
ρ	Radius of washer, in. (mm)
ϕ	Half cone angle, rad
ψ_0	Slope, rad

40
Load Transfer in Mechanical Connections

GENERAL COMMENTS

There is a significant number of structural interfaces where bolted joints described in Chapter 39 play a special role in design and performance of the mechanical systems [269–272, 274].

Several examples of application of stress analysis principles to joint and fitting design have been interspersed throughout the various chapters of this book. These included such topics as leaf springs, crane gear components, knuckle and swivel joints, complex pipe flanges, rolling diaphragms, nested and shrink-fit cylinders, and bolted connections.

Despite the obvious variety of applications of the mechanical fasteners in the general area of joints and fittings, a number of machine and structural elements may or may not require the use of the conventional fastening. For such reasons this chapter is devoted to a brief account of formulas and methods which may be helpful in assessing the critical loading and performance characteristics of clamps, couplings, pinned connections, and similar configurations not covered elsewhere in this volume.

The primary function of a mechanical joint is to transmit a given load from one piece of the material to the adjoining one. This process generates stresses and strains which are of interest to the designer and the user.

Special features of structural connections selected for this chapter fall largely in the province of machine design practice [177, 277]. While many components and systems can be held in place by means of rivets, bolts, welts, or adhesives, other applications may involve frictional effects, tapered surfaces, pins, keys, threads, or a combination of these load transfer elements. Every design consideration of a mechanical joint requires a combined knowledge of materials, geometry, and fabrication variables. This process is not easy because no single theoretical formula can satisfy the overall design conditions.

SPLIT-HUB DESIGN

The basic concept of a split-hub design is illustrated by the two views of the joint given in Figs. 40.1 and 40.2. The problem is concerned with the estimate of the bolt preload V to assure the frictional resistance against rotation and sliding of the shaft inside the hub. According to the symbols given in Fig. 40.1, F is the contact load and N denotes the number of bolts required to hold the split-hub assembly. The coefficient of friction between the hub and the shaft is denoted by f.

From the geometry in Fig. 40.1, the moment equilibrium gives $NV(x+x_1) = NFx_1$ so that $F = V(x+x_1)/x_1$. The frictional couple M_{tf} created by the clamping load is then

$$M_{tf} = \frac{VNfd(x+x_1)}{x_1} \tag{40.1}$$

In analyzing the axial equilibrium, the resistance W can be defined as

$$W = 2P_e L d f \tag{40.2}$$

In this equation P_e is the equivalent contact pressure acting on the two halves of the split hub. The exact distribution of the unit pressure on the contact surfaces is gen-

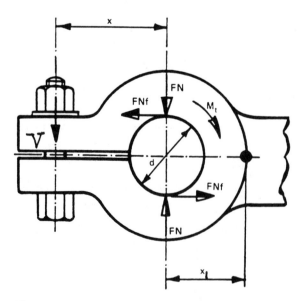

Fig. 40.1 Forces in split-hub joint.

Load Transfer in Mechanical Connections

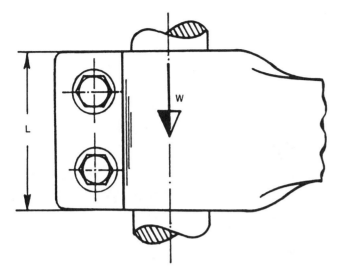

Fig. 40.2 Top view of split-hub joint.

erally unknown and it becomes necessary to make certain simplifying assumptions which can only be verified by experiment.

Consider, for instance, Fig. 40.3 defining the symbols for making a preliminary estimate of the contact pressure between the split hub and the shaft. The elementary torque resistance in this case can be taken as $P \times d\theta \times (d/2) \times L \times f \times (d/2)$. On the premise that $P = P_{\max} \sin\theta$, it can be shown by integration that the total frictional torque is

$$M_{\mathrm{tf}} = L f d^2 P_{\max} \tag{40.3}$$

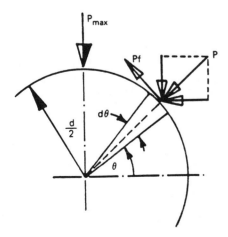

Fig. 40.3 Elements of contact pressure.

The design practice suggests that the frictional couple created by the bolt clamping force should be larger than the torque applied to the shaft. The first estimate of the corresponding contact pressure P_{max} can be obtained from Eqs. (40.1) and (40.3) as

$$P_{max} = \frac{VN(x + x_1)}{Ldx_1} \tag{40.4}$$

Assuming that the equivalent contact pressure $P_e = 0.5 P_{max}$, Eq. (40.2) yields the frictional drag force equal to $W = 2Ff$. When the calculation is based on the summation of the vertical components characterized by the term $P_{max} \sin^2 \theta$, the axially oriented frictional drag becomes $W = \pi F f$. It is evident that calculational variations can easily stem from the complex model required in solution of this type of problem.

PIPING SUPPORTS AND BRANCHES

The entire area of pipe support design methodology involves analytical complexities because of the pressure and temperature effects. This brief account is only concerned with examples of practical stress analysis pertinent to formulas and the related concepts of load distribution [200, 279, 280].

When local bending effects in the vessel walls are encountered, the theory of beams on elastic foundation is most likely to be used, as already discussed in Chapter 24. Consider the case of an external support consisting of a narrow ring fastened tightly around the outside of the vessel, as shown in Fig. 40.4. The sketch illustrates the two fundamental conditions. In the first case the pressure inside the cylindrical shell does not exist. If P_o is defined as the ring loading measured in pounds per inch of the circumference, then Y_r is the radial deflection, which can be calculated as

$$Y_r = 0.64 \left(\frac{P_o}{E}\right) \left(\frac{r}{t}\right)^{3/2} \tag{40.5}$$

In this particular case r denotes the mean radius and t is wall thickness of the pipe or vessel. The ring load is assumed to be sufficiently removed from the end closures or flanges.

The maximum stress produced in the vessel wall due to the external ring load is

$$S_b = \frac{1.17 P_o (r)^{1/2}}{t(t)^{1/2}} \tag{40.6}$$

When the cylindrical vessel under consideration contains internal pressure P_i, the corresponding ring load P_o resisting dilation is

$$P_o = \frac{0.85 A_r P_i (rt)^{1/2}}{0.64 A_r + t(rt)^{1/2}} \tag{40.7}$$

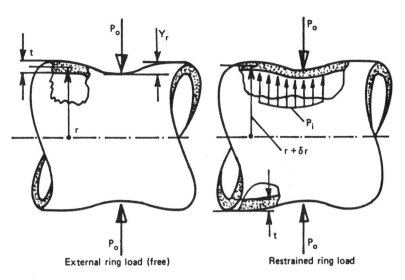

Fig. 40.4 Ring loading on cylindrical shell.

This constitutes the case of the restrained ring load. It is noted that the tightly assembled ring provides a means of support and restraint without any need for a weld or adhesion joint. The area of ring cross section here is denoted by A_r.

One of the more difficult design problems involves the interaction between the piping and the clamps which provide the attachment points and restraints. Recent success in the development of modeling techniques for the clamp analysis with linear and nonlinear characteristics has been due to the progress with finite-element codes involving many degrees of freedom and boundary conditions. An example of a standard industrial clamp used as one-way pipe restraint is illustrated in Fig. 40.5. Experience with this clamp shows that the pertinent load capacity can be related to a convenient dimensional parameter such as BH^2/D. The calculations also indicate that the highest clamp stress in this configuration is located at about 20° measured from the plane of the split as noted in the sketch. The load capacity factor for this clamp may be defined as the ratio W/S, where W is the external load on the clamp parallel to the plane of split and S defines the allowable clamp stress measured in the hoop direction. The distribution of the hoop stress in the clamp is nonlinear.

One of the basic questions related to a standard configuration is concerned with the clamp stiffness. The degree of difficulty in this area depends on the interaction effects between the pipe and the clamp. The analysis is simpler when the pipe is relatively thick. In the case of thin-walled piping the distribution of contact pressure between the pipe and the clamp is generally expected to follow a cosine function. However, according to some investigators, who used the finite-element modeling [278], the cosine distribution of contact pressure may not provide the most conservative estimate of hoop stress at the clamp interface of a thin-walled, high-temperature pipe.

The branched-pipe connections, such as those shown in Fig. 40.6, can be subjected to axial tension, axial compression, and in-plane as well as out-of-plane bending. The general failure criteria may be defined in terms of the maximum

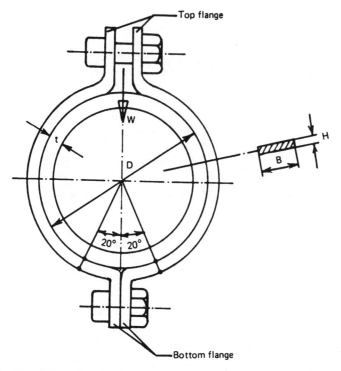

Fig. 40.5 Standard pipe clamp.

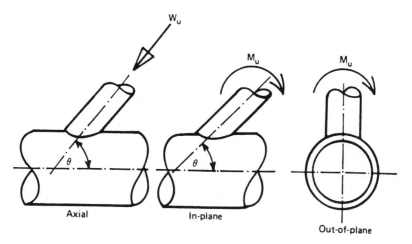

Fig. 40.6 Basic loading conditions.

Load Transfer in Mechanical Connections

load capacity, excessive deformation, or a crack growth. Since rigorous analysis of a tubular branch is very difficult, the industry has always relied on the empirical developments [279]. The practical design formulas summarized in this section are expressed in terms of the d/D ratio denoted by β and the sine function of the acute angle between the branch and the main pipe body. Elements of the basic notation are illustrated in Figs. 40.6 and 40.7.

For the case of axial tension or compression, directed along the axis of the branch, the maximum load capacity of the branch is

$$W_u = \frac{S_y T^2 (3.4 + 19\beta)}{\sin \theta} \tag{40.8}$$

The parameter $\beta = d/D$ can vary between 0.2 and 0.9 for the majority of typical industrial configurations. For instance when $\theta = \pi/2$ and $\beta = 0.5$, Eq. (40.8) gives $W_u = 12.9 S_y T^2$. This condition applies to a basic T-joint such as that shown in Fig. 40.7.

When the external bending moment is applied in the plane of the joint, as shown by the middle sketch of Fig. 40.6, the ultimate moment capacity becomes

$$M_u = \frac{S_y T^2 D \beta (2.7 + 15.2\beta)}{\sin \theta} \tag{40.9}$$

Equation (40.9) shows that for small values of β, the bending moment formula can be linearized. However, for the β values higher than about 0.6, the ultimate capacity transforms into a nonlinear effect.

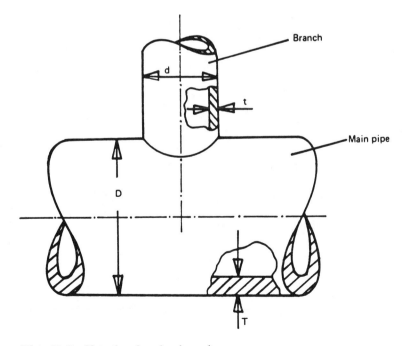

Fig. 40.7 Notation for pipe branches.

For the case of out-of-plane bending of the branched pipe, two separate formulas are recommended. When $\beta < 0.6$

$$M_u = \frac{S_y T^2 D \beta (2.7 + 5.6\beta)}{\sin \theta} \tag{40.10}$$

and for $0.6 < \beta < 1.0$ we have

$$M_u = \frac{S_y T^2 D (0.81 + 1.68\beta)}{\sin \theta (1 - 0.83\beta)} \tag{40.11}$$

In practice, D/T ratios for the main pipe shown in Fig. 40.7 vary between 20 and 100. The effect of this ratio on the ultimate capacity of the branched joint is reported to be relatively small [279]. By comparison with the standards of the American Petroleum Institute (API), the formulas given by Eqs. (40.8) to (40.11) are expected to predict lower joint capacities.

KEY AND PIN CONNECTIONS

Special mechanical joints, where rapid assembly and disassembly are of importance, involve various locking devices such as keys, splines, or pins. The designs are concerned with the bearing surfaces, geometrical transitions, stress patterns, taper criteria, and frictional characteristics of the joint material. The conventional range of taper lies within the limits of 0.01 to 0.20 for a cotter pin connection such as, for instance, that illustrated in Fig. 40.8 (A and B). In the case of a frequently disassembled joint the approximate range of taper is about 0.1 to 0.2. For a more permanent cottered fastening the taper should be on the order of 0.01 to 0.05.

When the external force F is applied to the shank as shown in Fig. 40.8A, we have

$$F = dS_t(0.7854d - b_o) \tag{40.12}$$

The compressive stress is

$$S_c = \frac{F}{b_o d} \tag{40.13}$$

For these calculations the compressive allowable can be taken as $1.5S_t$. Since there are two shear planes resisting the external load and since the allowable shear stress is $S_t/\sqrt{3}$, the appropriate formula is

$$F = 1.15 h_i d S_t \tag{40.14}$$

In order to avoid crushing of the socket wall the following upper boundary should be observed:

$$F = 2b_o t S_c \tag{40.15}$$

Load Transfer in Mechanical Connections 603

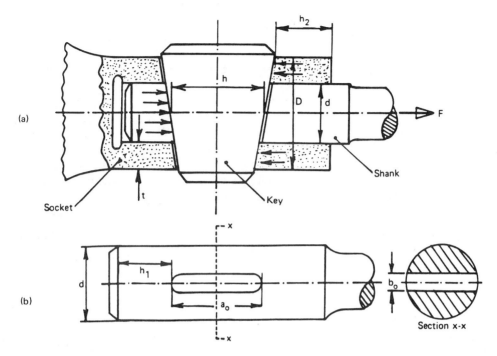

Fig. 40.8 Cotter pin connection.

Note that the thickness of the socket wall is $t = (D - d)/2$, so that the external socket diameter becomes

$$D = d + \frac{F}{b_o S_c} \tag{40.16}$$

A brief review of Eqs. (40.12) through (40.14) indicates that a balanced design can be obtained if $b_o = 0.3d$ and $h_i = 0.4d$. Also, Eqs. (40.13) and (40.16) are consistent when $D = 2d$. Using these criteria, the tensile strength of the socket can be defined as follows:

$$F = 2.4d(d - 0.25t)S_t \tag{40.17}$$

When $h_1 = h_2$ (see Fig. 40.8A and B) the shear-out strength of the socket edges becomes

$$F = 0.92 dt S_t \tag{40.18}$$

The strength of the cotter key may be calculated on the premise of beam bending provided the h/D ratio is smaller than 0.5. This leads to the following criterion:

$$h = 0.87 \left(\frac{FD}{b_o S_b} \right)^{1/2} \tag{40.19}$$

The foregoing design equations are intended for a cotter pin connection in an unstrained condition. Special allowance should be made when the joint is subjected to residual stresses.

Many machines involve a combination of the shaft and wheel-type assemblies in such applications as gears, pulleys, discs, and hubs. The problem of joining such members can be solved with the help of keys and splines [277] designed to transmit the required torque. The resulting forces cause compressive and shear stresses. A typical example of a key joint is illustrated in Fig. 40.9. The dimension not shown in the sketch is L, which defines the length of the key. This dimension is governed by the two basic criteria involving a crushing force on the side of the key and a direct shearing action. These criteria can be expressed in terms of the tensile strength of the key material as follows:

$$L = \frac{2.67 M_t}{h D S_t} \tag{40.20}$$

and

$$L = \frac{3.46 M_t}{b_o D S_t} \tag{40.21}$$

The normal routine here is to select the larger of the two values provided L is not greater than the hub length. When the length of the hub is insufficient it may be necessary to use two keys. Also, Eqs. (40.20) and (40.21) indicate that for a

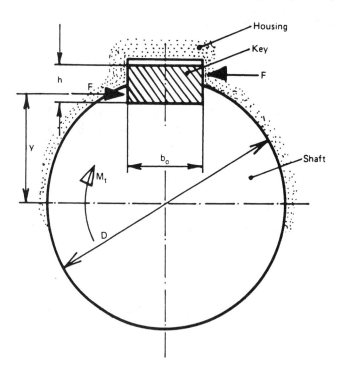

Fig. 40.9 Forces in a typical key joint.

balanced design $b_o = 1.3h$. Many interesting examples of this nature can be found in the machine design literature.

When the connection is made by means of a tapered key, a radial thrust is created through the interaction between the hub keyway and the shaft. The principle of a strained-key joint is shown in Fig. 40.10, which also serves as a summary of symbols for a brief analysis that follows. The magnitude of the normal force N_f created by driving the key during the joint assembly is

$$N_f = \frac{P_c}{\tan \rho + \tan(\alpha + \rho)} \tag{40.22}$$

In the foregoing equation α denotes the taper angle of the key and ρ is the angle of friction. The allowable force P_c is limited by the yield strength of the material according to the criterion

$$P_c = b_o h S_y \tag{40.23}$$

The equilibrium of forces acting on an element of the key, shown in Fig. 40.10, is represented by a model in Fig. 40.11. Taking moments about the key center yields $\epsilon = fh$, where f is the coefficient of friction. The forces reacting to the torque M_t applied to the shaft are indicated in Fig. 40.12. Assuming cosine distribution, the interaction between the external load N_f and the maximum unit pressure [277] leads to

$$N_f = 0.7854 L D q_{max} \tag{40.24}$$

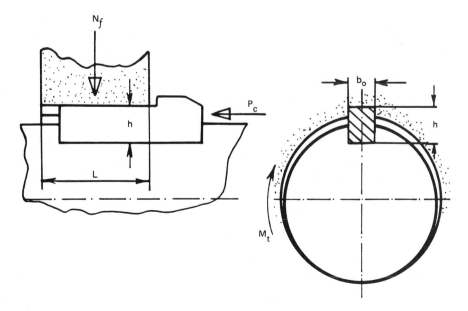

Fig. 40.10 Strained-key joint.

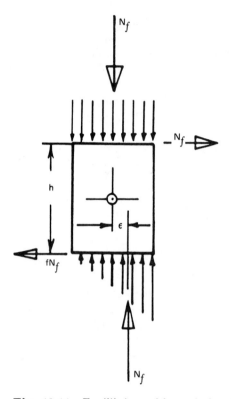

Fig. 40.11 Equilibrium of forces in key joint.

On the premise that $q_f = qf$, the total frictional moment M_f becomes

$$M_f = \frac{2N_f f D}{\pi} \tag{40.25}$$

The sum of all the moments taken about the center of the shaft is equal to the applied torque M_t, so that

$$M_t = M_f + \frac{f N_f (D - h)}{2} + N_f \epsilon \tag{40.26}$$

Hence, recalling that $\epsilon = fh$ and using Eq. (40.25) gives the design criterion for selecting the appropriate limit for N_f in order not to crush the keyway in the hub.

$$M_t = f N_f (1.14 D + 0.5 h) \tag{40.27}$$

In addition to emphasizing the role of practical stress analysis in machine design, the discussion is intended to make a point that even the simplest joint configuration cannot be described by a single design formula. The calculations made so far did not even begin to address the complex issue of torsional stresses in a shaft with a keyway [280].

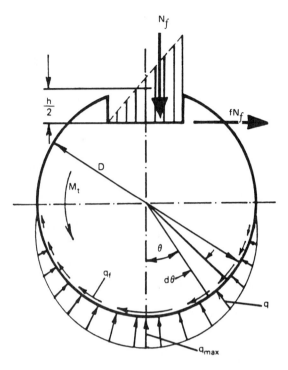

Fig. 40.12 Surface forces reacting to torque.

The analysis of an infinitely long groove in a circular shaft was first attempted close to 90 years ago. Several attempts have been made over the years using experiments and two-dimensional models, which, however, could not accurately describe the real stress distribution. It was not until 1975 that this problem was actually tackled with the help of a three-dimensional photoelastic technique. The experimental approach still persists because of insurmountable mathematical difficulties in obtaining a reliable solution to the boundary value problem posed by a keyway of a finite length. The failures traced to the presence of a keyway seem to be due to a circumferential crack, subsurface peeling-type fracture on a cylindrical surface, or a 45° helix crack propagation observed in certain spline connections. The proper documentation of failure is still badly needed for the assessment of realism implied by the various experimental results.

STRUCTURAL PIN DESIGN

The analysis of eyebar connections, discussed in Chap. 29, raises the question of the complementary problem of pin design. The primary function of a structural pin is to transmit the shear force. This machine element has very wide application in such devices as swivels, clevis links, shackles, hooks, sockets, sheaves, and construction blocks, to mention a few. Because of the unavoidable local deformation of the pin and the parts in contact, questions often arise as to the manner according to which pin loading can be defined. Consider, for instance, the typical pin connection shown in Fig. 40.13. The customary machine design approach is to adopt a model of a four-point loading for the purpose of calculating the bending stresses in the pin. The average compressive stress on the projected pin area is [177]

$$S_c = \frac{W}{Ld} \qquad (40.28)$$

The bending stress can be defined in terms of the compressive stress as

$$\frac{S_{bmax}}{S_c} = 1.273 \frac{(k+2)m^2}{k} \qquad (40.29)$$

where $k = L/a_0$ and $m = L/d$, with the dimensions given in Fig. 40.13. The design chart based on Eq. (40.29) is shown in Fig. 40.14. This family of curves applies to a solid pin only. The concept of average compressive stress used in Eq. (40.29) is hard to justify. However, it is a convenient model for the preliminary design and experimental purposes.

Using S_c as a basis, the maximum shear stress in the pin can be expressed as

$$S_{smax} = 0.849 m S_c \qquad (40.30)$$

In the majority of practical situations, the individual results obtained from Eqs. (40.29) and (40.30) are sufficient to define the critical stresses. This is not surprising because the maximum bending stress is found at the extreme fiber while the maximum shear stress is located at the centerline of the pin. It may be recalled from the fundamental strength theory that shear stresses equilibrate on four sides of an element simultaneously. The shear stress must also be zero on a free surface where only one pair of shear stresses can exist and therefore it does not satisfy the conditions of equilibrium.

At times, questions may arise as to the state of the combined stress at any point along the radius y shown in Fig. 40.13. The ratio of the maximum principal

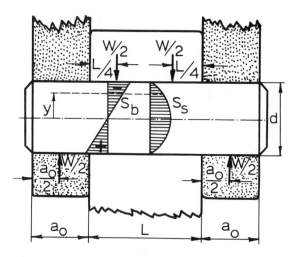

Fig. 40.13 Notation and geometry for a pinned connection.

Load Transfer in Mechanical Connections

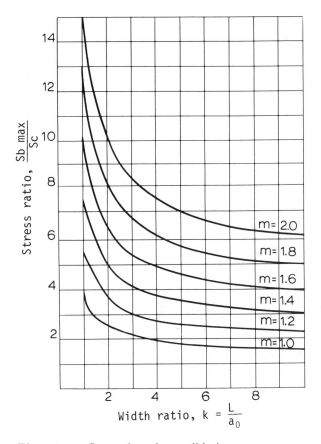

Fig. 40.14 Stress chart for a solid pin.

stress to the maximum bending stress can be calculated from the expression

$$\frac{S}{S_{\text{bmax}}} = n + \frac{[9m^2n^2(k+2)^2 + 4k^2(1-4n^2)^2]^{1/2}}{3m(k+2)} \qquad (40.31)$$

where nondimensional ratio y/d is denoted by n, which can vary between 0 and 0.5. When $n = 0$, the maximum stress is equal to that represented by Eq. (40.30). Also, when $n = 0.5$, $S = S_{\text{bmax}}$, given by Eq. (40.29).

The form of Eq. (40.31) indicates that as long as the pin length is equal to or greater than its diameter, combining the stresses does not cause the maximum principal stress to numerically exceed the bending stress at the outer fiber.

WIRE ROPE CONNECTION

One of the more intriguing cases of joint technology is concerned with the task of providing a reliable end fitting for the industrial wire rope. Apparently the mechanics of a tensile joint is known well only to nature while the designer of a wire rope fitting must struggle with limited tests and experience. This brief section can only touch on the elementary aspects of joint behavior applicable to wire ropes, cables, tie-bars, and similar configurations.

Pulling or jacking heavy loads suspended from cables often require unique gripping systems or swage fittings found in conventional and offshore applications. Imagine, for instance, the forces involved in controlling the tension of a suspension cable on the Golden Gate Bridge in San Francisco or on an offshore rig having 3.5 in.-diameter rope rated at 1400-ton capacity. Sophisticated pulling equipment is now available in industry for performing various functions on the premise that the wire rope can be threaded through a puller and grip system during the operation. Such a system is designed to hold the rope in suitable wedges at any point along the rope's length, until a permanent end fitting can be secured for the specified loading conditions.

Galvanized steel wire strand in rope design is generally used for guying poles, overhead support members, railroad hardware, and similar structures. For more critical overhead duties involving trolley systems and special anchors, improved grades of steel are required to assure greater strength and toughness. For the production of a splicing or fitting, the wire rope is cut and properly sized to assure uniform tension in the socket after the pouring of molten zinc. Some of the typical end fittings are illustrated in Fig. 40.15. Type A in Fig. 40.15 is referred to as a swaged socket, formed by applying external pressure or other metalworking techniques. Sketch C shows the basic elements of a pin-type shackle, the design of which involves curved-beam theory and pin analysis. The stress characteristics of the U-shape portion of the shackle are similar to those of the chain link analyzed in Chapter 29. The sizing of the pin can be performed with the help of the conventional beam theory. Type B shown in Fig. 40.15 can be sized on the basis of the available taper and the frictional adhesion between the body of the fitting and the appropriate filler material of a metallic or nonmetallic compound.

Experience shows that properly fabricated end fittings for the wire rope are expected to have a high efficiency. It appears, for instance, that swaged and zinc-filled sockets can be rated as high as the rope for which the fitting is intended. Other configurations, such as the shackle or thimble type, are likely to have 80% efficiency based on the rated capacity of the rope.

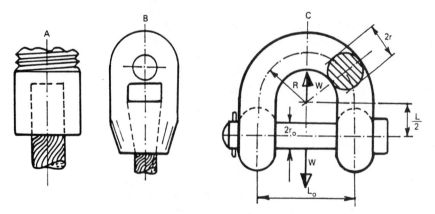

Fig. 40.15 Examples of rope fittings.

Load Transfer in Mechanical Connections

Suppose we consider the analysis of the shackle-type fitting shown in Fig. 40.15C. In order to assure equal-strength design of the pin and the U-shape part on the basis of bending, the assumption can be made that the pin behaves as a simply supported beam of length L_o subjected to a central load W. For the notation indicated in Fig. 40.15C, the section modulus of the pin is $\pi r_o^3/4$. The relevant bending moment at the center of the beam for $L_o = 2R$, is $WR/2$. Hence, the bending stress in the pin is

$$S_b = \frac{2WR}{\pi r_o^3} \qquad (40.32)$$

The maximum bending moment for the U-shape portion of the fitting can be described with the aid of Eq. (29.10). Hence, for the section modulus of $\pi r^3/4$, the critical bending stress becomes

$$S_b = \frac{2WR(2R+L)}{\pi(\pi R+L)r^3} \qquad (40.33)$$

For the condition of equal bending strength, Eqs. (40.32) and (40.33) yield

$$\frac{r_o}{r} = \left(\frac{\pi R + L}{2R + L}\right)^{1/3} \qquad (40.34)$$

Interpretation of Eq. (40.34) indicates that the effect of U-shape length on the shackle-pin size is, for most practical purposes, rather small.

The wire rope socket shown in Fig. 40.15B involves a tapered interface which provides a degree of resistance to pull-out of the rope, cable, or a rod depending on the cone angle and the degree of frictional adhesion. The normal pressure acting on the interface can be very high when the cone angle and the frictional adhesion strength are both low. As a conservative estimate, the conventional friction factor between the body of the socket and the filler material can be used. However, this type of analysis can, at best, be approximate and contingent upon the resistance of individual strands of the wire rope to pull-out. This fact is well known in industry and the manufacturer of rope and cable terminations has to rely on the qualification testing. It should also be emphasized that even the best theoretical prediction of the strength of a wire rope fitting should not be used in lieu of the performance test.

A relatively crude mechanical model of the end fitting is shown in Fig. 40.16. The half-angle of the cone is exaggerated for the purpose of clarity and the filler end does not show any specific details of strands or the filler compound. The term N_f in this derivation is regarded as a total interface force reacting with the filler. The corresponding resistance to pull-out is assumed to be directly proportional to the friction factor f. From the equilibrium of forces we get

$$W = N_f(\sin\alpha + f\cos\alpha) \qquad (40.35)$$

When $\alpha = 0$, $W = fN_f$, representing only the frictional effect if the joint is assembled with sufficient compression between the wire rope matrix and the socket

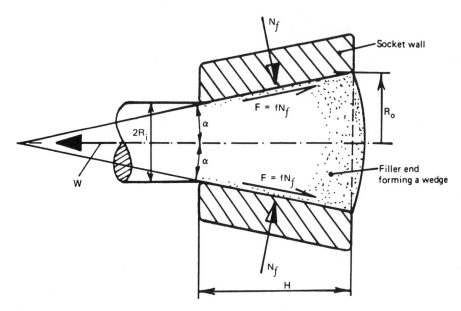

Fig. 40.16 Tapered end fitting.

wall. In reality, the external load has to overcome additional adhesion forces. As the cone angle α increases to the theoretical limit of $\pi/2$, the end fitting forms a shoulder and the effect of interface friction, according to Eq. (40.35), must vanish. This is, indeed, a gross oversimplification and a trivial solution which suggests that $W = N_f$. Hence any reasonable interpretation of Eq. (40.35) should be restricted to the values of α sufficiently removed from the two extremes of zero and $\pi/2$.

The approximate contact pressure q between the socket and the filler can be estimated as follows. From the geometry depicted in Fig. 40.16 the contact area is

$$A = \pi(R_o + R_i)[(R_o - R_i)^2 + H^2]^{1/2} \tag{40.36}$$

Hence, postulating that $q = N_f/A$ and using Eqs. (40.35) and (40.36) gives

$$q = \frac{W}{\pi(\sin\alpha + f\cos\alpha)(R_o + R_i)[(R_o - R_i)^2 + H^2]^{1/2}} \tag{40.37}$$

Observing that the above trigonometric functions can be expressed in terms of R_o, R_i, and H, Eq. (40.37) can be simplified as follows:

$$q = \frac{W}{\pi(R_o + R_i)(R_o - R_i + fH)} \tag{40.38}$$

The contact pressure q can now be used to estimate the stresses in the socket on the premise that the bond between the socket and the filler is maintained with the help of friction. The reliable magnitude of this effect can only be established experimentally.

Load Transfer in Mechanical Connections

SYMBOLS

A	Area of contact, in.^{2} (mm^2)
A_r	Area of ring cross-section, in.^{2} (mm^2)
a_o	Length of pin support, in. (mm)
B	Width of section, in. (mm)
b_o	Width of groove or key, in. (mm)
D	Outer diameter of main pipe (also diameter of shaft or socket), in. (mm)
d	General symbol for diameter, in. (mm)
E	Modulus of elasticity, psi (N/mm^2)
F	General symbol for external force, lb (N)
f	Coefficient of friction
H	Depth of section, in. (mm)
h	Depth of key, in. (mm)
h_1	Distance to end of shank, in. (mm)
h_2	Limit distance, in. (mm)
k	Length ratio
L	Length, in. (mm)
L_o	Shackle distance, in. (mm)
M_f	Frictional moment, lb-in. (N-mm)
M_t	Torsional moment, lb-in. (N-mm)
M_{tf}	Frictional torque, lb-in. (N-mm)
M_u	Ultimate bending moment, lb-in. (N-mm)
m	Length ratio
N	Number of bolts
N_f	Normal force, lb (N)
n	Dimensionless ratio
P	Shrink-fit pressure, psi (N/mm^2)
P_c	Concentrated load, lb (N)
P_e	Equivalent contact pressure, psi (N/mm^2)
P_i	Internal pressure, psi, (N/mm^2)
P_o	Ring loading, lb/in. (N/mm)
P_{max}	Maximum contact pressure, psi (N/mm^2)
q	Local contact pressure, psi (N/mm^2)
q_f	Frictional drag, psi (N/mm^2)
q_{max}	Maximum local pressure, psi (N/mm^2)
R	Radius of eyebar or curved beam, in. (mm)
R_i	Small radius of cone, in. (mm)
R_o	Large radius of cone, in. (mm)
r	Mean radius of tube (also general symbol), in. (mm)
r_o	Pin radius in shackle, in. (mm)
S	Stress, psi (N/mm^2)
S_b	Bending Stress, psi (N/mm^2)
S_{bmax}	Maximum bending stress, psi (N/mm^2)
S_c	Compressive stress, psi (N/mm^2)
S_s	Shear stress, psi (N/mm^2)
S_{smax}	Maximum shear stress, psi (N/mm^2)
S_t	Tensile stress, psi, (N/mm^2)

S_y	Yield strength, psi (N/mm²)
T	Thickness of main pipe, in. (mm)
t	Wall thickness of tubing, in. (mm)
V	Bolt force, lb (N)
W	External load, lb (N)
W_u	Ultimate axial load, lb (N)
x, x_1	Arbitrary distances, in. (mm)
Y_r	Radial deflection, in. (mm)
y	Coordinate, in. (mm)
α	Angle of taper, deg
β	Diameter ratio
ϵ	Distance to center, in. (mm)
θ	Arbitrary angle, rad
ρ	Angle of friction, deg

41
Mechanical Springs

INTRODUCTION

The technology of conventional mechanical springs has been developed over the past hundred years and sufficient data are currently available on materials, design, and manufacture so that the designer can size a spring for a particular application. For instance, an excellent summary of the mechanical properties of spring materials, working formulas, and fabrication variables has been compiled by Carlson [281].

The term "conventional" is difficult to defend because many structural and machine elements, discussed throughout the various sections of this book, can easily be put into the general category of springs. Also the term, "spring constant" enters many design calculations under static and dynamic conditions involving such configurations as straight beams, curved members, complex-shape springs, arched cantilevers, conical disks, or multiple-leaf springs. Probably the closest type to a traditional spring concept is obtained when the externally applied loads create a couple, which turns the spring while winding or holds it in place when wound up. Hence the geometry of a helically coiled spring seems to fit the "conventional" or "traditional" description best.

The basic derivation of spring formulas and some of the more complicated equations can be found in a classical text on mechanical springs by Wahl [145]. This chapter is limited to conventional formulas and the elementary principles of stress analysis based on the design office experience.

The principal variables of importance to spring design are the force, deflection, and stress. Once the preliminary design for a given material and environment is accomplished, the appropriate production techniques are best selected with the help of the established companies specializing in spring manufacture.

The methodology of spring calculation has evolved through the use of special slide rules, charts, tables, and finally electronic computers, which have considerably reduced the laborious process of repetitive trial-and-error computations. The principal design relationships, however, remained the same since they evolved from the theoretical concepts of stress and strain.

The design, manufacture, and application of mechanical springs result in a customary subdivision of this topic into compression, extension, and torsion categories. The basic stress and deflection formulas are essentially the same for the compression and extension springs. The only difference between these two categories lies in the fact that in most applications extension springs have initial tension wound into them. In the case of extension springs the errors in both stress and deflection calculations will result when the helix angle exceeds $12\frac{1}{2}°$. Furthermore, if the extension spring were pulled out far enough, the tensile and not the torsional component of stress would become more significant.

COMPRESSION SPRING

The compression springs in general are open-coil, helically wound springs which may be cylindrical, conical, barrel-shaped, or concave in form. Since the solid height and end conditions are important factors in design, it is necessary to know whether a compression spring under consideration is expected to have plain or squared ends with or without ground end coils. These conditions determine the appropriate corrections developed by the industry for the total number of coils, load tolerances, operational characteristics, and other features affecting the performance and cost. The compression springs are most popular and represent 80 to 90% of all springs produced by the industry.

The basic design parameters in the case of a round wire cylindrical spring include the number of active coils N, wire diameter d, mean coil diameter D, and the applied force P, which can be either compressive or tensile. The appropriate symbols are given in Fig. 41.1.

From the point of view of stress analysis the helical compression spring is somewhat of a misnomer because the compressive force produces the torsional fiber

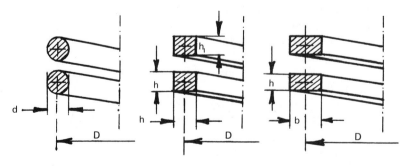

Fig. 41.1 Typical symbols for helical springs.

Mechanical Springs

stress. When using a round wire, the basic design formula for a helical compression spring is

$$\tau = \frac{2.55PD}{d^3} \tag{41.1}$$

This formula is based on the assumption that the spring wire responds as a straight bar of circular cross section when acted upon by a torsional moment. The simplified model cannot yield the correct value for the fiber stress because the coiled wire behaves as a curved member loaded out of the plane of curvature. The actual stresses should vary in a hyperbolic fashion, with the higher stress level existing on the inside surface of the spring.

The approximate formula for the torsional stress given by Eq. (41.1) is sufficiently accurate for most practical purposes where the static loads are involved. However, when the helical springs operate under severe dynamic conditions the additional stresses caused by the curvature of the wire and the shear load should be accounted for. The required stress correction factor, first proposed by Wahl [145], was shown to vary with the ratio of mean to wire diameter. Utilizing the Wahl formula for the correction factor and rewriting Eq. (41.1) gives

$$\tau_{\max} = PDK_0/d^3 \tag{41.2}$$

where K_0 is plotted in Fig. 41.2 as a function of the ratio D/d. Since the Wahl correction applies to square as well as to round wire, the corresponding maximum

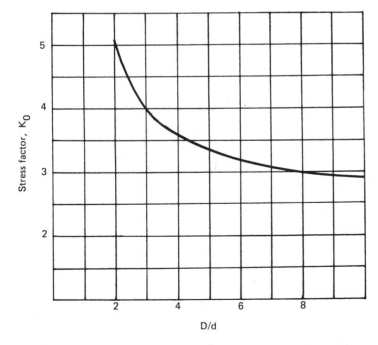

Fig. 41.2 Stress correction factor (based on the Wahl theory).

torsional stress for the square wire may be expressed as

$$\tau_{max} = \frac{0.94PDK_0}{h^3} \tag{41.3}$$

In this instance h denotes the side of a square wire before coiling. Other symbols are the same for the round and square wire designs. Since the most efficient cross section for the compression springs is round wire, the square wire design should be used only when it is necessary to assure the maximum load capacity within a given spring envelope. The total elastic energy which the spring can contain depends upon the cross section and length of the wire. Hence for $h = d$, a compression spring using square wire should have a greater energy-storing capacity than the one using round wire on the premise that all other dimensions are equal. It should be noted that when square wire is coiled into a spring the cross-section becomes trapezoidal in shape. The result is that the inner side of the square increases. Although this effect is relatively small, the solid height of a square wire spring should be corrected. Industry recommends the following expression for estimating the enlarged thickness of the inner edge [208]:

$$h_1 = \frac{0.48h(2D + h)}{D} \tag{41.4}$$

The initial step in spring sizing is to estimate the proper wire diameter for a given external load P, the approximate mean diameter D, and the allowable torsional stress τ. The convenient formula for starting the calculational procedure is given by Eq. (41.1). The number of coils can now be obtained from the standard spring formula

$$Y = \frac{8PD^3N}{Gd^4} \tag{41.5}$$

Here P, D, and d have the same meaning as before. In addition, N defines the number of active coils and G is the modulus of rigidity, or torsional modulus as it is known in the spring industry. If required, this property can be calculated from Eq. (2.2) provided the elastic modulus and Poisson's ratio of the spring material are known.

It is noted that the solution of Eq. (41.5) for N depends on the assumption of the desired deflection Y consistent with the external load P. The magnitude of N obtained from Eq. (41.5) should be increased by the appropriate number of "dead" coils established by the industrial practice. According to the type of spring ends, this allowance may vary between one and two coils. The solid height of the spring is taken as the total number of coils times the wire diameter. However, the compression spring should never be designed to deflect in its working travel until the coils actually contact each other. Therefore in selecting the working range Y, certain allowance should be made so that the spring can carry the load above its solid height.

The spring industry in general is not expected to recommend the use of square or rectangular wire. If required, however, the deflection of the spring made out of the square wire may be calculated as follows:

$$Y = \frac{5.58PD^3N}{Gh^4} \tag{41.6}$$

When the ratio D/d is larger than 12 the effect of coiling on the shape of the square wire can be neglected.

SPRING DEFORMATION AND DYNAMICS

This section relates to some of the more involved characteristics of spring behavior, such as the diametral change under load, buckling, suddenly applied loads, and vibrational effects, which normally go beyond the elementary aspects of spring design. The purpose here is to bring some of these topics to the attention of spring designers and users with special regard to any methods of practical stress analysis which might bear on spring engineering.

It is at times necessary to estimate the change in the outer diameter of the compression spring under load. Allowance for this change should be made when the spring is housed, for instance, in a tube. Unnecessary contact friction and binding can cause additional stresses and malfunction of the machinery. The increase in the outer spring diameter under the compressive load can be defined as

$$\Delta = [D^2 - 0.1(e^2 - d^2)]^{1/2} - D \tag{41.7}$$

In the foregoing expression D and d are the customary spring symbols. The amount of the diametral increase measured in inches is Δ. The pitch e is defined in terms of the free length of the spring, corrected for the type of spring ends, and divided by the number of active coils.

Whenever a relatively long compression spring is used, buckling may occur if the free length, for instance, is four or more times the mean diameter [281]. The dominating variables in this case include the deflection Y and the slenderness ratio L/D. The design curve in Fig. 41.3 illustrates the buckling problem for a compression spring with end coils closed and ground. The chart shows, for example, that when the slenderness ratio is 0.6, the spring will buckle if the working deflection is equal to 40% of the free length. It is desirable that such springs are guided over a rod or in a tube. For unusually long helical springs in compression it may be necessary to use several shorter springs assembled over a rod with the appropriate washers in between.

Rigorous analysis of dynamic behavior of the individual springs and spring systems is not a mundane task. However, certain elementary formulas and rules, supported by the industrial experience, can go a long way toward better understanding of spring design and performance.

In a more general sense, the loads for which springs are designed can be applied gradually, suddenly, or in a manner of high-velocity impact. The most direct approach to this problem is to model the spring as a weightless body on the premise that the spring surge effects can be neglected. In the case of a slowly applied load

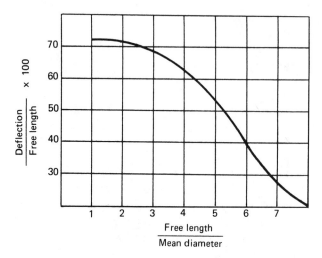

Fig. 41.3 Buckling chart for compression springs.

we have

$$Y = \frac{P}{k} \qquad (41.8)$$

Here k denotes the conventional spring constant in pounds per inch. The assumption is that the work done on the spring is equal to $PY/2$, where the load increases uniformly from zero to P. When this load is acting through the entire deflection the work done is PY and the formula for the condition of "instantly applied load" is

$$Y = \frac{2P}{k} \qquad (41.9)$$

Note that the factor of 2 appearing in Eq. (41.9) is consistent with the case of a sudden loading discussed in Chapter 11, where the change in potential energy is made equal to the elastic work done on the structural beam. When the load is suddenly dropped on the compression spring from a height a, the energy absorbed is equal to $P(Y + a)$, leading to the formula generally recognized by the spring industry [208, 281] as

$$Y^2 = \frac{2P(Y + a)}{k} \qquad (41.10)$$

If, instead of a free fall represented by Eq. (41.10), the horizontal impact velocity V is specified, then the corresponding design formula becomes

$$Y = V \left(\frac{P}{kg}\right)^{1/2} \qquad (41.11)$$

In this formula V is given in in./sec if g, denoting acceleration of gravity, is 386.4 in./sec². If we assume that the dynamic stress in the spring is equal to the

Mechanical Springs

static stress multiplied by the specific ratio, obtained from dividing Eq. (41.11) by Eq. (41.10), then the result can be directly correlated with the stress ratio derived from Eq. (11.6) in Chapter 11.

The design formulas represented by Eqs. (41.8) through (41.11) have been derived on the assumption that the spring is weightless. To include the effect of the spring's own weight use the term $P+(W_s/3)$ instead of P in the foregoing equations, with W_s denoting the spring weight.

When the compression spring is subjected to vibration and surging loads the stresses can be at times increased as much as 40% [281]. For this reason it is good practice to estimate the natural frequency of the spring so that it can be compared with the forcing frequency of the mechanical system for which this spring is intended. A typical formula for calculating the resonant frequency of an unloaded compression spring with both ends fixed is

$$n = \frac{2.18d}{D^2 N}\left(\frac{G}{\gamma}\right)^{1/2} \tag{41.12}$$

In this expression γ denotes the weight density of the spring material in pounds per cubic inch and the frequency has the dimensions of cycles per second. Other spring symbols in Eq. (41.12) are the same as those employed throughout this chapter. When the spring is compressed initially the approximate formula for estimating the frequency at the same end conditions becomes

$$n = 187.6(Y)^{-1/2} \tag{41.13}$$

It is important to emphasize that the fundamental frequency values obtained from Eqs. (41.12) and (41.13) are far removed from the operating frequencies of the machine, by at least one order of magnitude, so that the resonance and surging conditions can be avoided. This is known to be especially serious in the case of automobile engine valve springs and similar applications where high-speed force oscillations and structural fatigue cannot be tolerated.

EXTENSION SPRING

Extension springs are closely coiled helixes that offer a finite resistance to a pulling force. The load buildup obtained by the coiling process is defined as the initial tension. The extension springs are normally made out of round wire, although in special cases square or shaped wires can also be found. From the design point of view, initial tension means the presence of a definite stress in the spring wire at the original undeflected condition. The initial tensile stress should be added to the subsequent working stress of the spring. When there is no initial tension the compression spring formulas quoted in this chapter are directly applicable to extension spring design.

Experience shows that extension springs represent about 10% of all springs produced by the various manufacturers. The design lead time for these springs is longer because of the additional requirements for the control of initial tension, special coiling methodology, and stress analysis of end hooks. There is quite a variety of standard and special extension spring ends, designed according to the military and industrial applications, to which the designer is referred during the process of selecting the appropriate configurations. This is not a mundane problem because of the end curvature and the stress concentration phenomena involved. A typical geometry of hooks with sharp bends on the extension springs is illustrated in Fig. 41.4. Because of the geometrical transition from the plane of the end hook to the regular coil geometry, four different radii of curvature may be involved, denoted as R_1 through R_4 in Fig. 41.4. It appears that in general the maximum bending and torsional moments can be taken equal to PR. The exact solution of the maximum bending and torsional stresses may require certain design corrections due to the sharp curvature. However, the relevant industrial practice suggests that in a well-proportioned spring end the probability of a structural failure due to the torsional moment is rather low. Hence the simplified formula for bending is often all that might be needed [281].

$$S_b = \frac{5PD^2}{(D-d)d^3} \tag{41.14}$$

Another form of the approximate design equation for stresses due to bending utilizes the two radii of curvature R_1 and R_3 [208] as shown in Fig. 41.4.

$$S_b = \frac{10.19 PRR_1}{R_3 d^3} \tag{41.15}$$

The approximate formulas given by Eqs. (41.14) and (41.15) can be used for the majority of spring configurations including regular end hooks over center, crossover

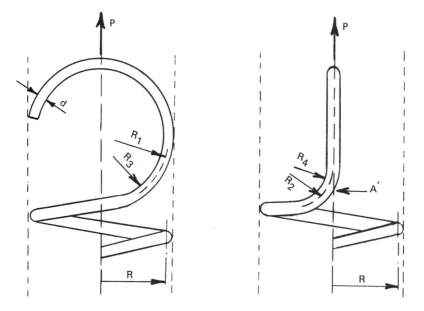

Fig. 41.4 Typical end hook for extension spring.

Mechanical Springs

center hooks, extended hooks, as well as machine or double hooks, to mention a few. The exceptions may involve half and side hooks subjected to additional stresses. It is also easy to shown that when $R_1 = R_2$, and when wire dimension d is rather small compared to coil diameter D, both formulas, Eqs. (41.14) and (41.15), give identical numerical results.

In the region of A' shown in Fig. 41.4, where the bend joins the helical portion of the spring wire, the stress is caused by twisting. The corresponding torsional shear stress can be obtained by multiplying the result from Eq. (41.1) by the ratio R_2/R_4. This approximation should be sufficiently conservative for most practical needs.

The amount of stretch in an extension spring has no well-defined limit similar to the solid-height limit found in a compression spring. Hence the amount of extension has to be governed by the maximum fiber stress dependent on the wire size and the diameter ratio. From the point of view of production, the extension springs can be made with a definite space between coils, with zero space and zero initial tension, or close-wound with a given initial tension. In the latter case, after the initial tension has been broken to create a space between the coils, the extension has the same rate as that which can be calculated from the deflection formula for the compression spring. The specific extension spring which requires a uniform rate of load from zero to the maximum allowable deflection must be wound with zero initial tension.

TORSION SPRING

The apparent confusion of terms reigns supreme when a torsion spring, exerting pressure along a path of circular arc, is subjected to bending stresses while the compression and extension springs carry torsional stresses. Although the torque has been known to be stored in such configurations as clock springs and motor and power springs, the common term of "torsion spring" employed in industry relates essentially to a helical coil spring that exerts a torque or rotary force. In this type of design the spring winds up from the free position, causing reduction in the coil diameter, and it is often supported over a rod. Although most of such springs in commercial use have been fabricated from round wire, the most efficient cross-sectional shape in this particular case must be rectangular, or the so called flat coil. The basic principle of a torsion spring is shown in Fig. 41.5, where the applied torque is not equal to the force times the radial distance of this force from the center of the rod $(P \times L_r)$ but $(P \times L_0)$.

There are numerous working tables, formulas, and recommended design steps for the analysis of torsion-type springs [145, 208, 281]. The underlying principles here follow the conventional beam theory on the premise that a constant moment acts on the entire wire cross section.

The bending stress for the round-wire torsion spring is

$$S_b = \frac{10.2M}{d^3} \qquad (41.16)$$

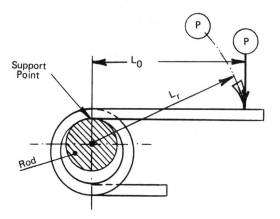

Fig. 41.5 Torsion spring.

where M denotes the applied torque in lb-in. and d is the spring wire diameter in inches as before. The corresponding stress for a rectangular section, having width b and depth h, is

$$S_b = \frac{6M}{bh^2} \tag{41.17}$$

The relationship for a round wire between the applied torsional moment M and the number of turns T is

$$M = \frac{0.093 E T d^4}{ND} \tag{41.18}$$

The number of active coils of the wire is denoted by N, while E and D are the elastic modulus and mean coil diameter, respectively. Finally, the relationship between the moment and the number of turns of the rectangular wire gives

$$M = \frac{0.152 E b T h^3}{ND} \tag{41.19}$$

In the foregoing expression the thickness h must always be the radial dimension and the width b the dimension parallel to the axis of the coil.

From the point of view of practical stress analysis, the coiling procedure and the degree of springback of a torsion spring depend on the D/d ratio. In the case of smaller D/d ratios, the tensile stresses induced by the coiling process can be in excess of the proportional limit of the material so that the calculations can only indicate the apparent rather than real stresses. The basic concept of the apparent stress is discussed in Chapter 33.

When a flat-wire torsion spring is wound on itself in the form of a spiral the governing bending stress can be estimated from Eq. (41.17). The applied moment

Mechanical Springs

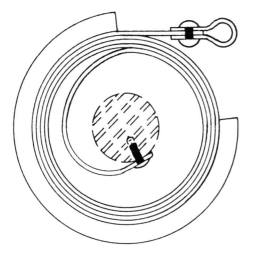

Fig. 41.6 Spiral power spring.

M_a per one turn of this spring can be expressed as

$$M_a = \frac{0.52 E b h^3}{L} \tag{41.20}$$

in which L denotes the total length of the active material. The spiral type of a spring shown in Fig. 41.6 may be used as a lock mechanism in automobiles or brush holders for electric motors and generators when it is required to sustain a relatively low pressure over a longer period of time. The concept of a constant-force spring is similar to that of a flat-wire torsion spring with the inner end usually not fastened. Hence, by pulling on the outer end outward, the spring can uncoil over a snugly fitted roller at a constant force. The actual design and stress analysis of this type of a spring can be avoided because there are literally thousands of constant-force springs available on the market.

BUCKLING COLUMN SPRING

When a flat and initially straight piece of steel is subjected to a compressive end load, large deflections can develop. The spring formed in this way is essentially a special curved member characterized by what is known as a *zero rate response*. The basic notation and geometry pertaining to this case are shown in Fig. 41.7. The analysis of this type of a structure requires solution of the following differential equation:

$$EI \left(\frac{d\psi}{ds} \right) = -Px \tag{41.21}$$

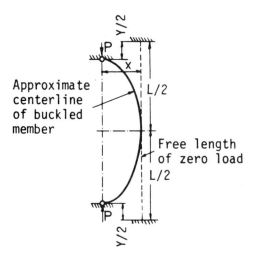

Fig. 41.7 Buckling column spring.

where ψ denotes slope at a point of the deflection curve and s is measured along the length of the curved member. Solution of this differential equation involves elliptic integrals, which can, however, be simplified to give design formulas and charts suitable for practical use [176].

The relevant working formulas are

$$P = \frac{h^2 A E F_1}{L^2} \tag{41.22}$$

and

$$S_b = \frac{h E F_2}{L} \tag{41.23}$$

The required area of the cross section for sizing the special spring is then

$$A = \frac{P E F_2^2}{F_1 S_b^2} \tag{41.24}$$

The design factors F_1 and F_2 are featured in Fig. 41.8. When the amount of vertical deflection Y is found from Eq. (41.22) and Fig. 41.8, the corresponding value of lateral displacement can also be determined from Fig. 41.8. Since relatively thin strips in axial compression cannot support substantial loads, parallel arrangement of a number of strips may be required for a particular design. The controlling factor here is the stress given by Eq. (41.23) and the type of load-deflection characteristics required. For best approximation to the zero rate spring response, the Y/L ratio should be kept somewhere between 0.1 and 0.3, while L/h should not be appreciably smaller than about 200.

Mechanical Springs

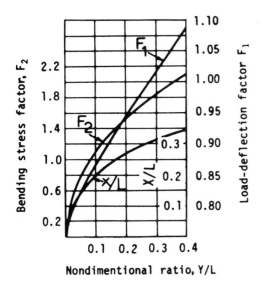

Fig. 41.8 Design factors for buckling column spring.

SYMBOLS

A	Area of cross section, in.^{2} (mm^2)
a	Length of vertical drop, in. (mm)
b	Width of section, in. (mm)
D	Mean coil diameter, in. (mm)
d	Wire diameter, in. (mm)
E	Modulus of elasticity, psi (N/mm^2)
e	Spring pitch, in. (mm)
F_1, F_2	Design factors
G	Modulus of rigidity, psi (N/mm^2)
g	Acceleration of gravity, in./sec^2 (mm/sec^2)
h, h_1	Depths of sections, in. (mm)
I	Moment of inertia, in.^{4} (mm^4)
K_0	Stress correction factor
k	Spring constant, lb/in. (N/mm)
L	Length of spring or flat stock, in. (mm)
L_0	Arbitrary length, in. (mm)
L_r	Radial distance, in. (mm)
M	Applied torque, lb-in. (N-mm)
M_a	Moment per one turn of torsion spring, lb-in. (N-mm)
N	Number of active coils
n	Frequency, cycles/sec.
P	Spring force, lb (N)
R	Mean radius of coil, in.
$R_1 \ldots R_4$	Radii of curvature, in. (mm)
S_b	Bending stress, psi (N/mm^2)

s	Length of deflection curve, in. (mm)
T	Number of turns
V	Impact velocity, in./sec, (mm/sec)
W_s	Weight of spring, lb (N)
x	Horizontal deflection, in. (mm)
Y	Deflection, in. (mm)
γ	Weight density, lb/in.^{3} (N/mm^3)
Δ	Diameter increase, in. (mm)
τ	Torsional stress, psi (N/mm^2)
τ_{max}	Maximum torsional stress, psi (N/mm^2)
ψ	Slope, rad

42
Application of Metric Units to Stress Analysis

INTRODUCTION

A modernized version of the old metric system, evolved during the past several years, is known as the International System of Units (SI). As far as solid mechanics is concerned, the system is built upon three base units and two supplementary units. The base units are the meter, kilogram, and second; the radian and steradian are classified as supplementary.

As a matter of general record, the three base units can be defined as follows:

The unit of length (meter) is equal to 1,650,763.73 wavelengths of the radiation of the atom of krypton-86 under specified conditions.

The unit of mass (kilogram) is the mass of a particular cylinder of platinum-iridium alloy preserved in a vault at Sèvres, France.

The unit of time (second) is the exact fraction 1/31,556,925.9747 of the tropical year 1900 (January).

As far as the mechanical design engineer, structural analyst, and hardware manufacturer are concerned, the most important immediate choice of the units would be that for the stress, pressure, elastic constants, weight, and physical dimensions. Before giving a few dimensional interpretations of some typical formulas used in this book, it should be noted that the SI system differentiates rather strongly between the concepts of weight and mass. *Weight* is defined as the force of gravity, and it is expressed in newtons. The newton, which is then a unit of force, is that magnitude of a force which gives to a mass 1 kilogram an acceleration of 1 meter per second per second. On this basis the *stress*, defined as the force divided by the area, would be expressed as newtons divided by meters squared. In the SI system 1 pascal (Pa) is equal to 1 newton per square meter, which is the formal SI unit for pressure, stress, material strength, or elastic constants. Long-overdue emphasis on the basic distinction between the concepts of mass and weight teaches us, then, that the unit of force is newtons and the acceleration of gravity g is now 9.8 m/s^2.

The choice of units for pressure, stress, and elastic constants is, unfortunately, not so straightforward, and it has been the subject of numerous discussions in engineering circles. Although the use of "pound," "kilogram," or "newton" does not present any particular calculational difficulties, and we can get used to thinking in terms of a different scale, the proposed unit "pascal" is not only too small for practical purposes, but it is also quite superfluous. It would be much more reasonable to use newtons per area instead of pascals. The material related to the SI system in this book is expressed in newtons per square millimeter.

There are several practical reasons why the use of N/mm^2 should be promoted. In the first place, the dimensions of countless machine components and structural elements, found worldwide, are expressed in millimeters. Secondly, the absolute magnitude of the newton force appears to be compatible with the area of a square millimeter and in such a case 1 atmosphere, for example, is approximately equal to 0.1 N/mm^2. This number is relatively easy to recall and many other similar numbers can be developed which can certainly represent a convenient practical range. For example, the modulus of elasticity of steel will work out to be 2×10^5 N/mm^2 and the approximate yield strength of a typical mild steel could be remembered as 250 N/mm^2. Numerically, of course, 1 megapascal is equal to 1 newton per square millimeter, as shown by the following simplistic transformation:

$$\frac{N}{mm^2} = \frac{N}{(m \times 10^{-3})^2} = \frac{N}{m^2 \times 10^{-6}} = \frac{10^6 N}{m^2} = 1 \text{ MPa}$$

The use of N/mm^2 should automatically eliminate our present lack of uniformity caused by the open choice of pascal, kilopascal, megapascal, and other potential multiples of this small and cumbersome unit.

The dual system of units adopted throughout the book wherever numerical calculations are involved utilizes the following basic conversions:

1 lb = 4.4482 N
1 psi = $\frac{lb}{in.^2} = \frac{4.4482 \text{ N}}{(0.0254 \text{ m})^2}$ = 6895 Pa
1 in. = 25.4 mm
1 psi = 0.006895 N/mm^2
1 lb/in. = 4.4482 N $\times$ 25.4 mm = 112.9842 N-mm
1 lb/in. = 4.4482 N/25.4 mm = 0.1751 N/mm
1 MPa = 1 N/mm^2 = 145 psi
1 kg = 9.8066 N

These values have been rounded off slightly for practical reasons.

The great majority of engineering design formulas presented in the book are homogeneous, so that a consistent set of units can be employed. The mechanics of interpretation and conversion of some of the dimensional expressions used will be illustrated with reference to a selected number of problems.

Application of Metric Units to Stress Analysis

SHEAR STRESS

$$\tau = \frac{M_t r}{I_p}$$

where M_t = twisting moment, lb-in.
r = radius, in.
I_p = polar moment of inertia, in.^{4}

Hence, assuming unit values, we get

$$\tau = \frac{4.4482 \text{ N} \times 0.0254 \text{ m} \times 0.0254 \text{ m}}{(0.0254 \text{ m})^4}$$

Suppose that we have

$$M_t = 500 \text{ lb-in.} = 500 \times 4.4482 \text{ N} \times 0.0254 \text{ m} = 56.49 \text{ Nm}$$
$$r = 2 \text{ in.} = 2 \times 0.0254 \text{ m} = 0.0508 \text{ m}$$
$$I_p = 0.4 \text{ in.}^4 = 0.4(0.0254)^4 \text{ m}^4 = 1.6649 \times 10^{-7} \text{ m}^4$$

Then using the English system,

$$\tau = \frac{500 \times 2}{0.4} = 2500 \text{ psi}$$

and the SI system gives

$$\tau = \frac{56.49 \times 0.0508}{1.6649 \times 10^{-7}} = 1.7236 \times 10^7 \text{ Pa}$$

or

$$\tau = \frac{500 \times 112.9842 \text{ (N-mm)} \times 2 \times 25.4 \text{ (mm)}}{0.4 \times 25.4^4 \text{ (mm)}^4} = 17.24 \text{ N/mm}^2$$

Note that $2500/17.24 = 145$. This is the conversion factor between MPa and N/mm^2, as shown previously. The resulting shear stress can be expressed as 2500 psi or 17.24 MPa = 17.24 N/mm^2, depending on the system adopted. Clearly, however, 17.24 is more convenient to use and remember than 1.724×10^7.

BEAM DEFLECTION

The general deflection formula for a cantilever beam can be stated as follows:

$$Y = \frac{W(2L^3 - 3aL^2 + a^3)}{6EI}$$

Let us now assume the following metric quantities:

$L = 2$ m
$a = 1$ m
$W = 100$ kg
$E = 2.069 \times 10^{11}$ Pa
$I = 1.5 \times 10^{-6}$ m^4

Since

$$1 \text{ kg} = 9.8066 \text{ N}$$

the deflection formula gives

$$Y = \frac{9.8066 \times 100(2 \times 2^3 - 3 \times 1 \times 2^2 + 1)}{6 \times 2.069 \times 10^{11} \times 1.5 \times 10^{-6}} = 0.002633 \text{ m}$$

or

$$\frac{0.002633 \times 10^3}{25.4} = 0.104 \text{ in.}$$

Direct conversion of the basic input into the English system of units, involving pounds and inches, leads to an identical result as long as the units are consistent and the formula is homogeneous. A dimensional check for the deflection formula follows directly, and there are no conversion factors involved unless we wish to express inches in terms of meters or millimeters.

$$\frac{\text{lb} \times \text{in.}^3}{\frac{\text{lb}}{\text{in.}^2} \times \text{in.}^4} = \text{in.}$$

$$\frac{\text{N} \times \text{mm}^3}{\frac{\text{N}}{\text{mm}^2} \times \text{mm}^4} = \text{mm}$$

or

$$\frac{\text{N} \times \text{m}^3}{\frac{\text{N}}{\text{m}^2} \times \text{m}^4} = \text{m}$$

BUCKLING STRESS

A simple illustration can be taken from Table 10.1, which summarizes the formulas for the critical buckling stress of structural columns. For instance,

$$S_{\text{CR}} = 15{,}000 - 0.325 \left(\frac{L}{\rho}\right)^2$$

where L = length of column, in.
$\rho = (I/A)^{1/2}$ = radius of gyration, in.

Suppose that

Application of Metric Units to Stress Analysis

$\frac{L}{\rho} = 100$
$\rho = 0.5$ in.
$\frac{I}{A} = 0.25$
$A = 4$ in.^{2}
$I = 0.25 \times 4 = 1$ in.^{4}

Then

$$S_{\text{CR}} = 15{,}000 - 0.325 \times 100 \times 100 = 11{,}750 \text{ psi}$$

Using metric data,

$$L = 50 \text{ in.} \times 0.0254 \frac{\text{m}}{\text{in.}} = 50 \times 0.0254 \text{ m}$$

$$\rho = \frac{0.5 \text{ in.} \times 25.4 \text{ mm/in.}}{1000 \text{ mm/m}} = 0.0127 \text{ m}$$

so that

$$\frac{L}{\rho} = \frac{50 \times 0.0254 \text{ m}}{0.0127 \text{ m}} = 100$$

Since this ratio is dimensionless, the resultant equation for S_{CR} still gives the critical buckling stress in psi, which in this case is 11,750 psi. To convert the formula to N/mm^2, we can multiply by 0.006895, to give

$$S_{\text{CR}} = 103.425 - 0.00224 \left(\frac{L}{\rho}\right)^2$$

This yields the critical stress equal to 81.03 N/mm^2, which is the same as 11,750 psi.

FRACTURE TOUGHNESS

In English units the term defining the plane strain fracture toughness is given as follows:

$$K_{\text{IC}} = A(\text{psi})(\text{in.})^{1/2} = A(\text{lb-in.}^{-3/2})$$

where A is a number representing the numerical value of toughness. To convert this to a metric form, we get

$$K_{\text{IC}} = A(0.006895 \text{ N/mm}^2)(25.4 \text{ mm})^{1/2}$$
$$= A \left(\frac{0.03475 \text{ N}}{\text{mm}^2}\right)(\text{mm})^{1/2}$$
$$= A \times 0.03475 \text{ (N/mm}^{-3/2})$$

Suppose that in English units $K_{IC} = 200{,}000$ psi (in.)$^{1/2}$. Then the equivalent fracture toughness in the metric system becomes

$$200{,}000 \times 0.03475 = 6950 \text{ N-mm}^{-3/2} = 219.8 \, \frac{\text{MN}}{\text{m}^2}(\text{m})^{1/2}$$

The last number represents a conversion factor of 1.098×10^{-3} in going from psi (in.)$^{1/2}$ to meganewtons per meter squared and multiplied by the square root of meter.

STRESS PROPAGATION

In estimating the stress level in the case of impact, we often use the following equation:

$$\sigma = \rho C V$$

where ρ = materials density, $\frac{\text{lb-sec}^2}{\text{in.}^4}$
C = sonic velocity, in./sec
V = particle velocity, in./sec

Hence, the dimension of stress follows directly:

$$\frac{\text{lb} \times \text{sec}^2}{\text{in.}^4} \times \frac{\text{in.}}{\text{sec}} \times \frac{\text{in.}}{\text{sec}} = \frac{\text{lb}}{\text{in.}^2}$$

since

$$\text{lb} = 4.4482 \text{ N}$$

and

$$\text{in.} = 25.4 \text{ mm}$$

The formula to give the metric equivalent becomes

$$\sigma = \frac{4.4482 \text{ N}}{(25.4 \text{ mm})^2} \times \rho C V$$

or

$$\sigma = 0.006895 \rho C V \; (\text{N/mm}^2)$$

where $\rho = \frac{4.4482 \text{ N} \times \text{sec}^2}{(25.4)^4 \text{mm}^4} = 0.000010686 \frac{\text{N sec}^2}{\text{mm}^4}$
$C = \frac{25.4 \text{ mm}}{\text{sec}}$
$V = \frac{25.4 \text{ mm}}{\text{sec}}$

$$\frac{4.4482 \text{ N} \times \text{sec}^2}{(25.4)^4 \text{ mm}^4} \times \frac{25.4 \text{ mm}}{\text{sec}} \times \frac{25.4 \text{ mm}}{\text{sec}} = 0.006895 \frac{\text{N}}{\text{mm}^2}$$

Application of Metric Units to Stress Analysis

For example, taking

$$\rho = \frac{0.283}{386.4} \frac{\text{lb sec}^2}{\text{in.}^4} = 0.0007324 \frac{\text{lb sec}^2}{\text{in.}^4}$$
$$C = 201{,}600 \text{ in./sec}$$
$$V = 120 \text{ in./sec}$$

we get

$$\sigma = 0.0007324 \times 201{,}600 \times 120 = 17{,}718 \text{ psi}$$

or

$$\sigma = 0.006895 \times 17{,}718 = 122.17 \text{ N/mm}^2$$

In converting these data into metric units, we have

$$\rho = \frac{0.0007324 \times 4.4482 \text{ N} \times \text{sec}^2}{(25.4)^4 \text{mm}^4} = 7.827 \times 10^{-9} \frac{\text{N sec}^2}{\text{mm}^4}$$
$$C = 201{,}600 \times 25.4 \times \frac{\text{mm}}{\text{sec}} = 5{,}120{,}640 \text{ mm/sec}$$
$$V = 120 \times 25.4 \times \frac{\text{mm}}{\text{sec}} = 3048 \text{ mm/sec}$$

Then

$$\sigma = 7.827 \times 10^{-9} \frac{\text{N sec}^2}{\text{mm}^4} \times 5{,}120{,}640 \frac{\text{mm}}{\text{sec}} \times 3048 \frac{\text{mm}}{\text{sec}} = 122.17 \text{ N/mm}^2$$

Hence, any consistent set of units can be used with the basic formula, although the numerical data in the metric system appear to be more cumbersome to use.

The examples selected at random from this book indicate that the numerically speaking metric unit of stress defined as N/mm^2 is rather convenient. It also reminds us of the old metric system, practiced in many corners of the world, where the measure of pressure, stress, or a mechanical property was kg/mm^2. In this old notation, kg usually stood for weight or force. The weight of a good-sized man is, say, 80 kg or about 176 lb. In the not too distant future, then, the SI man will weigh 782.9 newtons. However, since $N = mg$ and $g = 9.8 \text{ m/sec}^2$, solving for the mass m^* gives

$$m^* = \frac{N}{g} = \frac{782.9}{9.8} = 79.89 \text{ kg} \cong 80 \text{ kg}$$

The reader is left to ponder again the meaning of mass and weight.

Appendix
Selection of Practical Stress Formulas

The purpose of this appendix is to give the reader a quick overview of the principal formulas and topics which are most likely to occur in mechanical engineering design. Although the choice of the formulas presented is largely subjective, the reader will find that their field of application is very extensive.

Some of the quantities defined in this appendix have been denoted by more than one symbol. This is due to the fact that each chapter has its own list of symbols dictated by the diversity of the topics discussed. The equation numbers refer to the appropriate chapters where the original formulas first appeared. In addition to the brief selection of the design equations, the book contains a number of tables and more complex formulas not represented here.

Hooke's Law [Eq. (1.3)]

$$E = \frac{S}{\epsilon}$$

Poisson's ratio [Eq. (1.4)]

$$\nu = \frac{u}{\epsilon}$$

Volumetric strain [Eq. (1.6)]

$$\frac{\Delta V}{V} = \frac{\Delta L}{L}(1 - 2\nu)$$

True strain [Eq. (1.22)]

$$\epsilon_t = \ln(1 + \epsilon)$$

Modulus of rigidity [Eq. (2.2)]

$$G = \frac{E}{2(1+\nu)}$$

Angle of twist [Eq. (2.15)]

$$\phi = \frac{M_t L}{GI_p}$$

Torsional stress in circular tube [Eq. (2.18)]

$$\tau = \frac{M_t}{2\pi r^2 t}$$

Bending stress [Eq. (3.1)]

$$S_b = \frac{M_b C}{I}$$

Maximum principal stress [Eq. (4.4)]

$$S_1 = \frac{1}{2}(S_x + S_y) + \frac{1}{2}\sqrt{(S_x - S_y)^2 + 4\tau_{xy}^2}$$

Minimum principal stress [Eq. (4.5)]

$$S_2 = \frac{1}{2}(S_x + S_y) - \frac{1}{2}\sqrt{(S_x - S_y)^2 + 4\tau_{xy}^2}$$

Maximum shear stress [Eq. (4.6)]

$$\tau_{max} = \frac{S_1 - S_2}{E}$$

Principal strains [Eqs. (4.10) and (4.11)]

$$\epsilon_1 = \frac{S_1 - \nu S_2}{E}$$

$$\epsilon_2 = \frac{S_2 - \nu S_1}{E}$$

Formula for ductile response [Eq. (5.4)]

$$S_0 = (S_x^2 + 3\tau_{xy}^2)^{1/2}$$

Selection of Practical Stress Formulas

Unit strain energy (resilience) [Eq. (6.1)]

$$U' = \frac{S^2}{2E}$$

Strain energy in tension [Eq. (6.2)]

$$U = \frac{W^2 L}{2AE}$$

Uniaxial change in length [Eq. (6.3)]

$$\Delta L = \frac{WL}{AE}$$

Castigliano hypothesis [Eq. (6.5)]

$$\frac{\partial U}{\partial W} = \frac{WL}{AE}$$

Strain energy in bending [Eq. (6.6)]

$$U = \frac{M_b^2 L}{2EI}$$

Castigliano equation for straight bar [Eq. (6.10)]

$$\frac{\partial U}{\partial W} = \int_0^L \frac{M_b}{EI} \left(\frac{\partial M_b}{\partial W} \right) dx$$

Castigliano equation for curved bar [Eq. (6.11)]

$$\frac{\partial U}{\partial W} = \int_0^\theta \frac{M_b}{EI} \left(\frac{\partial M_b}{\partial W} \right) R \, d\theta$$

Shear strain energy [Eq. (6.12)]

$$U' = \frac{\tau^2}{2G}$$

Strain energy for curved bar [Eq. (7.3)]

$$U = \int_0^\theta \left(\frac{M_b^2}{2AER\delta} + \frac{N^2}{2AE} + \frac{\xi Q^2}{2AG} - \frac{M_b N}{AER} \right) R \, d\theta$$

Deflection curve of beam [Eq. (7.5)]

$$EI \frac{d^2 y}{dx^2} = M_b$$

Deflection of simple cantilever beam [Eq. (7.8)]

$$Y = \frac{W}{EI}\left(\frac{x^3}{6} - \frac{L^2 x}{2} + \frac{L^3}{3}\right)$$

Stress near circular hole [Eq. (9.2)]

$$S_{\max} = S''\left[1 + \frac{1}{8}\left(\frac{d}{x}\right)^2 + \frac{3}{32}\left(\frac{d}{x}\right)^4\right]$$

Inglis stress formula [Eq. (9.3)]

$$S_{\max} = S\left[1 + 2\left(\frac{L}{r}\right)^{1/2}\right]$$

Buckling load on column [Eq. (10.1)]

$$W_{\mathrm{CR}} = K_c \frac{EI}{L^2}$$

Radius of gyration [Eq. (10.3)]

$$\rho = \left(\frac{I}{A}\right)^{1/2}$$

Tangent modulus [Eq. (10.5)]

$$E_t = \frac{dS}{d\epsilon}$$

Buckling stress in plates (elastic) [Eq. (10.7)]

$$S_{\mathrm{CR}} = K_p E \left(\frac{t}{b}\right)^2$$

Plastic buckling of plates [Eq. (10.8)]

$$S_{\mathrm{CR}} = K_p \eta E \left(\frac{t}{b}\right)^2$$

Shear buckling of panels [Eq. (10.11)]

$$\tau_{\mathrm{CR}} = K_s \eta E \left(\frac{t}{b}\right)^2$$

Selection of Practical Stress Formulas

Buckling of I beam [Eq. (10.12)]

$$S_{\text{CR}} = \frac{0.7Ebt}{hL}$$

Column of variable cross section [Eq. (10.13)]

$$W_{\text{CR}} = \frac{\phi EI}{L^2}$$

Column's own weight [Eq. (10.14)]

$$q_{\text{CR}} = \frac{7.89EI}{L^3}$$

Natural period of vibration [Eq. (11.1)]

$$T = 2\pi \left(\frac{\delta_{\text{st}}}{g}\right)^{1/2}$$

Fundamental frequency [Eq. (11.3)]

$$f = \frac{3.1}{(\delta_{\text{st}})^{1/2}}$$

Free-fall impact stress [Eq. (11.5)]

$$S_{\text{dyn}} = S_{\text{st}} \left[1 + \left(1 + 2\frac{a}{\delta_{\text{st}}}\right)^{1/2}\right]$$

Horizontal impact stress [Eq. (11.6)]

$$S_{\text{dyn}} = S_{\text{st}} \frac{0.051V}{(\delta_{\text{st}})^{1/2}}$$

Drop of a packaged item [Eq. (11.7)]

$$\Omega = 0.113 \frac{(a)^{1/2}}{T_r}$$

Shock mitigation formula [Eq. (11.8)]

$$n = \frac{2(a + \delta_c)}{\delta_c}$$

Frequency of cantilever [Eq. (11.13)]

$$f = \frac{11.04}{L^2} \left(\frac{EI}{\gamma A}\right)^{1/2}$$

Frequency of built-in beam [Eq. (11.15)]

$$f = \frac{70.19}{L^2}\left(\frac{EI}{\gamma A}\right)^{1/2}$$

Seismic load [Eq. (12.1)]

$$Q_b = WCK_d$$

Seismic coefficient [Eq. (12.2)]

$$C = \frac{0.05}{T^{1/3}}$$

Seismic displacement [Eq. (12.3)]

$$\delta_{max} = \frac{0.49}{f^{5/3}}$$

Cumulative damage [Eq. (13.1)]

$$\sum \frac{n}{N} = 1$$

Notch sensitivity factor [Eq. (13.2)]

$$q = \frac{K_f - 1}{K - 1}$$

Low-cycle stress [Eq. (13.5)]

$$\sigma_a = \sigma_e + \frac{CE}{2N^{1/2}}$$

Cycles to failure at yield [Eq. (13.8)]

$$N = \frac{C^2 E^2}{4\sigma_y^2}$$

Stress at fracture [Eq. (14.1)]

$$\sigma = \frac{K_{IC}}{(\pi a)^{1/2}}$$

Effect of crack shape of stress [Eq. (14.2)]

$$\sigma = \frac{K_{IC}(\psi)^{1/2}}{\left(3.77b + 0.21\frac{K_{IC}^2}{\sigma_y^2}\right)^{1/2}}$$

Selection of Practical Stress Formulas

Correlation of impact energy [Eq. (14.3)]

$$K_{IC} = A(CVN)^n$$

Correlation of CVN and DT [Eq. (15.1)]

$$(CVN) = 0.12(DT) + 15$$

Yield criterion [Eq. (15.4)]

$$B \leq \left(\frac{K_{IC}}{S_y}\right)^2$$

Plane strain limit [Eq. (15.5)]

$$B \geq 2.5 \left(\frac{K_{IC}}{S_y}\right)^2$$

Sonic velocity [Eq. (16.2)]

$$C = \left(\frac{E}{\rho}\right)^{1/2}$$

Velocity of plastic wave [Eq. (16.3)]

$$C_p = \left[\left(\frac{d\sigma}{d\epsilon}\right)/\rho\right]^{1/2}$$

Dynamic elastic stress [Eq. (16.7)]

$$\sigma = \rho C V$$

Attenuation of impact stress [Eq. (16.11)]

$$\sigma_0 = \sigma e^{-(A\sqrt{\rho E}/M)t}$$

Fundamental period of pipe in breathing mode [Eq. (16.12)]

$$T = 2\pi r \left[\frac{\rho(1-\nu^2)}{E}\right]^{1/2}$$

Thermal stress [Eq. (17.1)]

$$\sigma = E\alpha\Delta T$$

Thermal volumetric strain [Eq. (17.2)]

$$\left(\frac{\Delta v}{v}\right)_T = 3\alpha \Delta T$$

Thermal stress index [Eq. (17.3)]

$$\text{TSI} = \frac{\sigma_u k}{\alpha E}$$

Thermal shock [Eq. (17.4)]

$$T_{\max} = \frac{\sigma_u(1-\nu)}{\alpha E}$$

Temperature drop through pipe wall [Eq. (17.5)]

$$\Delta T = \frac{Q \ln(R_o/R_i)}{2\pi k}$$

Elongation of bar in tension in terms of resilience [Eq. (19.1)]

$$\Delta L = L \left(\frac{2U}{E}\right)^{1/2}$$

Uniaxial stress in terms of resilience [Eq. (19.2)]

$$S = (2UE)^{1/2}$$

Elongation of tapered bar [Eq. (19.4)]

$$\Delta L = \frac{4WL}{\pi D d E}$$

Stress in a truncated cone under own weight [Eq. (19.5)]

$$S = \frac{\gamma L}{3D^2}(Dd + D^2 + d^2)$$

Elongation of a truncated cone under own weight [Eq. (19.8)]

$$\Delta L = \frac{\gamma L^2}{6D(D-d)^2 E}(D^3 + 2d^3 - 3Dd^2)$$

Deflection of cantilever with intermediate load [Eq. (20.8)]

$$y = \frac{W}{6EI}\left[x^2(x - 3a_0) + 3xL(2a_0 - L) + L^2(2L - 3a_0)\right]$$

Selection of Practical Stress Formulas

Constant beam depth and stress [Eq. (21.4)]

$$b = \left(\frac{6Wx}{\sigma_m h^2}\right)$$

Constant beam width and stress [Eq. (21.5)]

$$h = \left(\frac{6W}{b\sigma_m}\right)^{1/2} x^{1/2}$$

Constant beam proportions and stress [Eq. (21.6)]

$$h = \left(\frac{6W}{K\sigma_m}\right)^{1/3} x^{1/3}$$

Stress in flat springs [Eq. (21.11)]

$$S = \frac{6PL}{Bh^2}$$

Deflection of a tapered flat spring [Eq. (21.12)]

$$y = \frac{6PL^3}{EBh^3}$$

Deflection of a trapezoidal flat spring [Eq. (21.14)]

$$y = \frac{PL^3 \beta_T}{EBh^3}$$

Deflection of a cantilever spring with parabolically varying depth [Eq. (21.16)]

$$y = \frac{8PL^3}{EBh^3}$$

Deflection of multiple-leaf cantilever spring [Eq. (21.18)]

$$y = \frac{6PL^3}{Ebnh^3}$$

Stress in a multiple-leaf cantilever spring [Eq. (21.19)]

$$S = \frac{6PL}{bnh^2}$$

Deflection of a multiple-leaf simple beam [Eq. (21.21)]

$$y = \frac{3PL^3}{8Ebnh^3}$$

Stress in a multiple-leaf simple beam [Eq. (21.22)]

$$S = \frac{3PL}{2bnh^2}$$

Deflection of unsymmetric multiple-leaf spring [Eq. (21.24)]

$$y = \frac{6P(L-a)^2 a^2}{EbnLh^3}$$

Stress in unsymmetric multiple-leaf spring [Eq. (21.25)]

$$S = \frac{6P(L-a)a}{bnLh^2}$$

Deflection of beam with one sinking support [Eq. (23.1)]

$$Y = \frac{WL^3}{12EI}$$

Bending stress in clevis with single pin [Eq. (23.3)]

$$S_b = \frac{3WL}{BH^2}$$

Pull-up deflection in clevis [Eq. (23.4)]

$$U = \frac{WL^3}{EBH^3}$$

Maximum principal stress for a deep section beam [Eq. (24.1)]

$$S_{max} = \frac{S}{2} + \frac{1}{2}(S^2 + 4\tau^2)^{1/2}$$

Moment ratio for composite beam [Eq. (24.4)]

$$\frac{M_1}{M_2} = \frac{E_1 I_1}{E_2 I_2}$$

Bending stiffness for composite beam [Eq. (24.9)]

$$EI = E_1 I_1 + E_2 I_2$$

Bending moment for beam column [Eq. (24.12)]

$$M_{max} = \frac{W \tan \frac{L}{2} \left(\frac{P}{EI}\right)^{1/2}}{2 \left(\frac{P}{EI}\right)^{1/2}}$$

Selection of Practical Stress Formulas

Maximum bending stress for beam column [Eq. (24.13)]

$$S_{max} = \frac{P}{A} + \frac{M_{max}C}{I}$$

General equation for beams on elastic foundation [Eq. (24.16)]

$$EI\frac{d^4y}{dx^4} = -ky$$

Maximum deflection for semi-infinite beam on elastic foundation [Eq. (24.23)]

$$y_{max} = \frac{2\beta}{k}(W^* - \beta M_0^*)$$

Maximum slope for semi-infinite beam on elastic foundation [Eq. (24.24)]

$$\theta_{max} = -\frac{2\beta^2}{k}(W^* - 2\beta M_0^*)$$

Flexural rigidity of plate [Eq. (24.27)]

$$EI = \frac{ET^3}{12(1-\nu^2)}$$

Elastic foundation parameter for analysis of shells [Eq. (24.28)]

$$\beta = \frac{1.285}{(RT)^{1/2}}$$

Vertical deflection of arched cantilever under vertical load (for K_1, see Figs. 25.3 and 25.4) [Eq. (25.6)]

$$Y = \frac{PR^3 K_1}{EI}$$

Horizontal deflection of arched cantilever under vertical load (for K_2, see Figs. 25.3 and 25.4) [Eq. (25.7)]

$$X = \frac{PR^3 K_2}{EI}$$

Vertical deflection of arched cantilever under horizontal load (for K_2, see Figs. 25.3 and 25.4) [Eq. (25.13)]

$$Y = \frac{HR^3 K_2}{EI}$$

Horizontal deflection of arched cantilever under horizontal load (for K_3, see Figs. 25.3 and 25.4) [Eq. (25.14)]

$$X = \frac{HR^3 K_3}{EI}$$

Vertical deflection of arched cantilever under uniform load (for K_4, see Fig. 25.7) [Eq. (25.24)]

$$Y = \frac{qR^4}{EI} K_4$$

Horizontal deflection of arched cantilever under uniform load (for K_5, see Fig. 25.7) [Eq. (25.25)]

$$X = \frac{qR^4}{EI} K_5$$

Vertical deflection of arched cantilever under end couple (for K_6, see Fig. 25.10) [Eq. (25.30)]

$$Y = \frac{M_0 R^2}{EI} K_6$$

Horizontal deflection of arched cantilever under end couple (for K_7, see Fig. 25.10) [Eq. (25.31)]

$$X = \frac{M_0 R^2}{EI} K_7$$

Stress in snap ring [Eq. (26.2)]

$$S_b = \frac{6PR(1 + \cos \alpha)}{bh^2}$$

Snap ring deflection [Eq. (26.7)]

$$Y = \frac{3\pi PR^3}{EI}$$

Piston ring opening under diametral load [Eq. (26.13)]

$$Y = \frac{PR^3(4 + \pi)}{2EI}$$

Stress in three-quarter circular spring [Eq. (26.15)]

$$S_b = \frac{6PR}{bh^2}$$

Selection of Practical Stress Formulas

Deflection of three-quarter circular spring [Eq. (26.16)]

$$Y = \frac{3\pi P R^3}{4EI}$$

Deflection of spring clip [Eq. (26.28)]

$$Y = \frac{2PR^3}{EI}\left(9k + \frac{19\pi}{4}\right)$$

Stress in U spring [Eq. (26.37)]

$$S_b = \frac{6PR(1+k)}{bh^2}$$

Deflection of wave spring [Eq. (26.43)]

$$Y = \frac{PR^3}{EI}(1.25\phi + \phi \sin^2\phi - \sin\phi + 0.375\sin 2\phi)$$

Diametral compression (or tension) of thin ring [Eq. (27.2)]

$$Y = \frac{PR^3(\pi^2 - 8)}{4\pi EI}$$

Increase (or decrease) in horizontal diameter of thin ring [Eq. (27.3)]

$$X = \frac{PR^3(4-\pi)}{2\pi EI}$$

Vertical deflection of thick ring [Eq. (27.4)]

$$Y = \frac{P\chi}{bE}(1.7856\chi^2 + 0.7854 + 2.0453)$$

Horizontal deflection of thick ring [Eq. (27.5)]

$$X = \frac{P\chi}{bE}(1.6392\chi^2 - 0.5000 + 1.3020)$$

Opening of piston ring [Eq. (27.14)]

$$Y = \frac{3\pi R^2 S_b}{Eh}$$

Diametral opening of piston ring [Eq. (27.16)]

$$Y = \frac{6(\pi+4)PR^3}{Ebh^3}$$

Stress in uniformly loaded piston ring [Eq. (27.21)]

$$S_b = \frac{YEh}{3\pi R^2}$$

Depth of constant-stress piston ring [Eq. (27.22)]

$$h = 2.45R \left[\frac{q(1 - \cos\theta)}{bS_b}\right]^{1/2}$$

Initial gap for constant-stress piston ring [Eq. (27.24)]

$$Y = \frac{24qR^4}{bEh^3}$$

Mean diameter of piston ring [Eq. (27.26)]

$$d = 2.048R + 0.318Y$$

Stress in rotating ring [Eq. (27.28)]

$$S = \frac{\gamma v^2}{g}$$

Deflection of simply supported arch (for G_1, see Fig. 27.11) [Eq. (27.31)]

$$Y = \frac{PR^3}{EI}G_1$$

Thrust in pin-jointed arch (for G_2, see Fig. 27.13) [Eq. (27.36)]

$$H_p = PG_2$$

Central deflection in a pin-jointed arch [Eq. (27.41)]

$$Y = \frac{0.0189PR^3}{EI}$$

Thrust in a semicircular built-in arch [Eq. (27.58)]

$$H_b = \frac{P(4 - \pi)}{\pi^2 - 8}$$

Fixed moment in a semicircular built-in arch [Eq. (27.59)]

$$M_f = \frac{PR(\pi^2 - 2\pi - 4)}{2(\pi^2 - 8)}$$

Selection of Practical Stress Formulas

Deflection of a semicircular built-in arch [Eq. (27.61)]

$$Y = \frac{PR^3}{EI}\left[\frac{\pi^3 - 20\pi + 32}{8(\pi^2 - 8)}\right]$$

Deflection of a pin-jointed arch with uniform loading [Eq. (27.64)]

$$Y = \frac{0.0135qR^4}{EI}$$

Winkler–Bach formula for curved beams [Eq. (28.1)]

$$S = \frac{M}{AR}\left[1 + \frac{c}{\lambda(R + c)}\right]$$

General stress formula for curved beams [Eq. (28.7)]

$$S = \phi_0\left(\frac{P}{A} + \frac{Mc}{I}\right)$$

Tensile stress in hook [Eq. (28.10)]

$$S_t = \frac{Pc\cos\theta}{A\lambda(R - c)}$$

Depth of circular section for hook design [Eq. (28.15)]

$$R_o - R_i = 0.023(P)^{1/2} + 0.18R_i$$

Depth of trapezoidal or bull-head section for hook design [Eq. (28.16)]

$$R_o - R_i = 0.026(P)^{1/2} + 0.20R_i$$

Deflection of quarter-circle curved beam under horizontal load [Eq. (28.21)]

$$X = \frac{\pi HR(4R + 27\delta)}{16AE\delta}$$

Deflection of quarter-circle curved beam under vertical load [Eq. (28.27)]

$$Y = \frac{PR}{4A}\left[\frac{R(3\pi - 8) + \delta(8 - \pi)}{\delta E} + \frac{\pi\xi}{G}\right]$$

Fixing moment in chain link [Eq. (29.10)]

$$M_0 = \frac{WR(2R + L)}{2(\pi R + L)}$$

Fixing moment in studded link [for C_1, see Eq. (29.14)] [Eq. (29.12)]

$$M_0 = \frac{WRC_1}{2}$$

Force in link stud [for C_2, see Eq. (29.15)] [Eq. (29.13)]

$$H = \frac{WC_2}{2}$$

Stress in eyebar with zero clearance [Eq. (29.23)]

$$S = \frac{W\phi}{BR}$$

Clearance-corrected eyebar stress [Eq. (29.36)]

$$S_{tmax} = \frac{W\psi}{\lambda r B}$$

Factor of safety for eyebar section in shear [Eq. (29.39)]

$$F_{ss} = \frac{S_y}{\sqrt{3}S_{max}}$$

Factor of safety for eyebar section in tension [Eq. (29.40)]

$$F_{st} = \frac{S_y}{S_{tmax}}$$

Shear stress in eyebar under inclined load [for G, see Eq. (29.43)] [Eq. (29.42)]

$$S_s = \frac{W \cos \alpha}{2A_c} \left[(1 + G \tan \alpha)^2 + 4 \tan^2 \alpha\right]^{1/2}$$

Torsional strength of bar [Eq. (30.1)]

$$\tau = \frac{T}{K_s}$$

Deflection of ring under diametral twist [Eq. (30.15)]

$$Y = \frac{T_0 R^2}{8EI}(1 + \lambda)(2\theta \cos \theta - \pi \cos \theta + \pi - 2 \sin \theta)$$

Slope of ring in diametral twist [Eq. (30.16)]

$$\psi = \frac{T_0 R}{8EI}(1 + \lambda)(\pi - 2\theta) \sin \theta$$

Selection of Practical Stress Formulas

Angle of twist of ring [Eq. (30.17)]

$$\eta = \frac{T_0 R}{8EI}[2(1-\lambda)\sin\theta + (1+\lambda)(\pi-2\theta)\cos\theta]$$

Plate equations [Eq. (31.6)]

$$D\left(\frac{\partial^2 w}{\partial x^2} + \nu\frac{\partial^2 w}{\partial y^2}\right) = -M_1$$

[Eq. (31.7)]

$$D\left(\frac{\partial^2 w}{\partial y^2} + \nu\frac{\partial^2 w}{\partial x^2}\right) = -M_2$$

Radius of spherical shape [Eq. (31.8)]

$$\zeta = \frac{D(1+\nu)}{M}$$

Thermal stress in plate [Eq. (31.11)]

$$S = \frac{\alpha E \Delta T}{2(1-\nu)}$$

Rectangular plate equation [Eq. (31.12)]

$$\frac{\partial^4 w}{\partial x^4} + 2\frac{\partial^4 w}{\partial x^2 \partial y^2} + \frac{\partial^4 w}{\partial y^4} = \frac{q(x,y)}{D}$$

General displacement equation for circular plate [Eq. (31.23)]

$$w = \frac{qx^4}{64D} + \frac{wx^2}{8\pi D}(\log_e x - 1) - \frac{C_1 x^2}{4} - C_2 \log_e x + C_3$$

Deflection of circular diaphragm [Eq. (31.29)]

$$\frac{\delta}{t} + 0.583\left(\frac{\delta}{t}\right)^3 = 0.176\frac{q}{E}\left(\frac{R_o}{t}\right)^4$$

Large deflection of plate [Eq. (31.30)]

$$\delta = 0.662 R_o \left(\frac{qR_o}{Et}\right)^{1/3}$$

Tensile membrane stress at plate center [Eq. (31.31)]

$$S_t = 0.423\left(\frac{Eq^2 R_o^2}{t^2}\right)^{1/3}$$

Tensile membrane stress at built-in edge of plate [Eq. (31.32)]

$$S_t = 0.328 \left(\frac{Eq^2 R_o^2}{t^2}\right)^{1/3}$$

Deflection of flat plate with central opening [Eq. (32.19)]

$$w = \frac{M_o R_o^2}{2D \left[R_o^2(1+\nu) + R_i^2(1-\nu)\right]} \left[R_i^2 - x^2 + 2R_i^2(\log_e x - \log_e R_i)\right]$$

Bending stress in pipe or cylinder [Eq. (33.4)]

$$S_b = \frac{6M_o}{T^2}$$

Parameter for beams on elastic foundation [Eq. (33.7)]

$$\beta_s = \frac{1.285}{(R_i T)^{1/2}}$$

Hub stress in flanges of simple geometry [Eq. (33.10)]

$$\frac{S_b T^2}{W} = \frac{0.614(k+1)(k^2-1)(m)^{1/2}}{(0.77+1.43k^2)[2.57(m)^{1/2}+1.65n] + n^3(k^2-1)}$$

Bending stress at flange junction in German practice [Eq. (33.12)]

$$S_b = \frac{W(k-1)[0.228 + 0.035n \sin^2(\pi/n)]}{nT^2}$$

Waters–Taylor formula for flanges [Eq. (33.13)]

$$S_b = \frac{W}{T^2} \left\{ \frac{0.25(m)^{1/2}[k^2(1+8.55\log k)-1]}{(1.05+1.94k^2)[(m)^{1/2}+0.64n] + 0.53n^3(k^2-1)} \right\}$$

Flange ring stress ratio [Eq. (33.14)]

$$\frac{S_F}{S_b} = \frac{n(2m+1)}{3.64mn + 4m(2m+1)^{1/2}}$$

Plastic stress [Eq. (33.16)]

$$S_p = S\frac{E_0}{E} + S_y\left(1 - \frac{E_0}{E}\right)$$

Equivalent depth of ribbed flange [Eq. (33.19)]

$$H_e = 2.29 \left(\frac{I_x}{B_r + T}\right)^{1/3}$$

Selection of Practical Stress Formulas

Elastic foundation parameter in ribbed flange analysis [Eq. (33.22)]

$$\phi(\beta, L) = \frac{\cosh \beta L \sinh \beta L - \cos \beta L \sin \beta L}{2\beta(\cosh^2 \beta L - \cos^2 \beta L)}$$

Load-sharing formula for ribbed flanges [Eq. (33.24)]

$$f = \frac{0.16 N R T_r B_r^3}{(B_r + T)h^3 \phi(\beta, L)}$$

Rib bending stress [Eq. (33.26)]

$$S_{bR} = \frac{6fW(R-r)}{NT_r(1+f)(B_r+T)^2}$$

Total stress in flange rib [Eq. (33.28)]

$$S_{TR} = \frac{6fW(R-r)}{NT_r(1+f)(B_r+T)^2} + \frac{W}{NB_rT_r + 2\pi rT}$$

Torsional stress in flange ring [Eq. (33.29)]

$$S_t = \frac{0.96W(R-r)R}{h^2 r^2 (1+f) \log_e k}$$

Bending stress in flange which supports ribs [Eq. (33.32)]

$$S_b = \frac{6VWB_r}{Na_0 h^2}$$

General formula for normal stress in fillet weld [Eq. (33.43)]

$$S_n = \frac{P \cos \theta (\sin \theta + \cos \theta)}{2Bh}$$

Bending stress in angle bracket [Eq. (33.44)]

$$S_b = \frac{6W(d-T)}{BT^2}$$

Shear stress in angle bracket [Eq. (33.45)]

$$S_s = \frac{W}{BT}$$

Tensile stress in heavy plate bracket [Eq. (33.52)]

$$S_t = \frac{W \sin \phi}{BT \cos \phi}$$

Compressive stress in heavy plate bracket [Eq. (33.53)]

$$S_c = \frac{W}{BT \cos \phi}$$

Critical buckling stress in bracket plate [Eq. (33.56)]

$$S_{Cr} = \frac{3.62ET^2}{H^2}$$

Bending stress in tapered bracket plate [Eq. (33.58)]

$$S_b = \frac{6WL^2(x-e)}{T[aL + x(H-a)]^2}$$

Edge buckling stress [Eq. (33.61)]

$$S_{Cr} = K_b E \left(\frac{T}{H}\right)^2$$

Plastic buckling of bracket edge [Eq. (33.62)]

$$S_{Cr} = K_b E \eta \left(\frac{T}{H}\right)^2$$

Allowable load on triangular bracket [Eq. (33.66)]

$$W = S_{max}(0.60H - 0.21L)\frac{TL}{H}$$

Plastic load on triangular bracket [Eq. (33.71)]

$$W_{pl} = TS_y \cos^2 \alpha \left[(L^2 + 4e^2)^{1/2} - 2e\right]$$

Ligament efficiency [Eq. (34.1)]

$$\frac{P_0^2 - A}{P_0^2}$$

Rigidity of plate with orthogonal ribs [Eq. (34.2)]

$$D = \frac{Et^3}{12(1-\nu^2)} + \frac{EI}{d}$$

Empirical formula for deflection of cover plate with radial stiffeners [Eq. (34.4)]

$$Y = \frac{21.6qa^{10}\lambda^3}{EW^3}$$

Selection of Practical Stress Formulas

Plate bearing stress ratio on pin [Eq. (34.5)]

$$K_b = \frac{2rtS}{F}$$

Standard formula for stress in Belleville washer [Eq. (34.6)]

$$S_c = \frac{YE}{2(1-\nu^2)a^2 C_1}[C_2(2h-Y)+2tC_3]$$

Load on Belleville washer [Eq. (34.7)]

$$Q = \frac{YE}{(1-\nu^2)a^2 C_1}\left[(h-Y)\left(h-\frac{Y}{2}\right)t+t^3\right]$$

Stress for flattened Belleville washer [Eq. (34.13)]

$$S_{hc} = \frac{CEh(h+2t)}{a^2}$$

Load for flattened Belleville washer [Eq. (34.14)]

$$Q_h = \frac{GEht^3}{a^2}$$

Parametric ratio for load-deflection characteristics [Eq. (34.15)]

$$\frac{Q}{Q_h} = \frac{km^2}{2}(1-k)(2-k)+k$$

Stress ratio for Belleville washer [Eq. (34.16)]

$$\frac{S_{hc}}{S_c} = \frac{m+2}{k(2+2m-km)}$$

Tangential stress in cylinder Eq. (35.3)]

$$S_t = mP$$

Meridional stress in cylinder [Eq. (35.5)]

$$S_l = \frac{mP}{2}$$

Radial growth of cylinder [Eq. (35.6)]

$$\delta = \frac{(2-\nu)PR^2}{2ET}$$

Radial growth of sphere [Eq. (35.7)]

$$\delta = \frac{(1-\nu)PR^2}{2ET_s}$$

Tangential stress in conical shell [Eq. (35.9)]

$$S_t = \frac{mP}{\cos\alpha}$$

Meridional stress in conical shell [Eq. (35.10)]

$$S_l = \frac{mP}{2\cos\alpha}$$

Meridional stress in ellipsoid [Eq. (35.11)]

$$S_l = \frac{Pa}{2T}$$

Tangential stress in ellipsoid [Eq. (35.12)]

$$S_t = \frac{Pa(2b^2 - a^2)}{2Tb^2}$$

Radial growth of conical shell [Eq. (35.14)]

$$\delta = \frac{(2-\nu)PR^2}{2ET\cos\alpha}$$

Equatorial dilation of ellipsoid [Eq. (35.15)]

$$\delta = \frac{PR^2}{ET}\left(1 - \frac{a^2}{2b^2} - \frac{\nu}{2}\right)$$

Hoop stress in toroidal vessel [Eq. (35.17)]

$$S_t = \frac{PR}{2T}\left(\frac{2\zeta + R\cos\theta}{\zeta + R\cos\theta}\right)$$

Laplace equation [Eq. (35.20)]

$$\frac{S_l}{R_1} + \frac{S_t}{R_2} = \frac{P}{T}$$

Radial stress in thick cylinder [Eq. (35.22)]

$$S_r = \frac{R_i^2 P_i - R_o^2 P_o}{R_o^2 - R_i^2} - \frac{R_i^2 R_o^2 (P_i - P_o)}{r^2(R_o^2 - R_i^2)}$$

Selection of Practical Stress Formulas

Tangential stress in thick cylinder [Eq. (35.23)]

$$S_t = \frac{R_i^2 P_i - R_o^2 P_o}{R_o^2 - R_i^2} + \frac{R_i^2 R_o^2 (P_i - P_o)}{r^2 (R_o^2 - R_i^2)}$$

Radial displacement in thick cylinder [Eq. (35.25)]

$$u = \frac{1-\nu}{E} \frac{R_i^2 P_i - R_o^2 P_o}{R_o^2 - R_i^2} r + \frac{1+\nu}{E} \frac{R_i^2 R_o^2 (P_i - P_o)}{(R_o^2 - R_i^2) r}$$

Axial stress in rigid closed cylinder [Eq. (35.30)]

$$S_l = \frac{2\nu P_i R_i^2}{R_o^2 - R_i^2}$$

Radial stress in sphere [Eq. (35.35)]

$$S_r = \frac{P_i R_i^3 - P_o R_o^3}{R_o^3 - R_i^3} - \frac{R_i^3 R_o^3 (P_i - P_o)}{r^3 (R_o^3 - R_i^3)}$$

Tangential stress in sphere [Eq. (35.36)]

$$S_t = \frac{P_i R_i^3 - P_o R_o^3}{R_o^3 - R_i^3} + \frac{R_i^3 R_o^3 (P_i - P_o)}{2 r^3 (R_o^3 - R_i^3)}$$

Radial displacement in sphere [Eq. (35.37)]

$$u = \frac{(1-2\nu) r (P_i R_i^3 - P_o R_o^3)}{E (R_o^3 - R_i^3)} + \frac{(1+\nu) R_i^3 R_o^3 (P_i - P_o)}{2 E r^2 (R_o^3 - R_i^3)}$$

Burst pressure of thick cylinder [Eq. (35.44)]

$$P_c = S_y \psi B_1$$

Burst pressure of thick sphere [Eq. (35.45)]

$$P_s = S_y \psi B_2$$

Burst pressure of thin cylinder [Eq. (35.46)]

$$P_c = \frac{S_y}{m} B_3$$

Burst pressure of thin sphere [Eq. (35.47)]

$$P_s = \frac{S_y}{m} B_4$$

Shrink-fit formula [Eq. (35.48)]

$$P = \frac{E\delta}{2R_i^3} \frac{(R_i^2 - R_s^2)(R_o^2 - R_i^2)}{(R_o^2 - R_s^2)}$$

Shrink fit of cylinders with different properties [Eq. (35.50)]

$$P = \frac{\delta}{\frac{R_i}{E_o}\left(\frac{R_i^2+R_o^2}{R_o^2-R_i^2}+\nu_o\right) + \frac{R_i}{E_i}\left(\frac{R_s^2+R_i^2}{R_i^2-R_s^2}-\nu_i\right)}$$

Thinness factor [Eq. (36.1)]

$$\lambda = 1.2(m)^{1/4}(k\phi)^{-1/2}$$

Pressure to yield at midbay [Eq. (36.2)]

$$P_y = \frac{0.9 S_y}{m}$$

Collapse pressure of long elastic cylinder [Eq. (36.3)]

$$P_c = \frac{0.275 E}{m^3}$$

Collapse pressure of long ductile cylinder [Eq. (36.4)]

$$P_c = \frac{\phi S_y}{m(\phi + 3.64 m^2)}$$

Classical formula for short cylinder buckling pressure [Eq. (36.5)]

$$P_c = \frac{0.887 E k}{m^{3/2}}$$

Collapse pressure of imperfect short cylinder [Eq. (36.6)]

$$P_u = S_y F(k) F(m)$$

Definition of out-of-roundness parameter [Eq. (36.13)]

$$n = \frac{D_{max} - D_{min}}{4T} = \frac{e}{T}$$

Collapse pressure of long, out-of-round cylinder [Eq. (36.14)]

$$P_{ul} = S_y A_m A_n$$

Selection of Practical Stress Formulas

Effective out-of-roundness parameter [Eq. (36.19)]

$$n_e = n \sin^2(36\lambda)$$

Empirical formula for collapse pressure of cylinders [Eq. (36.20)]

$$P_{CR} = S_y Z_1 (Z_2 - n Z_3)$$

Tensile correction factor for collapse of cylinders [Eq. (36.30)]

$$F_a = (1 - 0.91 a^2)^{1/2} - 0.3a$$

Theoretical buckling pressure for spheres (elastic) [Eq. (36.31)]

$$P_{CR} = \frac{1.21 E}{m^2}$$

Buckling pressure for imperfect spheres (elastic) [Eq. (36.32)]

$$P_{CR} = \frac{0.37 E}{m^2}$$

Plastic buckling of spheres [Eq. (36.33)]

$$P_{CR} = \frac{0.84 (E_s E_t)^{1/2}}{m^2}$$

Shallow cap parameter [Eq. (36.39)]

$$\lambda_0 = \frac{1.82 a_0}{T(m)^{1/2}}$$

Approximate formula for buckling of spherical caps [Eq. (36.42)]

$$P_{CR} = 0.075 E n_0^{-4} \lambda_0^{4.15} e^{-0.095 \lambda_0}$$

Strength of thick sphere under external pressure [Eq. (36.47)]

$$S = \frac{3 P_0 R_o^3}{2(R_o^3 - R_i^3)}$$

Compression of solid sphere [Eq. (36.50)]

$$u_o = \frac{P_0 R_o (1 - 2\nu)}{E}$$

Stress in out-of-round cylinder [Eq. (36.52)]

$$S = Pm(1 + 6n)$$

Moment of inertia for thin tube [Eq. (37.2)]

$$I = \pi R^3 T$$

Section modulus for thin tube [Eq. (37.3)]

$$Z = \pi R^2 T$$

Buckling column load on pipe [Eq. (37.6)]

$$P = \frac{\pi^3 E R^3 T (1 + 2\alpha)^2}{L^2}$$

Buckling of pipe on elastic foundation [Eq. (37.9)]

$$P_{CR} = \frac{\pi^3 E R^3 T}{L_e^2}$$

Buckling mode of pipe on elastic foundation [Eq. (37.13)]

$$\alpha = \frac{L - L_e}{2 L_e}$$

Critical pipe load in one-way buckling [Eq. (37.15)]

$$P_{CR} = 4.49 \left(\frac{QEI}{\delta} \right)^{1/2}$$

Axial pipe stress in one-way buckling [Eq. (37.16)]

$$S_{CR} = 1.27 \left(\frac{QER}{\delta T} \right)^{1/2}$$

Donnell's formula for axial stress in buckling of cylinders [Eq. (37.19)]

$$S_{CR} = E \frac{0.605 - 10^{-7} m^2}{m(1 + 0.004\phi)}$$

Number of half-wave buckles for cylinders in elastic compression [Eq. (37.20)]

$$n_e = \frac{0.58 L}{(RT)^{1/2}}$$

Number of half-waves in plastic buckling of cylinders [Eq. (37.21)]

$$n_p = \frac{0.58 L (E)^{1/4}}{(RT)^{1/2} (E_r)^{1/4}}$$

Selection of Practical Stress Formulas

Plastic stress in axial compression of cylinders [Eq. (37.22)]

$$S_p = \frac{0.605 E_r T}{R}$$

Reduced modulus [Eq. (37.23)]

$$E_r = \frac{4EE_t}{\left(E^{1/2} + E_t^{1/2}\right)^2}$$

Ultimate axial load on pipe in axial compression [Eq. (37.26)]

$$W = 3.80 E_r T^2$$

Total stress in pipe under eccentric compression [Eq. (37.30)]

$$S = \frac{W}{T^2}\left(\frac{0.16}{m} + 0.32\frac{n}{m^2}\right)$$

Load eccentricity to cause local buckling of pipe [Eq. (37.31)]

$$e = 1.1\frac{E}{W}mT^3 + \frac{mT}{2} - 1.57\frac{P_e}{W}m^3T^3$$

Bending moment on pipe to initiate local buckling [Eq. (37.33)]

$$M_b = 1.1 E m T^3$$

Buckling stress in pipe due to bending [Eq. (37.33a)]

$$S_{CR} = \frac{0.35 E}{m}$$

Actuating pressure for cylindrical diaphragms [Eq. (37.36)]

$$\Delta P = \frac{(k+m)(0.55k + 4m^2 + 2m)S_y}{mk^3}$$

Approximate actuating pressure for annular diaphragms [Eq. (37.39)]

$$\Delta P_a = \frac{[0.55(k_0 + n_0) + 8m^2 + 4m]S_y}{m(k_0^2 - n_0^2)}$$

Radial growth for closed-end tube under pressure [Eq. (38.1)]

$$\delta = \frac{PD(2-\nu)(1+k)^2}{16E(1-k)}$$

Attenuation ratio for nested cylinders [Eq. (38.4)]

$$\psi = \frac{2(1 - k_1^2)}{(1.3 + 0.7k_2^2)(1 - k_1^2) + (0.7 + 1.3k_1^2)(1 - k_2^2)}$$

Buckling pressure for elastic ring [Eq. (38.5)]

$$q_{CR} = \frac{3EI}{R^3}$$

Tightening torque [Eq. (39.2)]

$$M_t = \frac{W_i d}{5}$$

Actual bolt load [Eq. (39.3)]

$$W = W_i + \frac{W_e}{1 + \left(\frac{K_c}{K_B}\right)}$$

Practical strength formula for bolts [Eq. (39.12)]

$$S = \frac{4W}{d(2.2d - 1)}$$

Maximum bolt spacing for tight joint [Eq. (39.14)]

$$B = 2d + \frac{6h}{m + 0.5}$$

Boardman formula for number of bolts [Eq. (39.15)]

$$W = \frac{\pi P R_i^2 (R_o - R_i)}{N(R_o - R)}$$

Practical criterion for flange thickness [Eq. (39.16)]

$$h = 1.75 R_i \left[\frac{P(R - R_i)}{RS_r}\right]^{1/2}$$

Frictional torque in split hub [Eq. (40.3)]

$$M_{tf} = L f d^2 P_{max}$$

Stress due to external ring load on pipe [Eq. (40.6)]

$$S_b = \frac{1.17 P_o (r)^{1/2}}{t(t)^{1/2}}$$

Selection of Practical Stress Formulas

Axial tension on pipe branch [Eq. (40.8)]

$$W_u = \frac{S_y T^2 (3.4 + 19\beta)}{\sin \theta}$$

Out-of-plane load on pipe branch [Eq. (40.10)]

$$M_u = \frac{S_y T^2 D \beta (2.7 + 5.6\beta)}{\sin \theta}$$

Balanced socket design [Eq. (40.17)]

$$F = 2.4 d (d - 0.25 t) S_t$$

Torque criterion for key joint [Eq. (40.20)]

$$L = \frac{2.67 M_t}{h D S_t}$$

Crush limit for hub keyway [Eq. (40.27)]

$$M_t = f N_f (1.14 D + 0.5 h)$$

Stress ratio for structural pin [Eq. (40.29)]

$$\frac{S_{bmax}}{S_c} = 1.273 \frac{(k+2) m^2}{k}$$

Shear stress for pin [Eq. (40.30)]

$$S_{max} = 0.849 m S_c$$

Bending of U-Shape fitting [Eq. (40.33)]

$$S_b = \frac{2 W R (2R + L)}{\pi (\pi R + L) r^3}$$

Contact pressure in socket [Eq. (40.38)]

$$q = \frac{W}{\pi (R_o + R_i)(R_o - R_i + f H)}$$

Stress in helical compression spring [Eq. (41.1)]

$$\tau = \frac{2.55 P D}{d^3}$$

Deflection of helical spring [Eq. (41.5)]

$$Y = \frac{8PD^3N}{Gd^4}$$

Impact criterion for helical spring [Eq. (41.10)]

$$Y^2 = \frac{2P(Y+a)}{k}$$

Frequency of spring under load [Eq. (41.13)]

$$n = 187.6(Y)^{-1/2}$$

Bending stress in extension spring

$$S_b = \frac{5PD^2}{(D-d)d^3}$$

Bending stress in torsion spring [Eq. (41.16)]

$$S_b = \frac{10.2M}{d^3}$$

Moment on torsional spring [Eq. (41.18)]

$$M = \frac{0.093ETd^4}{ND}$$

Load on buckling column spring [Eq. (41.22)]

$$P = \frac{h^2 AEF_1}{L^2}$$

Bending stress in buckling column spring [Eq. (41.23)]

$$S_b = \frac{hEF_2}{L}$$

References

1. Case, J., and A. H. Chilver, *Strength of Materials*, Edward Arnold, London, 1959.
2. Timoshenko, S., and J. N. Goodier, *Theory of Elasticity*, McGraw-Hill, New York, 1951.
3. Popov, E. P., *Mechanics of Materials*, Prentice-Hall, Englewood Cliffs, N. J., 1957.
4. Nash, W. A., *Strength of Materials*, Schaum, New York, 1957.
5. Timoshenko, S., *Strength of Materials*, D. Van Nostrand, New York, 1956.
6. Durelli, A. J., E. A. Phillips, and C. H. Tsao, *Introduction to the Theoretical and Experimental Analysis of Stress and Strain*, McGraw-Hill, New York, 1958.
7. Boyd, J. E., and S. B. Folk, *Strength of Materials*, McGraw-Hill, New York, 1950.
8. Castigliano, A., Nuova teoria intorno dell'equilibrio dei sistemi elastici, *Atti Accad. sci., Torino*, 1875.
9. Andrews, E. S., *Elastic Stresses in Structures*, Scott, Greenwood, London, 1919.
10. Sokolnikoff, I. S., and E. S. Sokolnikoff, *Higher Mathematics for Engineers and Physicists*, McGraw-Hill, New York, 1941.
11. Blake, A., *Design of Curved Members for Machines*, Robert E. Krieger, Huntington, N. Y., 1979.
12. Timoshenko, S. P., and D. H. Young, *Theory of Structures*, McGraw-Hill, New York, 1965.
13. Shermer, C. L., *Fundamentals of Statically Indeterminate Structures*, Ronald Press, New York, 1957.

14. Neuber, H., *Theory of Notch Stresses*, J. W. Edwards, Ann Arbor, Mich., 1946.
15. Hetenyi, M., *Handbook of Experimental Stress Analysis*, John Wiley, New York, 1950.
16. Peterson, R. E., *Stress Concentration Design Factors*, John Wiley, New York, 1953.
17. Battelle Memorial Institute, *Prevention of Fatigue of Metals*, John Wiley, New York, 1941.
18. Heywood, R. B., *Designing Against Fatigue of Metals*, Van Nostrand Reinhold, New York, 1962.
19. Forrest, P. G., *Fatigue of Metals*, Pergamon Press, Elmsford, N. Y./Addison-Wesley, Reading, Mass., 1962.
20. Sines, George, and J. L. Waismau (Eds.), *Metal Fatigue*, McGraw-Hill, New York, 1959.
21. Faupel, J. H., *Engineering Design*, John Wiley, New York, 1964.
22. Coker, E. G., and L. N. G. Filon, *Treatise on Photoelasticity*, Cambridge Univ. Press, London, 1931.
23. Jeffery, J. B., Plane Stress and Plane Strain in Bipolar Coordinates, *Philos. Proc. R. Soc. Lond.*, Ser. A, 221: 265, 1921.
24. Dumont, C., *Stress Concentration Around an Open Circular Hole in a Plate Subjected to Bending Normal to the Plane of the Plate*, NACA, Tech. Note 740, 1939.
25. Frocht, M., Factors of Stress Concentration Photoelastically Determined, *J. Appl. Mech.*, 2 (2), 1935.
26. Peterson, R. E., and A. M. Wahl, Two and Three-Dimensional Cases of Stress Concentration, and Comparison with Fatigue Tests, *J. Appl. Mech.*, 3 (1): A-15, 1936.
27. Sass, F., Ch. Bouche, und A. Leitner, *Dubbels Taschenbuch für den Maschinenbau*, Springer-Verlag, Berlin, 1966.
28. Gaylord, E. H., and C. N. Gaylord, *Design of Steel Structures*, McGraw-Hill, New York, 1957.
29. Shanley, F. R., *Strength of Materials*, McGraw-Hill, New York, 1957.
30. Hardrath, H. F., and L. Ohman, *A Study of Elastic and Plastic Stress Concentration Factors Due to Notches and Fillets in Flat Plates*, NACA Tech. Note 2566, 1951.
31. Timoshenko, S., *Theory of Elastic Stability*, Engineering Societies Monograph, McGraw-Hill, New York, 1936.
32. Bleich, F., *Buckling Strength of Metal Structures*, McGraw-Hill, New York, 1952.
33. Gerard, G., and H. Becker, *Handbook of Structural Stability*, NACA Tech. Notes 3781-3786 and D163, 1957–1959.
34. Royal Aeronautical Society, *Data Sheets on Structures*, London, 1963.
35. Roark, R. J., *Formulas for Stress and Strain*, McGraw-Hill, New York, 1965.
36. Final Report, *Trans. ASCE*, 98, 1933.
37. Priest, H. M., and J. A. Gilligan, *Design Manual for High-Strength Steels*, U. S. Steel Corp., Pittsburgh, Pa., 1956.
38. Flügge, W., *Handbook of Engineering Mechanics*, McGraw-Hill, New York, 1962.
39. Shanley, F. R., *Weight-Strength Analysis of Aircraft Structures*, Dover, New York, 1960.

References

40. Tall, L., L. S. Beedle, and T. V. Galambos, *Structural Steel Design*, Ronald Press, New York, 1964.
41. deVries, K., Strength of Beams as Determined by Lateral Buckling, *Trans. ASCE*, 112, 1947.
42. American Society of Testing and Materials, *Symposium on Speed of Testing of Non-metallic Materials*, ASTM Spec. Tech. Publ. No. 185, Philadelphia, Pa., 1956.
43. Rinehart, J. S., and J. Pearson, *Behavior of Metals Under Impulsive Loads*, American Society for Metals, Cincinnati, Ohio, 1949.
44. Skewmon, P. G., and V. F. Zackay, *Response of Metals to High Velocity Deformation* (Vol. 9, Metallurgical Society Conference), Interscience, New York, 1961.
45. Biggs, J. M., *Introduction to Structural Dynamics*, McGraw-Hill, 1964.
46. Crede, C. E., *Vibration and Shock Isolation*, John Wiley, New York, 1957.
47. Wiegel, R. L., *Earthquake Engineering*, Prentice-Hall, Englewood Cliffs, N. J., 1970.
48. Newmark, N. M., and E. Rosenblueth, *Fundamentals of Earthquake Engineering*, Prentice-Hall, Englewood Cliffs, N. J., 1971.
49. Biot, M. A., Analytical and Experimental Methods in Engineering Seismology, *Trans. ASCE*, 108, 1943.
50. Anderson, A. W., et al., Lateral Forces of Earthquake and Wind, *Trans. ASCE*, 117, 1952.
51. Blume, J. A., N. M. Newmark, and L. H. Corning, *Design of Multistory Reinforced Concrete Buildings for Earthquake Motions*, Portland Cement Assn., Chicago, 1961.
52. Housner, G. W., Intensity of Earthquake Ground Shaking near the Causative Fault, *Proc. Third World Conf. Earthquake Eng.*, 1, 1965.
53. Orowan, E., Theory of the Fatigue of Metals, *Proc. Lond. R. Soc., Ser. A*, 171, 1939.
54. Freudenthal, A. M., and T. J. Dolan, *The Character of the Fatigue of Metals*, 4th Prog. Rep., ONR, Task Order IV, 1948.
55. Machlin, E. S., *Dislocation Theory of the Fatigue of Metals*, NACA Tech. Note 1489, 1948.
56. Coffin, L. F., The Resistance of Material to Cyclic Thermal Strains, *ASME Trans.*, Paper 57-A-286, 1957.
57. Coffin, L. F., The Stability of Metals Under Cyclic Plastic Strain, *ASME Trans.*, Paper 59-A-100, 1959.
58. Grover, H. J., S. A. Gordon, and L. R. Jackson, *Fatigue of Metals and Structures*, U. S. Government Printing Office, Washington, D. C., 1954.
59. Miner, M. A., Cumulative Damage in Fatigue, *J. Appl. Mech.*, 12, 1945.
60. Manson, S. S., Cumulative Fatigue Damage, *Mach. Des.*, 1960.
61. Metcut Research Associates, Inc., *Machining Data Handbook*, Cincinnati, Ohio, 1972.
62. Horger, O. J., and H. R. Neifert, Effect of Surface Condition on Fatigue Properties, *Surface Treat. Met. ASM Symp.*, 1941.
63. Low, A. C., Short Endurance Fatigue, *Int. Conf. Fatigue Met.*, New York, 1956.
64. Noll, G. C., and C. Lipson, Allowable Working Stresses, *SESA Proc.*, 3 (11), 1946.

65. Field, M., and J. F. Kahless, *The Surface Integrity of Machined-and-Ground High—Strength Steels*, DMIC Rep. 210, Defense Metals Information Center, Battelle Memorial Institute, Columbus, Ohio, 1964.
66. Moore, H. F., *Shot Peening and the Fatigue of Metals*, American Foundry Equipment Co., Tulsa, Okla., 1944.
67. Tapsell, H. J., Fatigue at High Temperatures, *Symposium on High Temperature Steels and Alloys for Gas Turbines*, Iron and Steel Institute, London, 1950.
68. Travernelli, J. F., and L. F. Coffin, *Experimental Support for Generalized Equation Predicting Low Cycle Fatigue*, Appl. Ser. M-3, Instron Engineering Corp., Canton, Mass., 1959.
69. Shannon, J. L., Fracture Mechanics—The Search for Safety in Numbers, Part 1, *Mach. Des.*, Sept. 1967.
70. Shannon, J. L., Fracture Mechanics—Reducing Theory to Practice, Part 2, *Mach. Des.*, Oct. 1967.
71. Shannon, J. L., and W. F. Brown, Progress in Fracture Mechanics, *Mach. Des.*, Mar. 1970.
72. Inglis, C. E., Stresses in a Plate Due to the Presence of Cracks and Sharp Corners, *Proc. Inst. Nav. Archit.*, 60, 1913.
73. Griffith, A. A., The Phenomena of Rupture and Flow in Solids, *Trans. R. Soc. Lond.*, 221, 1920.
74. Westergaard, H. M., Bearing Pressures and Cracks, *J. Appl. Mech.*, 6, 1939.
75. Sneddon, I. N., The Distribution of Stress in the Neighborhood of a Crack in an Elastic Solid, *Proc. R. Soc. Lond., Ser. A*, 187, 1946.
76. Irwin, G. R., A Critical Energy Rate Analysis of Fracture Strength, *Weld. J. (Res. Suppl.)*, 1954.
77. Irwin, G. R., Analysis of Stresses and Strains near the End of a Crack Traversing a Plate, *J. Appl. Mech.*, 24 (3), 1957.
78. Irwin, G. R., The Crack Extension Force for a Part Through Crack in a Plate, *J. Appl. Mech.*, 1962.
79. McClintock, F. A., and G. R. Irwin, *Plasticity Aspects of Fracture Mechanics*, ASTM Spec. Tech. Pub. No. 381, Philadelphia, Pa., 1965.
80. Paris, P. C., and G. C. M. Sih, *Stress Analysis of Cracks*, ASTM Spec. Tech. Pub. No. 381, Philadelphia, Pa., 1965.
81. Srawley, J. E., and B. Gross, Stress Intensity Factors for Crackline-Loaded Edge-Crack Specimens, *Mater. Res. Stand.*, 7(4), 1967.
82. Wilson, W. K., and D. G. Thompson, On the Finite Element Method for Calculating Stress Intensity Factors for Cracked Plates in Bending, *Engineering Fracture Mechanics*, Vol. 3, Pergamon Press, Elmsford, N. Y., 1971.
83. Steigerwald, E., What You Should Know About Fracture Toughness, *Met. Prog.*, Nov. 1967.
84. Sippel, G. R., Processing Affects Fracture Toughness, *Met. Prog.*, Nov. 1967.
85. American Society for Testing and Materials, *Plane Strain Crack Toughness Testing of High Strength Metallic Materials*, ASTM 410, Philadelphia, Pa., Dec. 1967.
86. Pellini, W. S., and P. P. Puzak, *Fracture Analysis Diagram Procedures for the Fracture-Safe Engineering Design of Steel Structures*, NRL Rep. 5920, U. S. Naval Research Lab., 1963.

References

87. Pellini, W. S., *Evolution of Engineering Principles for Fracture-Safe Design of Steel Structures*, NRL Rep. 6957, U. S. Naval Research Lab., 1969.
88. Pellini, W. S., *Integration of Analytical Procedures for Fracture-Safe Design of Metal Structures*, NRL Rep. 7251, U. S. Naval Res. Lab., 1971.
89. Pellini, W. S. Principles of Fracture-Safe Design, *Weld. J. (Suppl.)*, 1971.
90. Wells, A. A., Brittle Fracture Strength of Welded Steel Plates, *Br. Weld. J.*, 1961.
91. Kolsky, H., *Stress Waves in Solids*, Dover, New York, 1963.
92. Von Karman, T., *On the Propagation of Plastic Deformation in Solids*, Rep. No. A-29, National Defense Research Council, 1943.
93. Hoppmann, W. H., The Velocity Aspect of Tension—Impact Testing, *Proc. Am. Soc. Test. Mater.*, 1947.
94. Kornhauser, M., *Structural Effects of Impact*, Spartan Books, New York, 1964.
95. Jacobsen, L. S., and R. S. Ayre, *Engineering Vibrations*, McGraw-Hill, New York, 1958.
96. Merritt, J. L., and N. M. Newmark, *Nuclear Geoplosics*, DASA-1285, 1964.
97. Wheeler, V. E., and R. G. Preston, *Scaled Free-Field Particle Motions from Underground Nuclear Explosions*, Res. Lab. Rep. 50563, University of California, 1968.
98. Rodean, H. C., *Nuclear-Explosion Seismology*, U. S. Atomic Energy Commission, Oak Ridge, Tenn., 1972.
99. Freudenthal, A. M., *Introduction to the Mechanics of Solids*, John Wiley, London, 1966.
100. Corrucini, R. J., Chem. Eng. Prog., 53: 262, 342, 397, 1957.
101. Scott, R. B., *Cryogenic Engineering*, D. Van Nostrand, New York, 1959.
102. Barron, R. F., Low-Temperature Properties of Engineering Materials, *Mach. Des.*, 1960.
103. Bussard, R. W., The Engine Internal Environment, *Aerosp. Eng.*, 1963.
104. Ramke, N. G., and J. D. Latva, Refractory Ceramics and Intermetallic Compounds, *Aerosp. Eng.*, 1963.
105. Kelble, J. M., and J. E. Bernados, High Temperature Nonmetallic Materials, *Aerosp. Eng.*, 1963.
106. Manson, S. S., *Thermal Stress and Low-Cycle Fatigue*, McGraw-Hill, New York, 1966.
107. Gatewood, B. E., *Thermal Stresses*, McGraw-Hill, New York, 1957.
108. Goodier, J. N., Thermal Stress, *J. Appl. Mech.*, 1937.
109. Maulbetsch, J. L., Thermal Stresses in Plates, *J. Appl. Mech.*, 1935.
110. Kent, C. H., Thermal Stresses in Spheres and Cylinders Produced by Temperatures Varying with Time, *ASME Trans.*, 54(18), 1932.
111. Ayres, F., *Differential and Integral Calculus*, Schaum, New York, 1950.
112. Large, G. E., *Basic Reinforced Concrete Design: Elastic and Creep*, Ronald Press, New York, 1957.
113. Blake, A., *Blade Vibrations—A General Survey*, Lab. Rep. No. 581, British Electricity Research Laboratories, London, 1955.
114. Blake, A., *Design of Vibration Test Rig for Frequency Analysis of Turbine Blades and Other Machine Components*, Lab. Rep. No. 640, British Electricity Research Laboratories, London, 1956.
115. Blake, A., and L. Scherff, Cantilever Deflection-Factors, *Prod. Eng.*, 1960.

116. Hall, I. H., *Deformation of Solids*, Thomas Nelson, London, 1968.
117. Stephens, R. C., *Strength of Materials Theory and Examples*, Edward Arnold, London, 1970.
118. Marguerre, K., *Technische Mechanik*, Springer-Verlag, Berlin, 1967.
119. Ickovitch, G. M., *Resistance of Materials*, (in Russian), College Publishers, Moscow, 1968.
120. American Institute of Steel Construction, *Manual of Steel Construction*, New York, 1970.
121. Darkov, A., and V. Kuznetsov, *Structural Mechanics*, Peace Publishers, Moscow, 1970.
122. Tuma, J. J., *Theory and Problems of Structural Analysis*, McGraw-Hill, New York, 1969.
123. Brooks, W. H., *Strength and Elasticity of Materials*, MacDonald & Co., London, 1956.
124. Blake, A., and G. Kurasz, Beams with Partial Uniform Loading, *Mach. Des.*, 1960.
125. Saelman, B., and L. C. Coombs, Minimizing Maximum Beam Moments, *Mach. Des.*, 1959.
126. Saelman, B., Minimizing Maximum Beam Deflection, *Mach. Des.*, 1959.
127. Holl, D. L., *Analysis of Thin Rectangular Plates Supported on Opposite Edges*, Bull. 129, Iowa Eng. Exp. Sta., Iowa State College, 1936.
128. Westergaard, H. M., Computation of Stress Due to Wheel Loads, *Public Roads*, 11, 1930.
129. Holl, D. L., Cantilever Plate with Concentrated Edge Load, *J. Appl. Mech.*, 4(1), Paper A-8, 1937.
130. MacGregor, C. W., Deflection of Long Helical Gear Tooth, *Mech. Eng.*, 57, 1935.
131. Zimmermann, H., *Die Berechnung des Eisenbahnoberbaues*, Ernst U. Korn, Berlin, 1888.
132. Hetenyi, M., *Beams on Elastic Foundation*, University of Michigan Press, Ann Arbor, Mich., 1946.
133. Maxwell, J. C., On the Calculation of the Equilibrium and Stiffness of Frames, *Phil. Mag.*, 27, 1864.
134. Betti, E., *Nuovo Cimento*, Ser. 2, 7 and 8, 1872.
135. Blake, A., Arched Cantilever Beams, *Mach. Des.*, 1958.
136. Biezeno, C. B., and R. Grammel, *Technische Dynamik*, Springer-Verlag, Berlin, 1953.
137. Pippard, A. J. S., *Studies in Elastic Structures*, Arnold & Co., London, 1952.
138. Love, A. E. H., *A Treatise on the Mathematical Theory of Elasticity*, Cambridge University Press, Cambridge, 1927.
139. Seely, F. B., and S. O. Smith, *Advanced Mechanics of Materials*, John Wiley, New York, 1966.
140. Blake, A., Analysis of a Curved Cantilever, *Design News*, 1974.
141. Gascoyne, J., *Analysis of Pipe Structures for Flexibility*, John Wiley, New York, 1959.
142. Gaylord, E. H., and C. N. Gaylord, *Structural Engineering Handbook*, McGraw-Hill, New York, 1968.

References

143. Palm, J., and K. Thomas, Berechnung gekrümmter Biegefedern, *VDI Z.* 101(8), 1959.
144. Blake, A., Curved-End Cantilevers, *Mach. Des.*, 1959.
145. Wahl, A. M., *Mechanical Springs*, McGraw-Hill, New York, 1963.
146. Blake, A., *Flexibility and Strength Considerations for Engine Lines*, Tech. Memo., Aerojet General Corp., Sacramento, Calif., 1964.
147. Blake, A., Complex Flat Springs, *Prod. Eng.*, 1961.
148. Blake, A., Design of U-Beams, *Design News*, 1975.
149. Fleckenstein, J. E., *U-Springs–Stress and Deflection Calculations, ASME Trans.*, Paper 50-WA-72, 1960.
150. Griffel, W., *Handbook of Formulas for Stress and Strain*, Frederick Ungar, New York, 1966.
151. Meck, H. R., Three Dimensional Deformation and Buckling of a Circular Ring of Arbitrary Section, *J. Eng. Ind.*, 1969.
152. Blake, A., Rings and Arcuate Beams, *Prod. Eng.*, 1963.
153. Blake, A., Circular Arches, *Mach. Des.*, 1958.
154. Leontovich, V., *Frames and Arches*, McGraw-Hill, New York, 1959.
155. Winkler, E., *Die Lehre von der Elastizität und Festigkeit*, Prague, 1867.
156. Goodenough, G. A., and L. E. Moore, *Strength of Chain Links*, Bull. 18, Eng. Exp. Sta., University of Illinois, Urbana, Ill., 1907.
157. Morley, A., Bending Stresses in Hooks and Other Curved Beams, *Engineering (Lond.)*, 98, 1914.
158. Winslow, A. M., and R. H. G. Edmonds, Tests and Theory of Curved Beams, *ASME Trans.*, 1926.
159. Gough, H. J., H. L. Cox, and D. C. Sopwith, Design of Crane Hooks and Other Components of Lifting gear, *Proc. Inst. Mech. Eng. (Lond.)*, 1934.
160. Wilson, B. J., and J. F. Quereau, *A Simple Method of Determining Stress in Curved Flexural Members*, Circ. 16, Eng. Exp. Sta., University of Illinois, Urbana, Ill., 1927.
161. Blake, A., Stress in Curved Beams, *Mach. Des.*, 1974.
162. Lofgren, K. E., Calculating Deflection of Curved Beams, *Mach. Des.*, 1948.
163. Leeman, E. R., *Stresses in a Circular Ring*, Engineering Ltd., London, 1956.
164. Blake, A., Deflection of a Thick Ring in Diametral Compression by Test and by Strength of Materials Theory, *J. Appl. Mech.*, 1959.
165. Pippard, A. J. S., and C. V. Miller, The Stresses in Links and Their Alternation in Length Under Load, *Proc. Inst. Mech. Eng. (Lond.)*, 1923.
166. Timoshenko, S., On the Distribution of Stresses in a Circular Ring Compressed by Two Forces Acting Along a Diameter, *Philos. Mag.*, 44(263), 1922.
167. Hall, A. S., A. R. Holowenko, and H. G. Laughlin, *Theory and Problems of Machine Design*, McGraw-Hill, New York, 1961.
168. Blake, A., Stresses in Eye Bars, *Design News*, 1974.
169. American Petroleum Institute, *API Specification for Drilling and Production Hoisting Equipment*, API STD 8A, Dallas, Tex., 1965.
170. Korkut, M. D., Pin Joint Design, *Mach. Des.*, 1961.
171. Hogan, M. B., Circular Beams Loaded Normal to the Plane of Curvature, *J. Appl. Mech.*, 5, 1938.
172. Moorman, R. B. B., *Stresses in a Curved Beam Under Loads Normal to the Plane of Its Axis*, Bull. 145, Iowa Eng. Exp. Sta., Iowa State College, 1940.

173. Tabakman, H. D., and H. P. Valentijn, Distortion of Circular Rings, *Mach. Des.*, 1964.
174. McGuiness, H. D., *Solution of a Circular Ring Structural Problem*, Tech. Rep. 32-178, California Institute of Technology, Pasadena, Calif., 1961.
175. Sadin. S. R., E. E. Ungar, and B. W. Shaffer, *Out-of-Plane Bending of a Relatively Stiff, Elastically Supported Ring*, New York University, 1956.
176. Blake, A., Analysis of Buckling Column Spring with Pivoted Ends and Uniform Rectangular Cross-Section, *ASME Trans.*, Paper 60-SA-10, 1960.
177. Blake, A., Structural Pin Design, *Design News*, 1974.
178. Timoshenko, S., *Theory of Plates and Shells*, McGraw-Hill, New York, 1940.
179. Nadai, A., *Theory of Flow and Fracture of Solids*, McGraw-Hill, New York, 1963.
180. Jeffrey, A., *Mathematics for Engineers and Scientists*, Barnes & Noble, New York, 1969.
181. Filolenko-Borodich, M., *Theory of Elasticity*, Peace Publishers, Moscow, 1963.
182. Holmberg, E. O, and K. Axelson, Analysis of Stresses in Circular Plates and Rings, *ASME Trans.*, Paper APM-54-2, 1931.
183. Waters, E. O., and J. H. Taylor, The Strength of Pipe Flanges, *Mech. Eng.*, 49, 1927.
184. Wahl, A. M., and G. Lobo, Stresses and Deflections in Flat Circular Plates with Central Holes, *ASME Trans.*, 52, 1930.
185. Blake, A., Simplified Approach to Panel Design, *Mach. Des.*, 1974.
186. Presscott, J., *Applied Elasticity*, Longmans, Green, 1924.
187. Prager, W., and P. G. Hodge, *Theory of Perfectly Plastic Solids*, John Wiley, New York, 1951.
188. Hodge, P. G., *Plastic Analysis of Structures*, McGraw-Hill, New York, 1959.
189. Waters, E. O., D. B. Wesstrom, D. B. Rossheim, and F. S. G. Williams, Formulas for Stresses in Bolted Flanged Connections, *ASME Trans.*, 1937.
190. Bernhard, H. J., Flange Theory and the Revised Standard B. S. 10: 1962— Flanges and Bolting for Pipes, Valves and Fittings, *Proc. Inst. Mech. Eng. (Lond.)*, 178(5), pt. 1, 1963-64.
191. Lake, R. L., F. W. DeMoney, and R. J. Eiber, Burst Tests of Pre-Flawed Welded Aluminum Alloy Pressure Vessels at $-220°F$, *Advances in Cryogenic Eng.*, 13, Plenum Press, 1968.
192. Johns, R. H., and T. W. Orange, *Theoretical Elastic Stress Distributions Arising from Discontinuities and Edge Loads in Several Shell Type Structures*, NASA TRR-103, Lewis Research Center, Cleveland, Ohio, 1961.
193. Waters, E. O., and F. S. G. Williams, Stress Conditions in Flanged Joints for Low-Pressure Service, *ASME Trans.*, 1952.
194. Phillips, A., *Introduction to Plasticity*, Ronald Press, New York, 1956.
195. Blake, A., Flanges that Won't Fail, *Mach. Des.*, 1974.
196. Blake, A., Stress in Flanges and Support Rings, *Mach. Des.*, 1974.
197. Blake, A., Design of Welded Brackets, *Mach. Des.*, 1975.
198. American Society of Mechanical Engineers, *ASME Boiler and Pressure Vessel Code Section VIII*, New York, 1971.
199. Slot, T., and W. J. O'Donnell, Effective Elastic Constants for Thick Perforated Plates with Square and Triangular Penetration Patterns, *J. Eng. Ind.*, 1971.

200. Harvey, J. F., *Theory and Design of Modern Pressure Vessels*, Van Nostrand Reinhold, New York, 1974.
201. Nash, W. A., Effect of a Concentric Reinforcing Ring of Stiffness and Strength of a Circular Plate, *J. Appl. Mech.*, 14, Paper 47-A-15, 1947.
202. Harvey, J., and J. P. Duncan, the Rigidity of Rib-Reinforced Cover Plates, *Proc. Inst. Mech. Eng. (Lond.)*, 177(5), 1963.
203. Frocht, M. M., and H. N. Hill, Stress Concentration Factors Around a Central Circular Hole in a Plate Loaded Through Pin in the Hole, *J. Appl. Mech.*, 7, 1940.
204. Gurney, D. A., Tests on Belleville Springs by the Ordnance Department, U. S. Army, *ASME Trans.*, APM-51-2, 1929.
205. Almen, J. O., and A. Laszlo, The Uniform-Section Disk Spring, *ASME Trans.*, RP-58-10, 1936.
206. Ashworth, G., The Disk Spring or Belleville Washer, *Proc. Inst. Mech. Eng. (Lond.)*, 155, 1946.
207. Wempner, G. A., The Conical Disk Spring, *Proc. 3rd U. S. Natl. Cong. Appl. Mech.*, 1958.
208. Associated Spring Corporation, *Handbook of Mechanical Spring Design*, Bristol, Conn., 1964.
209. Fortini, E. T., Conical-Disk Springs, *Mach. Des.*, 1958.
210. Flügge, W., *Stresses in Shells*, Springer-Verlag, Berlin, 1960.
211. Gill, S. S., *The Stress Analysis of Pressure Vessels and Pressure Vessel Components*, Pergamon Press, Oxford, 1970.
212. American Society of Mechanical Engineers, *Pressure Vessel and Piping Design*, ASME Collect. Pap., New York, 1960.
213. Jürgensonn, H., *Elastizität und Festigkeit im Rohrleitungsbau*, Springer-Verlag, Berlin, 1953.
214. Schwaigerer, S., *Festigkeitsberechnung von Bauelementen des Dampfkessel Behälter und Rohrleitungsbaues*, Springer-Verlag, Berlin, 1970.
215. Brownell, L. E., and E. H. Young, *Process Equipment Design*, John Wiley, New York, 1959.
216. Adachi, J., and M. Benicek, Buckling of Tori-Spherical Shells Under Internal Pressure, *J. Soc. Exp. Stress Anal.*, 1964.
217. Warnock, F. V., and P. P. Benham, *Mechanics of Solids and Strength of Materials*, Isaac Pitman, London, 1965.
218. Svensson, N. L., The Bursting Pressure of Cylindrical and Spherical Vessels, *J. Appl. Mech.*, 1957.
219. Montague, Q., Experimental Behavior of Thin-Walled Cylindrical Shells Subjected to External Pressure, *J. Mech. Eng. Sci.*, 11, 1969.
220. Windenburg, D. F., and C. Trilling, Collapse by Instability of Thin Cylindrical Shells Under External Pressure, *ASME Trans.*, 56, 1934.
221. von Sanden, K., and K. Günther, Über das Festigkeitsproblem querversteiffer Hohlzylinder unter Allseitig Gleichmässigem Aussendruck, *Werft, Reederei*, 1, 1920; 2, 1921.
222. Blake, A., Formulas for Canister and Pipe Design in Underground Nuclear Emplacement, *J. Pressure Vessel Technol.*, 1974.
223. American Petroleum Institute, *Performance Properties of Casing and Tubing*, API Bull. 5C2, Dallas, Tex., 1970.

224. Edwards, S. H., and C. P. Miller, Discussion on the Effect of Combined Longitudinal Loading and External Pressure on the Strength of Oil-Well Casing, *Drilling and Production Practice*, American Petroleum Institute, Dallas, Tex., 1939.
225. Armco Steel Corporation, *Oil Country Tubular Products Engineering Data*, Middletown, Ohio, 1966.
226. von Kármán, Th., and H. S. Tsien, The Buckling of Spherical Shells by External Pressure, *J. Aeronaut. Sci.*, 1939.
227. Biezeno, C. B., Über die Bestimmung der Durchschlagkraft einer schwach gekrümmten kreisförmigen Platte, *AAMM*, 15, 1938.
228. Bijlaard, P. P., Theory and Tests on the Plastic Stability of Plates and Shells, *J. Aeronaut. Sci.*, 16(9), 1949.
229. Gerard, G., Plastic Stability Theory of Thin Shells, *J. Aeronaut. Sci.*, 24(4), 1957.
230. Krenzke, M. A., Tests of Machined Deep Spherical Shells Under External Hydrostatic Pressure, Rep. 1601, David Taylor Model Basin, Department of the Navy, 1962.
231. Krenzke, M. A., and R. M. Charles, The Elastic Buckling Strength of Spherical Glass Shells, Rept. 1759, David Taylor Model Basin, Department of the Navy, 1963.
232. Kloppel, K., and O. Jungbluth, Beitrag zum Durchschlagproblem dünnwandiger Kugelschalen, *Stahlbau*, 1953.
233. Weydert, J. C., Stresses in Oval Tubes Under Internal Pressure, *Soc. Exp. Stress Anal.*, 12(1), 1954.
234. Allen, T., Experimental and Analytical Investigation of the Behavior of Cylindrical Tubes Subject to Axial Compressive Forces, *J. Mech. Eng. Sci.*, 10, 1968.
235. Stephen, R. M., *Compressive Testing of Coupled Pipe Sections*, Rep. No. 71-6, Richmond Field Station, University of California, 1971.
236. Blake, A., Soaking Up Shock with Rolling Diaphragms, *Mach. Des.*, 1974.
237. Dean, J. S., Approximate Calculations for the Radial Expansion of Thin Tubes Used in Pressure Transducers, *J. Strain Anal.*, 8(4), 1973.
238. Kiesling, E. W., R. C. DeHart, and R. K. Jain, Testing of Ring-Stiffened Cylindrical Shells Encased in Concrete—Instrumentation and Procedures, *Exp. Mech.*, June 1970.
239. Wiegand, H., and K. H. Illgner, *Berechnung und Gestaltung von Schraubenverbindungen*, Springer-Verlag, Berlin, 1962.
240. Dobrovolsky, V., K. Zablonsky, S. Mak, A. Radchik, and L. Erlikh, *Machine Elements*, Mir, Moscow, 1968.
241. Roehrich, R. L., Torquing Stresses in Lubricated Bolts, *Mach. Des.*, 1967.
242. Almen, J. O., Tightening Is a Vital Factor in Bolt Endurance, *Mach. Des.*, 1944.
243. Faires, V. M., *Design of Machine Elements*, Macmillan, New York, 1955.
244. Osgood, C. C., How Elasticity Influences Bolted Joints, *Mach. Des.*, 1972.
245. Hooper, A. G., and G. V. Thompson, How to Calculate and Design for Stress in Preloaded Bolts, *Prod. Eng.*, 1964.
246. Oberg, E., F. D. Jones, and H. L. Horton, *Machinery's Handbook*, 20th ed., Industrial Press, New York, 1978.

References

247. Radzimovksy, E. I., *Schraubenverbindungen bei veranderlicher Belastung*, Manu-Verlag, Augsburg, West Germany, 1949.
248. Rothbart, H. A., *Mechanical Design and Systems Handbook*, McGraw-Hill, New York, 1964.
249. Goodier, J. N., The Distribution of Load on Threads of Screws, *J. Appl. Mech.*, 6, 1939.
250. Wasley, R. J., *Stress Wave Propagation in Solids*, Marcel Dekker, New York, 1973.
251. Shigley, J. E., *Mechanical Engineering Design*, McGraw-Hill, New York, 1963.
252. Moltrecht, K. H., written communication, Columbus, Ohio, 1977.
253. Scott, R. G., and J. C. Stone, *The Effects of Design Variables on the Critical Stresses of Eye Bars Under Load: An Evaluation by Photoelastic Modeling*, UCRL-85805, Lawrence Livermore National Laboratory, Livermore, Calif., 1981.
254. Blake, A., Ed., *Handbook of Mechanics, Materials, and Structures*, John Wiley, New York, 1985.
255. Rooke, D. P., and D. J. Cartwright, *Compendium of Stress Intensity Factors*, H. M. Stationery Office, London, 1976.
256. Lange, E. A., Fracture Toughness Measurements and Analysis for Steel Castings, *AFS Trans.*, 1978.
257. Sih, G. C., The Role of Fracture Mechanics in Design Technology, *ASME J. Eng. Ind.*, 1976.
258. Lange, E. A., and L. A. Cooley, Fracture Control Plans for Critical Structural Materials Used in Deep-Hole Experiments, NRL Memorandum Report 2497, U. S. Naval Res. Lab., 1972.
259. Barsom, J. M., Development of the ASSHTO Fracture Toughness Requirements for Bridge Steels, *Engineering Fracture Mechanics*, 7, Pergamon Press, London, 1975.
260. Puzak, P. P., and E. A. Lange, Significance of Charpy-V Test Parameters as Criteria for Quenched and Tempered Steels, NRL Report 7483, U. S. Naval Res. Lab., 1972.
261. Robertson, T. S., Propagation of Brittle Fracture in Steel, *J. Iron Steel Inst.*, 175, 1953.
262. Rolfe, S. T., and J. M. Barsom, Fracture and Fatigue Control in Structures, Prentice-Hall, Englewood Cliffs, N. J., 1977.
263. Wilson, A. D., High Strength Weldable Precipitation Aged Steels, *J. Metals*, 39(3), 1987.
264. Maier, K. W., Springs That Store Energy Best, *Prod. Eng.*, 1958.
265. Collins, J. A., *Failure of Materials in Mechanical Design: Analysis, Prediction, Prevention*, John Wiley, New York, 1981.
266. Neudecker, J. W., M. W. Gragg, and R. E. Grace, written communication, Los Alamos National Laboratory, Los Alamos, N. M., 1987.
267. Blake, A., Design Considerations for Rib-Stiffened Flanges, UCRL-50756, Lawrence Livermore Laboratory, Livermore, Calif., 1969.
268. Werne, R. W., Theoretical and Experimental Stress Analysis of Rib-Stiffened Flanges, EG&G Technical Report, EGG-1183-4054, San Ramon, Calif., 1972.
269. Fisher, J. W., and J. H. A. Struik, *Guide to Design Criteria for Bolted and Riveted Joints*, John Wiley, New York, 1974.

270. Bickford, J. H., An Introduction to the Design and Behavior of Bolted Joints, Marcel Dekker, New York and Basel, 1981.
271. Webjörn, J. Die Moderne Schraubenverbindung, VDI-Zeitschrift 130, No. 1, Düsseldorf, FRG, 1988.
272. Roberts, I., Gaskets and Bolted Joints, *ASME J. Appl. Mech.*, July 1950.
273. Blake, A., *What Every Engineer Should Know About Threaded Fasteners*, Marcel Dekker, New York and Basel, 1986.
274. Bickford, J. H., New Twists in Bolting, *Mech. Eng.*, May, 1988.
275. Newmark, N. M. and W. J. Hall, Dynamic Behavior of Reinforced and Prestressed Concrete Buildings Under Horizontal Loads and the Design of Joints (Including Wind, Earthquake, Blast Effects), Proc. Eighth Congress, International Assoc. for Bridge and Structural Engineering, New York, 1968.
276. Newmark, N. M., and W. J. Hall, Seismic Design Criteria for Nuclear Reactor Facilities, Proc. Fourth World Conf. Earthquake Engineering, Santiago, Chile, 1969.
277. Blake, A., *Design of Mechanical Joints*, Marcel Dekker, New York and Basel, 1985.
278. American Society of Mechanical Engineers, *Effects of Piping Restraints on Piping Integrity*, PUP-40, New York, 1980.
279. Yura, J. A., Zettlemoyer, N., and I. F. Edwards, Ultimate Capacity Equations for Tubular Joints, OTC 3690, Offshore Technology Conference, Houston, 1980.
280. Orthwein, W. C., A New Key and Keyway Design, *J. Mech. Des.*, Paper No. 78-WA/DE-7, 1979.
281. Carlson, H., *Spring Designer's Handbook*, Marcel Dekker, New York and Basel, 1978.

Index

Anticlastic surface, 403
Apparent stress, 349, 440–444
Arches, circular (*see also* Curved
    members), 328–335
  built-in, 332–334
  pin-jointed, 329–331
  simply supported, 328–329
  under uniform load, 334–335
Area properties, 385–386, 546–547
Axial stress, 5, 493
  in piping, 546–562
  Kern limit, 207–210

Beam columns, 254–259
  approximate rules, 255
  bending of, 255
  design of, 256–259
Beam theory, 29–32
  central axis in, 32
  fiber stress in, 33
  in one-way buckling, 551–555
  in pipe on elastic foundation, 549–551
  limitations of, 32–33
  neutral axis in, 32
  shearing stress in, 33

[Beam theory]
  theoretical equations in, 234–235
Beams, 233–234
  basic assumptions, 233–234
  buckling of, 91–92
  built-in, 246
  cantilever, 211–232
  clevis type, 242–244
  composite, 253–254
  curved, 337–343
  deep section, 249–250
  mathematical concepts in, 234–235
  on elastic foundation, 259–267
    applications, 262–264
    bearing pressure, 265
    damping factor, 260
    formulas, 262
    in pressure vessels, 263–264, 266–267
    modeling discontinuity, 263–264
    modulus of foundation, 259
    semi-infinite model, 260–262
  on simple supports, 233–239
    graphical solution, 236–239
    tables, 235–237
  partial loading on, 245

[Beams]
   plastic strength of, 192
   wide section, 250–253
      effective width for, 250–251
      formulas for, 251
   with constraint, 240–248
      end conditions, 241
      formulas, 242
      sinking supports, 240–241
   with three supports, 247
Belleville washer, 482–486
   conventional formulas, 483
   design charts, 484
   elastic-plastic effect in, 482
   elastic stress in, 482
   experiments on, 482–483
   load-deflection response, 485
   notation for, 482
   permanent set in, 485
   simplified formulas, 483–485
   stress at solid, 485
   stress ratios, 485–486
Bending stress, 29–33
   in brackets, 454–457, 462–468
   in flanges, 436–439
   maximum, 30
   moment of inertia, 30
   moment of internal resistance, 29
   section modulus, 30
   with plastic limit, 192
Bolted assembly, 203–205
   Hooke's method, 203–204
   stiffness method, 205
Bolted joints, 579–593
   bracketing conditions, 583
   effective area in, 582
   elastic response of, 583
   external load, 580–581
   friction in, 579–580
   leakage control in, 588–590
      criteria for, 589–590
   load tables for, 592
   materials in, 590–591
   performance of, 586
   quality control of, 592–593
   spring constants in, 581
   stresses in, 583–586
      effect of friction, 584
      interaction method, 583
   torque formula, 579–580
   thread recession, 587
Brackets, 459–472

[Brackets]
   buckling parameters in, 469
   edge compression, 470–471
   formulas for, 466–468
   inverse strain parameter, 468
   plastic buckling of, 469–470
      tangent modulus in, 469–470
   stability of, 467–472
      elastic, 468–469
   typical configurations, 460–462
   weld stresses in, 462–466
Breathing mode, 171
Bredt's formula, 24–25
Brittle failure, 46–47, 135
   heat affected zone (HAZ), 159
Buckling
   of beams, 91–92
   of brackets, 467–471
   of circular arch, 96–97
   of columns, 81–87
   of cylinders, 555–559, 561–562
   of long pipe, 547–549
   of panels, 90–91
   of plates, 87–89
   of spherical shells, 532–537, 538–539
Buckling resistance, 80–81
   inelastic column, 85
      Engesser's theory, 85
      tangent modulus, 85
      ultimate load, 85
   inelastic plate, 89–90
      reduction factor, 89
      secant modulus, 89–90
      tangent modulus, 89–90
   in bending, 91–92, 561–562
   in brackets, 467–471
   in higher modes, 549
   shear of panel, 90–91
      approximate correction, 90–91
      critical response, 90
      parallel wrinkles in, 91
      secant modulus in, 90
Buckling (column) spring, 625–627
Bulk modulus, 14
Burst pressure, 506–509
   general criteria, 506
      for thick wall, 506
      offset yield, 507
      strain hardening, 506
   Ludwik theory, 506
   of cylinders, 507–509
   of spheres, 507–509

Index

[Burst pressure]
  plastic interface, 506
  Svensson's theory, 507

Cantilever beams, 211–232
  design charts, 215–216
  effect of shear, 216–217
  intermediate loading, 212–215
    by double integration, 212–214
    by energy method, 214–215
  of variable cross-section, 222–226
    analysis, 223–226
    constant parameters, 223
Castigliano theory, 51–53, 66–67
  in straight members, 214–215
  in curved members, 271–317, 328–363, 383–395
CAT curve, 153
Columns (*see also* Elastic stability), 81–87, 92–96
Combined stresses, 12–14, 34–42
  components, 35
  in three dimensions, 14
  in two dimensions, 12–13, 35–36
  maximum shear stress, 36
  Mohr's circle, 37–41
  other methods, 39, 41
  von Mises–Hencky criterion, 47
Complex shape springs (*see also* Curved members), 292–309
Composite beams, 253–254
  equivalent rigidity, 254
  strength of, 253–254
Conical shell (*see also* Membrane stress)
  dilation of, 495
  stress in, 494
Curved members, 271–396
  arched cantilevers, 272–282
    horizontal load, 273–275
    uniform load, 278–280
    vertical load, 272–273
  assumptions, 271–272
  chain link, 359–363
    reinforced, 360–363
  circular arches, 328–335
    built-in, 332–334
    design charts for, 329, 331, 334
    horizontal reaction, 330
    on simple supports, 328–329
    pin-jointed, 329–331
    redundant forces in, 330–334
    under uniform load, 334–335

[Curved members]
  complex shape cantilever, 282–291
    applications, 283
    design charts, 285–286
  curved beam theory, 337–343
    apparent stress in, 349
    approximate formula, 341–344
    developments in, 337–338
    elasticity in, 349–350
    experimental design, 340–343
    hook failure mode, 348–349
    neutral axis factor, 338–340
    optimum hook shape, 347–348
    strain energy method, 350–353
    stresses in hooks, 343–350
    variable cross-section, 353–354
    Winkler–Bach formula, 338
  curved-end cantilever, 287–290
  eyebars, 364–381
    API standard, 371–372
    experiments with, 377–378
    failure modes, 363–364, 378–379
    sling loads, 379–381
    thick-ring method, 367–369
    with finite clearance, 372–376
    with zero clearance, 364–367
  gimbals, 392, 394–395
  in-plane couple on, 282–283
  knuckle joint, 363–364
  of complex shape, 292–309
    clip spring, 300
    definition, 292–293
    double U beam, 303
    formulas and charts, 303–307
    frame spring, 308
    general U spring, 300–302
    instrument applications, 302–307
    precurved cantilever, 296–298
    S-spring, 298
    snap ring, 293–296
    three-quarter wave, 298–300
    wave spring, 307–308
  open ring, 280–281
  out-of-plane loading, 383–395
    basic equations, 384, 386
    cantilever model in, 387–389
    ring in, 390–393
    torsional factors, 383–387
  piston rings, 317–326
    allowable opening, 319–320
    assembly of, 318
    design allowance, 324–325

[Curved members]
  [piston rings]
    features of, 317–318
    materials for, 318–319
    of uniform thickness, 319–322
    of variable thickness, 322–325
    radial pressure on, 321–322
    residual stress in, 319
    theory of, 319–326
  proving ring formula, 362
  rotating ring, 326
  special ring, 326–327
    under radial forces, 327
  thick rings, 356–358
  thin rings, 310–312
    design chart, 313
    exact formula, 311–312
    superposition in, 312–315
    with constraint, 315–317
Cylinders (*see also* Pipe string)
  axial stress in, 527–531
  buckling due to bending, 561–562
  buckling of, 555–559
    bellows type, 558–559
    diamond shape, 555
    plastic, 556–558
  column behavior of, 547–549
  dilation of, 493–494
  Donnell equation for, 556
  membrane stress in, 492–493
  on elastic foundation, 549–551
  out-of-roundness in, 519–521
  under external pressure, 513–531
  thick, 498–502
Crack propagation, 145
Crack shape parameter, 139–140
CVN (Charpy V-notch), 145

Deflection analysis, 59–65
  bracketing assumptions, 62
  by double integration, 63–64
  by strain energy, 59
  fictitious force method, 64–65
  flexural rigidity in, 60
  methods compared, 64–65
  of clevis, 242–244
  of curved members, 61–63, 271–396
  of frame, 244–245
  of redundant members, 65
  of straight bars, 59–60
  total energy formulas, 62
    for thick members, 62

[Deflection analysis]
  [total energy formulas]
    neutral axis, 62
    shear distribution factor, 62
Design, 3–4
  Bredt's formula in, 24–25
  elastic strain energy in, 50–58
  for dynamic loads, 101–114
  for fatigue, 125–128
  fracture mechanics in, 135–150
  fracture safe, 159
  fracture tough, 159
  Kern limit in, 207–210
    definition of, 207–208
    for compressive members, 209
  lower bound stress in, 156
  nominal stress in, 137, 158–159
  of beam columns, 254–259
  of bolted joints, 579–593
  of brackets, 459–472
  of built-in beams, 246
  of cantilever beams, 211–220
    design charts for, 215–216
    effect of shear, 216–219
    maximum deflection, 218
    shear deflection, 220
    variable cross-section, 222–231
  of chain links, 359–363
  of circular arches, 328–335
  of circular openings, 422–427
  of circular plates, 404–406, 427–428
  of complex-shape springs, 292–309
  of composite bars, 205–207
  of composite-material beams, 253–254
  of crane hooks, 343–350
  of curved members, 271–396
  of deep-section beams, 249–250
  of elliptical plates, 407–408
  of eyebars, 364–381
  of flanges, 431–459
  of gimbals, 392, 394–395
  of hemispheres, 535–537
  of indeterminate structures, 66–70
  of instrument springs, 300–307
  of mechanical connections, 595–612
  of mechanical springs, 615–627
  of multiple-leaf springs, 229–231
  of multiple-support beams, 247
  of nested cylinders, 571–573
  of open rings, 280–281
  of panels, 417–419
  of piston rings, 317–326

Index

[Design]
    of pressure vessels, 491–545
    of rectangular plates, 403–404
    of ring stiffeners, 573–575
    of shallow caps, 537–539
    of shrink-fit, 509–511
    of simply-supported beams, 233–239
    of straight members, 189–268
        downhole string, 201–203
        round bar, 193–194
        stepped bar, 196–197
        tapered bar, 197–199
        under own weight, 197–199
    of tapered cantilevers, 211–232
    of thick cylinders, 482–502, 504–505
    of thick spheres, 502–505
    of thin cylinders, 492–493
    of thin rings, 310–312
        in-plane loads, 311–317
        out-of-plane loads, 313–317, 390–395
    of thin spheres, 532–537
    of thick rings, 356–358
    of wide-section beams, 250–253
    Poisson's ratio in, 8–10
    seismic, 115–124
    stress concentration in, 71–79
    stress criteria in, 189–191
    stress propagation in, 163–176
Dilation, 493
    of closed cylinders, 569–571
        application, 569–570
    of conical shell, 495
    of elliptical shell, 495
    of open cylinder, 493–494
    of spherical vessel, 494
    thickness criteria, 494
Distortion energy, *see* Ductile failure
Ductile failure, 47–49
    cup-and-cone, 47
    distortion energy theory, 47
    ductility, 45, 47
    shear lip, 47
Dynamic analysis, 102
    beam impact, 105–106
    formulas for, 113
    free fall, 109–110
    frequency of vibration, 111–113
    impact theory, 111
    methodology, 111–113
    vibrational stress, 111
Dynamic response, 101–114
    criteria of, 102

[Dynamic response]
    [criteria of]
        intermediate range, 102
        impact, 102
        static loading, 102
    in natural mode, 103–104
    kinematics, 113
    of crushable support, 108
Dynamic strength, 102
    dynamic yield, 137
    effect of rapid loads, 103
    of structural materials, 102–103
Dynamic stress, 104
    effect of loading rate, 106
    energy loss factors, 107
    in free fall, 104–107

Elastic constants, 17
    ratio of, 17
Elastic modulus, 7, 44
Elastic curves, 44
Elastic stability, 80–81
    circular arch, 96–97
        buckling factor for, 97
    circular ring, 96–97
    comparison of theories, 86
    effect of bending on,
        for rectangular bars, 91–92
        shape factor in, 91–92
        torsional rigidity, 91
    Euler buckling, 85–86
        deflection curve, 87
        end effects in, 85–86
        weight comparison, 86–87
    flat plates, 87–89
        as columns, 89
        between rivets, 89
        between welds, 89
        buckling stress coefficients, 87
        edge loading, 88–89
        local buckling in, 87
        natural stress limit in, 87–88
    slender column, 81
        column factor, 81
        eccentricity in, 83
        in steel design, 81–82
        radius of gyration, 81
        secant formula, 81
    special columns, 92–96
        approximate solution, 94
        buckling coefficients, 93–94
        on elastic foundation, 96

[Elastic stability]
  [special columns]
    stepped column, 93–94
    under own weight, 94–96
Elastic strain energy, 50–58
  for unit volume, 13, 14, 50
  form factors (storage), 56–57
    in Belleville washer, 56
    in cantilever, 57
    in ring-spring, 56
    in shear, 56
    in shear mount, 56
    in shock cord, 56
    in solid bar, 56
    in tube torsion, 56
  in bending, 53–55
    of beams, 55
    of curved members, 55
  in frequency analysis, 111–113
  storage of, 55–57
    volumetric efficiency, 55–56
  theory of Castigliano, 51–53
    limitations, 53
    proof, 52
  total, 50–51
Ellipse of plasticity, 527–528
Ellipsoidal shell, 494–495
  dilation of, 495
  hoop stress in, 494
  longitudinal stress in, 494
  shear stress in, 495
Elliptical plates, 407–408, 418–421
Endurance limit, 125, 130
Engesser theory, 85
Euler theory, 84
External pressure, 513–544
  design formulas, 525
  effect of axial stress, 527–531
    design charts, 531
    ellipse of plasticity, 527–528
    limits of testing, 528–529
    long cylinder criterion, 529
    theoretical correction, 530
    wall thinning, 527
  empirical data, 534
  numerical correlation, 526
  on cylinders, 513–531
    in mixed mode, 516–517
    inverse strain, 517
    in plastic mode, 515
    long cylinder formula, 515
    midbay collapse, 514

[External pressure]
  [on cylinders]
    modified formula, 518
    short cylinder formula, 517
    thinness ratio, 514
    with out-of-roundness, 519–520
  on hemispheres, 535–537
    discontinuity effect, 537
    empirical curve, 536
  on shallow caps, 537–539
    buckling criteria, 537
    design chart, 539
    empirical formula, 538
    inelastic behavior, 539
    limitations, 538
    of small thickness, 539
  on thick cylinders, 540–542
    approximate criteria, 543
    deflection factors, 540–541
    limitations, 544
    out-of-roundness effect, 543–544
    stress factor, 540
  on thick spheres, 542–543
  on thin spheres, 532–537
    basic theory, 532
    Durchschlag, 533
    effect of imperfections, 534–535
    elastic buckling, 532
    formula corrections, 532–533
    local curvature, 534–535
    plastic strength, 533–534
Eyebars, 364–381

Fatigue, 125–134
  basic diagram, 126
  creep combined with, 131
  cumulative damage, 125–127
  endurance limit, 125, 130
  influence on, 128–130
    corrosion, 129–130
    creep, 129
    size, 130–131
    surface finish, 128–129
  low cycle formula, 131
    reduction of area in, 132
  Neuber effect, 127
  notch sensitivity factor, 127
  pseudo-elastic limit in, 133
  Soderberg's law, 128
  strength ratios, 132
  stress cycle in, 126
  stress concentration in, 127

Index

Flanges, 430–459
  apparent stresses in, 440–441
  bending of, 454–457
    plate model in, 455–457
  circumferential stress, 439–440
  German code, 438–439
  hub stresses, 436–439
  plastic correction, 441–444
    application of, 443–444
    theory of, 442–443
  plate theory in, 435–436
  rotation of, 433–435
    moment arm, 435
  stress criteria, 431
  tapered gussets in, 457–458
  thick hub theory, 433
  thin hub theory, 432–433
  Waters–Taylor formula, 439
  with ribs, 445–459
    equivalent depth, 445–448
    load sharing, 448–452
    rib strength, 452–454
Flexural rigidity, 53, 60, 253–254, 385
Form factor, 217–219
  derivation of, 217–219
  in short cantilevers, 219
Fracture control, 151–162
  basic concepts of, 152–153
  CAT curve, 153, 155–157
  correlation of CVN and DT, 154
  crack arrest temperature (CAT), 152
  design rules in, 161
  ductile-to-brittle transition, 151
  DT (dynamic tear), 152
  fracture-safe design, 159
  fracture-tough design, 159
  fracture-tough steel, 160
  fracture transition elastic, 153, 156
  high-yield materials in, 151–152
  lower-bound design, 157
  lower-bound stress, 156
  NDT (nil-ductility transition), 152
  nominal stress, 158–159
  shift of CAT curve, 154
  shrinkage cracks, 159
  stress-temperature curve, 155–157
  thickness criteria, 157–158
  types of fracture, 153
Fracture mechanics, 135–150
  arrestable instability, 146–147
  classical theory of, 136–137
    strength limits, 136

[Fracture mechanics]
  conventional properties, 135–136
  crack shape parameter, 139–140
  crack tip blunting, 146
  critical crack length, 147
  CVN (Charpy V-notch), 145
    correlation with K_{1C}, 145
  design applications of, 138–144
  dynamic yield, 137
  in pressure vessel, 147–149
    failure stress, 148
    part-through flaw, 147–148
  K_C (plane stress toughness), 147
  K_{1C} (plane strain toughness), 137, 158
    related to strength, 139
  leak-before-break, 143
  linear elastic, 145
  nominal stress in, 137
  of glass, 143–144
  plane strain in, 137
  plane stress in, 137
  plane stress parameter, 145–147
  strain energy release, 143
    in glass panel, 144
  typical K_{1C} data, 146
Fracture toughness, 137, 146
Frame, 244–245
  deflection of, 308
  under lateral load, 244

Gimbal rings, 392–395
  concentrated load, 395
  uniform load, 394

Heat-affected zone (HAZ), 159
  metallurgical factors, 159–160
Heat flow in pipe, 183
Hemispherical vessels, 535–537
High-temperature materials, 181
Hooke's law, 4–5
  design implications, 12–14, 583
  in two dimensions, 12–13
  in three dimensions
Hooks, 343–350
Hoop stress, 493, 496–497, 514

Impact theory, 111, 167–169
  criteria for, 102
  in free fall, 109–110
  in stress propagation, 163–176
  particle velocity, 166
Implosion, 520

Inertia, moment of, *see* Moment of inertia
Inglis formula, 77
Internal pressure, 491–512
   hoop stress, 493
   in thin cylinders, 492–493
   on conical shell, 494
   on ellipsoidal shell, 494–495
   on toroidal vessel, 495–498
   on thick cylinder, 498–502
     design charts, 504–505
     Lamé theory, 501
     shrink fit, 509–511
     with closure, 502
   on thick sphere, 502–504
     general equations, 504
   radial growth, 493–494
Inverse strain parameter, 468
   in brackets, 468
   in cylinders, 517

Kern limit, 207–210
K_C (plane stress toughness), 147
K_{1C} (plane strain toughness), 137, 158

Lamé formulas, 501
Laplace equation, 497
Leak before break, 143
Least work, principle of, 66–67
L'Hospital's rule, 406
Ligament efficiency, 475–477
   chart for, 477
   definitions of, 477
Links, 359–363
Lüder slip lines, 37
Ludwik theory, 506

Maxwell's theorem of reciprocity, 274
Mechanical properties, 44–46
   ductility, 45
   effect of composition, 45
   effect of heat treatment, 45–46
   effect of temperature, 45
   effect of metal working, 46
   elastic modulus, 7, 44
   high-yield material, 44
   nominal values of, 45
   of special materials, 180–182
   strain hardening, 44–45, 506
Mechanical strength, 43–49
   accuracy of, 43–44
   modes of failure, 43
   offset method, 44

[Mechanical strength]
   stress-strain curves, 44
     bilinear, 44
     generic, 44
     linear, 44
Mechanical connections, 595–613
   cotter pin, 602–604
   key joint, 604–607
     design of, 604–606
     surface forces in, 607
   pipe branches, 599–602
   piping supports, 598–600
     pipe clamp, 600
     ring loading in, 599
   split hub, 596–598
     contact pressure, 598
     frictional torque, 596–597
   structural pins, 607–609
     chart for, 609
     stresses in , 608–609
   wire rope fittings, 609–612
     applications, 610
     contact pressure in, 612
Mechanical springs, 615–627
   coil compression spring, 616–619
     buckling of, 620
     diameter change, 619
     dynamics of, 620–621
     round wire, 617
     square wire, 618
     typical geometry of, 616
     vibration of, 621
     Wahl factor, 617
   coil extension spring, 621–623
     end hook design, 622–623
   column (buckling) spring, 625–627
     design factors for, 627
     simplified design of, 626
     zero rate in, 625
   spiral power spring, 624–625
   torsion spring, 623–625
     coiling of, 624
     rectangular wire, 624
     round wire, 623
Megapascal (MPa), 6
Membrane stress, 492
   in conical shells, 494
   in ellipsoidal shells, 494–495
     maximum shear, 495
     radial growth, 495
   in spherical shells, 493
   in thin cylinders, 492–493

Index

Meridional stress, 493
Modulus of elasticity, 6–8
  bulk, 14
  equivalent, 475–476
  of special materials, 179–182
  of structural materials, 7
  reduced, 557–558
  secant, 90
  summary of concepts, 90
  tangent, 8, 90, 469–470
Modulus of foundation, 259
Modulus of rigidity, 16–17
  in experiments, 17
Mohr's theory, 37–41
  circle of stress, 37–41
  ductile yielding, 37
  Lüder slip lines, 37
  maximum shear, 37
  special cases, 40
Moment of inertia, 19, 30
  polar, 23
  second moment of area, 30
  section properties, 385–386
Moment ratio for beams, 192
Multiple-leaf springs, 229–231

NDT (nil-ductility transition), 152
Nadai theory of plates, 409–410
Nested cylinders, 571–573
  attenuation factor, 572
  design chart for, 573
  double wall in, 571
Neuber effect, 127
Neutral axis, 338–340
Nominal stress, 137
Notch sensitivity, 127

Offset method, 44
One-way buckling, 551–555
Out-of-plane loading, 383–395
Out-of-roundness, 519
  effective, 520–521
  effect on collapse, 521–523
    in long cylinders, 421
    in short cylinders, 520
  elliptical mode, 523
  relation to thinness factor, 521

Panels, 416–421
  edge conditions, 417
  elliptical, 418
  of other shapes, 419–421

[Panels]
  rectangular, 417–418
  under uniform load, 418–419
Pins, 602–604, 607–609
Pin-loaded plate, 480–482
Pipe string, 201–203
  as column, 547–549
  experiments, 556–559
  load eccentricity, 559–561
  on elastic foundation, 549–551
  one-way buckling, 551–555
Piston rings (*see also* Curved members), 317–326
  allowable opening, 319–320
  design allowance, 324–325
  materials for, 318–319
  of uniform thickness, 319–322
  of variable thickness, 322–325
  theory of, 319–326
Plane stress, 41–42
  in fracture mechanics, 137
Plane strain, 41–42
  in fracture mechanics, 137
Plastic bending, 191
  moment ratio, 192
Plates, 399–414
  assumptions, 399–400
  basic theory, 400–401
  Belleville washer as plate, 482
  boundary conditions in, 403
  circular, 404–406
  elliptical, 418
  flexural rigidity of, 401
  in flanges, 408–409
    Waters formula, 408
    Holmberg–Axelson formula, 409
  large deflection in, 409–413
  limit loads on, 414
  perforated, 475–477
    ligament efficiency, 475–477
    limitations, 475
    square pattern, 476
    triangular pattern, 476
  pin-loaded, 480–482
    bearing stress, 480–481
    effect of clearance, 481
    empirical data, 481
    stress concentration, 480–481
  rectangular, 403–404
    anticlastic surface, 403
    Sophie Germain equation, 403
  reinforced, 477–480

[Plates]
  [reinforced]
    concentric stiffening, 478
    deflection criterion, 478
    deflection formula, 479
    radial ribs, 478–480
    rigidity of, 478
    toroidal stiffness, 478
  spherical deformation, 402
  thermal gradient in, 402
  with central openings, 422–427
Poisson's ratio, 8–10
  equivalent, 475
  in bulk modulus, 14
  of various materials, 10
  theoretical limits, 10
Polar moment of inertia, 23
Pressure, collapse, see External pressure
Pressure vessels, 491–545
Principal planes, 36
Principal strain, 37
Principal stresses, 35

Radius of gyration, 81
Rankine theory, 46
Reciprocity, see Maxwell's theorem
Rectangular plates, 403–404
  panels, 416–418
Resilience, 193–194
  of solid bars, 193
Richter scale, 117–118
Rigidity, 53
  flexural, 53, 60, 385
  torsional, 385
  ratio, 385
Rings (see also Curved members)
  thin, 310–317
  thick, 356–358
Ring stiffeners, 573–575
Rolling diaphragm, 562–567
  annular version, 565–566
  comparison of configurations, 566
  cylindrical, 565
  theory of, 560–565

Secant modulus, 90
Section modulus, 30–31
  in bending, 30
  in torsion, 25–26
    by experiment, 26
  table of properties, 385–386
Seismic design, 115–124

[Seismic design]
  building code, 115–116
  fundamental mode, 116
  of frame, 244–245
  plastic response, 116, 123
  Richter scale, 117–118
    energy equivalent, 118
  rigidity in, 117
  rule of thumb, 117
  shear load, 116
  spectral velocity method, 118
    design charts for, 121–123
    determination of g level, 121–123
    example of, 118–121
    formulas in, 121
  structural damping, 123–124
Shear, 16–17
  average, 16–17
  distribution, 18–19
  displacement in, 17
  in closed sections, 20
  in open sections, 20
  in shafts, 22–23
  in tubes, 21–22
  maximum, 21–23
  moment of resistance in, 22–23
  polar moment in, 23
Shear distribution factor, 20–21, 62
Shock mitigation, 109–111
  dynamic force in, 109–111
  of packaged item, 109
Shrink-fit design, 509–511
  two cylinders, 509
  with different properties, 510–511
Soderberg's law, 128
Sonic velocity, 164–166
  dilatational, 167
  formulas for, 166
Sophie Germain equation, 403
Spall, 169
Special materials, 180–182
Spherical caps, 537–539
Spherical shells, 532–537, 542–543
Spring members (see also Mechanical Springs), 227–231, 292–309, 615–627
  Belleville washer, 482–486
  of complete shape, 292–309
  of multiple-leaf design, 229–231
    design formulas, 230–231
    unsymmetrical, 231
  tapered cantilever, 227–229

Index

[Spring members]
  [tapered cantilever]
    taper factor, 227–229
Static indeterminacy, 66–70
  analysis of, 66–70
    by theorem of least work, 66–67
    by superposition, 66
  of pin-jointed arch, 68–69
  of propped cantilever, 67–68
  of thin ring, 310–312
  of thick ring, 356–358
Stepped columns, 93–94
Stiffeners, in brackets, 459–472
  in circular plates, 477–480
  in cylinders, 573–575
  in flanges, 445–459
Strain, 6
  inverse, 468, 517
  lateral, 8
  longitudinal, 6
  plane, 41–42
  principal, 37
  shearing, 37
  three-dimensional, 13
  true, 15
  two-dimensional, 36
  volumetric, 9
Stress-strain curves (generic), 44
Stress, 4
  apparent, 349, 440–441
  axial (longitudinal), 5, 493
  bending, 29–33
  combined, 12–14, 34–42
  dimensions of, 6
  hydrostatic, 9
  in fatigue, 128
  in vibration, 111
  meridional, 493
  nominal, 43, 137
  plane, 41–42
  principal, 35
  shearing, 16, 18–27
  superposition of, 66
  thermal, 177–186
  true, 14–15
Stress concentration, 71–79
  distribution of, 74
  effect of secant modulus, 78
  elastic, 72
  factor of, 71
  in bending, 76
  in brittle material, 72

[Stress concentration]
  in circular shafts, 73–75
  in cycling loading, 72
  in fillets, 73
  in flange holes, 75, 77
  in rectangular bars, 73–75
  in sharp notches, 73
  in tension, 74
  in torsion, 76
  in welded joints, 74, 76
  Inglis formula, 77
  local effects in, 72
  macroscopic, 71
  near holes, 73
  net section in, 71–72
  plastic correction, 77
  source of, 78
  theoretical, 74–75
Stress propagation, 163–176
  analogy of, 164
  behavior of stress waves, 164
  breathing mode, 171
  critical impact velocity, 165
  dilatational velocity, 167
  elastic impact, 166
  elastic stress, 167
  frequency data for, 172
    axial mode, 172
    radial mode, 172
  impact theory, 167–169
  in granular media, 173
    adiabatic, 173
    isothermal, 173
    mechanical model, 174
    spherical wave, 173
  in machinery, 174–175
    dynamic stress, 175
  lumped mass model, 171
  sonic velocity, 164–166
    elastic, 164
    plastic, 165
  spall, 169
  theory of, 164–166
Svensson theory, 507

Tangent modulus, 90
Tensile test, 4–7
  gage dimension, 4–5
  Hooke's law, 4–6
  load-deflective diagram, 6
  stress-strain diagram, 7
Thermal stress, 177–186

[Thermal stress]
  basic theory of, 178
  causes of, 177
  design tables for, 184–185
  material properties, 178
    coefficient of expansion, 178
    effect of temperature, 179
    in special applications, 180–182
    specific heat, 178, 186
    thermal conductivity, 178
    thermal diffusion, 178
Thermal stress fatigue, 183
Thermal stress index (TSI), 182
  for various materials, 182
  in piping, 183
Thermal shock, 182–183
  limiting temperature, 182
Toroidal vessel, 495–498
  hoop stress in, 496–497
  Laplace equation for, 497
  meridional stress in, 497
Torsion, 23–27
  angle of twist, 23–24
  Bredt's formula in, 24–25
  of tubes, 24–25
  of rectangular sections, 25–26
  of structural shapes, 26
  parameters in, 26
  section modulus in, 23
  shape factor for, 25
Torsional rigidity, 385

Torsional shape factor, 25–26, 385–386
  by experiment, 26

Ultimate strength, 190–192
  bending moment ratio, 192
  of typical beams, 192
  plastic moment, 191

Variable section,
  of cantilevers, 211–232
  of flange gussets, 457–458
  of curved beams, 353–354
  of piston rings, 322–325
  of leaf springs, 227–229
Velocity of impact, 165
Vibration frequency, 111–113
Volumetric strain, 9
von Mises–Hencky theory, 47

Weld stresses, 462–466
  theory of fillet weld, 464–466
    applications to bracket design, 466–468
Winkler theory, 337–338
  Winkler–Bach formula, 338

Yield point, 6–7
  offset method, 44
Young's modulus, *see* Modulus of elasticity